CONSTRUCTION

PROFESSIONAL REFERENCE

MASTER EDITION

RESIDENTIAL AND LIGHT COMMERCIAL

William Spence

Published by:

www.DEWALT.com/guides

OTHER TITLES AVAILABLE

Trade Reference Series

- Blueprint Reading
- Construction
- Construction Estimating
- Construction Safety/OSHA
- Datacom
- Electric Motor
- Electrical – 2008 Code
- HVAC – Master Edition
- HVAC Estimating
- Lighting & Maintenance
- Plumbing
- Plumbing Estimating
- Residential Remodeling & Repair
- Security, Sound & Video
- Spanish/English Construction Dictionary – Illustrated
- Wiring Diagrams

Exam and Certification Series

- Building Contractor's Licensing Exam Guide
- Electrical Licensing Exam Guide
- HVAC Technician Certification Exam Guide
- Plumbing Licensing Exam Guide

Code Reference Series

- Building
- Electrical
- HVAC/R
- Plumbing

For a complete list of The DEWALT® Professional Reference Series visit **www.dewalt.com/guides**.

This Book Belongs To:

Name: ____________________

Company: ____________________

Title: ____________________

Department: ____________________

Company Address: ____________________

____

____

Company Phone: ____________________

Home Phone: ____________________

DEWALT® Construction Professional Reference, Master Edition:
Residential and Light Commercial
William Spence

Vice President, Technology and
Trades Professional Business Unit: .... Gregory L. Clayton
Product Development Manager: .... Robert Person
Editorial Assistant: .... Nobina Chakraborti
Director of Marketing: .... Beth A. Lutz
Executive Marketing Manager: .... Taryn Zlatin
Marketing Manager: .... Marissa Maiella
Production Director: .... Carolyn Miller
Production Manager: .... Andrew Crouth
Content Project Manager: .... Brooke Greenhouse
Art Director: .... Benjamin Gleeksman

Library of Congress Control Number: 2009928007

ISBN-13: 978-1-4180-6632-1
ISBN-10: 1-4180-6632-X

Delmar
5 Maxwell Drive
Clifton Park, NY 12065-2919
USA

Cengage Learning is a leading provider of customized learning solutions with office locations around the globe, including Singapore, the United Kingdom, Australia, Mexico, Brazil, and Japan. Locate your local office at: **international.cengage.com/region**.

Cengage Learning products are represented in Canada by Nelson Education, Ltd.

Visit us at **www.InformationDestination.com**
For more learning solutions, please visit our corporate website at **www.cengage.com**

Notice to the Reader
Cengage Learning and DEWALT® do not warrant or guarantee any of the products described herein or perform any independent analysis in connection with any of the product information contained herein. Cengage Learning and DEWALT® do not assume, and expressly disclaim, any obligation to obtain and include information other than that provided to it by the manufacturer. The reader is expressly warned to consider and adopt all safety precautions that might be indicated by the activities described herein and to avoid all potential hazards. By following the instructions contained herein, the reader willingly assumes all risks in connection with such instructions. Cengage Learning and DEWALT® make no representations or warranties of any kind, including but not limited to, the warranties of fitness for particular purpose or merchantability, nor are any such representations implied with respect to the material set forth herein, and Cengage Learning and DEWALT® take no responsibility with respect to such material. Cengage Learning and DEWALT® shall not be liable for any special, consequential, or exemplary damages resulting, in whole or part, from the readers' use of, or reliance upon, this material.

Printed in the United States of America.

8 9 10 11 12 13 14 23 22 21 20 19

Preface

This book is broad in scope and contains a series of carefully designed chapters giving detailed information about various materials used in the construction industry. Technical information is rapidly changing providing new and improved materials and construction processes. This handbook provides current information on widely used construction materials. In some areas, various illustrations are given plus a short descriptive paragraph to help clarify the data shown in a table or chart.

This book is designed to serve as a guide in the construction company office as decisions are being made and also as a field manual to assist with on-site issues which often occur. In addition, it serves as a supplement to a library of technical books as found in architectural, engineering, and contractors offices.

The careful, precise selection of the widely used construction materials and related construction activities condensed in a series of chapters, makes it an easy to use source of technical information.

Special attention should be given to Chapter 1, The Construction Specifications Institute MasterFormat™, because of the influence it has on all written manuals prepared for the construction industry. Chapter 2, Safety on the Construction Site, is also a chapter you should observe.

William Spence

CONTENTS

CHAPTER 1

The Construction Specifications Institute MasterFormat™

MasterFormat™ is a system of numbers and descriptive titles used to organize the extensive array of construction information in a standard order that facilitates the retrieval of information and serves as a means for communicating within the construction industry. The MasterFormat system was developed by the Construction Specifications Institute, 99 Canal Center Plaza, Suite 300, Alexandria, VA 22314-1588, 1-800-689-2900; 1-703-684-0300. SINetURL: http://www.csinet.org and Construction Specifications Canada, 100 Lombard St., Suite 200, Toronto, Ontario, Canada M5C 1M3, http://www.csc.dcc.ca. It is widely used by governmental organizations, construction product manufacturers, and professional and technical publications.

The numbers and titles in MasterFormat are divided into 48 basic groupings called **divisions**. Each division has a title and identifying number. Within each division, specifications are identified in numbered **sections**, each of which covers a particular part of the work of the division. Each section is identified by a six-digit number.

INTRODUCTION *(cont.)*

For a more in-depth explanation of MasterFormat, please contact:

The Construction Specifications Institute
99 Canal Center Plaza, Suite 300
Alexandria, VA 22314
800-689-2900, 703-684-0300
CSINet URL: http://www.csinet.org

THE MASTERFORMAT™ — LEVEL 2 NUMBERS AND TITLES

PROCUREMENT AND CONTRACTING REQUIREMENT GROUP

Division 00 – Procurement and Contracting Requirements

INTRODUCTORY INFORMATION

PROCUREMENT REQUIREMENTS

00 10 00 Solicitation

00 20 00 Instructions for Procurement

00 30 00 Available Information

00 40 00 Procurement Forms and Supplements

CONTRACTING REQUIREMENTS

00 50 00 Contracting Forms and Supplements

00 60 00 Project Forms

00 70 00 Conditions of the Contract

THE MASTERFORMAT™ – LEVEL 2 NUMBERS AND TITLES *(cont.)*

00 90 00 Revisions, Clarifications, and Modifications

SPECIFICATIONS GROUP

General Requirements Subgroup

Division 01 – General Requirements

01 10 00 Summary

01 20 00 Price and Payment Procedures

01 30 00 Administrative Requirements

01 40 00 Quality Requirements

01 50 00 Temporary Facilities and Controls

01 60 00 Product Requirements

01 70 00 Execution and Closeout Requirements

01 80 00 Performance Requirements

01 90 00 Life Cycle Activities

FACILITY CONSTRUCTION SUBGROUP

Division 02 – Existing Conditions

02 20 00 Assessment

02 30 00 Subsurface Investigation

02 40 00 Demolition and Structure Moving

02 50 00 Site Remediation

02 60 00 Contaminated Site Material Removal

02 70 00 Water Remediation

02 80 00 Facility Remediation

THE MASTERFORMAT™ — LEVEL 2 NUMBERS AND TITLES *(cont.)*

Division 03 – Concrete

03 10 00	Concrete-Forming and Accessories
03 20 00	Concrete Reinforcing
03 30 00	Cast-in-Place Concrete
03 40 00	Precast Concrete
03 50 00	Cast Decks and Underlayment
03 60 00	Grouting
03 70 00	Mass Concrete
03 80 00	Concrete Cutting and Boring

Division 04 – Masonry

04 20 00	Unit Masonry
04 40 00	Stone Assemblies
04 50 00	Refractory Masonry
04 60 00	Corrosion-Resistant Masonry
04 70 00	Manufactured Masonry

Division 05 – Metals

05 10 00	Structural Metal Framing
05 20 00	Metal Joists
05 30 00	Metal Decking
05 40 00	Cold-Formed Metal Framing
05 50 00	Metal Fabrications
05 70 00	Decorative Metal

THE MASTERFORMAT™ — LEVEL 2 NUMBERS AND TITLES *(cont.)*

Division 06 – Wood, Plastics, and Composites

06 10 00	Rough Carpentry
06 20 00	Finish Carpentry
06 40 00	Architectural Woodwork
06 50 00	Structural Plastics
06 60 00	Plastic Fabrications
06 70 00	Structural Composites
06 80 00	Composite Fabrications

Division 07 – Thermal and Moisture Protection

07 10 00	Dampproofing and Waterproofing
07 20 00	Thermal Protection
07 25 00	Weather Barriers
07 30 00	Steep Slope Roofing
07 40 00	Roofing and Siding Panels
07 50 00	Membrane Roofing
07 60 00	Flashing and Sheet Metal
07 70 00	Roof and Wall Specialties and Accessories
07 80 00	Fire and Smoke Protection
07 90 00	Joint Protection

Division 08 – Openings

08 10 00	Doors and Frames
08 30 00	Specialty Doors and Frames

THE MASTERFORMAT™ – LEVEL 2 NUMBERS AND TITLES *(cont.)*

08 40 00	Entrances, Storefronts, and Curtain Walls
08 50 00	Windows
08 60 00	Roof Windows and Skylights
08 70 00	Hardware
08 80 00	Glazing
08 90 00	Louvers and Vents
Division 09 – Finishes	
09 20 00	Plaster and Gypsum Board
09 30 00	Tiling
09 50 00	Ceilings
09 60 00	Flooring
09 70 00	Wall Finishes
09 80 00	Acoustic Treatment
09 90 00	Painting and Coating
Division 10 – Specialties	
10 10 00	Information Specialties
10 20 00	Interior Specialties
10 30 00	Fireplaces and Stoves
10 40 00	Safety Specialties
10 50 00	Storage Specialties
10 70 00	Exterior Specialties
10 80 00	Other Specialties

THE MASTERFORMAT™ — LEVEL 2 NUMBERS AND TITLES *(cont.)*

Division 11 – Equipment

11 10 00	Vehicle and Pedestrian Equipment
11 15 00	Security, Detention, and Banking Equipment
11 20 00	Commercial Equipment
11 30 00	Residential Equipment
11 40 00	Food Service Equipment
11 50 00	Educational and Scientific Equipment
11 60 00	Entertainment Equipment
11 65 00	Athletic and Recreational Equipment
11 70 00	Healthcare Equipment
11 80 00	Collection and Disposal Equipment
11 90 00	Other Equipment

Division 12 – Furnishings

12 10 00	Art
12 20 00	Window Treatments
12 30 00	Casework
12 40 00	Furnishings and Accessories
12 50 00	Furniture
12 60 00	Multiple Seating
12 90 00	Other Furnishings

THE MASTERFORMAT™ — LEVEL 2 NUMBERS AND TITLES *(cont.)*

Division 13 – Special Construction

- 13 10 00 Special Facility Components
- 13 20 00 Special Purpose Rooms
- 13 30 00 Special Structures
- 13 40 00 Integrated Construction
- 13 50 00 Special Instrumentation

Division 14 – Conveying Equipment

- 14 10 00 Dumbwaiters
- 14 20 00 Elevators
- 14 30 00 Escalators and Moving Walks
- 14 40 00 Lifts
- 14 70 00 Turntables
- 14 80 00 Scaffolding
- 14 90 00 Other Conveying Equipment

Divisions 15 to 19 – Reserved for Future Expansion

FACILITY SERVICES SUBGROUP

Division 20 – Reserved for Future Expansion

Division 21 – Fire Suppression

- 21 10 00 Water-Based Fire-Suppression Systems
- 21 20 00 Fire-Extinguishing Systems
- 21 30 00 Fire Pumps
- 21 40 00 Fire-Suppression Water Storage

THE MASTERFORMAT™ — LEVEL 2 NUMBERS AND TITLES *(cont.)*

Division 22 – Plumbing

22 10 00	Plumbing Piping and Pumps
22 30 00	Plumbing Equipment
22 40 00	Plumbing Fixtures
22 50 00	Pool and Fountain Plumbing Systems
22 60 00	Gas and Vacuum Systems for Laboratory and Healthcare Facilities

Division 23 – Heating, Ventilating, and Air Conditioning (HVAC

23 10 00	Facility Fuel Systems
23 20 00	HVAC Piping and Pumps
23 30 00	HVAC Air Distribution
23 40 00	HVAC Air Cleaning Devices
23 50 00	Central Heating Equipment
23 60 00	Central Cooling Equipment
23 70 00	Central HVAC Equipment
23 80 00	Decentralized HVAC Equipment

Division 24 – Reserved for Future Expansion

Division 25 – Integrated Automation

25 10 00	Integrated Automation Network Equipment
25 30 00	Integrated Automation, Instrumentation, and Terminal Devices
25 50 00	Integrated Automation Facility Controls

THE MASTERFORMAT™ – LEVEL 2 NUMBERS AND TITLES *(cont.)*

25 90 00	Integrated Automation Control Sequences
Division 26 – Electrical	
26 10 00	Medium-Voltage Electrical Distribution
26 20 00	Low-Voltage Electrical Transmission
26 30 00	Facility Electrical Power Generating and Storing Equipment
26 40 00	Electrical and Cathodic Protection
26 50 00	Lighting
Division 27 – Communications	
27 10 00	Structured Cabling
27 20 00	Data Communications
27 30 00	Voice Communications
27 40 00	Audio-Video Communications
27 50 00	Distributed Communications and Monitoring Systems
Division 28 – Electronic Safety and Security	
28 10 00	Electronic Access Control and Intrusion Detection
28 20 00	Electronic Surveillance
28 30 00	Electronic Detection and Alarm
28 40 00	Electronic Monitoring and Control
Division 29 – Reserved for Future Expansion	

THE MASTERFORMAT™ — LEVEL 2 NUMBERS AND TITLES *(cont.)*

SITE AND INFRASTRUCTURE SUBGROUP

Division 30 – Reserved for Future Expansion

Division 31 – Earthwork

31 10 00 Site Clearing

31 20 00 Earth Moving

31 30 00 Earthwork Methods

31 40 00 Shoring and Underpinning

31 50 00 Excavation Support and Protection

31 60 00 Special Foundations and Load-Bearing Elements

31 70 00 Tunneling and Mining

Division 32 – Exterior Improvements

32 10 00 Bases, Ballasts, and Paving

32 30 00 Site Improvements

32 70 00 Wetlands

32 80 00 Irrigation

32 90 00 Planting

Division 33 – Utilities

33 10 00 Water Utilities

33 20 00 Wells

33 30 00 Sanitary Sewerage Utilities

33 40 00 Storm Drainage Utilities

33 50 00 Fuel Distribution Utilities

THE MASTERFORMAT™ – LEVEL 2 NUMBERS AND TITLES *(cont.)*

33 60 00 Hydronic and Steam Energy Utilities

33 70 00 Electrical Utilities

33 80 00 Communications Utilities

Division 34 – Transportation

34 10 00 Guideways/Railways

34 20 00 Traction Power

34 40 00 Transportation Signaling and Control Equipment

34 50 00 Transportation Fare Collection Equipment

34 70 00 Transportation Construction and Equipment

34 80 00 Bridges

Division 35 – Waterway and Marine Construction

35 10 00 Waterway and Marine Signaling and Control Equipment

35 20 00 Waterway and Marine Construction and Equipment

35 30 00 Coastal Construction

35 40 00 Waterway Construction and Equipment

35 50 00 Marine Construction and Equipment

35 70 00 Dam Construction and Equipment

Divisions 36 to 39 – Reserved for Future Expansion

THE MASTERFORMAT™ — LEVEL 2 NUMBERS AND TITLES *(cont.)*

PROCESS EQUIPMENT SUBGROUP

Division 40 – Process Integration

- 40 10 00 Gas and Vapor Process Piping
- 40 20 00 Liquids Process Piping
- 40 30 00 Solid and Mixed Materials Piping and Chutes
- 40 40 00 Process Piping and Equipment Protection
- 40 80 00 Commissioning of Process Systems
- 40 90 00 Instrumentation and Control for Process Systems

Division 41 – Material Processing and Handling Equipment

- 41 10 00 Bulk Material Processing Equipment
- 41 20 00 Piece Material Handling Equipment
- 41 30 00 Manufacturing Equipment
- 41 40 00 Container Processing and Packaging
- 41 50 00 Material Storage
- 41 60 00 Mobile Plant Equipment

Division 42 – Process Heating, Cooling, and Drying Equipment

- 42 10 00 Process Heating Equipment
- 42 20 00 Process Cooling Equipment
- 42 30 00 Process Drying Equipment

THE MASTERFORMAT™ — LEVEL 2 NUMBERS AND TITLES *(cont.)*

Division 43 – Process Gas and Liquid Handling, Purification, and Storage Equipment

- 43 10 00 Gas Handling Equipment
- 43 20 00 Liquid Handling Equipment
- 43 30 00 Gas and Liquid Purification Equipment
- 43 40 00 Gas and Liquid Storage

Division 44 – Pollution Control Equipment

- 44 10 00 Air Pollution Control
- 44 20 00 Noise Pollution Control
- 44 40 00 Water Treatment Equipment
- 44 50 00 Solid Waste Control

Division 45 – Industry-Specific Manufacturing Equipment

Divisions 46 to 47 – Reserved for Future Expansion

Division 48 – Electrical Power Generation

- 48 10 00 Electrical Power Generation Equipment
- 48 70 00 Electrical Power Generation Testing

Division 49 – Reserved for Future Expansion

Courtesy The Construction Specifications Institute and Construction Specifications Canada.

CHAPTER 2
Safety on the Construction Site

The construction industry is one of the more dangerous workplaces. The wide range of the types of operations being performed and the mixture of workers from various trades set the stage for accidents. It is important that construction management have safety plans in effect and see that all involved observe them. Following are several sources where additional information about recommended and required safety practices can be secured.

Occupational Safety and Health Administration (OSHA)
www.osha.gov

National Fire Protection Association (NFPA)
http://catalog.nfpa.org

American National Standards Institute (ANSI)
www.ansi.org

National Institute for Occupational Safety and Health (NIOSH)
www.cdc_gov.niosh

National Association of Home Builders (NAHB)
www.nahb.org

National Safety Council (NSC)
www.nsc.org

EMPLOYER-ESTABLISHED PERSONAL SAFETY RULES

Safety rules for those on the site should be clearly specified, and supervisors should see that they are observed. These should be reviewed regularly by giving toolbox safety talks. Following are examples of rules frequently required.

1. Do not wear loose clothing or jewelry.
2. Wear required safety equipment. Hard hats and safety glasses must always be worn on the job site. Other items such as respirators, hearing protection, fall harnesses, and gloves must be worn when circumstances require them.
3. Smoking on the site is discouraged, and prohibited in areas where flammable and combustible materials are present or where chemicals are stored or are being used.
4. Alcohol and drugs must not be brought onto the construction site. Suspected users should be reported to a supervisor.
5. Only authorized persons should enter restricted areas.
6. If anyone is not certain how to do a particular procedure, that person must ask a supervisor for assistance.
7. Report any observed unsafe conditions to a supervisor.
8. Any injury, even if minor, should be reported to a supervisor.

EMPLOYER-ESTABLISHED PERSONAL SAFETY RULES *(cont.)*

9. Personal behavior must be controlled and courteous at all times. No horseplay or tricks should be allowed.
10. Expect disciplinary action if safety rules are violated. This could begin with a written warning, and additional violations could lead to a hearing with management and possible termination.

EMERGENCY ACTION PLAN

State and local laws specify requirements for emergency action plans. OSHA requires general contractors with more than eleven employees to have such plans. Those with fewer employees should still inform all employees of the emergency procedures to be followed. Examples of things to be on an emergency action plan:

1. If someone is injured, specify who on the site is to be immediately contacted.
2. Specify who on site is to provide emergency first aid and maintain control until outside help arrives.
3. Specify the location of the first aid equipment. On a big job, this could be a first aid station. On others, it could be the job site trailer.
4. Have telephone communications established and telephone numbers of local fire, ambulance, and rescue teams available.

EMERGENCY ACTION PLAN *(cont.)*

5. If there is a major occurrence, such as a fire requiring evacuation, specify the signal that is the evacuation alarm, such as a loud bell or air horn.
6. Establish escape routes and update them as construction continues and the situation changes.
7. Provide a procedure for accounting for all who are in the building when an evacuation is required.
8. If a job has unusual conditions, such as excavations, specify the procedure to be used to rescue trapped or unconscious workers.
9. If a rescue is needed, specify if an on-site rescue team is expected to handle the situation or if a call for an off-site rescue team is required. Have phone communications available.

SAFEGUARDING DURING CONSTRUCTION

1. Construction materials and equipment must be stored so they present no danger to workers or the public.
2. During remodeling, sanitary facilities, fire protection, exits, and structural systems must be maintained at all times.
3. Fill materials and excavations must be constructed to prevent danger to property and persons.
4. Sanitary facilities must be available at all times.
5. Provisions to protect pedestrians must be installed and maintained. These include directional and warning signs, walkways, barricades, railings, and covered walkways.

SAFEGUARDING DURING CONSTRUCTION *(cont.)*

6. Adjoining property must be protected at all times.
7. Nothing should be placed or stored blocking fire protection facilities.
8. Fire extinguishers are required during construction. Portable fire extinguishers must be available on all floor levels. Multiple extinguishers are required in areas containing flammable materials. Remotely located storage buildings or trailers should have one fire extinguisher.
9. Buildings over four stories should have constructed at least one temporary stairway unless the permanent stairways are constructed as the building is erected.
10. Buildings over four stories must have one standpipe for fire department hose connections available during construction.
11. A water supply for fire protection must be made available as soon as combustible material arrives on the site.
12. Additional details are available in the International Building Code, Chapter 33, International Code Council.

SAFETY ON THE CONSTRUCTION SITE

Personal Attitudes

The construction site is a dangerous place, and workers must develop a positive attitude toward their safety and the safety of others. They need to alert others about dangerous situations so corrections can be made. Even a small act such as pulling protruding nails or putting waste in the dumpster is important. Finally, they need to always wear required personal safety equipment and keep all tools in proper condition.

General Housekeeping Rules

1. All walkways, runways, aisles and other work areas should be kept free of debris.
2. Tools, materials, and other supplies should be stored in a secure, safe location.
3. Remove any protruding items such as nails, reinforcing bars, and lumber bracing, so they do not cause injury.
4. Place all flammable materials, such as oily rags, finishing materials, and other combustibles, in nonflammable storage.
5. Keep snow and ice off all walkways, scaffolding, and ladders.
6. Remove any spilled liquids, such as oil or grease, from the walkways.
7. All trash and flammable waste must be removed from the site on a regular basis.
8. When dropping waste materials more than 20 feet, use an enclosed chute directing the flow into a waste container.

SAFETY ON THE CONSTRUCTION SITE *(cont.)*

9. When dropping waste materials a short distance, such as between floors, barricade the area where the material will land.
10. Keep the land around the building graded fairly level and free of rocks, stumps, and other obstacles.
11. Establish marked traffic ways around the site, including traffic control signs. Keep them smooth and level.
12. Provide enough lighting in all work areas so workers can adequately and safely perform their work.

SITE MAINTENANCE RECOMMENDATIONS

1. Remove all trash and flammable waste materials regularly.
2. Store combustible materials in nonflammable containers and mark FLAMMABLE.
3. Keep all walkways, stairs, and work areas free of spilled oil, water, grease, and other liquids. Regularly remove all scrap material.
4. Keep scaffolds and ladders clean. In winter, keep free of ice and snow.
5. Remove nails, screws, and other sharp projections from waste materials.
6. All materials must be stored in a safe manner.
7. When dropping material more than 20 feet, use an approved chute to direct it to a trash bin below.
8. When dropping scrap short distances through a hole in the floor, place a trash bin below to receive the material and put a barricade around the area.

SITE MAINTENANCE RECOMMENDATIONS *(cont.)*

9. Lay out roads around the site for use by trucks and other equipment. Set up stop signs if necessary. Maintain them in a level and smooth condition.
10. Keep the area on the construction site as level as possible, especially near the building under construction.
11. Maintain fences, gates, and other site security features.
12. Maintain all areas set aside to provide for surface water runoff.
13. Maintain all barriers used to control the flow of loose soil on to the neighboring property.

DEMOLITION SAFEGUARDS

Demolition activities may involve taking down an entire building or major reconstruction of several floors in a multistory building. These activities provide a wide range of dangerous situations. In many situations a special analysis by demolition experts is required. Following are some frequently occurring things to consider.

1. Removal of waste materials must be carefully planned and be a continuous operation. Some waste materials are hazardous and require special permits and disposal requirements.
2. Protection of the general population is required. This may be as simple as barriers keeping people away from the building, sheltered roofs

DEMOLITION SAFEGUARDS *(cont.)*

built over nearby sidewalks, or a substantial barrier blocking off the entire area.

3. Adjoining property must have adequate protection. Items such as footings, foundations, exterior walls, and roofs may require special protection. Also, surface water must be directed away and soil erosion controlled.
4. Provision must be made for the storing of equipment used for the demolition.
5. Since fire is a constant danger, adequate fire protection must be available. This can include portable fire extinguishers or maintaining a standpipe in multistory buildings.
6. In multistory buildings, provision for means to exit the building is mandatory. This could be temporary stairways that are adequately lighted.
7. Sanitary facilities must be provided.
8. All flammable materials should be removed before demolition starts.
9. If work is in only part of a building, fire protection such as maintaining an existing sprinkler system is required. This may require heating the area in cold weather to prevent it from freezing.
10. If explosives are used, fire protection must be actively maintained during the detonation.
11. Electrical service must be discontinued or reduced to a minimum in the affected area.

DEMOLITION SAFEGUARDS *(cont.)*

12. Gas service should be turned off and disconnected outside the building.
13. Waste water systems need to be capped off in the affected area.
14. Potable water can remain available, but provision for shutting it off if it is damaged is necessary.

SAFETY AROUND THE TRUCKS AND EARTHMOVING EQUIPMENT

When mobile equipment is in use on the site, observe the following recommendations.

1. Be aware of the backing alarm and move out of the way.
2. Do not hitch rides on the equipment.
3. Operator vision is limited, so stay alert any time a piece is moving.
4. Equip all off-road equipment with rollover protection.
5. Regularly check the braking system and lights.
6. When inspecting the area under a raised bed of a dump truck, insert blocking.
7. Do not overload a truck.
8. Verify the training and experience of truck operators.

SAFETY AROUND CRANES

Following are some safety actions needed when working on a site where cranes are in use.

1. Stay clear of the rotating cab.
2. Keep at least 10 feet clearance from overhead power lines.
3. Stay away from the area below a crane when it is in operation.
4. Never ride on a sling or on the load being lifted.
5. Keep watch for material that may fall off a load being moved. Give a warning if something appears to be loose.
6. Cranes are difficult to move around the site, so stay well away from them as they are moving.
7. Watch out for swinging loads, hooks, and booms.
8. Never serve as a crane signal person unless you have been trained in the use of standard signals.
9. Verify the rated capacity of the crane and do not exceed it.
10. Prepare the ground below the crane as needed to ensure its stability.
11. When moving loads, use a tag line to control the movement.
12. Verify the training and experience of the crane operator.

SEPARATION OF TEMPORARY CONSTRUCTION-RELATED STRUCTURES

Offices, Trailers, Sheds from Buildings under Construction

Temporary Structure Exposing Wall Length		Minimum Separation Distance	
m	ft.	m	ft.
6	20	9	30
9	30	11	35
12	40	12	40
15	50	14	45
18	60	15	50
>18	>60	18	60

Notes:

1. Where the separation distance between temporary structures is less than the minimum separation distance, then the exposing wall length shall be considered to be the sum of the individual exposing wall lengths of the temporary structure.

2. A 75% reduction in separation distances shall be permitted to be applied, provided automatic sprinkler protection is used in the exposing structure.

3. The separation distances apply to single-level structures only. This table does not apply to multilevel, unsprinklered structures. A level, where applying this table, is 3600 mm (144 in).

STANDARD FOR SAFEGUARDING CONSTRUCTION, ALTERATION AND DEMOLITION OPERATIONS

NFPA 241
http://catalog.nfpa.org

1. Temporary Construction, Equipment and Storage
2. Processes and Hazards
3. Utilities
4. Fire Protection
5. Safeguarding Construction and Alteration Operations
6. Safeguarding Roofing Operations
7. Safeguarding Demolition
8. Safeguarding Underground Operations

WORK ZONE TRAFFIC SAFETY

The construction project manager must establish the patterns for the flow of traffic to, around, and from the construction site. This will change as the project develops, so regular evaluation of the situation is necessary. Suggestions from those moving about the site are valuable. Following are some recommended procedures:

1. Traffic control signals, signs, and message boards must be used to direct the flow of traffic.
2. Various control devices such as cones, barrels, barricades, and posts should be used.
3. Plans for limiting motorists from accidentally entering the work area must be established. Control can be established using concrete, sand and earth barriers, collapsible crash barriers, and crash cushions.
4. Actual flow can be controlled by flaggers. They can control on-site equipment movement and that of the general public who must transverse the area.
5. Flaggers must wear approved high-visibility garments having a fluorescent background of a reflective material. The flaggers must be trained and certified.
6. When necessary, lighting must be provided for moving equipment and workers on the ground after dark.
7. All equipment must have seat belts and rollover protection.

RESPIRATORY PROTECTION

Whenever working in or near an area where the atmosphere is hazardous, an appropriate respirator must be worn. The type of device used depends upon the contaminates in the atmosphere and the protection factor required. Respirators must be NIOSH approved.

The inexpensive, single-strap comfort mask is not NIOSH approved. It is used to protect from pollen and some non-toxic dusts. Dispose after a short use.	
The filtering face piece is NIOSH approved and can be used when sanding wood, cutting and sanding drywall, cutting bricks and concrete, and installing fiberglass. It does not protect from asbestos or lead. Dispose when it begins to get dirty.	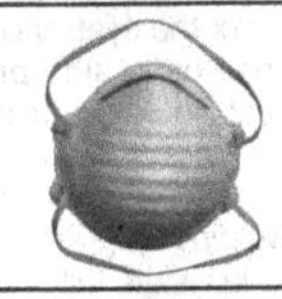
A comfort mask with replaceable filters is not NIOSH approved. It is used to protect from pollen and dust from sweeping, and when cutting grass.	
A half-face respirator protects against paint and pesticide spray and vapor, lead and asbestos fumes, fiberglass insulation, various solvents, adhesives, lacquers, enamels, and various dusts from sanding wood and drywall. It has various types of cartridges available for various atmospheres. It is NIOSH approved.	
A full-face respirator is just like the half-face respirator but has a clear plastic shield that covers the entire face. This helps protect from irritants and contaminants in the atmosphere. It is NIOSH approved.	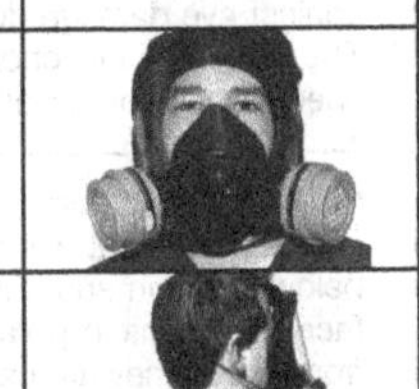
Other types of respirators have air-purifying packs connected to the face mask, providing breathing capacity from a battery-powered fan that pulls air through filters and circulates it through a hose to the hood. Another type has an air tank carrying oxygen so the contaminated air in the atmosphere is not involved.	

(Photos and information courtesy The Aearo Company)

EYE PROTECTION

Eye protection devices should fit properly, be comfortable, provide unrestricted vision, and be durable and easily cleaned. Standard prescription eyeglasses do not provide adequate eye protection.

Safety spectacles have durable frames and impact-resistant lenses. Side shields are available and are recommended for many situations.	
Goggles fit tightly against the face and completely cover the eyes and facial area surrounding them. They have side protection. They protect from dust, liquids, and impact. Some fit over standard eyeglasses.	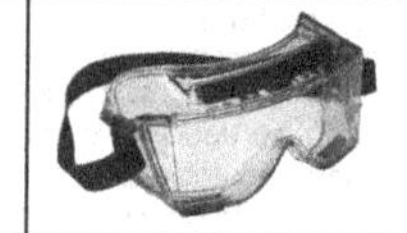
Welding goggles provide a durable lens, giving wide vision, fully cover the eyes and surrounding facial area, have the lens tinted to absorb infrared and ultraviolet radiation, and will fit over standard eyeglasses.	
Welding helmets have a front that can be lifted open, use standard filter plates, and have a drop down ratchet head gear which permits easy viewing of the work area. Some types accommodate standard eye glasses.	
Laser safety glasses and goggles protect against eye damage due to exposure to lasers. The specific one chosen depends upon the specifications of the laser equipment.	
Full-face shields are made using a transparent plastic shield that extends from the forehead to below the chin and covers the full width of the face. Some have polarized shields to protect from glare. They are used to protect from minor hazards such as dust and liquid sprays, but do not provide impact resistance. Safety goggles worn behind the shield are used to provide impact resistance if needed.	

(Photos and information courtesy The Aearo Company)

HEARING PROTECTION

Exposure to excessive noise depends upon:

1. Loudness in decibels (dB)
2. Duration of exposure to the noise
3. Workers moving between areas having different noise levels
4. Noise generated from several sources in the same area

Employees must be required to wear appropriate hearing protection. The hearing protectors worn should have their noise reduction rating on the packaging. Sound levels must be measured with a sound level meter. Hearing protectors must reduce the noise exposure to within the limits in the following chart.

PERMISSIBLE NOISE EXPOSURES

Duration per day, in hours	Sound level in dB*
8	90
6	92
4	95
3	97
2	100
1 1/2	102
1	105
1/2	110
1/4 or less	115

Source: OSHA 29 CFR 1910.95 Table G-16

SOUND INTENSITY			
Be aware of the intensity of sound in the working area and wear ear protection when necessary.			
Sound Pressure Level in Decibels	**Sensation**	**Source Example**	**Hearing Protection**
140	Deafening	Near Jet Aircraft Departing	Wear Hearing Protection
130	Threshold of Pain	Artillery Fire	
120	Threshold of Feeling	Siren, Rock Band	
110	Just Below Threshold of Pain	Riveting, Air Hammering	
100	Very Loud, Begin Hearing Damage	Power Mower	Work 1 to 2 Hours Unprotected
90	Very Loud	Symphony Orchestra	Work 6 to 8 Hours Unprotected
80	Loud	Noisy Industrial Plant, Diesel Truck Departing	Hearing Protection Not Needed
70	Loud	Average Radio or TV	
60	Moderately Loud	Normal Conversation	
50	Moderately Loud	Inside Large Office	

EAR PROTECTION DEVICES

Disposable ear plugs are a form of soft foam that compresses in the ear canal. They are used a few times and then discarded. They do not meet the requirements for high exposures over a long time.	
Reusable ear plugs have pre-formed pods that fit into the ear. This type has a triple flange, allowing a snug and comfortable fit. They should be wiped off frequently. They do not provide rated protection.	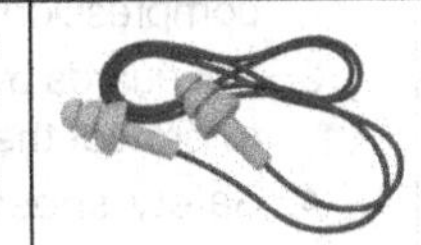
Ear muffs provide rated decibel protection and are used where noise levels require known protection levels. Glasses, heavy facial hair, and excessive facial movements such as constant talking or chewing can break the seal to the head and reduce effectiveness. This is especially important when the person wearing them is exposed to high-decibel sound over many hours.	

(Photos and information courtesy The Aearo Company)

HARD HATS

Hard hats are divided into three classes: **CLASS A.** They provide impact and penetration resistance and limited voltage protection. **CLASS B.** They provide the highest protection against electrical hazards, high voltage shock, and burn protection. **CLASS C.** They are lightweight and provide impact resistance but no electrical protection. Replace hard hats that have been penetrated, cracked, or deformed. If damaged by exposure to heat, chemicals, or ultraviolet light and show a loss of surface gloss, they should be replaced.	

LEG AND FOOT PROTECTION

1. Leggings on lower legs and feet are needed to protect against heat and hazards such as welding.
2. Metal, plastic, or fiberglass guards are placed over the toe of the shoe to protect from an impact or compression.
3. Toe guards over the toes of regular shoes are used to protect the toes from impact or compression.
4. Safety shoes with impact-resistant toe plates and heat-resistant soles protect from impact and excessive underfoot conditions.

PROTECTIVE GLOVES

Gloves fall into four groups:

1. Gloves made from leather, canvas, metal mesh
2. Gloves made from fabric and coated fabric
3. Gloves that are chemical and liquid resistant
4. Insulating rubber gloves

Protection provided by each type.

- Leather, canvas, metal mesh – cuts, abrasions, burns
- Leather, canvas – sustained heat
- Fabric – abrasions, slivers, dirt
- Coated fabric – slip and abrasion resistant, abrasion from jobs as carrying bricks or other construction materials
- Chemical-resistant – various chemicals, depending upon the type of material used to make the glove
- Insulating rubber – electrical shock

VISIBLE PROTECTION

Many jobs on the construction site require that those involved wear highly visible safety vests. Some are designed for daytime use, while others are approved for day and night use. They are lightweight and ventilated.

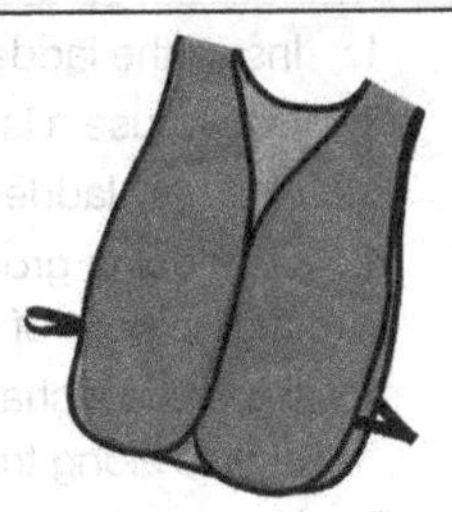

Photo courtesy The Aearo Company.

LADDER SAFETY

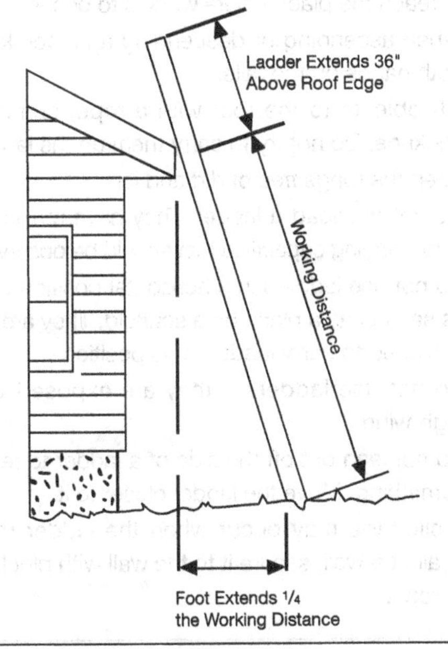

LADDER SAFETY *(cont.)*

1. Install the ladder as shown in the drawing.
2. Do not use a ladder that has a defect.
3. Use only ladders with approved nonskid feet.
4. On sloping ground, build up below the feet with large pieces of plywood.
5. If there is a chance the feet may slip, drive wood stakes along the bottom rung.
6. Keep debris away from the top and bottom of the ladder.
7. Adjust the length of the ladder so it is long enough to reach the place where work is to occur.
8. When ascending or descending a ladder, keep both hands on the rails.
9. Lift objects to the roof with a rope, crane, or backhoe. Do not try to carry them up the ladder.
10. Keep the rungs free of dirt and ice.
11. Do not overload a ladder. They have maximum load-carrying capacities that should be observed.
12. Do not use ladders in a horizontal position such as serving as a plank on a scaffold. They are not designed to carry loads in this position.
13. Do not use ladders if they are exposed to a high wind.
14. Do not lean out off the side of a ladder to reach something. Move the ladder closer to it.
15. If slippage may occur when the ladder rests against a wall, secure it to the wall with blocking or rope.

SITE-BUILT LADDERS

Site-built ladder material requirements.

Ladder Length	Min. Inside Width at the Base	Rung	Rails
Up to 12'	16"	1 × 3"	2 × 3"
Over 12' to 20'	18"	1 × 4"	2 × 4"
Over 20' to 30'	19"	1 × 4"	2 × 6"

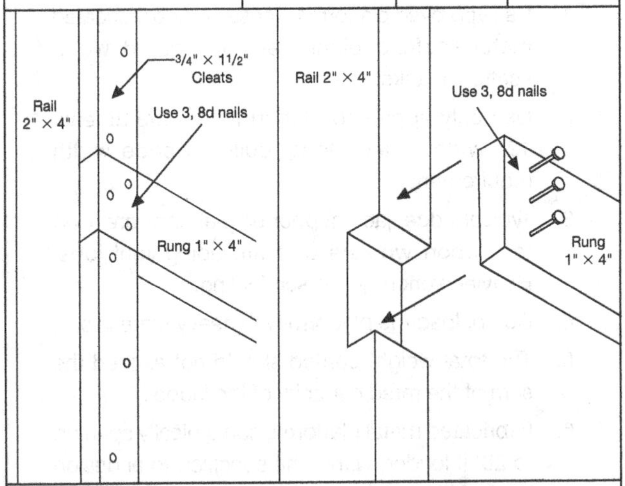

Site-built ladders should have the rungs recessed into the side rail or supported with wood cleats.

LADDER JACK SAFETY

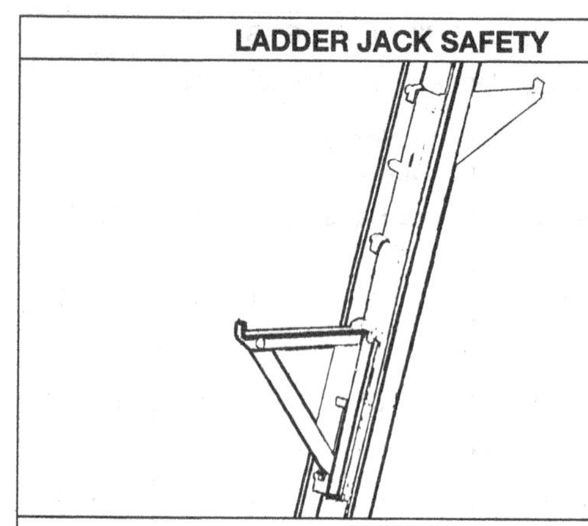

1. Use approved platforms. These include fabricated metal scaffold planks and laminated wood scaffold planks.
2. Use only approved platforms that are at least 12" wide. Check local building code width requirements.
3. Typical ladder jack-supported platforms are used to support workers who are doing light jobs. Heavier work requires scaffolding.
4. Do not load the platform with heavy materials.
5. The total weight carried should not exceed the sum of the rated capacity of the ladder.
6. Fabricated metal platforms can typically span up to 28' if loaded within the specified total design load of the ladders, ladder jacks, and platform. Check the manufacturer's specifications.

LADDER JACK SAFETY *(cont.)*

7. Do not use ladders, ladder jacks, or platforms that are damaged. Repair or replace the damage before using. Discard if repairs are not possible.
8. Observe the safety rules for installing ladders. Secure the ladders at the ground and the top to the wall.
9. Assemble the ladder jacks as shown in the manufacturer's instructions.
10. Use 250 or 300 pound Type 1 or 1A duty rated ladders. Manufacturers of jacks strongly recommend using the 300 pound type. Never use ladder jacks on 200 or 225 pound duty 1 ladders.
11. The work platform should never be more than 20' above the base of the ladder.
12. Secure the platform to the ladder jacks.
13. The platform should overhang the ladder jack by 12" but not more than 18".
14. It is recommended that not more than one person work on the platform unless it is certain the span between ladders is short enough to handle the added weight of the extra worker and the tools.
15. For additional information contact the American National Standards Institute, www.ansi.org and the Occupational Safety and Health Administration, www.osha.gov.

WALL SCAFFOLD BRACKET SAFETY

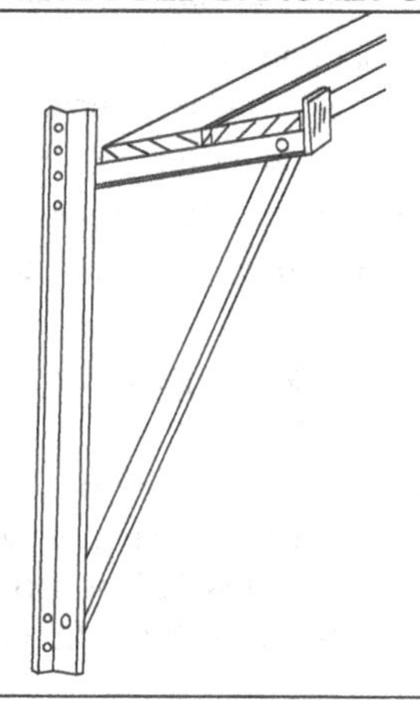

1. When possible, bolt the bracket to the wall framing.
2. Brackets can be nailed or screwed to the studs. Use 16d to 20d common nails or $3^{1}/_{2}$ to 4" wood screws.
3. Replace any nails that bend as they are installed.
4. Install an approved guard rail as specified for scaffolds.
5. Use only for light work. Do not overload with materials.
6. Use approved wood planks for the platform.

SCAFFOLD SAFETY

METAL TUBE SCAFFOLD

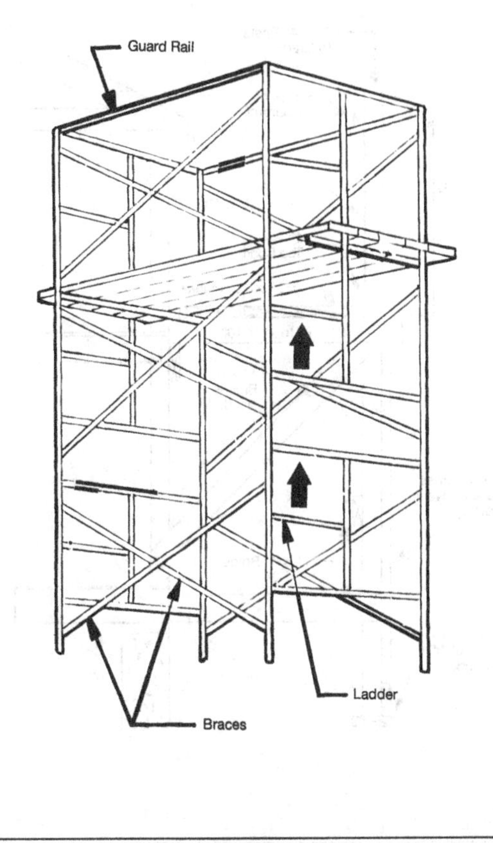

SCAFFOLD SAFETY *(cont.)*

SINGLE-POLE SCAFFOLD

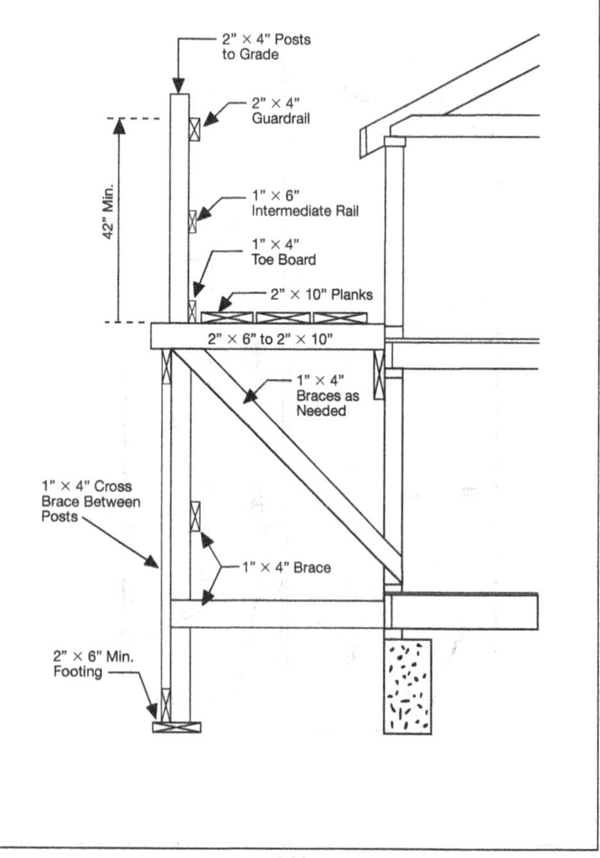

SCAFFOLD SAFETY *(cont.)*

DOUBLE-POLE SCAFFOLD

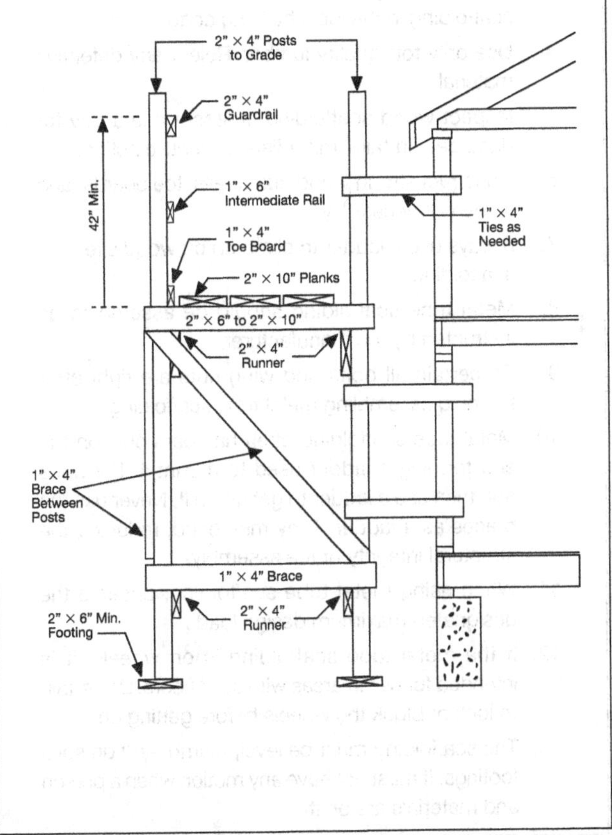

SCAFFOLD SAFETY *(cont.)*

1. Wood site-built scaffolds must be built under the supervision of an experienceD carpenter.
2. Wood scaffolding must meet OSHA standards.
3. Secure design regulations for wood site-built scaffolding in the local building code.
4. Use only top quality lumber. Reject any defective material.
5. Inspect wood scaffolding at least once a day for damage and take immediate corrective action.
6. Construct strong wood guard rails, toe boards, and screens if necessary.
7. Always use a ladder to climb up on wood site-built scaffolding.
8. Metal tube scaffolding should be assembled as instructed by the manufacturer.
9. Be certain all bolts and wing nuts are tight after finishing assembling metal tube scaffolding.
10. Metal tube scaffolding often has bars built on the end forming a ladder used to mount it. If it does not, then use a ladder to get up on it. Never use the braces as a ladder. They may bend, reducing the structural integrity of the assembly.
11. When using metal tube scaffolding, observe the designated maximum design loads.
12. If the metal tube scaffolding is on wheels, it is intended for use in areas with solid floors. Be certain to lock or block the wheels before getting on it.
13. The scaffolding must be level, plumb, and on solid footings. It must not have any motion when a person and materials are on it.

SCAFFOLD SAFETY *(cont.)*

14. When possible, secure the scaffolding to the wall.
15. Use only approved planking. It must be certified to carry the expected loads over the distance to be spanned.
16. Planking should overlap at least 12 inches on braces and extend 6 inches but no more than 12 inches beyond end supports. There should be no spaces between the boards.
17. Install guard rails as required by code.
18. Some codes require the installation of toe boards to keep tools and materials from falling off the platform.
19. If people must walk under scaffolding, cover the space below the wood toe board with screen to keep materials from falling on them.
20. The height of the platform should not exceed four times the smallest base dimension.
21. Keep the scaffold platform free of debris and materials no longer needed.
22. Never set a ladder on the platform to reach something higher.
23. Do not move scaffolding until all tools and materials have been removed from the platform.
24. If work is proceeding in an area above the scaffolding, some form of overhead protective covering is needed to protect those on it.
25. Beware of power lines near the work area. Contact the electric company for advice.

SINGLE-POLE WOOD SCAFFOLDS

Single-pole wood scaffolds for light duty.*

	Up to 20 ft. high	Up to 60 ft. high
Maximum Intended Load (lb./sq. ft.)	25	25
Poles	2 × 4"	4 × 4"
Maximum Pole Span—Longitudinal	6'	10'
Maximum Pole Span—Transverse	5'	5'
Runners	1 × 4"	1 1/4 × 9"
Bearers Size and Spacing	2 × 4" – 3'	2 × 4" – 5'
	2 × 6" – 5'	2 × 6" – 5'
Planking	1 1/4 × 9"	2 × 10"
Maximum Vertical Spacing of Horizontal Members	7'	9'
Bracing, Horizontal, Diagonal, and Tie-ins	1 × 4"	1 × 4"

*U.S. Department of Labor, Occupational Safety and Health Administrative publication, *Safety Standards for Scaffolds Used in the Construction Industry*; Final Rule, 1996.

DOUBLE-POLE WOOD SCAFFOLDS

Double-pole wood scaffolds for light duty.*

	Up to 20 ft. high	Up to 60 ft. high
Maximum Intended Load (lb./ft.^{2})	25	25
Poles	2 × 4"	4 × 4"
Maximum Pole Span—Longitudinal	6'	10'
Maximum Pole Span—Transverse	6'	10'
Runners	1 1/4 × 4"	1 1/4 × 9"
Bearers Size and Spacing	2 × 4" – 3'	2 × 4" – 3'
	2 × 6" – 6'	2 × 10" (rough) – 6'
Planking	1 1/4 × 9"	2 × 10"
Maximum Vertical Spacing of Horizontal Members	7'	7'
Bracing, Horizontal, Diagonal, and Tie-ins	1 × 4"	1 × 4"

*U.S. Department of Labor, Occupational Safety and Health Administrative publication, *Safety Standards for Scaffolds Used in the Construction Industry*; Final Rule, 1996.

MANUFACTURER FABRICATED SCAFFOLD PLANKS	
Loads used when fabricated planks are specified by the manufacturer.*	
Rated load capacity	**Intended load over the entire span area (lb./sq. ft.)**
Light duty	25
Medium duty	50
Heavy duty	75
No. of Persons	**Intended load by the number of persons on the scaffold (lb./sq. ft.)**
One	250 pounds at the center of the scaffold
Two	250 pounds 18" to the left and right of the center of the span
Three	One person at the center and one person 18" to the right and left of the center of the span. Each person equals 250 pounds.

**OSHA Report 29 CFR Part 126, Safety Standards for Scaffolds Used in the Construction Industry; Final Rule.*

SOLID WOOD GRADE-STAMPED SCAFFOLD PLANKS

Maximum spans for grade-stamped planks.*

Maximum Intended Nominal Load (lb./sq. ft.)	Maximum Permissible Span	
	Full Thickness Undressed Lumber (ft.)	**Nominal Thickness Undressed Lumber**
25	10	8
50	8	6
75	6	—

**OSHA Report 29 CFR Part 126, Safety Standards for Scaffolds Used in the Construction Industry; Final Rule.*

POWER TOOL SAFETY

General Safety Rules

1. Make certain that the tool is in good repair.
2. The tool should be sharp. Accidents are caused by dull tools.
3. The blades, cutters, belts, or drills should be firmly installed in the unit.
4. Be constantly on the alert. Someone may bump you. The tool may slip. Are you overtired and is your attention lacking?
5. Make certain that the material to be worked is solidly in place. It must not move or slip as it is being worked upon.
6. Keep all guards in place.
7. Wear safety glasses.
8. Do not overcrowd the tool. Let it do its job at the normal pace.
9. After finishing an operation, let the tool stop before you move on to something else.
10. Check the electric cord. It must have a third wire for a ground. It must not be worn.
11. Be certain your source of electrical power is wired so that the third wire (ground) is available. The third wire on your tools is of no use if the source is not grounded.

POWER TOOL SAFETY *(cont.)*

12. Be certain that the switch on the tool is operating correctly. Many tools have switches that turn off the motor when the operator releases pressure on them. These should always be in working order.
13. Do not wear rings, bracelets, necklaces, and so on. They tend to get tangled up in moving machinery.
14. Read the manufacturer's instructions before using a tool. Obey them.
15. When sawing boards, be certain that they are held securely.
16. When using portable electric tools, be certain that they have a grounding wire and that it is connected to an approved ground.
17. Use only high-quality extension cords. A 100-foot cord of No. 12 wire will safely carry 20 amperes (A). A No. 8 wire will carry 35 A. A No. 4 wire will carry 60 A.
18. Examine electric extension cords for cuts and breaks.
19. Unplug any electric tool before you adjust or replace the cutting device.
20. Use only the attachments recommended for the tool by the manufacturer.
21. If working near welders, do not look at the arc.

STATIONARY CIRCULAR SAW SAFETY

1. All adjustments should be made after the blade has stopped.
2. The blade should be the right one for the saw. The diameter and arbor hole should be as specified for the saw.
3. Keep the blade sharp. If it wobbles or is cracked, throw it away.
4. The blade should rise above the wood being cut no more than 1/8 to 1/4".
5. Always cut with the guards in place.
6. Use feather boards and push sticks to protect your fingers. Never, ever get your fingers near the saw blade.
7. Never stand directly behind the blade when cutting. It often kicks back a board, which will injure anyone behind the saw.
8. Use the miter gauge for crosscutting. Never use the fence for crosscutting.
9. When ripping, keep the smooth straight edge of the board next to the fence.
10. Always keep the table free of scraps. Push them off with a wood push stick. Do not use your fingers.
11. Before operating make certain that everything—including the fence, miter gauge, and depth control—is locked tightly.

STATIONARY CIRCULAR SAW SAFETY *(cont.)*

12. Warped boards will bind the blade and kick back. Do not cut warped boards.
13. The person who may be removing boards as they are cut should not pull on them. This might choke the saw and cause a kickback. They should only support the wood.
14. Never cut stock freehand. Always use the miter gauge for crosscutting and the fence for ripping.
15. Do not adjust the saw while it is running.

RADIAL ARM SAW SAFETY

1. Keep all guards in place.
2. After adjusting and before starting, make certain that all locking handles are secure.
3. Use the anti-kickback fingers. Keep them about 1/8" above the top of the board.
4. Make certain that the saw is set so that it is lower at the back. This tilts the table and arm to the rear. This keeps the yoke from moving forward on its own. It helps to return the saw to the rear position when a cut is finished.
5. The stock must be pressed firmly against the table and fence. The stock must rest flat on the table. If stock is bowed, it will bind the saw.

RADIAL ARM SAW SAFETY *(cont.)*

6. After crosscutting a board, return the saw to its rear position.
7. Let the saw reach its full speed before starting to cut.
8. When ripping, be absolutely certain to feed the stock into the blade so that the teeth are turning upward and toward you.
9. Before starting a cut, be certain that your fingers are out of the path of the blade. This is a difficult saw to guard. Do not depend on the guards to protect your fingers.
10. Keep the table free of scraps.
11. When crosscutting, the saw tends to feed itself into the work. As you move it toward you, control forward movement with the handle. It may be necessary to resist forward movement if the saw feeds too fast.
12. Do not attempt to rip stock unless it has one square edge to place against the fence.

POWER MITER SAW SAFETY

1. Keep all guards in place.
2. Make sure that the saw is tightly locked at the angle it is to cut.
3. Keep the stock pressed firmly against the table and fence. Be especially careful when cutting moldings.

POWER MITER SAW SAFETY *(cont.)*

4. Be certain that hands are clear of the cutting area before swinging the saw down for a cut.
5. Wear safety glasses or a face shield to keep eyes free of sawdust.
6. Keep the saw blade sharp.

SABER AND RECIPROCATING SAW SAFETY

1. Use sharp blades. Worn or bent blades should be discarded.
2. Be certain that the blade is securely mounted in the chuck. Follow the manufacturer's directions.
3. Unplug the machine before changing blades.
4. Never use an ungrounded saw.
5. Always let the motor get to full operating speed before starting a cut.
6. Be certain that the space below the board being cut is clear.
7. Do not let your fingers hang below the board. Remember that the blade is below the board.
8. Before cutting, make certain that the work is properly supported.
9. Do not walk around with the saw running. Let it stop before you move.

JOINTER SAFETY

1. Never make any adjustments when the machine is running.
2. Before starting, check the handles to make certain that everything is tight.
3. The knives should be replaced only by someone trained in doing this. If replacement is improperly done, they can fly out of the cutter head.
4. Always keep all guards in place.
5. Never pass your hands across the cutter head. Keep them clear on one side or the other.
6. When finishing a cut on small stock, use wood push sticks or push blocks.
7. Never run stock shorter than 12" or thinner than 3/8".
8. Never take too deep a cut. Follow the manufacturer's directions.
9. Do not joint stock with knots. They will chip the knives and often come loose and be thrown.
10. Always use sharp knives. Dull or improperly ground knives cause vibration and will kick the stock out of your hands.
11. Knives must be sharpened by someone having the proper equipment.

PORTABLE ELECTRIC PLANE SAFETY

1. Always hold the plane in two hands.
2. Follow the manufacturer's directions for adjusting the plane.
3. Clamp the work securely so that it will not kick back.
4. Disconnect the power plug before making adjustments or replacing cutters.

PORTABLE ROUTER SAFETY

1. Always unplug the router before changing cutters.
2. Make certain that the cutter is tight in the chuck.
3. Wear eye shields.
4. Clamp the work to be routed so that it will not be thrown.
5. When starting the router, torque from the motor tends to spin it. Hold it tightly.
6. When operating, hold the router with both hands.

PORTABLE ELECTRIC DRILL SAFETY

1. Use double-insulated drills or wireless drills in wet places.
2. Be certain that the drill is tight in the chuck. Remove the key before starting the motor.
3. Use eye protection.
4. When drilling, do not force the drill beyond its normal cutting speed.
5. Use sharp drills.

POWER NAILER AND STAPLER SAFETY

1. Since operation principles vary, study the manufacturer's operating manual.
2. Be certain to use the type of nail or staple recommended.
3. Be certain that electric units are grounded.
4. Use the recommended air pressure for pneumatic units.
5. Treat the machine as you would a gun. Do not point it at yourself or others.
6. When not in use, disconnect from the power source to prevent accidental releasing of fasteners.
7. Always keep the nailer or stapler tight against the surface to receive the staple or nail.

SIZE OF EXTENSION CORDS FOR PORTABLE TOOLS

	Full-Load Rating of the Tool in Amperes at 115 Volts					
	0 to 2.0	2.10 to 3.4	3.5 to 5.0	5.1 to 7.0	7.1 to 12.0	12.1 to 16.0
Cord Length, Feet	Wire Size (American Wire Gauge)					
25	18	18	18	16	14	14
50	18	18	18	16	14	12
75	18	18	16	14	12	10
100	18	16	14	12	10	8
200	16	14	12	10	8	6
300	14	12	10	8	6	4
400	12	10	8	6	4	4
500	12	10	8	6	4	2
600	10	8	6	4	2	2
800	10	8	6	4	2	1
1000	8	6	4	2	1	0

LASER SAFETY

Laser devices are widely used on construction sites. They can be dangerous to eyesight; however, most available are low-powered, reducing the possibility of danger. The Occupational Safety and Health Act (OSHA) lists various regulations related to laser use. Also, manufacturers issue information pertaining to safety procedures. Following are some frequently recommended considerations.

1. Post warning signs around the area where lasers are in use.
2. Do not enter areas marked off with warning signs.
3. Allow only those trained to operate the laser or make adjustments to it. Require that they carry their certified Operator's Card when using the laser.
4. Do not look directly into the beam. Be aware that the beam can be deflected by polished surfaces, such as polished metal, and can direct the beam in an unplanned direction.
5. Laser retinal burns can be painless, and those exposed may be unaware of possible damage. It is essential that all personnel in or around laser operations wear laser safety goggles. The goggles must be labeled with the laser wavelengths they will protect, the optical density of those wavelengths, and the visible light transmission. The laser should have these ratings on its label.

LASER SAFETY *(cont.)*

6. Do not play games with the beam such as pointing it at a fellow worker.
7. If there is a possible injury, seek medical care immediately.
8. When the laser is not in use, shut it off and cap it.

FIRE SAFETY

Classes of Fires

Class A fires involve solid combustible materials, such as wood, cloth, and paper, that are extinguishable by water or cooling or coating with a suitable chemical powder.

Class B fires involve flammable liquids, oil, or gasoline that are smothered with a cooling agent.

Class C fires involve electrical equipment where a malfunction such as a short will keep the source of the fire active until the current is cut off. The extinguishing agent must be nonconductive.

Class D fires involve combustible metals such as sodium and magnesium. These must be controlled with special powders and trained fire personnel.

Class K fires involve combustible cooking materials such as vegetable or animal oils or fats.

TYPES OF PORTABLE FIRE EXTINGUISHERS	
Type	**Class**
Water Type	Class A fires
Carbon Dioxide	Class B and C fires
Various Dry Chemicals	Class B and C fires
Wet Chemical	Class B, C, K fires
Halon	Class B, C fires

Many old extinguishers are no longer approved and should be removed from service. Check with the National Fire Protection Association, www.nfpa.org.

MARKING SYSTEM SYMBOLS

Indicates extinguisher suitability according to class of fire.

For Class A Types

For Class A, B Types

(1) AFFF
(2) FFFP

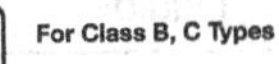

For Class B, C Types

(1) Carbon Dioxide
(2) Dry Chemical
(3) Halogenated Agents

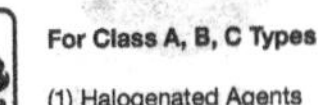

For Class A, B, C Types

(1) Halogenated Agents
(2) Multipurpose dry Chemicals

For Class K Types

(1) Wet Chemical-based
(2) Dry Chemical-based

LETTER-SHAPED FIRE EXTINGUISHER MARKINGS

Ordinary A Combustibles	Extinguishers suitable for Class A fires should be identified by a triangle containing the letter "A". If colored, the triangle is colored green.*
Flammable B Liquids	Extinguishers suitable for Class B fires should be identified by a square containing the letter "B". If colored, the square is colored red.*
Electrical C Equipment	Extinguishers suitable for Class C fires should be identified by a circle containing the letter "C". If colored, the circle is colored blue.*
Combustible D Metals	Extinguishers suitable for fires involving metals should be identified by a five-pointed star containing the letter "D". If colored, the star is colored yellow.

TYPES OF FIRE EXTINGUISHERS

Fire extinguishers are required on the construction site. Detailed information is available from the National Fire Protection Association publications NFPA 10 Standard for Portable Fire Extinguishers and NFPA 412, Standard for Safeguarding Construction, Alterations and Demolition Operations. www.nfpa.org and http://catalognfpa.org.

CLASS A	Class A includes fires involving typically available combustible materials such as wood, plastics, and fabrics. Water, water solutions, and dry chemicals are typical extinguishing agents.
CLASS B	Class B includes fires involving flammable liquids and pressurized gases. Dry chemical extinguishers are used.
CLASS C	Class C includes fires involving energized electrical equipment such as electric motors and power panels. A nonconductive extinguishing agent such as carbon dioxide or one using a halogenated agent such as Halon 1211 is used.
CLASS D	Class D includes fires involving combustible metals such as magnesium, zirconium, titanium, and phasium. These are in the form of a powder, flakes, shavings, or chips. Each fire extinguisher is labeled for use on a specific combustible metal hazard.

VISUAL ELECTRIC SAFETY INSPECTIONS

Following are references for making visual safety electrical inspections found in the National Electrical Code, ANSI/NFPA 1993. The chart indicates the sections and page numbers of this information.

NEC Section	Illustrated on Pg.	NEC Section	Illustrated on Pg.
110-7	19, 23	318-8 to 365-6	26
110-11	5	370-5	5
110-12	27	370-6	3
110-13(a)	28	370-16	3
110-14(b)	22	370-21	4
110-16	6	370-13	25
110-17	9, 12	370-15	7
110-8	8	370-18(c)	7
110-21	11	370-19	6
110-22	10, 11	384-2	6
110-31	12	400-3	19, 23
240-24	13	400-7	21
240-41	13	400-8	21
250-42	14	400-9	22
250-43	14	400-10	24
250-44	14	410-5	29
250-45	17, 19	410-6	29
250-45(d)	18	410-15	28
250-51	14, 15, 17	410-16(a)	28
250-59	7, 17, 19	410-42	24
250-92(b)	15	410-56(f)	8, 19
250-92(c)(1)	15	410-58	7
250-113	16	410-58(b)(3)	19
300-6	5	422-10	30
305-4	20	450-43(c)	12
305-4(d)	20	610-61	14
305-6	20	500-504	31

BLASTING ON THE CONSTRUCTION SITE

Explosives are sometimes used on the construction site to remove stumps, move earth, break rock, and excavate an area such as a ditch, trench, or pipeline. The handling, storage, and use of explosives are carefully regulated procedures. Manufacturers of explosives can provide useful information about their products. In addition, considerable information is available from organizations such as:

The Institute of Makers of Explosives
www.ime.org or e-mail: infor@ame.org

National Fire Protection Association
www.nfa.org

American National Standards Institute
www.ansi.org

U.S. Occupational Safety and Health Admin.
http://www.dot.gov

BLASTING SAFETY RECOMMENDATIONS

1. Clear the area of all personnel and post it as a danger area. Install barriers as needed to secure the area.
2. Post guards around the blast area.
3. Sound a warning indicating a detonation is about to occur.
4. Those who must be in close proximity to the blast area should wear personal protective equipment, including respiratory devices.

BLASTING SAFETY RECOMMENDATIONS *(cont.)*

5. When using explosives to fragment rock or disperse soil, there will be flying particles over a large area, so all equipment and workers must be moved a safe distance from the blast site.
6. If a misfire occurs, only experienced, qualified explosive personnel should attempt to handle the situation.
7. Occasionally, the ground displaced during a detonation will trap gasses below the surface. Excavate the blast area as soon as possible to release the trapped gasses.
8. All blasting operations must be supervised by a person who is legally licensed to use explosives and has the required permits.
9. The person with the BLASTING PERMIT must keep a record of all operations involving explosive materials and keep a daily record of explosive materials received, fired, and otherwise disposed.
10. Keep all explosive materials stored in code-approved magazines and keep them locked. Post the area warning of the presence of explosives.
11. Follow local, state, and federal laws regulating the transportation of explosives.
12. Explosives being transported should always be accompanied by an experienced employee. Never leave them unattended.
13. The motor vehicle to be used to move explosive materials must meet NFPA standards.

CURRENT/PRIOR DEPARTMENT OF TRANSPORTATION EXPLOSIVE CLASSIFICATION SYSTEMS	
Current Classification	**Prior DOT Classification**
Division 1.1	Class A Explosives
Division 1.2	Class A or Class B Explosives
Division 1.3	Class B Explosives
Division 1.4	Class C Explosives
Division 1.5	Blasting Agents
Division 1.6	No Applicable Hazard Class

CLASSIFICATION OF EXPLOSIVE MATERIALS

For transportation purposes, explosives are classified by DOT in accordance with 49 CFR, and under these regulations all explosives are listed as Hazard Class 1 materials. Class 1 materials are divided into six divisions to note the principal hazard of the explosive. These six divisions are as follows:

DIVISION 1.1	Explosives that have a mass explosion hazard. A mass explosion is one that affects the entire load instantaneously. Typical examples: dynamite, detonator (cap) sensitive emulsions, slurries, water gels, cast boosters, and mass detonating detonators.
DIVISION 1.2	Explosives that have a projection hazard but not a mass explosion hazard. Typical examples: certain types of ammunition, mines, and grenades.
DIVISION 1.3	Explosives that have a fire hazard and either a minor blast hazard or a minor projection hazard or both, but not a mass explosion hazard. Typical examples: certain types of fireworks, propellants, and pyrotechnics.

CLASSIFICATION OF EXPLOSIVE MATERIALS *(cont.)*	
DIVISION 1.4	Explosives that present a minor explosion hazard. The explosive effects are largely confined to the package, and no projection of fragments of appreciable size or range is to be expected. An external fire must not cause virtually instantaneous explosion of almost the entire contents of the package. Typical examples: safety fuse and certain electric, electronic, and nonelectric detonators.
DIVISION 1.5	Explosives that are very insensitive. This division is comprised of substances that have a mass explosion hazard but are so insensitive that there is very little probability of initiation or of transition from burning to detonation under normal conditions of transport. Typical examples: blasting agents, ANFO, non-cap-sensitive emulsions, blends, slurries, water gels, and other explosives that require a booster for initiation.
DIVISION 1.6	Extremely insensitive explosives that do not have a mass explosive hazard. This division is comprised of articles that contain only extremely insensitive detonating substances and that demonstrate a negligible probability of accidental initiation of propagation. The risk from articles of Division 1.6 is limited to the explosion of a single article. Generally, commercial explosives are not classified as Division 1.6.

FORMER EXPLOSIVE CLASSIFICATIONS	
Until October 1, 1991, DOT classified explosive materials as Explosives A, B, or Blasting Agent. Essentially these classes were as follows:	
Class A Explosives	Explosives that detonate or have maximum hazard. Typical examples: dynamites, cast boosters, cap-sensitive emulsions, slurries, water gels, and certain initiators and detonators.
Class B Explosives	Explosives that function by rapid combustion rather than detonation. Typical examples: pyrotechnics, certain propellants and fireworks, flash powders, and signal devices.
Class C Explosive	Explosives that contain Class A or B explosives as components but in restricted quantities that present minimum hazard. Typical examples: safety fuse, igniter cord, certain detonators, and specialty explosive devices.
Blasting Agents	Explosive materials that have been tested and found to be so insensitive that it is unlikely that they will initiate or detonate in a fire during normal transportation conditions. Typical examples: ANFO, blends, emulsions, slurries, and water gels that are not cap-sensitive.

U.S. BUREAU OF ALCOHOL, TABACCO, AND FIREARMS CLASSIFICATIONS	
High Explosives	Explosive materials that can be caused to detonate by means of a blasting cap when unconfined (for example, dynamite, flash powders, and bulk salutes). Typical examples: dynamites, cast boosters, and certain emulsions, slurries, and water gels.
Low Explosives	Explosive materials that can be caused to deflagrate when confined (for example, black powder, safety fuses, igniters, igniter cords, fuse lighters, and "display fireworks" classified as UN0333, UN0334, or UN0335 by the U.S. Department of Transportation regulations at 49 CFR 172.101, except for bulk salutes). Typical examples: black powder, safety fuse, igniters, and fuse lighters.
Blasting Agents	Any material or mixture, consisting of fuel and oxidizer, that is intended for blasting and not otherwise defined as an explosive; if the finished product, as mixed for use or shipment, cannot be detonated by means of a number 8 test blasting cap when unconfined. A number 8 test blasting cap is one containing 2 grams (g) of a mixture of 80 percent mercury fulminate and 20 percent potassium chlorate, or a blasting cap of equivalent strength. An equivalent strength cap comprises 0.40-0.45 grams of PETN base charge pressed in an aluminum shell with bottom thickness not to exceed to 0.03 of an inch (in.), to a specific gravity of not less than 1.4 g/cc, and primed with standard weights of primer depending on the manufacturer. Typical examples: ANFO, blends, and certain emulsions, slurries and water gels.
From *Safety in the Transportation, Storage, Handling and Use of Explosive Materials*, Courtesy The Institute of Makers of Explosives, www.ime.org	

EXPLOSIVE MATERIALS

1. Dynamite is available in several basic formulations: straight, ammonia, gelatin, extra-gelatin, and semi-gelatin
2. Water gels
3. Emulsions
4. Slurries
5. Ammonium nitrate and fuel oil
6. Black powder
7. Blasting caps
8. Safety fuses
9. Detonating cord
10. Igniter cord
11. Igniters

MONITOR THE SEISMIC EFFECT OF BLASTING

Blasting may cause damage to a building nearby or lead to alleged charges that the building was damaged. Some carefully examine and make notes about the condition of the building and take photos of existing damage to use to refute later charges of damage. Before blasting begins, seismic recording instruments can be placed in the vicinity of the blasting to monitor the magnitude of the effects of the blast. Often the insurance company covering the blasting activity will handle the monitoring.

CHAPTER 3
Architectural Design Information

ABBREVIATIONS USED ON ARCHITECTURAL DRAWINGS	
A	
Above Finished Counter	**AFC**
Above Finished Floor	**AFF**
Above Finished Grade	**AFG**
Acoustic	**AC**
Acoustic Plaster	**AC PL**
Acoustic Tile	**AC T**
Actual	**ACT**
Additional	**ADD**
Adhesive	**ADH**
Adjustable	**ADJ**
Aggregate	**AGGR**
Air Conditioning	**AIR COND**
Air Conditioning Unit	**ACU**
Alternating Current	**AC**
Aluminum	**AL or ALUM**
Amount	**AMT**
Ampere	**AMP or A**
Anchor Bolt	**AB**
Angle (in degrees)	∢
Angle (structural)	**L**
Approximate	**APPROX**
Architectural	**ARCH**
Area	**A**
Area Drain	**AD**
Asbestos	**ASB**
Asphalt	**ASPH**
Asphaltic Concrete	**ASPH CONC**
Assembly	**ASSEM**
At	**@**
Automatic	**AUTO**
Avenue	**AVE**
Average	**AVG**
B	
Balcony	**BALC**
Basement	**BSMT**
Baseplate	**BP**
Bathroom	**B**
Bathtub with Shower	**BTS**
Batten	**BATT**
Beam	**BM**

ABBREVIATIONS USED ON ARCHITECTURAL DRAWINGS *(cont.)*

Beam, Standard	**S BM**
Beam, Wide Flange	**W BM**
Bearing	**BRG**
Bearing Plate	**B PL**
Bedroom	**BR**
Bench Mark	**BM**
Between	**BET**
Beveled	**BEV**
Bidet	**BDT**
Block	**BLK**
Blocking	**BLKG**
Blower	**BLO**
Board	**BD**
Board Feet	**BD FT**
Both Sides	**BS**
Both Ways	**BW**
Bottom	**BOT**
Boulevard	**BLVD**
Bracket	**BRKT**
Brass	**BR**
Brick	**BRK**
British Thermal Unit	**BTU**
Broom Closet	**BC**
Building	**BLDG**
Building Line	**BL**
Built-in	**BLT-IN**
Built-up	**BU**
Buzzer	**BUZ**
By (used as 2 × 4)	×
C	
Cabinet	**CAB**
Candela	**cd**
Candlepower	**CP**
Carpet	**CPT**
Cast Iron	**CI**
Cast in Place	**CIP**
Catch Basin	**CB**
Caulking	**CLKG**
Ceiling	**CLG**
Ceiling Diffuser	**CD**
Celsius	**C**
Cement	**CEM**
Cement Plaster	**CEM PLAS**
Center	**CTR**
Center to Center	**C to C**
Centerline	**℄ or CL**
Centimeter	**cm**
Ceramic	**CER**
Ceramic Tile	**CT**
Chalkboard	**CHKBD**
Chamber	**CHAM**
Channel (structural)	**C**
Check	**CHK**
Cinder Block	**CLN BL**
Circle	**CIR**
Circuit	**CKT**

ABBREVIATIONS USED ON ARCHITECTURAL DRAWINGS *(cont.)*

Circuit Breaker	**CIR BKR**
Class	**CL**
Classroom	**CLRM**
Cleanout	**CO**
Clear	**CLR**
Closet	**CLO or CL**
Clothes Dryer	**CL D**
Cold Water	**CW**
Column	**COL**
Combination	**COMB**
Common	**COM**
Concrete	**CONC**
Concrete Block	**CONC B**
Concrete Masonry Unit (concrete block)	**CMU**
Construction	**CONST**
Continuous	**CONT**
Contractor	**CONTR**
Contractor Furnished	**CF**
Control Joint	**CJ**
Copper	**COP or CU**
Corridor	**CORR**
Counter	**CTR**
Countersink	**CSK**
Courses	**C**
Cover	**COV**
Cross Section	**X-SECT**
Cubic	**CU**
Cubic Feet	**CU FT**
Cubic Feet per Minute	**CFM**
Cubic Yard	**CU YD**
D	
Damper	**DMPR**
Decibel	**db**
Deep, Depth	**DP**
Degree	**° or DEG**
Department	**DEPT**
Detail	**DET**
Diagonal	**DIAG**
Diagram	**DIAG**
Diameter	**DIA**
Diffuser	**DIFF**
Dimension	**DIM**
Dining Room	**DIN RM**
Direct Current	**DC**
Dishwasher	**DW**
Disposal	**DISPL**
Distance	**DIST**
Ditto	**DO**
Divided or Division	**DIV**
Door	**DR**
Double	**DBL**
Double-hung	**DH**

ABBREVIATIONS USED ON ARCHITECTURAL DRAWINGS *(cont.)*

Double-strength (glass)	**DS**
Douglas Fir	**DF**
Dowel	**DWL**
Down	**DN**
Downspout	**DS**
Drain	**D or DR**
Drawing	**DWG**
Drinking Fountain	**DF**
Dryer	**D**
Drywall	**DW**
Duplicate	**DUP**
E	
Each	**EA**
Each Face	**EF**
Each Way	**EW**
East	**E**
Elbow	**ELL**
Electric(al)	**ELECT**
Electric Panel Board	**EPB**
Elevation	**EL or ELEV**
Elevator	**ELEV**
Enclosure	**ENCL**
Engineer	**ENGR**
Entrance	**ENT**
Equal	**EQ**
Equipment	**EQUIP**
Estimate	**EST**
Excavate	**EXC**
Exhaust	**EXH**
Existing	**EXIST'G**
Expansion Bolt	**EB**
Expansion Joint	**EXP JT**
Exposed	**EXPO**
Extension	**EXT**
Exterior	**EXT**
Exterior Grade	**EXT GR**
F	
Fabricate	**FAB**
Face Brick	**FB**
Face of Studs	**FCS**
Fahrenheit	**F**
Feet	**' or FT**
Feet per Minute	**FPM**
Fiberglass-reinforced Plastic	**FRP**
Figure	**FIG**
Finish(ed)	**FIN**
Finished All Over	**FAO**
Finished Floor	**FIN FL**
Finished Floor Elevation	**FFE**
Finished Grade	**FIN GR**
Finished Opening	**FO**
Firebrick	**FBRK**
Fire-extinguisher	**F EXT**

ABBREVIATIONS USED ON ARCHITECTURAL DRAWINGS *(cont.)*

Fire-extinguisher Cabinet	**FEC**
Fire Hose Cabinet	**FHC**
Fire Hydrant	**FH**
Fireproof	**FP**
Fitting	**FTG**
Fixture	**FIX**
Flammable	**FLAM**
Flange	**FLG**
Flashing	**FL**
Flexible	**FLEX**
Floor	**FLR**
Floor Drain	**FD**
Floor Sink	**FS**
Flooring	**FLG**
Fluorescent	**FLUOR**
Folding	**FLDG**
Foot	**' or FT**
Footing	**FTG**
Forward	**FWD**
Foundation	**FND**
Four-way	**4-W**
Frame	**FR**
Front	**FR**
Full Size	**FS**
Furnace	**FURN**
Future	**FUT**
G	
Gallon	**GAL**
Galvanized	**GALV**
Galvanized Iron (galvanized steel)	**GI**
Gauge	**GA**
Glass	**GL**
Glass block	**GL BL**
Glazed Structural Unit	**GSU**
Glue-laminated	**GLUELAM**
Government	**GOVT**
Grade	**GR**
Grade Beam	**GB**
Grating	**GRTG**
Gravel	**GVL**
Grille	**GR**
Ground	**GRND**
Grout	**GT**
Gypsum	**GYP**
H	
Hall	**H**
Hardboard	**HBD**
Hardware	**HDW**
Hardwood	**HDWD**
Head	**HD**
Header	**HDR**
Heater	**HTR**
Heating	**HTG**
Heating/Ventilating/ Air Conditioning	**HVAC**

ABBREVIATIONS USED ON ARCHITECTURAL DRAWINGS *(cont.)*

Heavy Duty	HD
Height	HT
Hexagonal	HEX
Highway	HWY
Hollow Core	HC
Hollow Metal	HM
Horizontal	HORIZ
Horsepower	HP
Hose Bibb	HB
Hospital	HOSP
Hot Water	HW
Hot Water Heater	HWH
Hour	HR
House	HSE
Hundred	C
I	
Illuminate	ILLUM
Incandescent	INCAND
Inch(es)	" or IN.
Inflammable	INFL
Information	INFO
Inside Diameter	ID
Inside Face	IF
Inspect(ion)	INSP
Install	INST
Insulate(d)(ion)	INS
Interior	INT
Interior Grade	INT GR
J	
Jamb	JMB
Janitor's Sink	JS
Janitor's Closet	JC
Joint	JT
Joist	JST
Joist and Plank	J & P
Junction	JCT
Junction Box	J-BOX
K	
Kelvin	K
Kiln Dried	KD
Kilogram	kg
Kilovolt	KV
Kilowatt	KW
Kitchen	KIT
Kitchen Cabinet	KCAB
Kitchen Sink	KSK
Knockout	KO
L	
Laboratory	LAB
Laminate(d)	LAM
Landing	LDG
Latitude	LAT
Laundry	LAU
Lavatory	LAV
Left	L
Length	LGTH

ABBREVIATIONS USED ON ARCHITECTURAL DRAWINGS *(cont.)*

Level	**LEV**
Library	**LIB**
Light (pane of glass)	**LT**
Linear Feet	**LIN FT**
Linen Closet	**L CL**
Linoleum	**LINO**
Live Load	**LL**
Living Room	**LR**
Location	**LOC**
Long	**LG**
Longitude	**LNG**
Lumber	**LBR**
M	
Manhole	**MH**
Manufacture(r)	**MFR**
Marble	**MRB**
Mark	**MK**
Masonry	**MAS**
Masonry Opening	**MO**
Material	**MAT**
Maximum	**MAX**
Mechanical	**MECH**
Medicine Cabinet	**MC**
Medium	**MED**
Membrane	**MEMB**
Metal	**MET**
Metal Lath and Plaster	**MLP**
Meter	**m**
Millimeter	**mm**
Minimum	**MIN**
Mirror	**MIRR**
Miscellaneous	**MISC**
Modular	**MOD**
Molding	**MLDG**
Mullion	**MULL**
N	
Noise Reduction Coefficient	**NRC**
Nominal	**NOM**
North	**N**
Not Applicable	**NA**
Not in Contract	**NIC**
Not to Scale	**NTS**
Number	**NO. or #**
O	
Oak	**O**
Office	**OFF**
On Center	**OC**
One-way	**1-W**
Open Web	**OW**
Opening	**OPG**
Opposite	**OPP**
Opposite Hand	**OPH**
Ounce	**OZ**
Outside Diameter	**OD**
Outside Face of Concrete	**OFC**

ABBREVIATIONS USED ON ARCHITECTURAL DRAWINGS *(cont.)*

Outside Face of Studs	**OFS**
Overhead	**OH**
P	
Painted	**PTD**
Pair	**PR**
Panel	**PNL**
Parallel	**PAR or II**
Partition	**PTN**
Passage	**PASS**
Pavement	**PVMT**
Penny (nail size)	**d**
Per	**/**
Percent	**%**
Perforate	**PERF**
Perimeter	**PERIM**
Perpendicular	**PERP or ⊥**
Pierce	**PC**
Plan	**PLN**
Plaster	**PLS**
Plasterboard	**PL BD**
Plastic	**PLAS**
Plastic Tile	**PLAS T**
Plate	**PL or ℞**
Plate Glass	**PL GL**
Platform	**PLAT**
Plumbing	**PLMB**
Plywood	**PLY**
Polished	**POL**
Polyethelyne	**POLY or PE**
Polystyrene	**PS**
Polyvinyl Chloride	**PVC**
Position	**POS**
Pound	**LB or #**
Pounds per Square Foot	**PFS**
Pounds per Square Inch	**PSI**
Precast	**PRCST**
Prefabricated	**PREFAB**
Preliminary	**PRELIM**
Premolded	**PRMLD**
Property	**PROP**
Public Address System	**PA**
Pull Chain	**PC**
Pushbutton	**PB**
Q	
Quantity	**QTY**
Quarry Tile	**QT**
Quart	**QT**
R	
Radiator	**RAD**
Radius	**RAD**
Random Length and Width	**RL&W**
Range	**R**
Receptacle	**RECP**

ABBREVIATIONS USED ON ARCHITECTURAL DRAWINGS *(cont.)*

Recessed	**REC**
Redwood	**RDWD**
Reference	**REF**
Refrigeration	**REF**
Refrigerator	**REFRIG**
Register	**REG**
Reinforced, Reinforcing	**REINF**
Reinforcing Bar	**REBAR**
Required	**REQ**
Resilient	**RES**
Resistance	**RES**
Return	**RET**
Revision	**REV**
Revolutions per Minute	**RPM**
Right	**R**
Right hand	**RH**
Riser	**R**
Road	**RD**
Roof	**RF**
Roof Drain	**RD**
Roofing	**RFG**
Room	**RM**
Rough	**RGH**
Rough Opening	**RO**
Round	**RD or Ø**
Rubber Base	**RB**
Rubber Tile	**RBT**

S

Schedule	**SCH**
Screw	**SCR**
Second	**s or SEC**
Section	**SECT**
Select	**SEL**
Select Structural	**SS**
Self-closing	**SC**
Service	**SERV**
Sewer	**SEW**
Sheathing	**SHTHG**
Sheet	**SHT**
Sheet Metal	**SM**
Shower	**SH**
Siding	**SDG**
Sill Cock	**SC**
Similar	**SIM**
Single-hung	**SH**
Single-strength (glass)	**SS**
Sink	**SK**
Slop Sink	**SS**
Socket	**SOC**
Soil Pipe	**SP**
Solid Block	**SLD BLK**
Solid Core	**SC**
South	**S**
Specifications	**SPEC**
Square	**□ or SQ**

ABBREVIATIONS USED ON ARCHITECTURAL DRAWINGS *(cont.)*

Square Feet	**SF or □'**
Square Inches	**SQ IN or □"**
Stainless Steel	**SST**
Stairs	**ST**
Stand Pipe	**ST P**
Standard	**STD**
Station Point	**SP**
Steel	**STL**
Stirrup	**STIR**
Stock	**STK**
Storage	**STO**
Storm Drain	**SD**
Street	**ST**
Structural	**STR**
Structural Clay Tile	**SCT**
Substitute	**SUB**
Supply	**SUP**
Surface	**SUR**
Surface Four Sides	**S4S**
Surface Two Edges	**S2E**
Suspended Ceiling	**SUSP CLG**
Switch	**S or SW**
Symbol	**SYM**
Symmetrical	**SYM**
Synthetic	**SYN**
System	**SYS**

T

Tack Board	**TK BD**
Tangent	**TAN**
Tar and Gravel	**T & G**
Technical	**TECH**
Tee	**T**
Telephone	**TEL**
Television	**TV**
Temperature	**TEMP**
Temporary	**TEMP**
Terra-cotta	**TC**
Terrazzo	**TZ**
Thermostat	**THERMO**
Thickness	**THK**
Thousand	**M**
Thousand Board Feet	**MBM**
Three-way	**3-W**
Threshold	**THR**
Toilet	**TOL**
Tongue and Groove	**T & G**
Top of wall	**TW**
Tread	**TR**
Two-way	**2-W**
Typical	**TYP**

U

Undercut Door	**UCD**
Underwriters' Laboratory, Inc.	**U.L.**

ABBREVIATIONS USED ON ARCHITECTURAL DRAWINGS *(cont.)*

Unfinished	**UNFIN**
Urinal	**UR**
Utility	**UTIL**
V	
V-joint	**VJ**
Vanishing Point	**VP**
Vanity	**VAN**
Vapor Barrier	**VB**
Vent Through Roof	**VTR**
Vent Stack	**VS**
Ventilation	**VENT**
Ventilator	**V**
Vertical	**VERT**
Vertical Grain	**VG**
Vestibule	**VEST**
Vinyl	**VIN**
Vinyl Base	**VB**
Vinyl Tile	**VT**
Vinyl Wall Covering	**VWC**
Vitreous Clay Tile	**VCT**
Volt	**V**
Volume	**VOL**
W	
Wainscot	**WSCT**
Wall Cabinet	**VCAB**
Wall Vent	**WV**
Waste Stack	**WS**
Water	**W**
Water Closet (toilet)	**WC**
Water Heater	**WH**
Waterproof	**WP**
Watt	**W**
Weatherproof	**WP**
Weephole	**WH**
Weight	**WT**
Welded Wire Fabric	**WWF**
West	**W**
Wet Bulb	**WB**
White Pine	**WP**
Wide Flange (structural)	**W**
Window	**WDW**
With	**w/**
Without	**WO**
Wood	**WD**
Working Point	**WPT**
Wrought Iron	**WI**
Y	
Yard	**YD**
Yellow Pine	**YP**
Z	
Zinc	**ZN**

ARCHITECTURAL SCALES

SCALES USED ON RESIDENTIAL AND COMMERCIAL DRAWINGS

U.S. Customary Scales (in.)	
Floor Plan	1/4" = 1'-0"
Foundation Plan	1/4" = 1'-0"
Elevations	1/4" = 1'-0"
Construction Details	3/4" to 1 1/2" = 1'-0"
Wall Sections	3/4" to 1 1/2" = 1'-0"
Cabinet Details	1/8" to 1/2" = 1'-0"
Site Plan	1" = 20' or 40'
ISO Metric Scales (mm)	
Floor Plan	1:50
Foundation Plan	1:50
Elevations	1:50
Construction Details	1:20 and 1:10
Wall Sections	1:20 and 1:10
Cabinet Details	1:50 and 1:25
Site Plan	1:100

ARCHITECTURAL SCALES *(cont.)*		
SCALES USED ON COMMERCIAL BUILDING DRAWINGS		
Drawing	**U.S. Customary (in.)**	**SI Metric (mm)**
Floor Plan	1/8" = 1'-0"	1:100
Foundation Plan	1/8" = 1'-0"	1:100
Elevations	1/8" = 1'-0"	1:100
Construction Details	1/2" to 1 1/2" = 1'-0"	1:25 or 1:10
Interior Details	1/2" = 1'-0"	1:25
Building Sections	1/4"= 1'-0"	1:50
Framing Details	1/8" = 1'-0"	1:100
Lighting, Electrical, Heating, Plumbing, Air Conditioning Plans	1/8" = 1'-0"	1:100
Site Plan	1" = 20' or 40'	1:200, or 1:500

GRAPHIC SYMBOLS USED ON ARCHITECTURAL DRAWINGS

Following are some of the most frequently used graphic symbols. Variations of these can be found on drawings made by those working in the various areas presented on such drawings.

STANDARD LINE SYMBOLS

Visible Line – Thick

Hidden Line – Thin

Center Line – Thin

Property Line – Thick

Long-Break Line – Thin

Short-Break Line – Thick

Cutting Plane Lines – Thick

Section Lining – Thin

Phantom Line – Thin

GRAPHIC SYMBOLS USED ON ARCHITECTURAL DRAWINGS *(cont.)*

CONTOURS

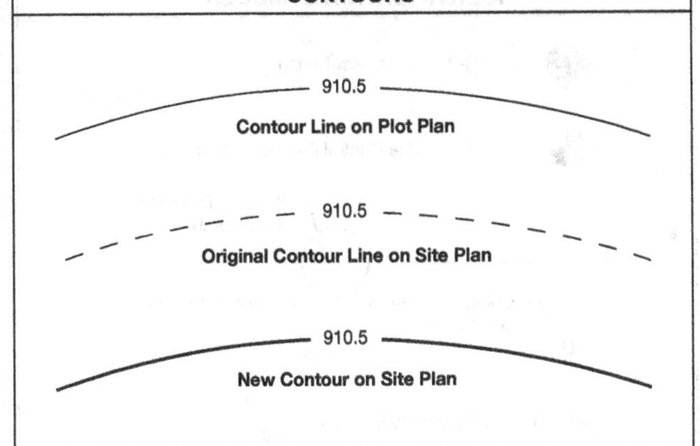

DIMENSIONS AND NOTES

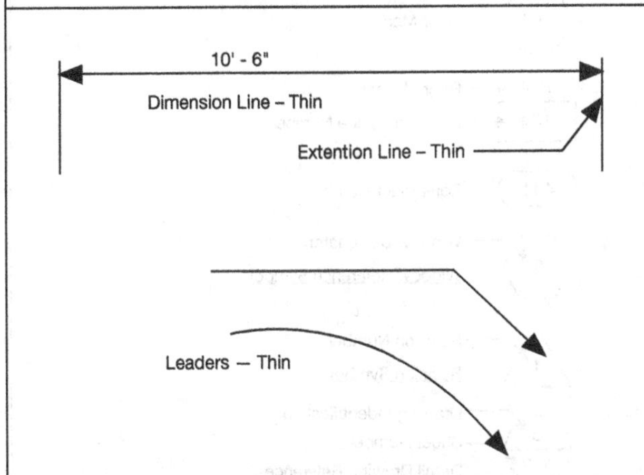

GRAPHIC SYMBOLS USED ON ARCHITECTURAL DRAWINGS *(cont.)*

IDENTIFICATION SYMBOLS

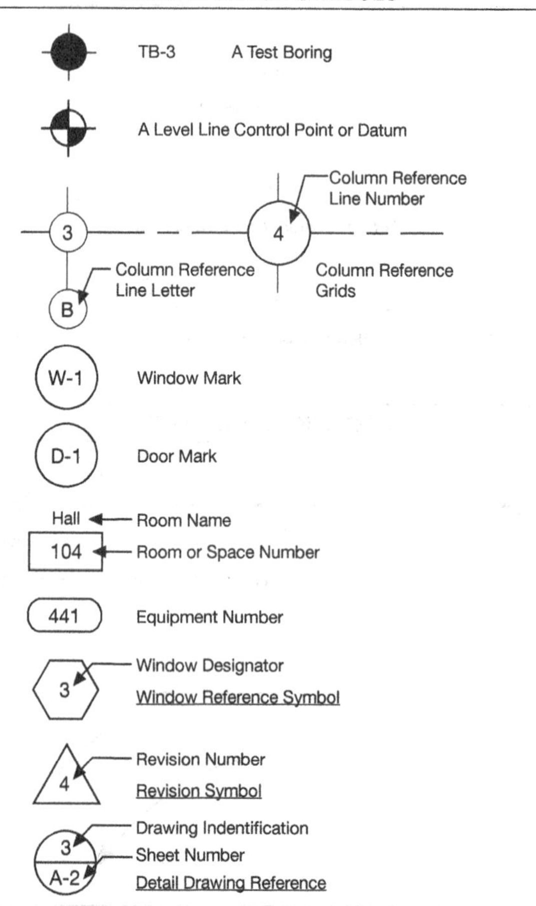

GRAPHIC SYMBOLS USED ON ARCHITECTURAL DRAWINGS *(cont.)*

IDENTIFICATION SYMBOLS *(cont.)*

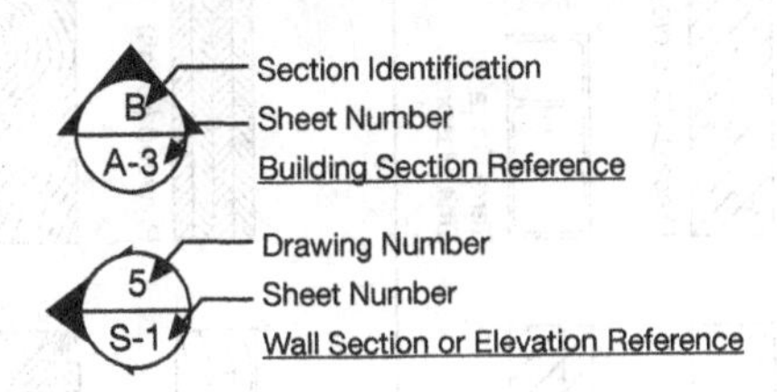

Drawing Reference Number Examples

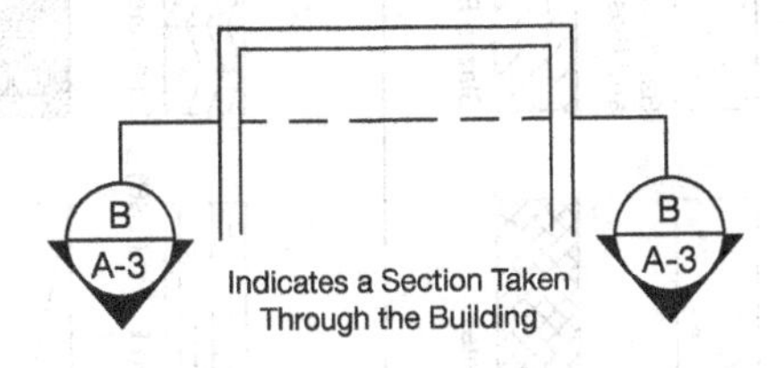

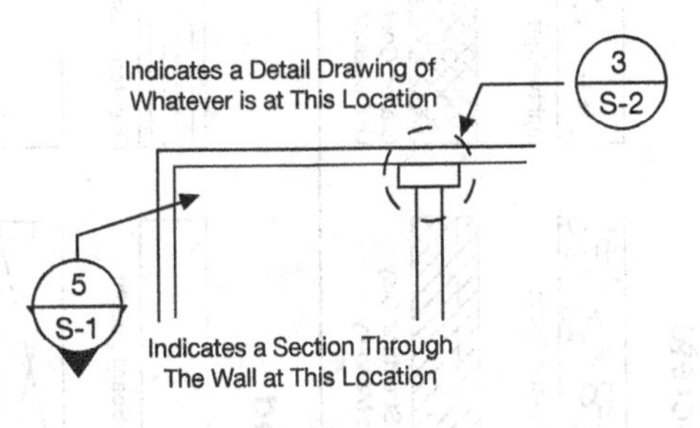

SYMBOLS FOR MATERIALS IN SECTIONS

Concrete

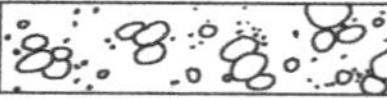

Cast-in-place/Precast

Lightweight

Sand, Mortar, Grout, and Plaster

Precast Concrete

Glazed Structural Clay Tile Unit Masonry

Glazed Concrete Unit Masonry

Glazed Brick

Terra Cotta Unit Masonry

Wood

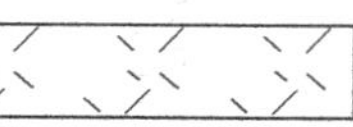

Particle Board (rough)

Oriented Strand Board (OSB)

Laminated Wood

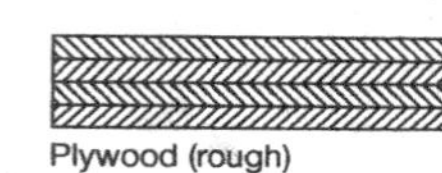

Plywood (rough)

End Grain, Construction Lumber

Blocking

Hardboard

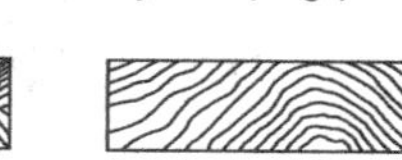

End Grain, Finish Lumber

SYMBOLS FOR MATERIALS IN SECTIONS *(cont.)*

Masonry

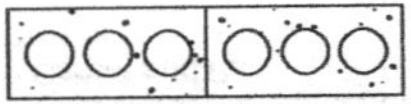

Gypsum Unit Masonry

Cast Stone

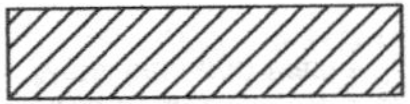

Common/Face

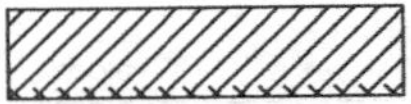

Fire Brick

Concrete Block

Structural Facing File

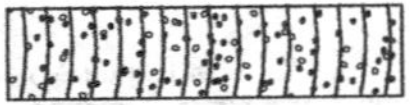

Tile Structural Clay

Clay Tile

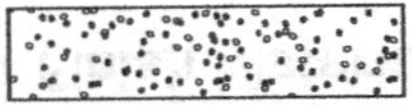

Rough Cut Stone

Metal

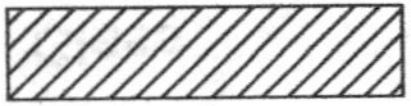

Cast Iron

Steel

Aluminum

Brass/Bronze

SYMBOLS FOR MATERIALS IN SECTIONS *(cont.)*

Glass

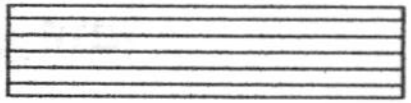

Glass

Structural

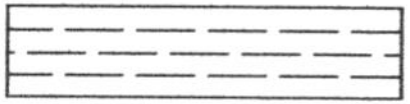

Plastic

Finish Materials

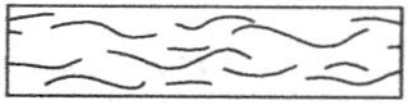

Gypsum Wallboard

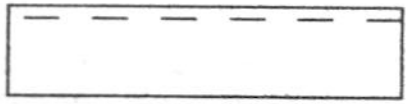

Lath and Plaster

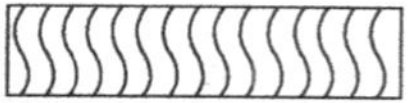

Ceramic Tile

Resilient Tile

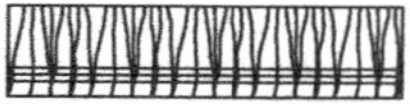

Carpet and Pad

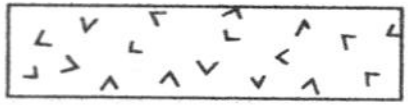

Terrazzo

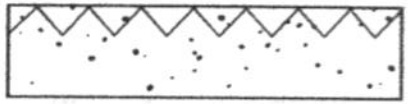

Metal Lath and Plaster

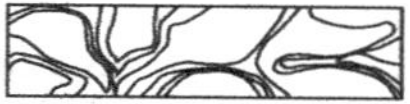

Marble

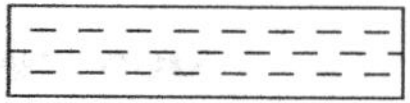

Slate, Bluestone, Flagging, Soapstone

SYMBOLS FOR MATERIALS IN SECTIONS *(cont.)*

Site Work

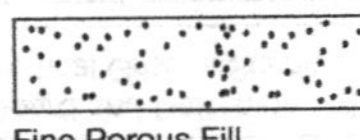
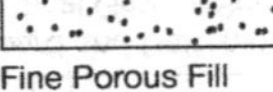

Fine Porous Fill

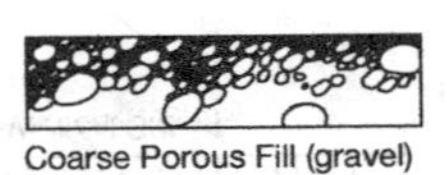

Coarse Porous Fill (gravel)

Earth

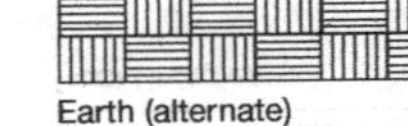

Earth (alternate)

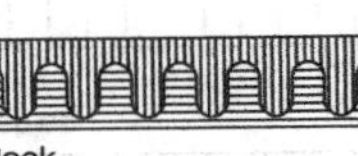

Rock

Rubble

Thermal Protection

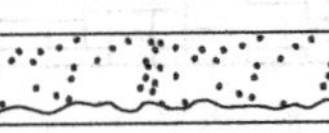

Insulation, Spray and Foam

Fibrous Fire Safing

Insulation, Rigid

Insulation Batts or Loose

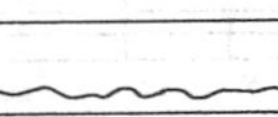

Foam Insulation

Exterior Insulation and Finish System (EIFS)

SYMBOLS FOR MATERIALS IN ELEVATION

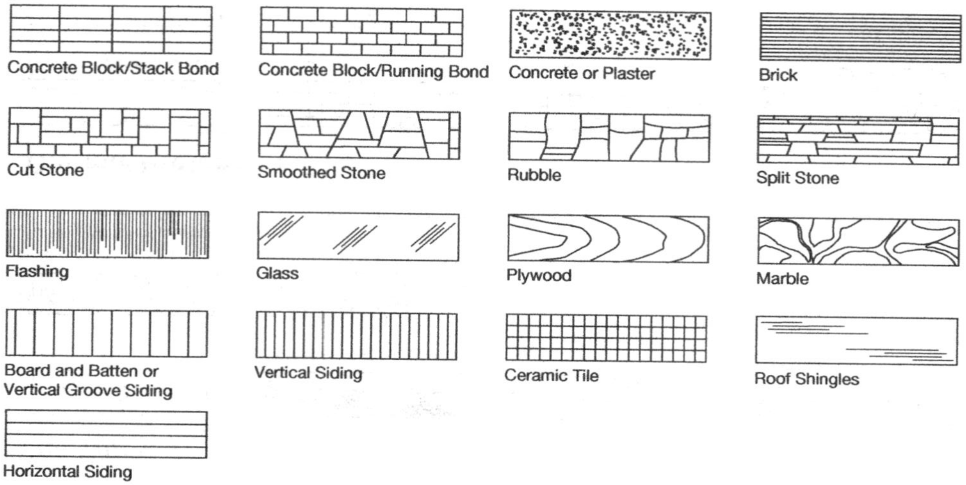

Concrete Block/Stack Bond

Concrete Block/Running Bond

Concrete or Plaster

Brick

Cut Stone

Smoothed Stone

Rubble

Split Stone

Flashing

Glass

Plywood

Marble

Board and Batten or Vertical Groove Siding

Vertical Siding

Ceramic Tile

Roof Shingles

Horizontal Siding

SYMBOLS FOR WALLS IN SECTION

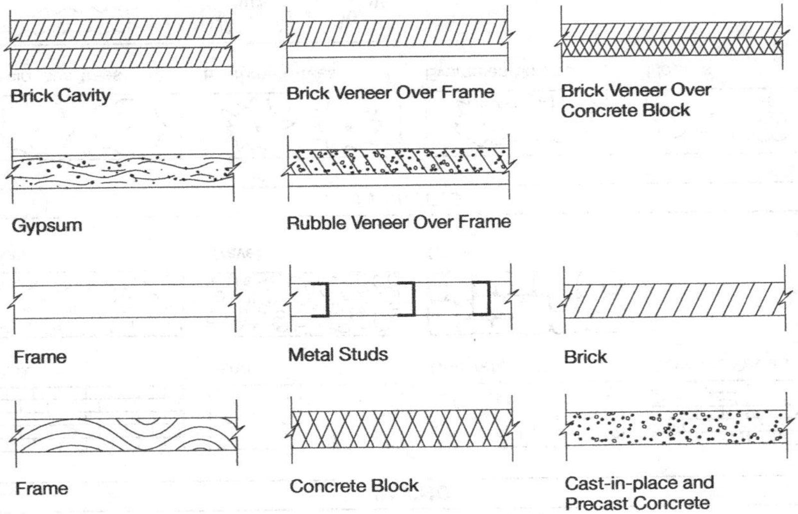

Brick Cavity

Brick Veneer Over Frame

Brick Veneer Over Concrete Block

Gypsum

Rubble Veneer Over Frame

Frame

Metal Studs

Brick

Frame

Concrete Block

Cast-in-place and Precast Concrete

LANDSCAPE SYMBOLS

PAVING

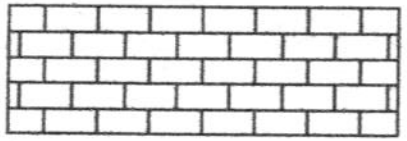

Brick

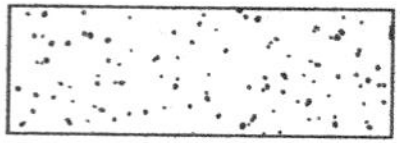

Sand

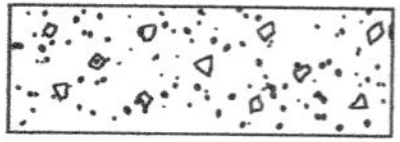

Concrete

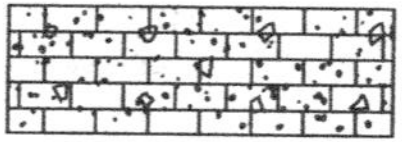

Concrete Pavers

Grass

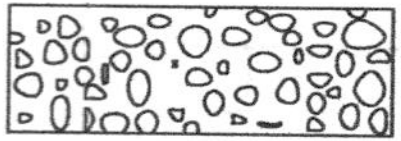

Gravel

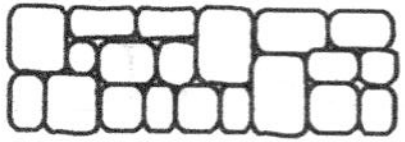

Stone

PLANTING

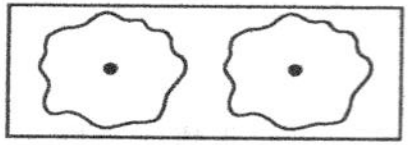

Deciduous Trees

Evergreen Trees

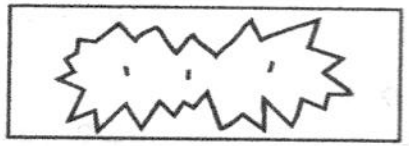

Evergreen Shrubs

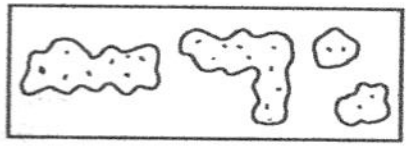

Flowers

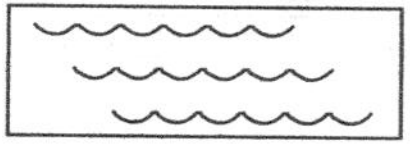

Water

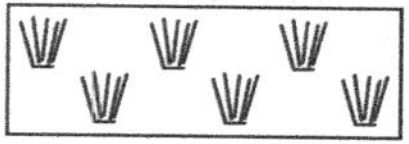

Wetlands

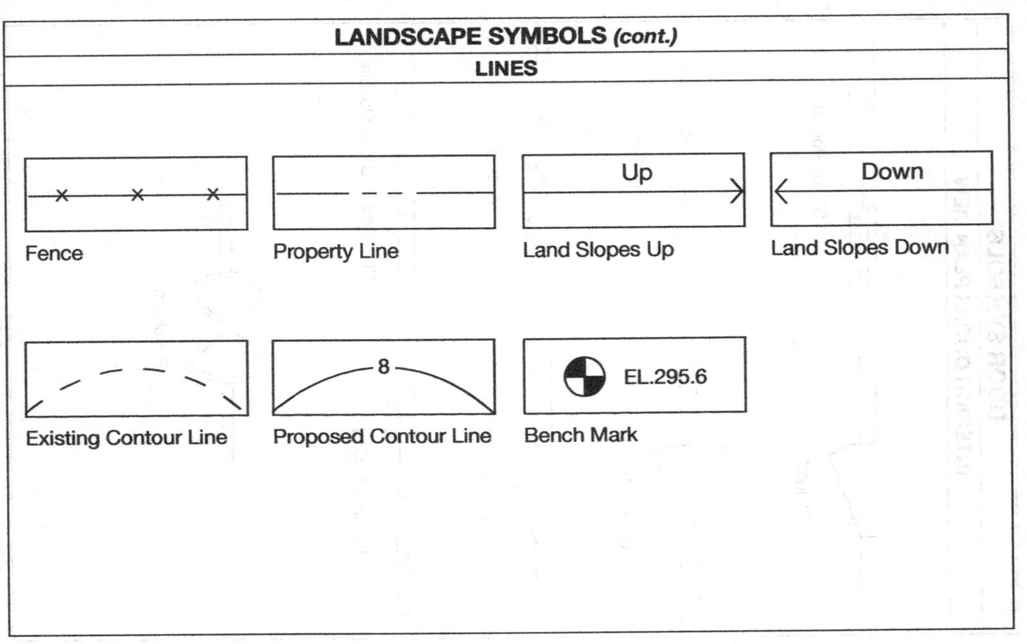
LANDSCAPE SYMBOLS (cont.)
LINES
Fence
Property Line
Up
Land Slopes Up
Down
Land Slopes Down
Existing Contour Line
8
Proposed Contour Line
EL.295.6
Bench Mark

DOOR SYMBOLS

INTERIOR DOORS PLAN VIEW

Hinged

Sliding Pocket

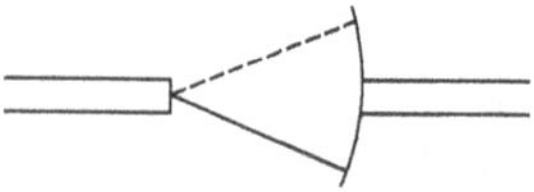

Double Action

Bi-fold

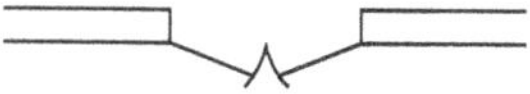

French

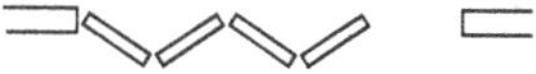

Accordion

By-pass Sliding

Plastered or Cased Opening

Revolving

DOOR SYMBOLS *(cont.)*

EXTERIOR DOORS PLAN VIEW

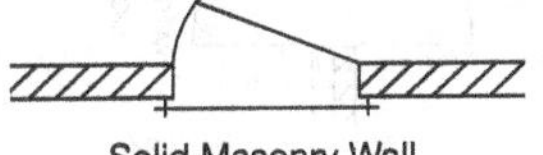

Solid Masonry Wall

Sliding Door in Wood Frame Wall

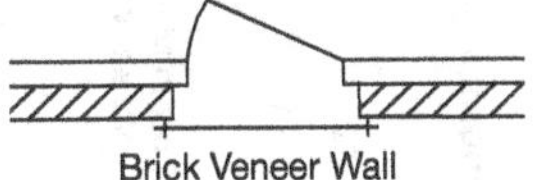

Brick Veneer Wall

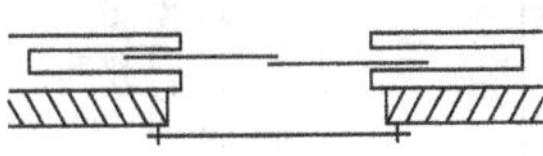

Sliding Door in Brick Veneer Wall

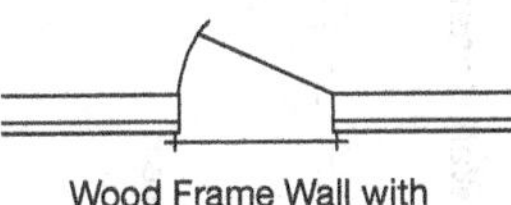

Wood Frame Wall with Stucco Surface

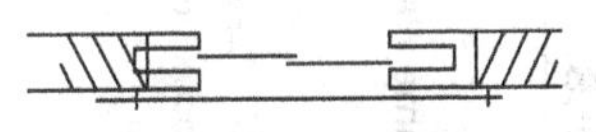

Sliding Door in Solid Masonry Wall

EXTERIOR DOORS PLAN VIEW

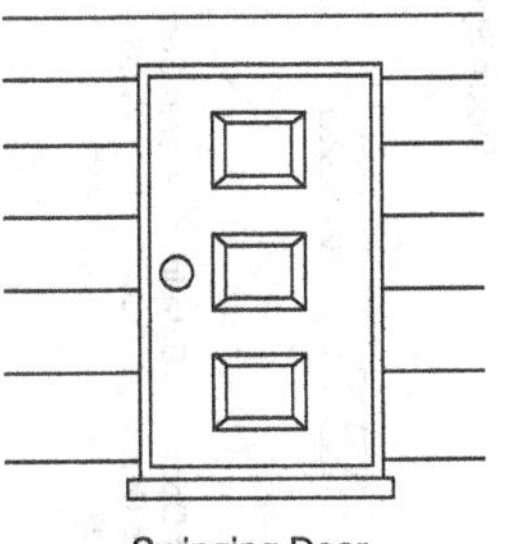

Swinging Door

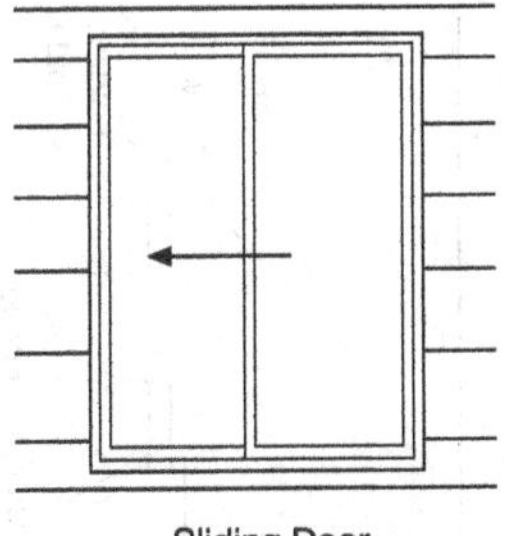

Sliding Door

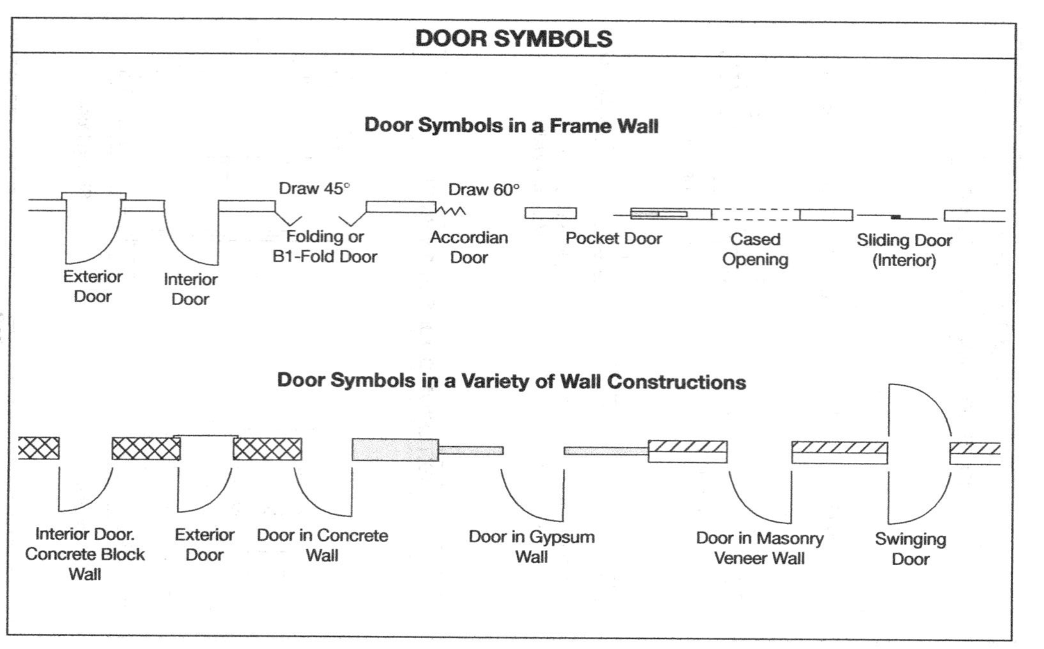
DOOR SYMBOLS
Door Symbols in a Frame Wall
Draw 45°
Draw 60°
Exterior Door
Interior Door
Folding or B1-Fold Door
Accordian Door
Pocket Door
Cased Opening
Sliding Door (Interior)
Door Symbols in a Variety of Wall Constructions
Interior Door. Concrete Block Wall
Exterior Door
Door in Concrete Wall
Door in Gypsum Wall
Door in Masonry Veneer Wall
Swinging Door

WINDOW SYMBOLS

Window Symbols in Masonry Wall

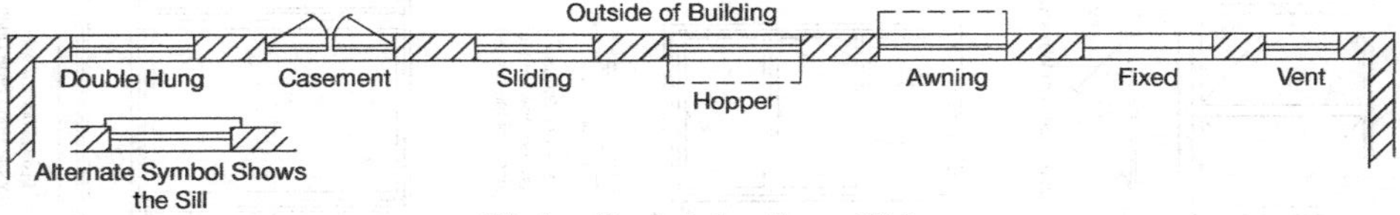

Outside of Building

Double Hung
Casement
Sliding
Hopper
Awning
Fixed
Vent

Alternate Symbol Shows the Sill

Window Symbols in a Frame Wall

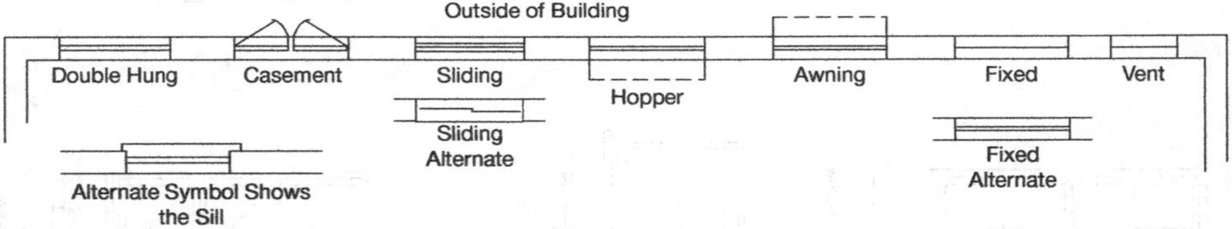

Outside of Building

Double Hung
Casement
Sliding
Hopper
Awning
Fixed
Vent

Sliding Alternate

Fixed Alternate

Alternate Symbol Shows the Sill

Window Symbols in Masonry Veneer Over Frame Wall

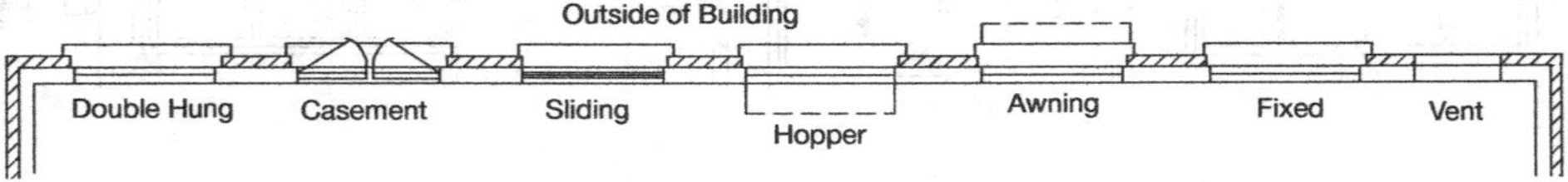

Outside of Building

Double Hung
Casement
Sliding
Hopper
Awning
Fixed
Vent

WINDOW SYMBOLS

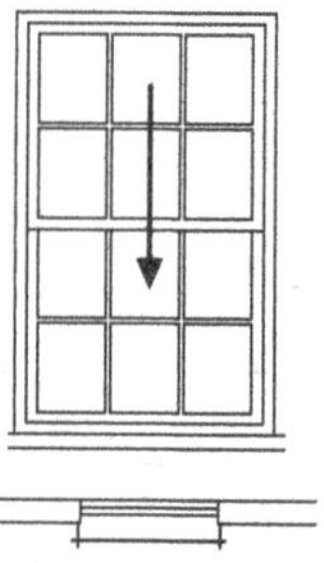

Double Hung Window

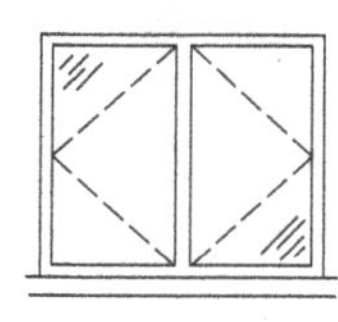

Double Casement Window

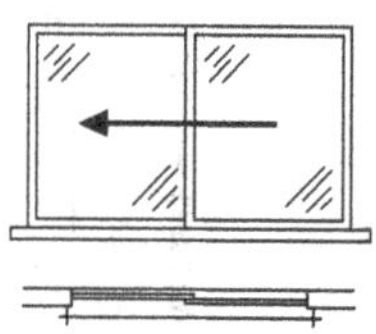

Sliding Window

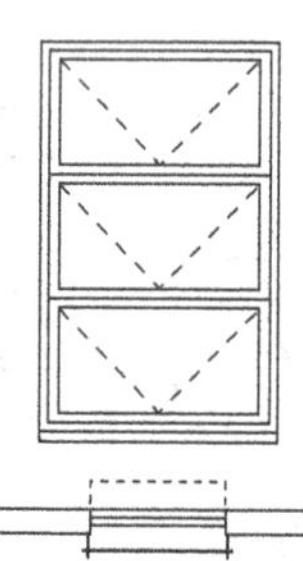

Hopper Window
(swings toward the
inside of the house)

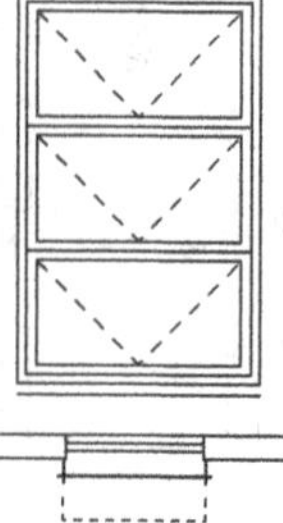

Awning Window

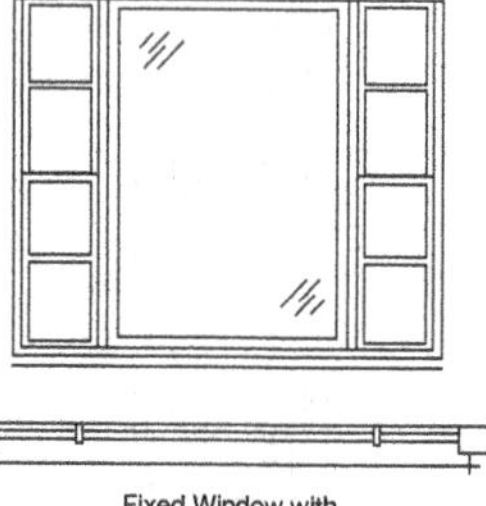

Fixed Window with
Double Hung Side Units

TYPICAL OPENINGS IN VARIOUS WALL CONTRUCTIONS

WOOD OR STEEL STUD EXTERIOR WALL

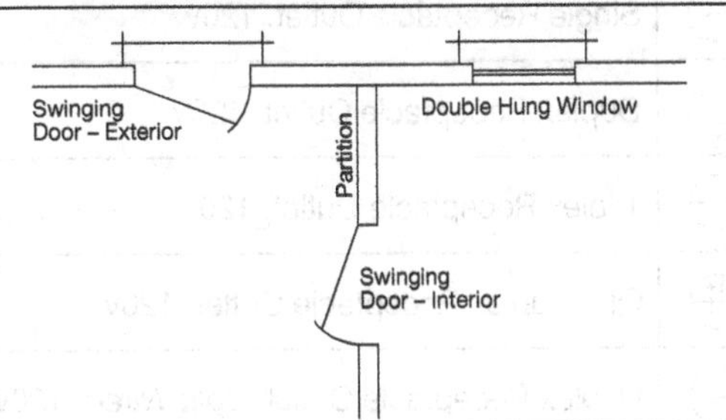

BRICK VENEER EXTERIOR WALL

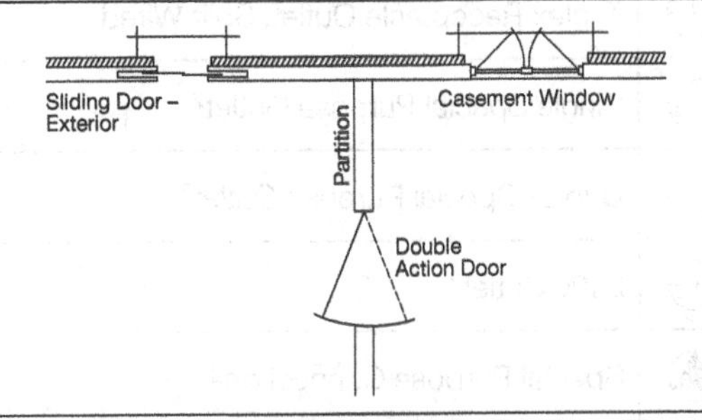

SOLID MASONRY EXTERIOR WALL

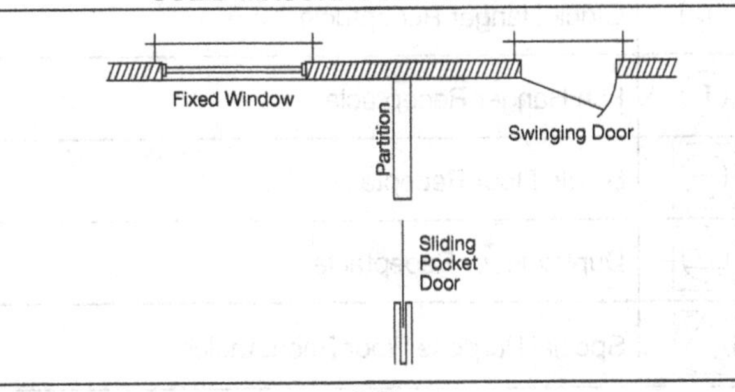

RECEPTACLE OUTLET SYMBOLS	
	Single Receptacle Outlet, 120v
	Duplex Receptacle Outlet, 120v
	Triplex Receptacle Outlet, 120v
	Quadruplex Receptacle Outlet, 120v
	Duplex Receptacle Outlet, Split Wired, 120v
	Triplex Receptacle Outlet, Split Wired
	Single Special Purpose Outlet[2]
	Duplex Special Purpose Outlet[2]
	240v Outlet[2]
	Special Purpose Connection[2]
C	Clock Hanger Receptacle
F	Fan Hanger Receptacle
	Single Floor Receptacle
	Duplex Floor Receptacle
	Special Purpose Floor Receptacle[2]

LIGHTING OUTLET SYMBOLS	
◯	Incandescent fixture, surface or pendant
Ⓡ	Recessed incandescent fixture
[○○○○]	Incandescent lighting track
Ⓓ	Drop cord
⊗	Exit light and outlet box, directional arrows to exit, shaded area is the front
Ⓙ	Junction box
Ⓛ PS	Lamp holder with pull switch
Ⓛ	Outlet controlled by low voltage switching when relay is in outlet box
B	Emergency battery pack and charger, has sealed beams
	Single fluorescent fixture, surface mounted
	Emergency service fluorescent fixture
R	Single recessed fluorescent fixture
	Continuous row fluorescent fixture, surface mounted
R	Recessed continuous row fluorescent fixture
	Single bare lamp fluorescent fixture
	Continuous bare lamp fluorescent fixture

1. Outlets requiring special identification may be indicated by lettering abbreviations beside the standard symbol, as WP for weatherproof or EP for explosion proof.
2. Use numeral or letter beside symbol keyed to a legend of symbols to indicate the type of receptacle or its use.

SWITCH SYMBOLS	
S	Single Pole Switch
S_2	Double Pole Switch
S_3	Three-way Switch
S_4	Four-way Switch
S_P	Switch with Pilot Lamp
S_K	Key-operated Switch
S_L	Switch – Low Voltage System
S_{LM}	Master Switch – Low Voltage System
S_T	Time Switch
S_D	Door Switch
S_{DM}	Dimmer Switch
S_D	Automatic Door Switch
(S)	Ceiling Pull Switch
⦵ S	Switch with Single Receptacle
⦵ S	Switch with Double Receptacle
S_{CB}	Circuit Breaker Switch

SIGNALING AND COMMUNICATIONS SYMBOLS

Symbol	Description	Symbol	Description
·	Pushbutton		Buzzer
	Bell	D	Electric Door Opener
CH	Chime		Bell and Buzzer
	Interconnection Box	BT	Bell-Ringing Transformers
R	Radio Outlet	TV	Television Outlet
	Data Communications		Telephone
	Floor Telephone	S	Speaker
F	Fire Alarm Pull Control	F	Fire Alarm Strobe/Horn
F	Fire Alarm Bell		

WIRE SYMBOLS

Wiring Concealed in Wall and Ceiling	Surface-mounted Wiring
Wiring Concealed in the Floor	Rigid Conduit
Flexible Conduit	Home Run to Panel
Wiring, Neutral	Wiring, Hot
Wiring, Ground	

MISCELLANEOUS ELECTRIC SYMBOLS

	Flush-mounted Service Panel
	Surface-mounted Service Panel
S	Smoke Alarm
Λ	Flame Detector
	Heat Detector
H	Humidistat
	Gas Detector
	Ground

HVAC SYMBOLS

REGISTER SYMBOLS

Sidewall Return, Exhaust, or Relief Register	
Floor Register	

HVAC SYMBOLS *(cont.)*	
DUCTWORK SYMBOLS	
Duct Size and Direction of Flow (indicate depth and width)	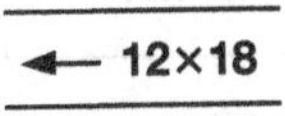 ← 12×18
Supply Duct	
Return Exhaust or Relief Duct	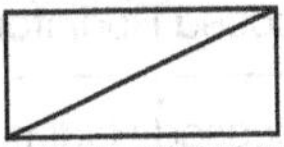
Flexible Duct	
Type of Duct (indicate shape, type, and size)	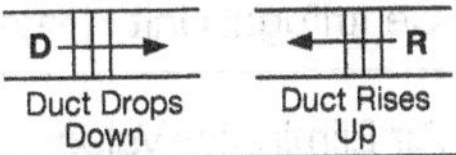 D → Duct Drops Down; ← R Duct Rises Up
DIFFUSER SYMBOLS	
Linear Diffuser	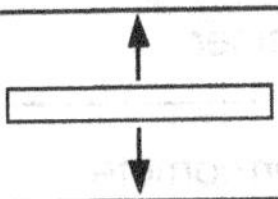
Round Diffuser	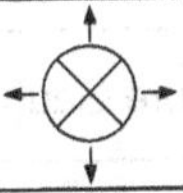
Rectangular Diffuser	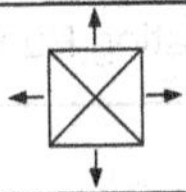
Sidewall Diffuser	
Indicate diffuser size, direction of flow, and cubic feet per minute of flow. Arrows indicate directions of flow required.	

OTHER HVAC SYMBOLS			
Thermostat	T	Damper	
Humidistat	H	Furnace	FURN

HOT WATER SYSTEM SYMBOLS	
Exposed Radiator	
Recessed Radiator	
Flush Enclosed Radiator	
Centrifugal Unit Heater	
Air Eliminator Value	
Strainer	
Thermometer	
Pressure Gauge and Cock	
Circulating Pump	C
Relief Valve	
Thermostat, Electric	T
Thermostat, Pneumatic	T

AIR CONDITIONING PIPING SYMBOLS	
Refrigerant Liquid	—— RL ——
Refrigerant Discharge	—— RD ——
Refrigerant Suction	—— RS ——
Condenser Water Supply	——CWS——
Condenser Water Return	——CWR——
Chilled Water Supply	——CHWS——
Chilled Water Return	——CHWR——
Make Up Water	——MU——
Drain	—— D ——
Brine Supply	—— B ——
Brine Return	—— BR ——
HEATING PIPING SYMBOLS	
Hot Water Heating Supply	—— HW ——
Hot Water Heating Return	——HWR——
High Pressure Steam	—— HPS ——
Medium Pressure Steam	—— MPS ——
Low Pressure Steam	—— LPS ——
High Pressure Return	—— HPR ——
Medium Pressure Return	—— MPR ——
Low Pressure Return	—— LPR ——
Boiler Blow Off	—— BD ——
Condensate or Vacuum Pump Discharge	—— VPD ——
Make-up Water	—— MU ——
Fuel Oil Suction	—— FOS ——
Fuel Oil Return	—— FOR ——
Fuel Oil Vent	—— FOV ——
Gas – Low Pressure	—— G ——
Gas – Medium Pressure	—— MG ——
Gas – High Pressure	—— HG ——
Arrow Indicates Direction of Flow	——▶——

PLUMBING PIPING SYMBOLS	
WASTE AND VENTS	
Soil, Waste, or Leader (above grade)	————————
Soil, Waste, or Leader (below grade)	— — — —
Vent	- - - - - - -
Combination Waste and Vent	——SV——
Acid Waste	——AW——
Acid Vent	- - -AV- - -
Indirect Drain	——IW——
Storm Drain	——S——
FIRE PIPING	
Fire Line	—F——F—
Wet Standpipe	——WSP——
Dry Standpipe	——DSP——
Combination Standpipe	——CSP——
Main Supplies Sprinkler	——S——
Branch and Head Sprinkler	—o——o—
WATER	
Cold Water	—-——-—
Soft Cold Water	——SW——
Industrialized Cold Water	——ICW——
Chilled Drinking Water Supply	——DWS——
Chilled Drinking Water Return	——DWR——
Hot Water	- - - —— - - -
Hot Water Return	- - - - —— - - - -
Sanitizing Hot Water Supply (180° F)	+- - -+- - -+
Sanitizing Hot Water Return (180° F)	++- - -++- - -
Industrialized Hot Water Supply	——IHW——
Industrialized Hot Water Return	——IHR——
Tempered Water Supply	——TWG——
Tempered Water Return	——TWR——

PLUMBING PIPING SYMBOLS *(cont.)*

OTHER PIPING

Compressed Air	——A——
Vacuum	——V——
Vacuum Cleaning	——VC——
Oxygen	——O——
Liquid Oxygen	——LOX——
Nitrogen	——N——
Liquid Nitrogen	——LN——
Nitrous Oxide	——NO——
Hydrogen	——H——
Helium	——HE——
Argon	——AR——
Liquid Petroleum Gas	——LPG——
Industrial Waste	——INW——

PLUMBING SYMBOLS

VALVES

Check Valve, Straightway	
Gate Valve	
Globe Valve	
Butterfly	
Solenoid	
Lock Shield	
2-way Automatic Control	
3-way Automatic Control	
Gas Cock	
Pressure Reducing Valve	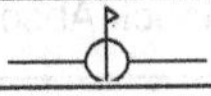

PLUMBING SYMBOLS *(cont.)*

FITTINGS	
90° Elbow	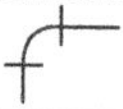
90° Elbow, Turned Down	
90° Elbow, Turned Up	
45° Elbow	
Straight Cross	
Straight Tee	
Straight Tee, Outlet Up	
Straight Tee, Outlet Down	
Union	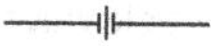
Concentric Reducer	
Eccentric Reducer	
Cap	
P-Trap	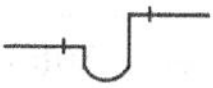
House Trap	
Shock Absorber	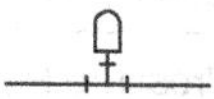

PLUMBING SYMBOLS *(cont.)*	
OTHER SYMBOLS	
Expansion Joint	
Expansion Loop	
Flexible Connection	
Flow Direction	
Pipe Pitch, Rise (R)/Drop (D)	P → R
Thermostat	T
Cleanout on End of Pipe	CO
Cleanout on Wall	CO
Floor, Roof, or Shower Drain	FD RD SD
Vent Through Roof	VTR
1 1/2" Waste Down (up)	← 1 1/2" W
1/2" Hot Water Down (up)	← 1/2" HW
1/2" Cold Water Down (up)	← 1/2" CW
Meter	M
Air Vent	
Vent Pipe	- - - - - - - -

SYMBOLS INDICATING TYPE OF PIPE MATERIAL	
Cast Iron	——CI——
Clay Tile	——CT——
Ductile Iron	——DI——
Reinforced Concrete	——RCP——
Polyvinyl Chloride	——PVC——
Acrilylonitrile Butadiene Styrene	——ABS——
Styrene Rubber Plastic	——SRP——

SYMBOLS INDICATING MATERIALS CARRIED BY PIPE	
Material	**Symbol**
Acid Waste	——AW——
Condensate	——C——
Compressed Air	——CA——
Carbon Dioxide	——CO2——
Cold Water	——CW——
Dry Standpipe	——DSP——
Fuel Oil	——FO——
Gasoline	——GAS——
Gas – Low Pressure	——G——
Gas – Medium Pressure	——MG——
Gas – High Pressure	——HG——
Hot Water	——HW——
Liquefied Propane Gas	——LPG——
Nitrogen	——N——
Natural Gas	——NG——
Nontoxic Industrial Waste	——NTW——
Oxygen	——O——
Refrigerant	——RFGT——
Steam	——STM——
Water	——W——
Wet Standpipe	——WSP——

SYMBOLS INDICATING FLANGED, SCREWED, WELDED, OR SOLDERED FITTINGS				
Fitting	Flanged	Screwed	Welded (X or •)	Soldered
90° Elbow			or	
90° Elbow, Turned Down				
90° Elbow, Turned Up				
45° Elbow				
Straight Cross				
Straight Tee				
Straight Tee, Outlet Up				
Straight Tee, Outlet Down				
Union				
Check Valve, Straightway				
Gate Valve				
Globe Valve				

FIRE PROTECTION SYSTEM SYMBOLS	
Fire Line Water Supply	——F——
Upright Sprinkler Head	——○——
Pendent Sprinkler Head	——●——
Wet Standpipe	——WSP——
Dry Standpipe	——DSP——
Combination Standpipe	——CSP——
Fire Hydrants	FH
Wall Mounted Fire Hydrant with Two Heads	
Recessed Fire Hose Cabinet	FHC
Surface Mounted Fire Hose Cabinet	FHC

PLUMBING FIXTURES

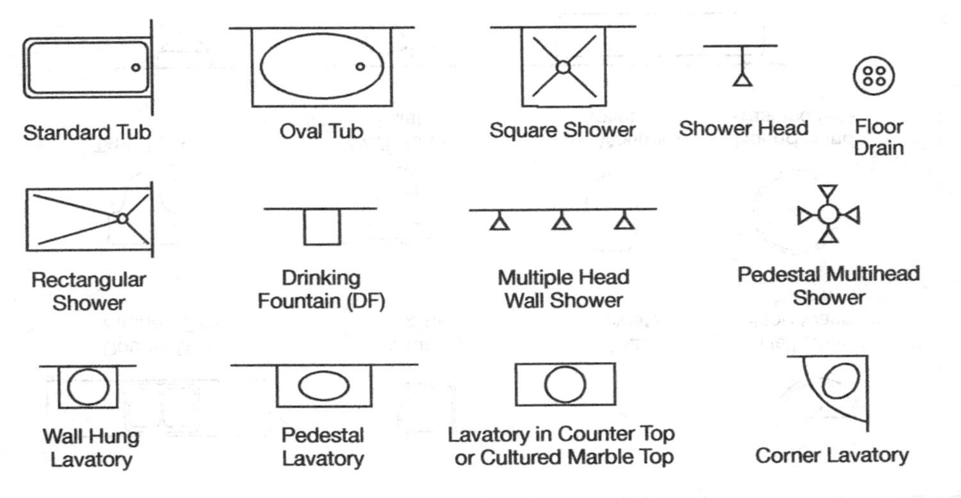

PLUMBING FIXTURES *(cont.)*

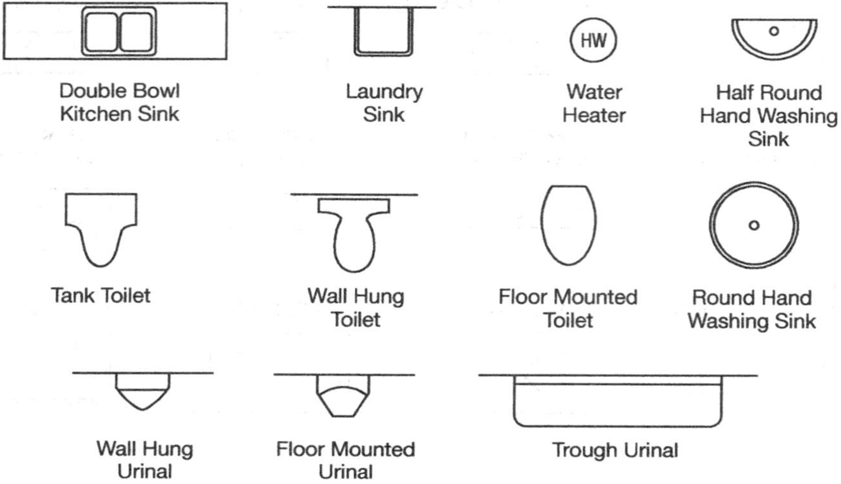

HW

Double Bowl Kitchen Sink

Laundry Sink

Water Heater

Half Round Hand Washing Sink

Tank Toilet

Wall Hung Toilet

Floor Mounted Toilet

Round Hand Washing Sink

Wall Hung Urinal

Floor Mounted Urinal

Trough Urinal

SCHEDULES

Complete architectural drawings contain schedules which give detailed information about various features of the building. They are important because they become part of the legal presentation of the design information. The information shown and how it is presented varies depending upon the extent of descriptive information wanted and the best way to record it. Small buildings will typically have short schedules as shown by the following examples. Large commercial buildings will have schedules requiring several pages of the drawings. Typical schedules include those for windows, doors, columns, lintels, electrical units, beams, plumbing units, heat and air-conditioning units, room for finish information, footings, and appliances.

TYPICAL WINDOW SCHEDULE FOR A SMALL RESIDENCE

Mark	No.	Unit Size	Rough Opening	Type	Material	Glazing	Remarks
W1	1	5'-0" × 6'0"	5'-1/2" × 6'-1/2"	Fixed	Pine	Double	Energy Efficient GL
W2	10	3'-1 5/8" × 5'-9 1/4"	3'-2 1/8" × 5'-9 5/8"	D. H.	Pine	Single	—
W3	2	4'-0" × 4'-6"	4'-1/2" × 4'-6 1/2"	Awning	Pine	Single	—

TYPICAL WINDOW SCHEDULE FOR A SMALL COMMERCIAL BUILDING

Mark	Unit Size	Type	Material	Glazing	Finish	Remarks
A	4'-0" × 5'6"	Fixed/Hopper	Aluminum	5/8" Insulation	Bronze	—
B	3'-0" × 5'-6"	DO	DO	DO	DO	—
C	4'-6" × 3'-0"	Awning	DO	DO	DO	—
D	4'-6" × 2'-6"	DO	DO	DO	DO	—
E	4'-0" × 5'-0"	Fixed/Hopper	DO	DO	DO	—

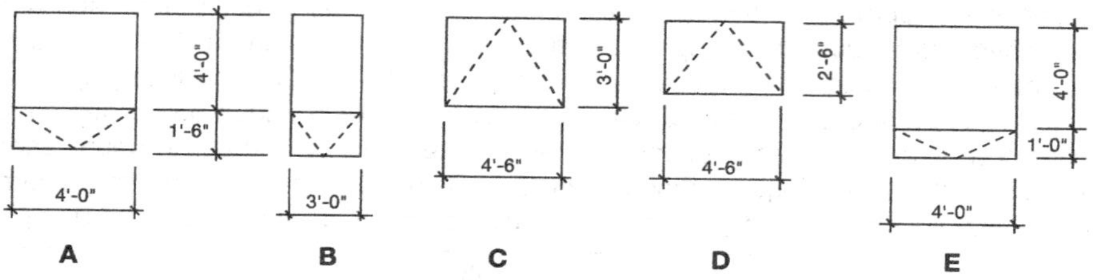

4'-0"
1'-6"
4'-0"
A

3'-0"
B

3'-0"
4'-6"
C

2'-6"
4'-6"
D

4'-0"
1'-0"
4'-0"
E

BEAM SCHEDULE				
Mark	Size	Splice Detail	End BRG Detail	Top BM Elevation
A	W10 × 49	1 / 5 | 10	2 / 5 | 10	100'-6"
B	W12 × 50	3 / 7 | 10	4 / 8 | 10	90'-5"
C	W14 × 48	5 / 9 | 10	6 / 3 | 10	101'-4"

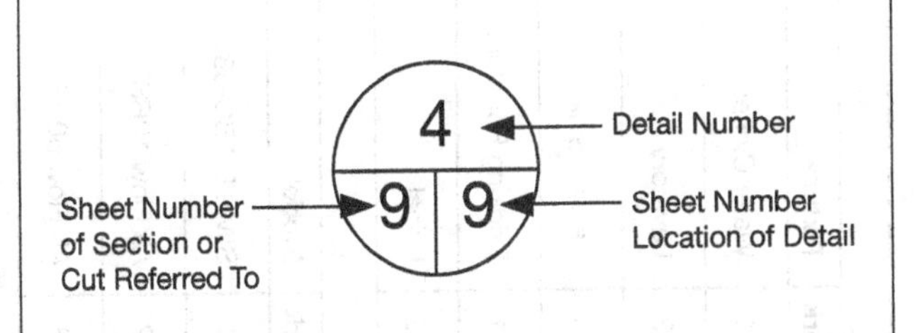

PLUMBING SCHEDULE

Mark	Fixture	No.	Mfr.	Mfg. No.	CW	HW	Vent	Waste	Trap
P-1	Water Closet	8	Crown	191-0900	1"	—	4"	4"	Int.
P-2	Lavatory	8	Frank	103-2214	1/2"	1/2"	1 1/4"	1 1/4"	1 1/4"
P-3	Service Sink	1	Frank	311-1100	3/4"	3/4"	1 1/2"	1 1/2"	1 1/2"
P-4	Drinking Fountain	3	Frank	091-5181	3/8"	—	1 1/4"	1 1/4"	1 1/4"
P-5	Urinal	4	Crown	151-9000	3/4"	—	2"	2"	2"

EXHAUST FAN SCHEDULE

Mark	Model	CFM	SP	Motor HP	Current	Remarks
E-1	SIMS E ABC-38	20,000	1/4"	2 1/2	208/3/60	Motorized Damper
E-2	Air Flow 10RS	550	1/8"	1/10	115/1/60	Gravity Damper
E-3	Air Flow 50RS	425	1/8"	1/10	115/1/60	Gravity Damper

METRIC INFORMATION

BASE UNITS

Quantity	Unit	Symbol
Length	Meter	m
Mass	Kilogram	kg
Time	Second	s
Electric Current	Ampere	A
Thermodynamic Temperature	Kelvin	K
Amount of Substance	Mole	mo
Luminous Intensity	Candela	cd

SUPPLEMENTARY SI UNITS

Quantity	Unit	Symbol
Plane Angle	Radian	rad
Solid Angle	Steradian	sr

DERIVED METRIC UNITS WITH COMPOUND NAMES

Physical Quantity	Unit	Symbol
Area	Square Meter	m^2
Volume	Cubic Meter	m^3
Density	Kilogram per Cubic Meter	kg/m^3
Velocity	Meter per Second	m/s
Angular Velocity	Radian per Second	rad/s
Acceleration	Meter per Second Squared	m/s^2
Angular Acceleration	Radian per Second Squared	rad/s^2
Volume Rate of Flow	Cubic Meter per Second	m^3/s
Moment of Inertia	Kilogram Meter Squared	$kg \bullet m^2$
Moment of Force	Newton Meter	N•m
Intensity of Heat Flow	Watt per Square Meter	W/m^2
Thermal Conductivity	Watt per Meter Kelvin	W/m•K
Luminance	Candela per Square Meter	cd/m^2

METRIC INFORMATION

SI PREFIXES

Multiplication Factor	Prefix	Symbol
1 000 000 000 000 000 000 = 10^{18}	exa	E
1 000 000 000 000 000 = 10^{15}	peta	P
1 000 000 000 000 = 10^{12}	tera	T
1 000 000 000 = 10^{9}	giga	G
1 000 000 = 10^{6}	mega	M
1000 = 10^{3}	kilo	k
100 = 10^{2}	hecto	h
10 = 10^{1}	deka	da
0.1 = 10^{-1}	deci	d
0.01 = 10^{-2}	centi	c
0.001 = 10^{-3}	milli	m
0.000 001 = 10^{-6}	micro	m
0.000 000 001 = 10^{-9}	nano	n
0.000 000 000 001 = 10^{-12}	pico	p
0.000 000 000 000 001 = 10^{-15}	femto	f
0.000 000 000 000 000 001 = 10^{-18}	atto	a

METRIC UNIT TO IMPERIAL UNIT CONVERSION FACTORS	
Metric Units	**Imperial Equivalents**
LENGTH	
1 Millimeter (mm)	= 0.0393701 Inch
1 Meter (m)	= 39.3701 Inches = 3.28084 Feet
1 Kilometer (km)	= 0.621371 Mile
LENGTH/TIME	
1 Meter per Second (m/s)	= 3.28084 Feet per Second
1 Kilometer per Hour (km/h)	= 0.621371 Mile per Hour
AREA	
1 Square Millimeter (mm^2)	= 0.001550 Square Inch
1 Square Meter (m^2)	= 19.7639 Square Feet
1 Hectare (ha)	= 2.47105 Acres
1 Square Kilometer (km^2)	= 0.386102 Square Mile
VOLUME	
1 Cubic Millimeter (mm^3)	= 0.0000610237 Cubic Inch
1 Cubic Meter (m^3)	= 35.3147 Cubic Feet = 1.30795 Cubic Yards
1 Milliliter (ml.)	= 0.0351951 Fluid Ounce
1 Liter (L)	= 0.219969 Gallon
MASS	
1 Gram (g)	= 0.0352740 Ounce
1 Kilogram (kg)	= 2.20462 Pounds
1 Tonne (t) (51,000 kg)	= 1.10231 Tons (2,000 lb.)
FORCE	
1 Newton (N)	= 0.224809 Pound-Force
STRESS	
1 Megapascal (MPa)	= 145.038 Pounds-Force psi
LOADING	
1 Kilonewton per Sq. Meter	= 20.8854 Pounds Force psf
1 Kilonewton per Meter	= 68.5218 Pounds Force per Ft.
MISCELLANEOUS	
1 Joule (J)	= 0.00094781 B.t.u.
1 Joule (J)	= 1 Watt-second
1 Watt (W)	= 0.00134048 Electric hp

IMPERIAL UNIT TO METRIC UNIT CONVERSION FACTORS	
Imperial Units	**Metric Equivalents**
LENGTH	
1 Inch	= 25.4 mm = 0.0254 m
1 Foot	= 0.3048 m
1 Mile	= 1.60934 km
LENGTH/TIME	
1 Foot per Second	= 0.3048 m/s
1 Mile per Hour	= 1.60934 km/h
AREA	
1 Square Inch	= 645.16 mm^2
1 Square Foot	= 0.0929030 m^2
1 Acre	= 0.404686 ha
1 Square Mile	= 2.58999 km^2
VOLUME	
1 Cubic Inch	= 16387.1 mm^3
1 Cubic Foot	= 0.0283168 m^3
1 Cubic Yard	= 0.764555 m^3
1 Fluid Ounce	= 28.4131 mL
1 Gallon	= 4.54609 L
MASS	
1 Ounce	= 28.3495 g
1 Pound	= 0.453592 kg
1 Ton	= 0.907185 t
FORCE	
1 Pound	= 4.44822 N
STRESS	
1 psi	= 0.00689476 MPa
LOADING	
1 psf	= 0.0478803 kN/m^2
1 plf	= 0.0145939 kN/m
MISCELLANEOUS	
1 B.t.u.	= 1055.06 J
1 Watt-second	= 1 J
1 Horsepower	= 746 W

FEET AND INCH CONVERSIONS

1 Inch	=	25.4 mm
1 Foot	=	304.8 mm
1 psi	=	6.89 kPa
1 psf	=	0.048 kPa

METRIC CONVERSIONS

1 mm	=	0.039 Inches
1 m	=	3.28 Feet
1 kPa	=	20.88 psf

mm	=	Millimeter
m	=	Meter
kPa	=	Kilopascal
psi	=	Pounds per Sq. Inch
psf	=	Pounds per Sq. Foot

INCHES TO MILLIMETERS AND CENTIMETERS								
in.	mm	cm	in.	mm	cm	in.	mm	cm
1/8	3	0.3	9	229	22.9	30	762	76.2
1/4	6	0.6	10	254	25.4	31	787	78.7
3/8	10	1.0	11	279	27.9	32	813	81.3
1/2	13	1.3	12	305	30.5	33	838	83.8
5/8	16	1.6	13	330	33.0	34	864	86.4
3/4	19	1.9	14	356	35.6	35	889	88.9
7/8	22	2.2	15	381	38.1	36	914	91.4
1	25	2.5	16	406	40.6	37	940	94.0
1 1/4	32	3.2	17	432	43.2	38	965	96.5
1 1/2	38	3.8	18	457	45.7	39	991	99.1
1 3/4	44	4.4	19	483	48.3	40	1016	101.6
2	51	5.1	20	508	50.8	41	1041	104.1
2 1/2	64	6.4	21	533	53.3	42	1067	106.7
3	76	7.6	22	559	55.9	43	1092	109.2
3 1/2	89	8.9	23	584	58.4	44	1118	111.8
4	102	10.2	24	610	61.0	45	1143	114.3
4 1/2	114	11.4	25	635	63.5	46	1168	116.8
5	127	12.7	26	660	66.0	47	1194	119.4
6	152	15.2	27	686	68.6	48	1219	121.9
7	178	17.8	28	711	71.1	49	1245	124.5
8	203	20.3	29	737	73.7	50	1270	127.0

TYPICAL R-VALUES FOR SELECTED CONSTRUCTION MATERIALS

R-value is a numerical value to indicate the resistance to the flow of heat. The higher the R-value, the greater the resistance. *(Consult manufacturers for specific values.)*

AIR FILMS

Type	R-values
Air, Inside	0.77
Heat Flow Up	0.61
Heat Flow Horizontal	0.68
Heat Flow Down	0.92
Air Outside	
Wind, 7 1/2 Miles per Hour	0.25
Wind, 15 Miles per Hour	0.17
Summer	0.25
Winter	0.17

CONCRETE AND MASONRY (PER INCH OR LISTED SIZE)

Type	R-values
Concrete Lightweight 1"	0.60
Concrete Sand and Gravel 1"	0.11
Concrete Block, Lightweight 4"	1.68
Concrete Block, Lightweight 6"	1.83
Concrete Block, Lightweight 8"	2.12
Concrete Block, Sand, Gravel 4"	1.17
Concrete Block, Sand, Gravel 6"	1.29
Concrete Block, Sand, Gravel 8"	1.46
Brick, Common 4"	0.78
Brick, Face 4"	0.40
Current Mortar 1"	0.20
Stone 1"	0.08
Stucco 3/4"	0.11
Clay Tile, Hollow, 3"	0.80
Clay Tile, Hollow, 4"	1.11
Clay Tile, Hollow, 6"	1.52

TYPICAL R-VALUES FOR SELECTED CONSTRUCTION MATERIALS *(cont.)*

DEAD AIR SPACE (LISTED SIZE)	
Type	**R-values**
1/2"	0.90
3/4"	0.90
1 1/2"	1.02
3 1/2"	1.20
ROOFING (AS LISTED)	
Type	**R-values**
Asphalt Shingles	0.45
Wood Shingles	0.95
Slate 1/2"	0.05
Concrete Tile	0.01
Builtup	0.33
Felt Underlayment	0.06
Metal Roofing	0.01
GYPSUM/PLASTER	
Type	**R-values**
Gypsum Wallboard 1/2"	0.45
Gypsum Wallboard 5/8"	0.56
Gypsum Lath and Plaster 1"	0.40
WOOD (PER INCH OR AS LISTED)	
Type	**R-values**
Hardwoods, Typical	0.91
Softwoods, Typical	1.25
Wood Shingles for Siding	0.90
Wood Bevel Siding, Typical 1/2"	0.80
Wood Bevel Siding, Typical 3/4"	0.93
Southern Pine 1 1/2"	1.88
Southern Pine 2"	2.50
Douglas Fir 3/4"	0.93
Hardboard siding 7/16"	0.68

TYPICAL R-VALUES FOR SELECTED CONSTRUCTION MATERIALS *(cont.)*

PANELS (PER INCH OR AS LISTED)

Type	R-values
Fiberglass	3.1 to 4.30
Extruded Polystyrene	5.00
Polyisocyanurate	7.00
Polyurethane	6.25
Polystyrene	4.0 to 5.4
Mineral Wood	4.0
Plywood 3/8"	0.48
Plywood 1/2"	0.64
Plywood 5/8"	0.79
Plywood 3/4"	0.96
Hardboard 1/4"	0.18
Hardboard 1/2"	0.36
Gypsum Sheathing 1/2"	0.45
Gypsum Sheathing 5/8"	0.55
Gypsum Wallboard 1/2"	0.35
Particleboard 1/2"	0.62
Wood Fiber Rigid 1"	2.80
Sound-deadening Board 1/2"	1.35

GLAZING MATERIALS (LISTED SIZES)

Type	R-values
Glass, Single Thickness 1/8"	0.91
Glass, Single Thickness 1/4"	0.96
Double Pane Insulating Glass	
with 3/16" Airspace	1.61
with 1/4" Airspace	1.72
with 1/2" Airspace	2.04
Insulating Glass with Three Panels	
with 1/4" Airspace	2.56
with 1/2" Airspace	3.23

TYPICAL R-VALUES FOR SELECTED CONSTRUCTION MATERIALS *(cont.)*	
GLAZING MATERIALS (LISTED SIZES) *(cont.)*	
Type	**R-values**
Glass block	
6" × 6" × 4"	1.67
8" × 8" × 4"	1.79
Single Plastic Pane 1/8"	0.94
Single Plastic Pane 1/4"	1.04
INSULATION (PER INCH)	
Type	**R-values**
Batts and Blankets	
Fiberglass	3.1 to 4.3
Rock Wool	2.9
Loose Fill (per inch)	
Cellulose Blown	3.1 to 4.0
Cellulose Poured	3.7 to 4.0
Vermiculite	2.1 to 3.0
Perlite	2.6 to 3.5
Fiberglass Poured	3.1 to 3.3
Fiberglass Blown	2.8 to 3.8
Rockwool Blown	3.0 to 3.3
Spray Applied (per inch)	
Polyurethane	6.3 to 7.2
Isocyanurate	6.8 to 8.0
Foamed Insulation (per inch)	4.5
Reflective Insulation	
Aluminum Multilayer with Four, 3/4"Airspaces	Winter 13 Summer 17
Multilayer Foil Batts with Six, One Inch Airspaces	Winter 18 Summer 30

TYPICAL R-VALUES FOR SELECTED CONSTRUCTION MATERIALS *(cont.)*	
INTERNAL WALLS AND FLOORING (LISTED SIZES)	
Type	**R-values**
Cork Tile 1/8"	0.28
Carpet with Fiber Pad	2.08
Carpet with Rubber Pad	1.23
Vinyl Tile	0.05
Terazzo	0.08
Slate	0.05
Hardwood Floors 3/4"	0.71
Softwood Floors	0.94
Maple Floors 3/4"	0.90
Gypsum Wallboard 1/2"	0.35
Gypsum Lath 1/2" and 3/4" Plaster	0.45
Metal Lath and 3/4" Plaster	0.15
Vermiculite Plaster 1"	0.59
TYPICAL EXTERIOR WALL R-VALUES	
UNINSULATED 2 × 4 STUD WALL	
Outside Air Film	0.17
1/2" Wood Siding	0.80
1/2" Plywood Sheathing	0.64
Vapor Barrier	0.00
1/2" Gypsum Wallboard	0.45
Inside Air Film	0.77
Total	**2.83**
INSULATED 2 × 4 STUD WALL	
As Above	2.83
31/2" Fiberglass Insulation	10.8
Total	**13.64**
INSULATED 2 × 6 STUD WALL	
As Above	2.84
51/2" Fiberglass Insulation	10.05
Total	**12.89**

R-VALUES FOR TYPICAL PARTITIONS AND WALLS

INTERIOR PARTITION

Layer	R	Layer	R
Inside Air Film	0.77	Inside Air Film	0.77
Outside Air Film	0.25	Outside Air Film	0.25
1/2" Gypsum	0.70	1/2" Sound Deadening Board	1.35
3 1/2" Air Space	1.20	1/2" Gypsum	0.35
	R 2.92	3 1/2" Air Space	1.20
			R 3.92

EXTERIOR WALL

Layer	R	Layer	R
Inside Air Film	0.77	Inside Air Film	0.77
Outside Air Film, Summer	0.25	Outside Air Film, Summer	0.25
4" Clay Face Brick	0.40	4" Concrete Block, Lightweight	1.68
	R 1.42		**R 2.95**

R-VALUES FOR TYPICAL PARTITIONS AND WALLS *(cont.)*

EXTERIOR WALL *(cont.)*

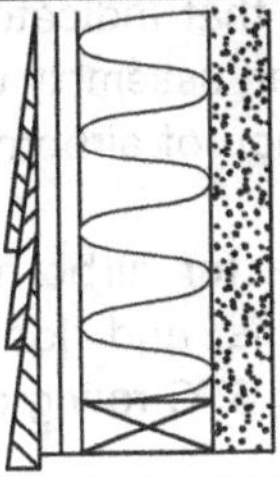

Material	R-value
Inside Air Film	0.77
Outside Air Film, Summer	0.25
1/2" Gypsum	0.35
3 1/2" Fiberglass	10.85
5/8" Plywood	0.79
1/2" Wood Bevel Siding	0.80
	R 13.81

U-VALUES

U is a measure of the amount of heat that will pass through an assembly of various materials, such as an exterior wall. The smaller the U-value, the greater the resistance to the transmission of heat. It is the coefficient of heat transfer. A U-value is the reciprocal of the R-value or U = 1/R-value. See the following example for an exterior wall.

EXTERIOR WALL

Type	R-value
Outside Air Film	0.17
Siding, Wood 1/2" × 8" Lapped	0.81
Sheathing, Plywood 1/2"	0.62
Insulation	11.00
Interior Finish Gypsum Board 1/2"	0.45
Inside Air Film	0.68
Total R-value	13.73
U-value (1/R)	0.073

SOUND TRANSMISSION CLASS RATINGS (STC)

STC is a single number rating that indicates the effectiveness of a material or an assembly of materials to reduce the transmission of airborne sound through it.

Typical STC recommendations for airborne sound for exterior walls, partitions, and floor/ceiling assemblies should have an STC rating of at least 50.

IMPACT ISOLATION CLASS (IIC)

IIC is a single number giving an approximate measure of the effectiveness of floor construction to provide isolation against sound transmission from impact.

Structure-bone sound such as floor/ceiling assemblies between dwelling units or public areas should have an IIC rating of at least 50.

SOUND TRANSMISSION CLASS (STC) RECOMMENDATIONS FOR VARIOUS OCCUPANCIES	
Room Occupancy*	**STC****
Conference Rooms, Doctor's Offices, High Privacy Areas	50-55
General Offices, Conference Rooms, Low Privacy Areas	45-50
Large Offices, Customer Areas as in Banks	40-50
Mechanical Equipment Rooms	50-60
Classrooms	50
School Music Rooms, Drama Rooms, Industrial Arts Shops	60
Apartment Buildings	—
Bedrooms	48-55
Bathrooms	55-60
Kitchens	50-55
Living Rooms	50-55
Halls	50-55
Single Family Residence Bedroom	40-50
Single Family Bathroom	45-50
Single Family Exterior Wall	45-50

*Typical examples. Refer to local codes for specific requirements.

**Sound Transmission Class

TYPICAL SOUND TRANSMISSION CLASS (STC) RATINGS FOR SELECTED MATERIALS	
Type	**STC**
1/4" Plate Glass	26
3/4" Plywood	28
1/2" Gypsum Board, Both Sides of 2 × 4 Studs	33
1/4" Steel Plate	36
6" Concrete Block Wall	42
8" Reinforced Concrete Wall	51
12" Concrete Block Wall	53
Cavity Wall, 6" Concrete Block, 2" Air Space	56

TYPICAL IMPACT ISOLATION CLASS (IIC) RATINGS FOR SEVERAL FLOOR/CEILING ASSEMBLIES

Type	ICC
WOOD-FRAMED FLOOR/CEILING ASSEMBLIES	
2 × 10" Joists, 1/2" Gypsum Board Ceiling, 1/2" Plywood Subfloor, Hardwood Flooring	30-35
Above with 3" Insulation	45-49
Above with Standard Carpet and Pad	58-60
Above with Insulation and Standard Carpet and Pad	75-80
STEEL BAR JOIST FLOOR/CEILING ASSEMBLIES	
Bar Joist, Gypsum Board Ceiling, Plywood Subfloor, 1 3/8" Concrete Topping, Standard Carpet and Pad	60-62
Above with Lath and Plaster Ceiling	60-62
Above with 2 1/2" Concrete Deck on Metal Decking, Vinyl Flooring	35-37

IIC 45 to 65 is typical for floor/ceiling assemblies in multifamily dwellings. Codes typically require IIC 50.

IIC is a single number giving an approximate measure of the effectiveness of floor/ceiling assemblies to provide isolation against the transmission of sound from impact.

NOISE REDUCTION COEFFICIENT (NCR)

A noise reduction coefficient (NCR) number is an indication of the amount of airborne sound energy absorbed by a material. The larger the number, the greater the efficiency of the material to absorb sound.

NOISE REDUCTION COEFFICIENTS (NRC) FOR SEVERAL MATERIALS

Material	NRC
Unpainted Brick Wall	0.02-0.05
Painted Brick Wall	0.01-0.02
Glazed Clay Tile	0.01-0.02
Concrete Wall	0.01-0.02
Lightweight Concrete Block	0.45
Heavyweight Concrete Block	0.27
Standard Plaster Wall	0.01-0.04
Gypsum Wallboard 1/2"	0.01-0.04
Acoustical Plaster Wall	0.21-0.75
Glass	0.02-0.03
Fiberglass 1" to 4"	0.75-0.95
Wood, 1" Thick	0.10-0.15
Mineral Wool	0.45-0.85
Acoustical Tile	0.55-0.65
Carpeting	0.45-0.75
Vinyl Floor Covering	0.01-0.05
Rubber Flooring	0.02-0.02
Stone	0.01-0.02
Terazzo	0.01-0.03
Wood Floor (solid)	0.015-0.07

NCR is an indicator of the amount of airborne sound energy absorbed by a material.

TYPICAL SOUND INTENSITY DECIBEL LEVELS		
Sound Levels (Decibels)	**Source of Sound**	**Sensation**
0-20	Whisper, Normal Breathing	Very Faint
20-40	Average Residence, Private Office, Quiet Conversation	Quiet
40-60	Conversation, Average Office, Quiet Radio or TV	Moderately Loud
60-80	Noisy Office, Average Radio or TV, Loud Conversation	Loud
80-100	Power Mower, Thunder Close By, Symphony Orchestra	Very Loud
110	Riveting, Air Hammering	Just Below Threshold of Feeling
120	Elevated Train, Rock Band, Siren	Threshold of Feeling
130	Artillery Fire	Threshold of Pain
140	Near a Jet Aircraft	Deafening

TYPICAL LIVE FLOOR DESIGN LOADS FOR RESIDENTIAL AND COMMERCIAL CONSTRUCTION	
Residential	**lb./sq. ft.²**
Living Areas	40
Sleeping Areas	30
Attic, No Storage	10
Attic, Limited Storage	20
Attic, with Pull-down Stair	30
Garage and Carports	50
Balconies and Porches	60
Stairs	60
Sidewalks and Driveways	250
Stairs and Corridors in Apartments	60
Rooms Open to Public	100
Commercial	**lb./sq. ft.²**
Assembly Areas and Balconies	50
Interior, Fixed Seating	100
Interior, Moveable Seating	100
Corridors	100
Dining Room, Restaurant	100
Garages, Auto Repair	100
Gymnasiums	100
Hospital	
Operating Room	60
Laboratories	60
Private Rooms	40
Wards	40
Corridors	80
Kitchens, Commercial	150

TYPICAL LIVE FLOOR DESIGN LOADS FOR RESIDENTIAL AND COMMERCIAL CONSTRUCTION *(cont.)*	
Commercial *(cont.)*	**lb./sq. ft.²**
Laboratories, Scientific	100
Laundry	150
Library	
Reading Room	60
Stacks	150
Office Buildings	
Offices	50
Business Machine Room	100
Lobbies	100
Corridors	80
Computer Rooms – Based on Analysis of Equipment	–
Hotels	
Guest Rooms	40
Public Rooms	100
Public Corridors	100
Restrooms, Public	60
Schools	
Classrooms	40
Corridors	80
Stairs and Exitways	100
Warehouses	
Light	125
Heavy	250
Stores	
Retail	75
Wholesale	100

TYPICALLY USED LIVE DESIGN LOADS FOR RESIDENTIAL CONSTRUCTION	
Type	**Live Loads (lb./sq. ft.)**
Floors	30, 40, 50, 60, 70, 80
Floor and Roof Beams	20, 30, 40
Ceiling Joists	10, 20
Rafters, Flat or Sloped, with or without Attic	20, 30
Rafters, Medium or High Slope, Light or Heavy Roofing	20, 30

FLAME-SPREAD REQUIREMENTS FOR INTERIOR WALL AND CEILING FINISH*

Use	Enclosed Vertical Exitways	Other Exitways	Rooms or Areas
Assembly Buildings	1	2	3
Storage and Sales Areas for Combustible Goods	1	2	3
Restaurants Less Than 50 Occupants, Retail Gasoline Stations	1	2	3
Schools	1	2	3
Factories, Warehouses, Not Using Highly Flammable Material	1	2	3
Hospital, Nursing Home	1	2	2
Hotels, Apartments	1	2	3
Residential, Single and Multifamily	3	3	3
Storage, Handling or Sale of Highly Flammable or Explosive Materials	1	2	3
Auto Repair Garages	1	2	3
Nurseries	1	2	2
Private Garages, Carports, Sheds, Agricultural Buildings	No Restrictions		

*Typical examples. Consult local code for specific requirements.

FIRE RATINGS AND SOUND TRANSMISSION CLASSES FOR SELECTED PARTITIONS

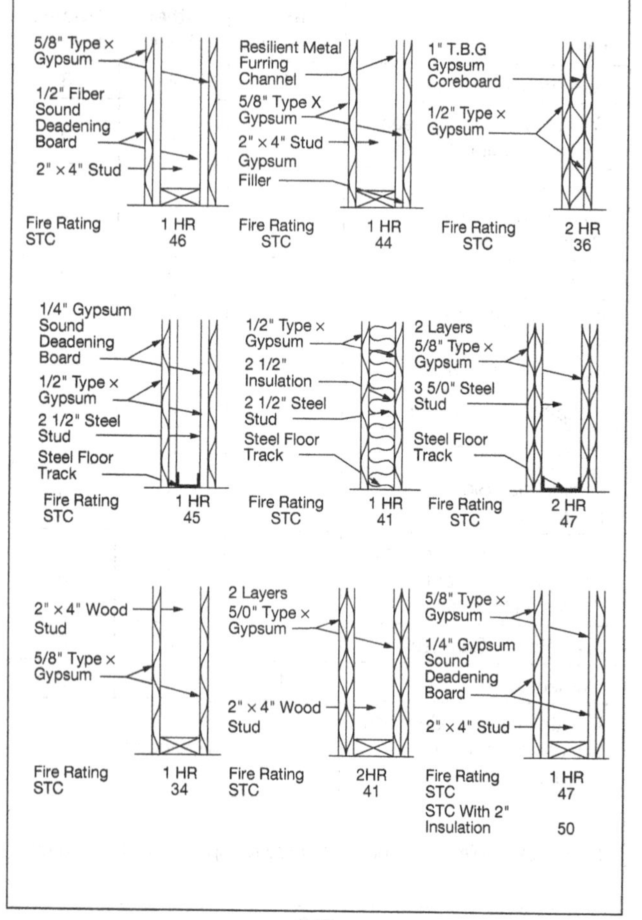

FIRE RATINGS AND SOUND TRANSMISSION CLASSES FOR SELECTED PARTITIONS *(cont.)*

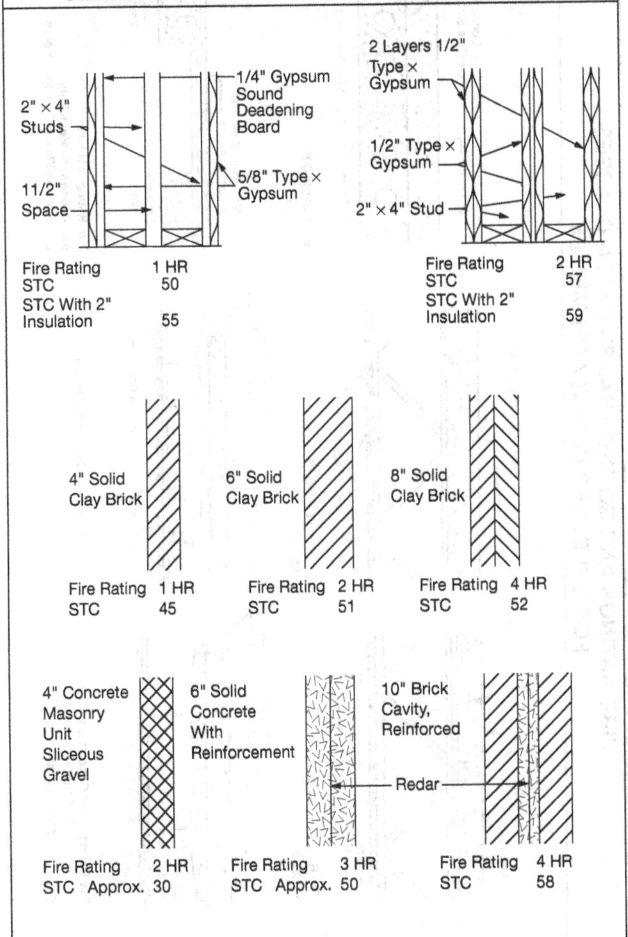

FIRE RATINGS AND SOUND TRANSMISSION CLASSES FOR SELECTED FLOOR ASSEMBLIES

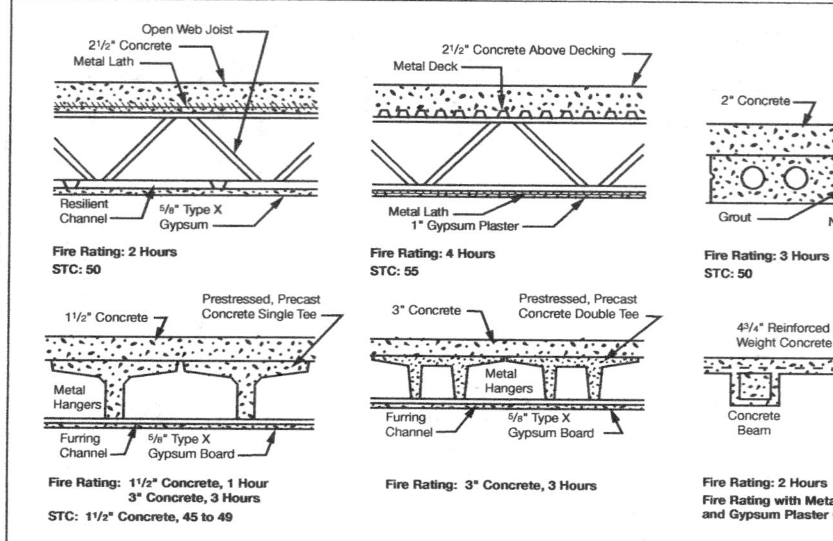

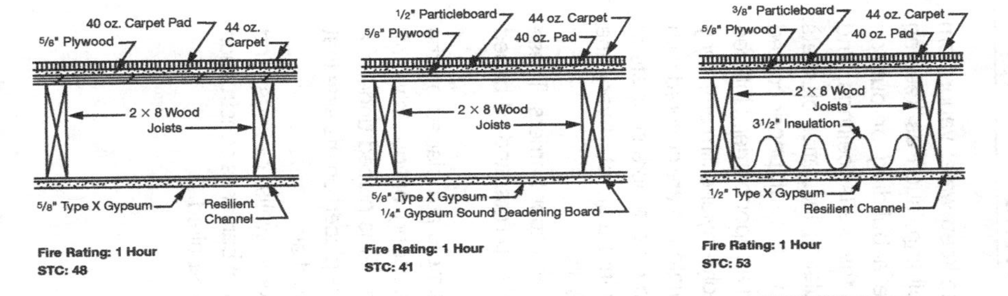

Fire Rating: 1 Hour
STC: 48

Fire Rating: 1 Hour
STC: 41

Fire Rating: 1 Hour
STC: 53

19/32 T.B.G. Plywood

2 Layer 5/8" Type X Gypsum Board

Wood Floor Truss

Fire Rating: 1 Hour
STC: Approx. 40

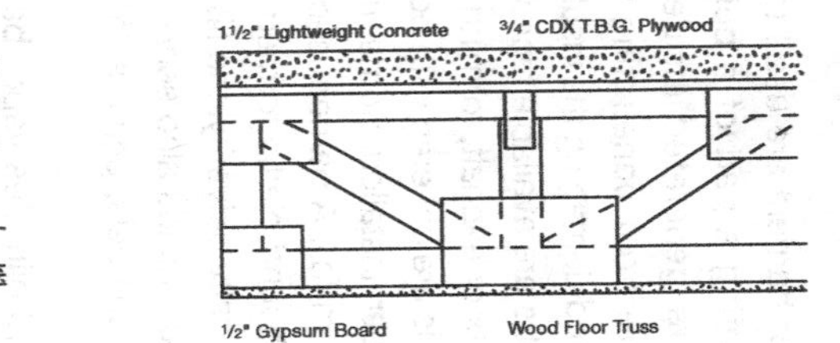

Fire Rating: 1 Hour
STC: Approx. 45

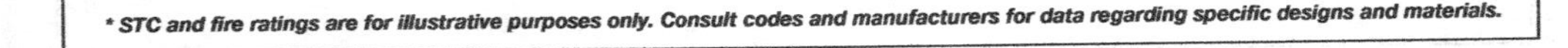

STC and fire ratings are for illustrative purposes only. Consult codes and manufacturers for data regarding specific designs and materials.

VAPOR BARRIERS

Vapor barriers are used to keep water vapor from penetrating a part of a building, such as keeping moisture generated inside a building or outside moisture from penetrating the walls, ceilings, and floors and damaging the insulation. Insulation blankets and batts are available faced with Kraft paper coated with wax or asphalt, forming a vapor barrier. Unfaced materials are covered with polyethylene film after they have been installed. Some materials are covered with aluminum foil. A barrier used on exteriors consists of a spunbound oleifin formed into a sheet of very fine high-density polyethylene fibers.

Various liquids also serve as vapor barriers. These include enamels, primers, latex paints, and oil-based paints.

In cold climates vapor barriers are placed on the inside (warm side) of the wall, ceiling, or floor. In warm, humid climates the vapor barrier is placed under the sheathing. In mild climates it is placed on the side that is most subject to warm, humid air.

Caulk all joints and cracks in walls, ceilings, and floors. Do not puncture the vapor barrier as construction continues. Breaks should be repaired.

TYPICAL PERM RATINGS OF VAPOR BARRIER MATERIALS	
Type	**Perm Rating**
Vapor-Retardant Paint	0.5
Latex Primer	6.0
15-lb. Building Felt	4.0
1" Expanded Polystyrene Bead Board	2 to 5.8
1" Extruded Polystyrene Board	1.2
Kraft Paper Insulation Facing	1.0
6-mil Polyethylene Sheet	.06
1-mil Aluminum Foil	0.0

A perm is a measure of the porosity of a material to the passage of water vapor. Vapor barriers must have a perm of 0.00 to 0.50.

WATERPROOFING MEMBRANES AND COATINGS

Areas where waterproofing is required include foundations, roofs, exterior wood and masonry walls, land exposed structural components, including steel. Methods for waterproofing include:

1. Applying a built-up bituminous membrane of felt and hot or cold tar pitch.
2. Applying a heavy coating, such as Portland cement plaster or a trowelable asphalt.
3. Bonding an elastomeric membrane to the wall.
4. Applying a thin film or coating to the exterior of the wall, such as liquid silicone or coal tar pitch.
5. Adding waterproofing admixtures to the concrete as it is mixed.
6. Applying a dry coating that will emulsify in place, such as bentonite clay.

WATERPROOFING MATERIALS IN GENERAL USE	
Sheet Membranes	**Composite Membranes**
Butyl Ethyene Propylene Neoprene Polyethylene Polyvinyl Chloride	Elastomeric, Backed Polyethylene and Rubberized Bitumen Polyvinyl Chloride Backed Saturated Felts and Bitumen Coated
Liquid Membranes	**Applied Coating**
Butyl Urethane Polychlorene (neoprene) Polyurethane, Coal Tar	Acrylic, Silicone Asphalt Emulsions, Cut Backs Cementitious with Admixtures Epoxy, Bitumen Urethane, Bitumen Bitumen, Rubberized
BUILT-UP MEMBRANES	
Hot Applied	**Cold Applied**
Asphalt, Type I, II, III Coal Tar Pitch, Type B Felts, Saturated and Coated	Bitumen Emulsion Bitumen, Fiberated Cement Felts, Coated Bentonite Clay Fabric, Saturated Glass Fiber Mesh, Saturated Cementitious Membrane

WATERPROOF TIPS

There are a number of ways to waterproof a foundation wall. The manufacturer of a system usually requires the contractor to employ a certified applicator if the manufacturer's guarantee is to be valid. Common systems include liquid membranes, sheet membranes, cementitious coating, built-up systems, and bentonite.

Liquid membranes are applied with a roller, trowel, or spray. The liquid solidifies into a rubbery coating. Different materials are available, such as polymer-modified asphalt and various polyurethane liquid membranes.

Sheet membranes tend to be self-adhering rubberized asphalt sheets, typically an assembly of multiple layers of bitumen and reinforcing materials. Some companies manufacture PVC and rubber butyl sheet membranes.

Cementitious products are available from building supply outlets. They are mixed on the site and are applied with a brush. Some have an acrylic additive available that improves bonding and makes the cementitious coating more durable. One disadvantage is that these coatings will not stretch if the foundation cracks, thus opening the possibility for leakage through cracks.

Built-up systems may be like the widely used hot tar and felt membrane. Alternate layers of hot tar and felt are bonded to the foundation. Usually at least three layers of felt are specified.

Bentonite is a clay material that expands when wet. It is available in sheets that are adhered to the foundation. As groundwater penetrates the clay, it swells many times its original volume, providing a permanent seal against water penetration.

Surface Preparation

Regardless of the type of system used, it is important to prepare the surface before application. This includes (1) drying the wall and footings, (2) removing the concrete form ties, making certain they break out inside the foundation so they do not penetrate the waterproof membrane, (3) cleaning the wall so it is free of all dirt or other loose material, and (4) sweeping the wall free of dust and mud film residue. A residue left when wet mud is wiped off and left to dry on the foundation can inhibit bonding. Finally, any openings around pipes or other items that penetrate the wall must be grouted.

Safety

Waterproofing presents some hazards that must be controlled. First is a possible cave-in of the soil, burying the workers. Normal shoring procedures should be observed. Many of the materials used are flammable and solvent-based, presenting a fire hazard. Workers should not smoke or use any tools that might cause ignition. Solvent fumes can be very harmful, and workers must wear respirators. Fumes are usually heavier than air and settle around the foundation in the excavated area. The solvents, asphalt, and other materials used may cause skin problems, so protective clothing, including gloves, is required with many products. As always, wear eye protection. When in doubt, consult the manufacturer of the product.

WEIGHTS OF CONSTRUCTION MATERIALS

BRICK AND MASONRY

Type	lb./sq. ft.
4" Brick Wall	40
4" Concrete Brick, Stone or Gravel	46
4" Concrete Brick, Lightweight	33
4" Concrete Block, Stone or Gravel	34
4" Concrete Block, Lightweight	22
6" Concrete, Stone or Gravel	50
6" Concrete Block, Lightweight	31
8" Concrete Block, Stone, or Gravel	55
8" Concrete Block, Lightweight	35
12" Concrete Block, Stone or Gravel	85
12" Concrete Block, Lightweight	55

CONCRETE

Type	lb./sq. ft.
Plain, Slag	132
Plain, Stone	144
Reinforced, Slag	138
Reinforced, Stone	150

GLASS

Type	lb./sq. ft.
Double Strength, 1/8"	26 ounces
Double Pane Insulating with 5/8" Air space	3.25
Glass Block, 4" Standard with Mortar	20
Glass Block, 3" Solid with Mortar	40
Glass Block, 3" Lightweight with Mortar	16
Wire Glass, 1/4"	18

WEIGHTS OF CONSTRUCTION MATERIALS *(cont.)*

SOFT WOODS

Type	lb./cu. ft. at 12% Moisture Content
Balsam Fir	23.8
Cedar, White	21.0
Cypress	33.0
Cedar, Western	24.5
Fir, Larch, Douglas	34.2
Hemlock	29.4
Pine, Yellow Southern	36.4
Pine, Northern	32.2
Pine, Ponderosa	29.4
Pine, White	25.9
Redwood, California	26.2
Spruce, Engleman	28.7

HARDWOODS

Type	lb./cu. ft. at 12% Moisture Content
Ash, White	40.5
Aspen	25.9
Birch	44.0
Cottonwood	28.0
Poplar, Yellow	30.1

WEIGHTS OF CONSTRUCTION MATERIALS *(cont.)*

INSULATION	
Type	**lb./sq. ft.**
Batts, Blankets, Blown Fiber per 1" Thickness	0.1 to 0.4
Corkboard per 1" Thickness	0.60
Foamed Insulation Board per 1" Thickness	2.60 ounces

METALS	
Type	**lb./cu. ft.**
Aluminum, Cast	165
Brass, Cast	534
Bronze, Commercial	550
Copper, Solid	555
Iron, Cast	450
Lead	710
Stainless Steel	490
Steel, Rolled	490
Tin	458

PARTITIONS	
Type	**lb./sq. ft.**
4 Metal Stud with Gypsum Wallboard, Both sides	8
2' × 4' wood Studs with Gypsum Wallboard, Both sides	8

WEIGHTS OF CONSTRUCTION MATERIALS *(cont.)*	
ROOFING	
Type	**lb./sq. ft.**
Asphalt Shingles	1.7 to 2.8
Built-up	6.5
Clay Tile	8 to 16
Copper	1.5 to 2.5
Corrugated Iron	2
Concrete Tile	9.5
Slate 3/16" to 1/4"	7 to 9
Slate 3/8" to 1/2"	14 to 18
Steel Deck Alone	2.5
Wood Shingles and Shakes	2 to 3
STONE VENEER	
Type	**lb./sq. ft.**
2" Granite with 1/2" Parging	30
4" Limestone with 1/2" Parging	36
4" Sandstone with 1/2" Parging	49
1" Marble	13
STRUCTURAL CLAY TILE	
Type	**lb./sq. ft.**
4" Hollow	23
6" Hollow	38
8" Hollow	45

WEIGHTS OF CONSTRUCTION MATERIALS *(cont.)*	
SUSPENDED CEILINGS	
Type	**lb./sq. ft.**
Mineral Fiberboard, 5/8" Acoustic	1.4
Plaster on Gypsum Lath	10 to 11
LIGHTWEIGHT CONCRETE	
Type	**lb./sq. ft.**
Concrete with Perlite	35 to 50
Concrete with Pumice	60-90
Concrete with Vermiculite	25-60
MORTAR AND PLASTER	
Type	**lb./cu. ft.**
Mortar Used for Masonry	115
Plaster, Gypsum, or Sand	105
WALL, CEILING, AND FLOOR	
Type	**lb./sq. ft.**
Acoustical Tile, 1/2"	0.8
Concrete, Reinforced, Stone 1"	12.5
Concrete, Plain, Stone 1"	12
Concrete, Reinforced, Lightweight 1"	6 to 10
Concrete, Plain, Lightweight 1"	3 to 9
Gypsum Wallboard, 1/2"	2
Gypsum Wallboard, 5/8"	2.5
Hardboard 1/2"	0.75
Hardwood Flooring 13/32"	4

WEIGHTS OF CONSTRUCTION MATERIALS *(cont.)*

WALL, CEILING, AND FLOOR *(cont.)*	
Type	**lb./sq. ft.**
Plaster, 2" Solid Partition	20
Plaster, 4" Solid Partition	34
Plaster, 1/2"	4.5
Plaster on Lath	10
OSB and Waferboard 1/2"	1.6
OSB and Waferboard 3/4"	2.5
Plank, 2" Cinder Concrete	15
Plank, 2" Gypsum	12
Plywood, 1/4"	0.75
Plywood, 1/2"	1.6
Plywood, 3/4"	2.25
Tile, Quarry 1/2"	5.8
Tile, Glazed 3/8"	3
Terrazzo, 1"	25
Vinyl Floor Tile	1.4
WALL, FLOOR, PARTITIONS, CEILINGS	
Type	**lb./sq. ft.**
Exterior Wall, 4' Wood Studs	10
Exterior Wall, 6" Wood Studs	14
Exterior Wall, Brick over 4" Wood Studs	50
Wood-Framed Floor	10
Ceiling with Drywall	10

SEISMIC CONSIDERATIONS

Certain sections of the country are known to have subsurface conditions that make them possible earthquake zones. Check local building codes because some areas will have special regulations for the design and construction of buildings and other structures. Masonry construction is usually required to meet the requirements set forth by the American Concrete Institute, URL: http://www.aci-int.org and the American Society of Civil Engineers, www.asci.org. Structural steel requirements may be those set by the American Institute of Steel Construction, www.aisc.org, while those for wood construction may follow the requirements of the National Forest and Paper Association, www.afandpa.org. Information is also available from the Building Seismic Safety Council and the National Institute of Building Sciences at www.nibs.org.

CHAPTER 4
Site Work

Site planning involves the services of building architects, landscape architects, engineers and other specialists. The on-site work includes a wide range of activities such as layout, grading, dewatering excavations, installing utilities, paving, surface drainage control during and after construction and site improvements such as walks, fences and landscaping. Local building codes have regulations pertaining to site work.

Refer to Chapter 2 for detailed safety rules which should be observed while performing activities on the construction site.

TYPICAL SITE PLAN WITH ORIGINAL AND FINISHED CONTOURS AND LAYOUT INFORMATION

Legend

———— - - - ————	Property Line
————G————	Gas Line
————W————	Water Line
————S————	Sanitary Sewer
————SS————	Storm Sewer
————P————	Electrical Power
————T————	Telephone
1002.0 (underlined)	Existing and Finish Grade
1002.0 / 1003.0	Existing Grade / Finish Grade
- - - - - - - -1002- - - - - - - -	Original Contour
————1002————	Finish Contour

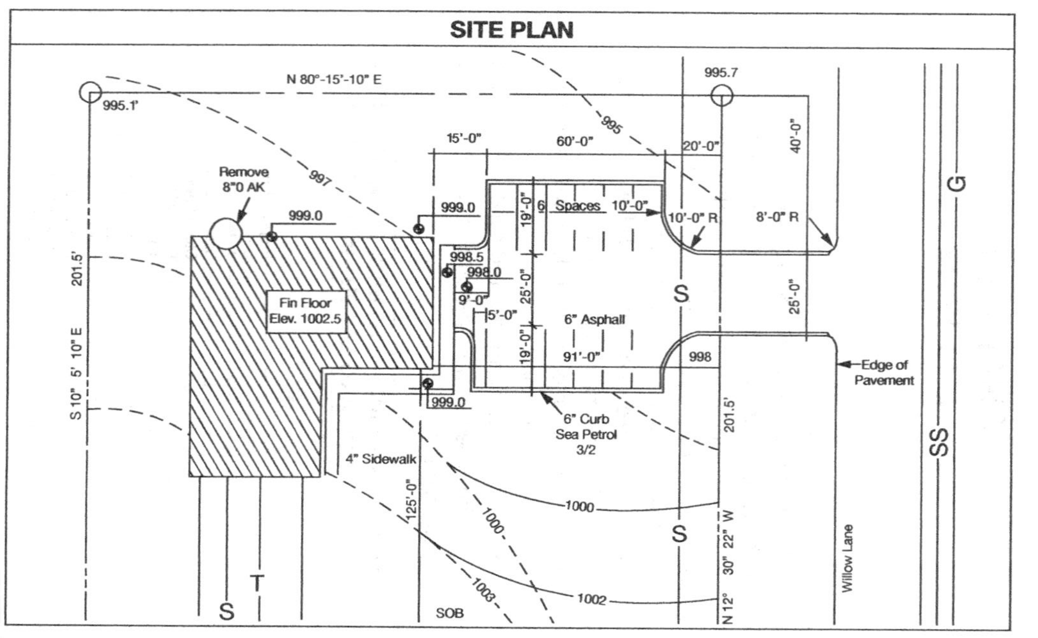

SITE PLAN
N 80°-15'-10" E
995.1'
995.7
995
15'-0"
60'-0"
20'-0"
40'-0"
Remove
8"0 AK
997
999.0
999.0
6 Spaces
10'-0"
10'-0" R
8'-0" R
G
19'-0"
998.5
998.0
9'-0"
25'-0"
5'-0"
S
25'-0"
201.5'
Fin Floor
Elev. 1002.5
6" Asphall
19'-0"
91'-0"
998
Edge of
Pavement
S 10" 5' 10" E
999.0
6" Curb
Sea Petrol
3/2
201.5'
SS
4" Sidewalk
125'-0"
1000
1000
S
N 12° 30" 22" W
Willow Lane
T
S
1003
1002
SOB

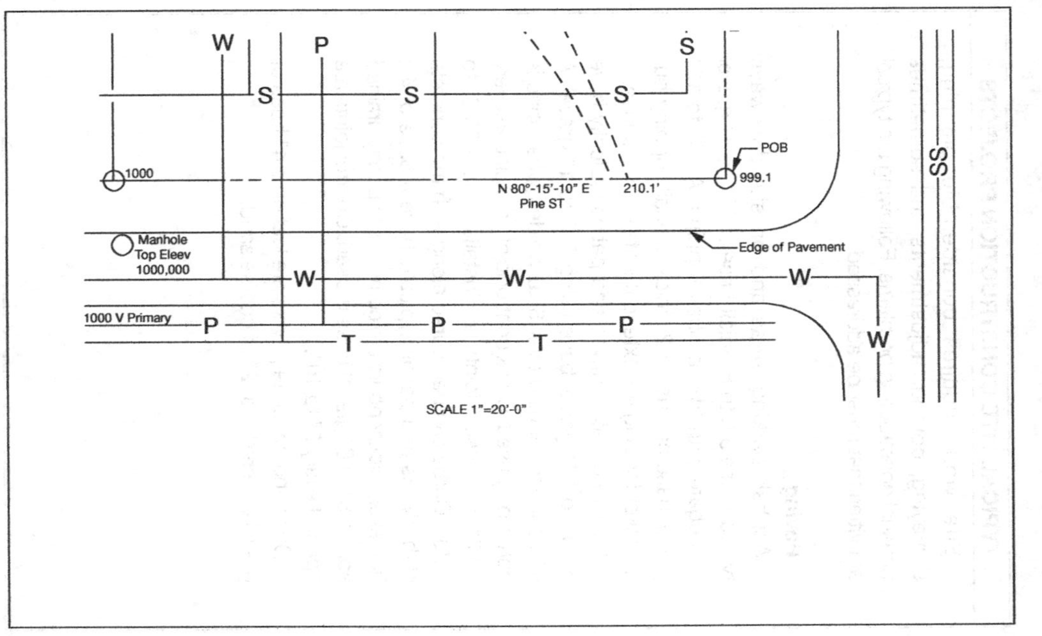
W
P
S
S
S
S
POB
1000
999.1
N 80°-15'-10" E
210.1'
Pine ST
Manhole
Top Eleev
1000,000
Edge of Pavement
W
W
W
W
1000 V Primary
P
P
P
T
T
SS
SCALE 1"=20'-0"

TYPICAL SITE CONSTRUCTION PROJECTS

Site work requires considerable planning, surveying, contour adjustments and sometimes unusual construction problems. Following are typical activities that must be addressed.

Paving

Asphalt parking areas and on site driveways typically use a 6 to 8" thick base of 3/4" to 1" gravel aggregate with 2½" asphalt topping. A 10" to 12" gravel base can use a 2" thick asphalt topping but a thicker topping will extend the life of the surface.

Concrete walks, drives and patios typically have an 8" to 10" gravel base using 3/4" to 1" gravel. A 4" concrete topping over 6×6 welded wire fabric is commonly used. Crack control joints are placed every 10'0". Expansion joints are installed every 20'0" to 30'0". Curbs can be poured along with the concrete slab. Walks and patios typically have a base of 4"; however, thicker pours are common. In poorly drained soils a 6" to 8" gravel base is used. Control joints are spaced every 8" to 10".

Brick and concrete pavers are laid on a 4" base of gravel covered with 2" of coarse sand.

TYPICAL SITE CONSTRUCTION PROJECTS *(cont.)*

Fences

Typically used fences include chain link, usually 4' to 6' high, wrought iron fences are typically 3' to 4' high, various types of wood fences are usually 3' to 6' high and brick and stone walls usually run in the range of 3' to 4' high.

Landscaping

The landscape plan identifies the types and location of shrubs, trees, both existing and those to be added, flowers, gardens, pools, arbors, fences and other such features. It may be a part of the plot plan or a separate drawing based on the plot plan data such as countours and site boundary information. Symbols are used to represent the various types of plants and other features. A listing of landscape symbols is in Chapter 3. Information about the species and size is included on the plan or in a landscape schedule.

Landscaping provides the attractive setting for the building. The size and type of plant must be carefully considered. Dense leaved plants can be used to shield the sunny side of a building, reducing air conditioning costs and blocking noise penetration. An irrigation system is often included. For lawns a spray type system is recommended, while a drip or bubble system feeds large shrubs and trees.

TYPICAL LANDSCAPE PLAN

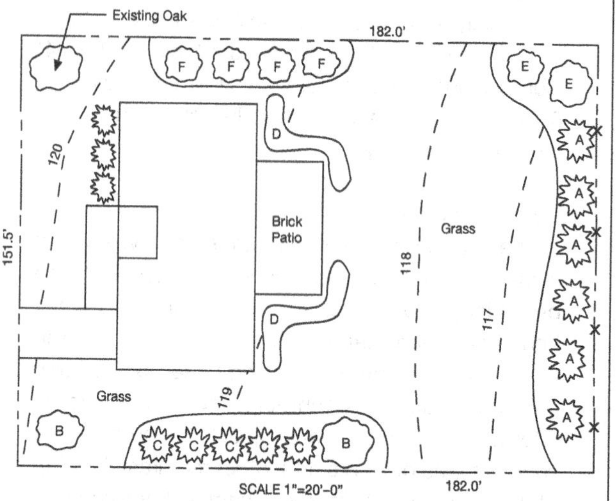

Existing Oak
182.0'
F
F
F
F
E
E
D
A
A
120
Brick
Patio
Grass
A
151.5'
118
A
D
117
A
Grass
119
A
B
C
C
C
C
C
B
A
SCALE 1"=20'-0"
182.0'

Key	Quantity	Plant
A	6	Leland Cypress
B	2	River Birch
C	8	Juniper
D	—	Flowers
E	2	Red Tips
F	4	Holly

PARKING AREAS

Parking areas require individual parking spaces sized to accommodate the largest automobiles available and various other vehicles as vans and make provisions for vehicles used by the disadvantaged. Access into and out of the parking area requires consideration as does the spacing of aisles of traffic within the area.

PARKING DATA

- Typical automobile parking stalls 8'6" to 10'0" wide by 19'0" to 20'0" long.
- Compact auto parking 7'0" to 8'0" wide by 15'0" by 16'0" long.
- Handicapped van parking 8'0" wide by 20'0" long, with an 8'0" wide side aisle.
- Regular vans 8'0" wide by 20'0" long with a 5'0" side aisle.
- One way traffic aisle between parking rows 12'0" to 14'0" wide.
- Two way traffic aisle between parking rows 18'0" to 24'0" wide.

TYPICAL PARKING STALL DATA

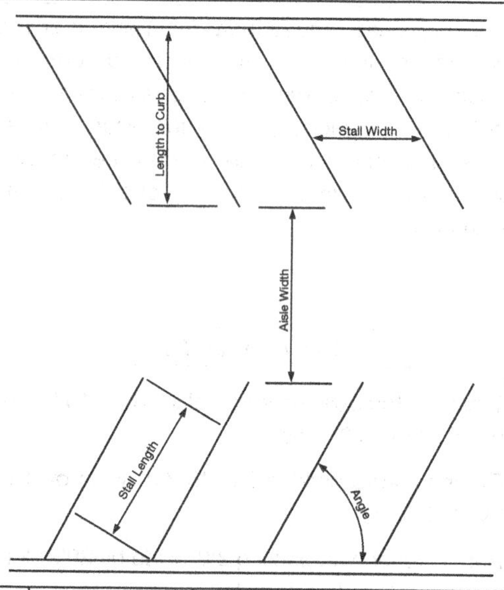

Stall Width	60° Parking[3] Stall Length to Curb (19' Stall)	90° Parking[3] Stall Length to Curb (19.0' Stall)	Aisle Width[1]	Aisle Width[2]
8'-6"	20.7	19.0'	18'-6"	25'-0"
9'-0"	21.0	19.0'	18'-0"	24'-0"
9'-6"	21.2	19.0'	18'-6"	24'-0"

[1] Aisle accommodates one-way traffic and provides room to back out of stall.

[2] Aisle accommodates two-way traffic and provides room to back into or out of stall.

[3] Other angles often used.

TYPICAL STREET AND DRIVEWAY DESIGN DATA

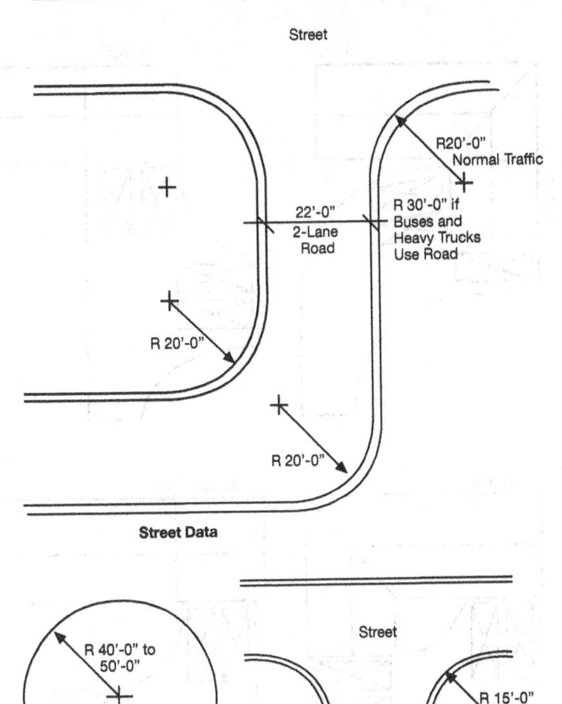

Street Data

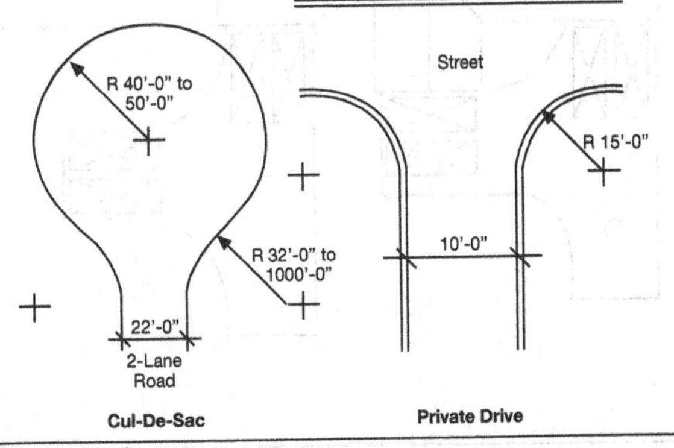

Cul-De-Sac **Private Drive**

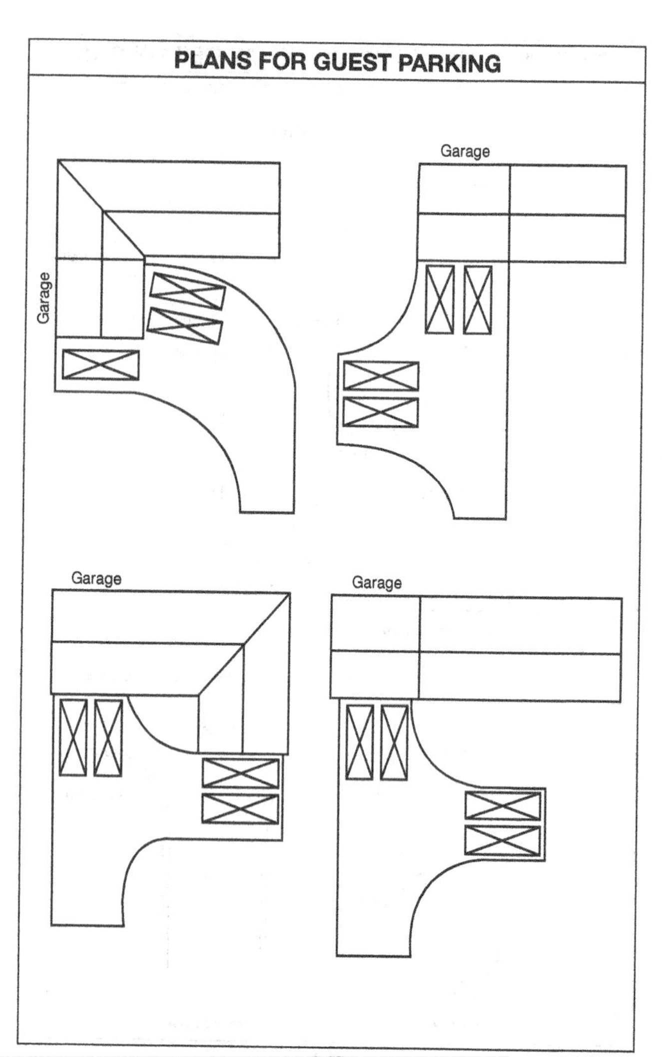
PLANS FOR GUEST PARKING
Garage
Garage
Garage
Garage

PLANS FOR GUEST PARKING *(cont.)*

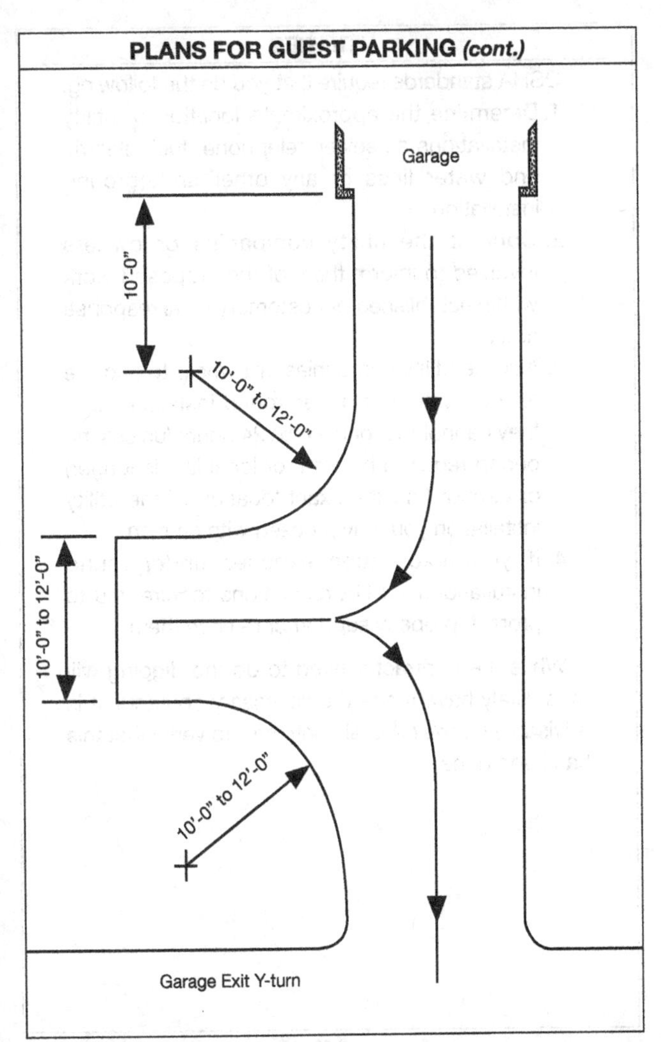

Garage Exit Y-turn

UTILITIES

OSHA standards require that you do the following:

1. Determine the approximate location of utility installations as sewer, telephone, fuel, electric and water lines or any other underground installation.
2. Contact the utility companies or owners involved to inform them of the proposed work within established or customary local response times.
3. Ask the utility companies or owners to find the exact location of underground installations. If they cannot respond within 24 hours (unless the period required by state or local law is longer) or cannot find the exact location of the utility installation you may proceed with caution.
4. If your excavation exposes underground installations, OSHA regulations require you to protect, properly support or remove them.

While the contractor hired to do the digging will most likely have made the necessary contacts, it is advisable for the general contractor to verify that this has been done.

LAYERED SOIL

As excavations get deeper there will usually be a change in the characteristics of the soil. This layering will require that the handling of the slope be adjusted to suit the soil encountered. The angle of slope will have to be changed and possibly some form of retaining wall may become necessary.

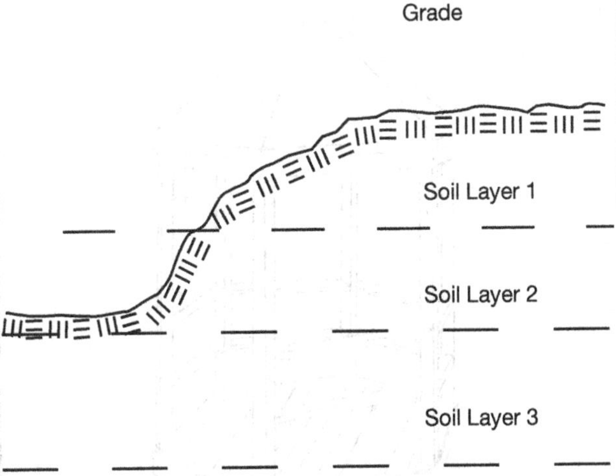

Grade

Soil Layer 1

Soil Layer 2

Soil Layer 3

TREE REMOVAL

Frequently large trees on a site must be removed. This can present dangers to those employed to remove them and others on the site. General safety rules must be enforced as covered in Chapter 2. It is always best to employ an experienced tree cutting firm that has the necessary insurance.

DECORATIVE STRUCTURES AND INSTALLATIONS

Part of the landscape plan will often include small structures as a garden arbor, trellis or a gazebo. Also decorative ponds with flowers and swimming pools often are a part of the site work activities.

TYPICAL RECOMMENDED SLOPES		
Type	% of Slope	
	Max.	Min.
Ramps with handrail	8.3	–
Curb ramps	5.0	–
Exitway ramps	10.0	–
Street, length	6 to 10	0.5
Streets, width	4	2
Parking area	5	1
Parking area crown or cross slope	5	1
Parking handicapped	2	–
Walks subject to freezing	5	–
Walks not subject to freezing	14	–
Stair landings	2	1
Slope away from foundations - Impervious surface	21	1
Slope away from foundations - Pervious surface	21	5
Swales and concrete on-grade drainage gutters	as needed	1

SOILS

Soils are the end result of mechanical and chemical weathering of rock. They include any earthen material other than bedrock. They are a complex material and have widely variable properties. The classification of soils requires that all these variables be considered. Even a change in the water content causes differences in physical properties. A knowledge of soils, their characteristics and other properties is essential when designing foundations, trenching and tunneling.

IMPORTANT SOIL PROPERTIES

Density.

Dry density is the amount of solid material in a unit volume of soil.

Cohesion.

This is the maximum tensile strength of the soil.

Compressibility.

As a soil is placed under stress it had a change in volume due to compression.

Permeability.

The ability of a soil mass to conduct fluid flow.

LABORATORY SOIL TESTS

Test	Findings
Plasticity Index	To find the quality of silty or clayey soils and predict their behavior.
Grain size analysis	Determining the added ingredients needed to bring fill material to the required bearing strength.
Direct Shear Test	Measurement of the relationships between compression and the shear strength at various pressures.
Field Density Test	To find the dry weight per cubic foot and the moisture content of soil samples.
Moisture Density Tests	Consolidation tests on soil samples having varying amounts of water added to find the maximum soil weight or most favorable density.
Consolidation Tests	Calculation of the actual settlement that will occur.

TERMS USED IN SOIL DESCRIPTIONS

- Swell is the volume increase of disturbed soil over the natural state.
- Shrinkage is the reduction in the volume of soil from the natural condition to a compacted condition.
- Bank volume is the volume of soil in its natural undisturbed state.
- Loose volume is the volume of soil after it has been excavated from its natural location.
- Compacted volume is the volume of soil after it has been subjected to mechanical compaction.
- Cohesiveness refers to the properties of soils that stick together dry, forming a solid mass and soften when liquid is added.
- Cohesionless soils are granular materials having large particles which do not bond.
- Plasticity refers to the condition of soils when wet that puts them in a state that is neither liquid or solid.
- Graduation refers to the distribution of particles of a range of sizes in a soil.
- Organic soils are those containing decaying plant or animal matter.
- Expansive clays are those which greatly expand when wet and shrink when dry.
- Engineered fill is a fill made using soil that has been carefully selected, placed and compacted as planned by an engineer.
- Permeability refers to the rate at which water passes through a soil.

TERMS USED IN SOIL DESCRIPTIONS *(cont.)*

- Elasticity designates the ability of a soil to return to this original shape after undergoing deformation under load.
- Shearing strength is the maximum shear stress which a soil is capable of sustaining.
- Compressibility is a volume change deformation that results from the expulsion of water from the soil.
- Compaction results when load is applied to a soil causing the soil to settle and compress.
- Grouting is the process of introducing chemical agents into the soil to stabilize it. Portland cement and special chemical grouting agents are used.
- Dewatering is a process used to lower the water table around a construction site to reduce the problem of working in a wet soil condition.
- Stabilization is a process used to stabilize unstable soils by adding special chemicals, blending in other soils or adding Portland cement, lime and salt.
- Liquid limit of a soil is the water content expressed as a percentage of the dry weight at which the soil will flow when tested with the shaking test.
- Plastic limit of a soil is the percent moisture content at which the soil begins to crumble when rolled into a 1/8" diameter thread.
- The shrinkage limit is the water content at which the soil volume is at its minimum.

TYPES OF SOILS

Soils are classified by the sizes of particles and their physical properties. The following five types of soils are those typically found. However, often soil is a mixture of these.

- Gravel is a hard rock material with particles larger than 1/4" (6.4 mm) in diameter but smaller than 3" (76 mm).
- Sand is fine rock particles smaller than 1/4" (6.4 mm) in diameter to 0.002" (0.05 mm).
- Silt is fine sand with particles smaller than 0.002" (0.05 mm) and larger than 0.00008" (0.002 mm).
- Clay is a very cohesive material with microscopic particles (less than 0.00008" or 0.002 mm).
- Organic matter is partly decomposed vegetable matter.
- Rock particles larger than 3" (76 mm) are called cobbles or boulders and are not classified as soil.

SOIL CLASSIFICATION – OSHA 29CFR 1926, CONSTRUCTION INDUSTRY

Type A

Cohesive soils with an unconfined compressive strength of 1.5 tons per square foot (144 kPa) or greater. Examples include clay, silty clay, sandy clay, clay loam, and cemented soils such as caliche and hardpan.

Type B

Cohesive soils with an unconfined compressive strength greater than 0.5 tons per square foot (48 kPa) but less than 1.5 tons per square foot (144 kPa). Examples include angular gravel, silt, silt loam, sandy loam, and often silty clay loam and sandy clay loam.

Type C

Cohesive soil with an unconfined compressive strength of 0.5 tons per square foot (48 kPa) or less. Examples include gravel sand, loamy sand, submerged soil or soil with seeping water and submerged rock.

SOIL CLASSIFICATION AND PROPERTIES DESCRIPTION USING THE UNIFIED CLASSIFICATION SYSTEM

Type	Letter Symbol	Description	Rating as Subgrad Material	Rating as Surfacing Material
Gravel and gravelly soils	GW	Well-graded gravel; gravel-sand mixture; little or no fines	Excellent	Good
	GP	Poorly graded gravel; gravel-sand mixture; little or no fines	Good	Poor
	GM	Gravel with silt; gravel-sand-silt mixtures	Good	Fair
	GC	Clayey gravels; gravelly sands; little or no fines	Good	Excellent
Sand and sandy soils	SW	Well-graded sands; gravelly sands; little or no fines	Good	Good
	SP	Poorly graded sands; gravelly sands; little or no fines	Fair	Poor
	SM	Silty sands; sand-silt mixtures	Fair	Fair
	SC	Clayey sands; sand-clay mixtures	Fair	Excellent
	ML	Inorganic salts; fine sands; rock flour; silty and clayey fine sands with slight plasticity	Fair	Poor
Silts and clays with liquid, limit greater than 50°	CL	Inorganic clays of low to medium plasticity; gravelly clays; silty clays; lean clays	Fair	Fair
	OL	Organic silts of low plasticity	Poor	Poor
Silts and clays with liquid, limit greater than 50°	MH	Inorganic silts; micaceous or diatomaceous fine sandy or silty soils; elastic silts	Poor	Poor
	CH	Inorganic clays of high plasticity	Very poor	Poor
	OH	Organic clays of medium to high plasticity; organic silts	Very poor	Poor
Highly organic soils	Pt	Peat and other highly organic soils	Unsuited for subgrade material	Unsuited for surfacing

ASTM D2487-06 Standard Practice for Classification of Soils for Engineering Purposes (Unified Soil Classification System), Courtesy American Society for Testing and Materials.

THE AASHTO SYSTEM OF SOIL CLASSIFICATION

	A-1 Sand and Gravel		A-2 Gravel, Silty, or Clayey Sand				A-3 Fine Sand	A-4 Silt	A-5 Silt	A-6 Clay	A-7 Clay
Typcial Material	A-1-a	A-1-b	A-2-4	A-2-5	A-2-6	A-2-7					
No. 10 sieve	50% max.	—	—	—	—	—	—	—	—	—	—
No. 40 sieve	30% max.	50% max.	—	—	—	—	51% min.	—	—	—	—
No. 200 sieve	15% max.	25% max.	35% max.	35% max.	35% max.	35% max.	10% max.	36% min.	36% min.	36% min.	36% min.
Fraction passing, No. 40 sieve, Liquid limit	6% max.	6% max.	40% max.	41% max.	40% max.	41% min.	—	40% max.	41% min.	40% max.	41% min.
Plasticity index	—	—	10%	10%	11%	11%	—	10%	10%	11%	11%

From Standard Specifications for Transportation Materials and Methods of Sampling and Testing, 2006, American Association of State Highway and Transportation Officials, Washington, D.C. www.transportation.org. Used by permission.

ENGINEERING PROPERTIES OF VARIOUS SOILS

Typical Names of Soil Groups	Important Engineering Properties			Compressibility When Compacted and Saturated	Workability as a Construction Material
	Group Symbols	Permeability When Compacted	Shear Strength When Compacted and Saturated		
Well graded gravels, gravel-sand mixtures, little or no fines	GW	Pervious	Excellent	Negligible	Excellent
Poorly graded gravels, gravel sand mixtures, little or no fines	GP	Very pervious	Good	Negligible	Good
Silty gravels, poorly graded gravel sand silt mixtures	GM	Semipervious to impervious	Good	Negligible	Good
Clayey gravels, poorly graded gravel clay mixtures	GC	Impervious	Good to fair	Very low	Good
Well graded sands, gravelly sands, little or no fines	SW	Pervious	Excellent	Negligible	Excellent
Poorly graded sands, gravelly sands, little or no fines	SP	Pervious	Good	Very low	Fair

ENGINEERING PROPERTIES OF VARIOUS SOILS *(cont.)*

Silty sands, poorly graded sand, silt mixtures	SM	Semipervious to impervious	Good	Low	Fair
Clayey sands, poorly graded sand clay mixtures	SC	Impervious	Good to fair	Low	Good
Inorganic silts and very fine sands, rock flour, silty or clayey fine sands with slight plasticity	ML	Semipervious to impervious	Fair	Medium	Fair
Inorganic clays of low to medium plasticity, gravelly clays, sandy clays, silty clays, lean clays	CL	Impervious	Fair	Medium	Good to fair
Organic silts and organic silt-clays of low plasticity	OL	Semipervious to impervious	Poor	Medium	Fair
Inorganic silts, micaceous or diatomaceous fine sandy or silty soils, elastic silts	MH	Semipervious	Fair to poor	High	Poor
Inorganic clays of high plasticity, fat clays	OH	Impervious	Poor	High	Poor
Organic clays of medium to high plasticity	OH	Impervious	Poor	High	Poor
Peat and other highly organic soils	PT	–	–	–	–

RELATIVE DESIRABILITY PROPERTIES OF VARIOUS SOILS

Typical Names of Soil Groups	Relative Desirability for Various Uses (No. 1 is considered the best)									
	Rolled Earthfill Dams			Canal Sections		Foundations		Roadways		
								Fills		
	Homogeneous Embankment	Core	Shell	Erosion Resistance	Compacted Earth Lining	Seepage Important	Seepage Not Important	Frost Heave Not Possible	Frost Heave Possible	Surfacing
Well graded gravels, sand mixtures, little or no fines	—	—	1	1	—	—	1	1	1	3
Poorly graded gravels, gravel sand mixtures, little or no fines	—	—	2	2	—	—	3	3	3	—
Silty gravels, poorly graded gravel sand-silt mixtures	2	4	—	4	4	1	4	4	9	5
Clayey gravels, poorly graded gravel sand clay mixtures	1	1	—	3	1	2	6	5	5	1
Well graded sands, gravelly sands, little or no fines	—	—	3 if gravelly	6	—	—	2	2	2	4
Poorly graded sands, gravelly sands, little or no fines	—	—	4 if gravelly	7 if gravelly	—	—	5	6	4	—
Silty sands, poorly graded sand silt mixtures	4	5	—	8 if gravelly	5 erosion critical	3	7	8	10	6
Clayey sands, poorly graded sand clay mixtures	3	2	—	5	2	4	8	7	6	2

RELATIVE DESIRABILITY PROPERTIES OF VARIOUS SOILS *(cont.)*

Inorganic silts and very fine sands, rock flour, silty or clayey fine sands with slight plasticity	6	6	—	—	6 erosion critical	6	9	10	11	—
Inorganic clays of low to medium plasticity, gravelly clays, sandy clays, silty clays, lean clays	5	3	—	9	3	5	10	9	7	7
Organic silts and organic silt clays of low plasticity	8	8	—	—	7 erosion critical	7	11	11	12	—
Inorganic silts, micaceous or diatomaceous fine sandy or silty soils, elastic silts	9	9	—	—	—	8	12	12	13	—
Inorganic clays of high plasticity, fat clays	7	7	—	10	8 volume change critical	9	13	13	8	—
Organic clays of medium to high plasticity	10	10	—	—	—	10	14	14	14	—
Peat and other highly organic soils	—	—	—	—	—	—	—	—	—	—

Source: SOILS MANUAL, U.S. Bureau of Reclamation

Soils are divided into sizes based on the size of the soil particles. The particles are sized using standard sieves.

STANDARD SIEVE SIZES

Standard Sieve Sizes	Opening Size	
	Inch	Millimeter
3 in.	3	76.2
1½ in.	1.50	38.1
¾ in.	0.75	19.0
No. 4	0.186	4.76
No. 10	0.078	2.00
No. 40	0.017	0.425
No. 100	0.006	0.150
No. 200	0.003	0.075

DESIGNATED SOIL SIZES

Type of Soil	Standard Sieve Size
Cobbles	Above 3"
Coarse gravel	3¼"
Fine gravel	¾" to No. 4
Coarse sand	No. 4 to No. 10
Medium sand	No. 10 to No. 40
Fine sand	No. 40 to No. 200
Silts and clays	Below 200

SOIL VOLUME CHARACTERISTICS

TYPICAL WEIGHTS, SWELL, AND SHRINKAGE FOR SEVERAL TYPES OF SOILS

	Weight Loose		Weight Bank Measure		Weight Compacted			
	lb/yd^3	kg/m^3	lb/yd^3	kg/m^3	lb/yd^3	kg/m^3	Swell	Shrinkage
Clay, natural	2300	1300	2900	1720	3750	2220	38%	20%
Common earth, dry	2100	1245	2600	1540	3250	1925	24%	10%
Sand with gravel, dry	2900	1720	3200	1900	3650	2160	12%	12%
Sand, dry	2400	1400	2700	1600	2665	1580	15%	12%

[3]Weights, swell, and shrinkage of actual samples may be larger or smaller.

SOIL CONSISTENCY IN TERMS OF WATER CONTENT

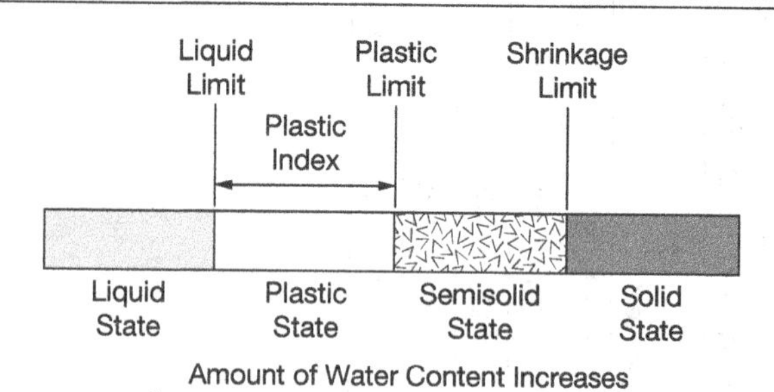

When water content in a soil increases the soil becomes more fluid

TYPICAL WEIGHTS OF SOIL, SAND, GRAVEL, AND STONE

SOIL, SAND, GRAVEL	lb per cu. yd. bank volume	lb per cu. yd. loose volume
Clay, natural bed	2900	2100
Clay and gravel, dry	3000	2200
Clay and gravel, wet	3700	2700
Earth, loam dry	2600	2100
Earth, loam wet	3400	2700
Gravel, to 2 inches	3300	2900
Sand, dry	3000	2700
Sand, wet	3000	3200
STONE	**lb per cu. ft.**	
Granite	165	
Limestone	160	
Marble	170	
Sandstone	145	
Slate	170	

SOIL LOAD BEARING VALUES

When footings are placed on undisturbed original soil, the known local load bearing capacities are usually accepted by the local building official. If there is a doubt about the soil at the site or if the soil has been disturbed and possibly compacted, tests will be required.

LATERAL SOIL VALUES

Lateral soil pressure occurs on the sides of foundations and retaining walls. The actual pressure increases with the depth of the foundation into the soil. The presence of water naturally in the soil can increase the lateral pressure and must be considered as the foundation or retaining wall is designed.

MAXIMUM ALLOWABLE FOUNDATION PRESSURE AND LATERAL PRESSURE RESISTANCE VALUES				
ALLOWABLE FOUNDATION AND LATERAL PRESSURE				
Class of Materials	**Allowable Foundation Pressure (psf)[d]**	**Lateral Bearing (psf/f Below Natural Grade)[d]**	**Lateral Sliding**	
			Coefficient of Friction[a]	**Resistance (psf)[b]**
1. Crystalline bedrock	12000	1200	0.70	—
2. Sedimentary and foliated rock	4000	400	0.35	—
3. Sandy gravel and/or gravel (GW and GP)	3000	200	0.35	—
4. Sand, silty sand, clayey sand, silty gravel and clayey gravel (SW, SP, SM, SC, GM and GC)	2000	150	0.25	—
5. Clay, sandy clay, silty clay, clayey silt, silt and sandy silt (CL, ML, MH and CH)	1500[d]	100	—	130

MAXIMUM ALLOWABLE FOUNDATION PRESSURE AND LATERAL PRESSURE RESISTANCE VALUES *(cont.)*

For SL 1 pound per square foot = 0.0479 KPa. 1 pound per square foot per foot = 0.157 kPa/m.

a. Coefficient to be multiplied by the dead load.

b. Lateral sliding resistance value to be multiplied by the contact area as limited by Section 1804.3.

c. Where the building official determines that in-place soils with an allowable bearing capacity of less than 1,500 psf are likely to be present at the site, the allowable bearing capacity shall be determined by a soils investigation.

d. An increase of one-third is permitted when using the alternate load combinations in Session 1605.3.2 that include wind or earthquake loads.

EXCAVATIONS

Excavations are required for footings, foundations, trenches, and other below ground projects. Some present considerable danger to the workers and are carefully regulated.

OSHA defines an excavation as any man-made cut, cavity, trench, or depression in the earth's surface formed by earth removal. A trench is defined as a narrow underground excavation that is deeper than it is wide and no deeper than 15 feet (4.5 meters). The Occupational Safety and Health Administration (OSHA) Excavation and Trenching standard, Title 29 of the Code of Federal Regulations (CFR) Part 1926.650, covers requirements for excavation and trenching operations.

OSHA requires that all excavations in which employees will work be protected by:

1. Sloping or benching the sides of the excavation
2. Supporting the sides of the excavation with some type of formwork
3. Placing a shield between the side of the excavation and the work area

PROTECTIVE SYSTEMS FOR TRENCHES

The Occupational Safety and Health Administration's (OSHA) Excavation and Trenching Standard, Title 29 of Code of Federal Regulation (CFR), Part 1926.650 and 1926.652 cover requirements for excavation and trenching operations.

OSHA Method One

For excavations less than 20 ft. deep slope the sides of the escavation at an angle of 34 degrees (1-1½:1). This slope is safe for any type of soil.

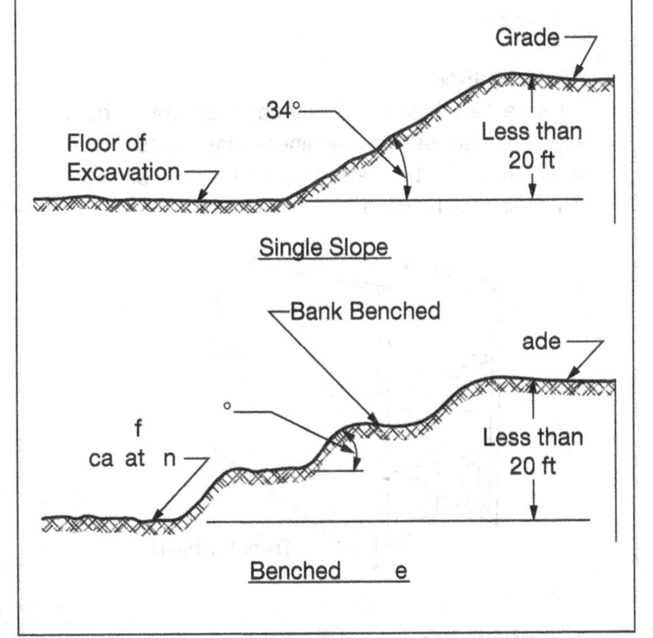

OSHA Method Two

OSHA recommends using tabulated data approved by a registered professional engineer. Data must be in writing and include enough explanatory information, including the criteria for making a selection and the limits on the use of the data, for the user to make a selection. A copy of the data and the identity of the registered professional engineer who approved it must be kept on the worksite during construction. A copy is provided to the Assistant Secretary of Labor for OSHA.

OSHA Method 3

Use a trench shield designed or approved by a registered professional engineer that meets OSHA standards. Timber, aluminum or other suitable materials may be used.

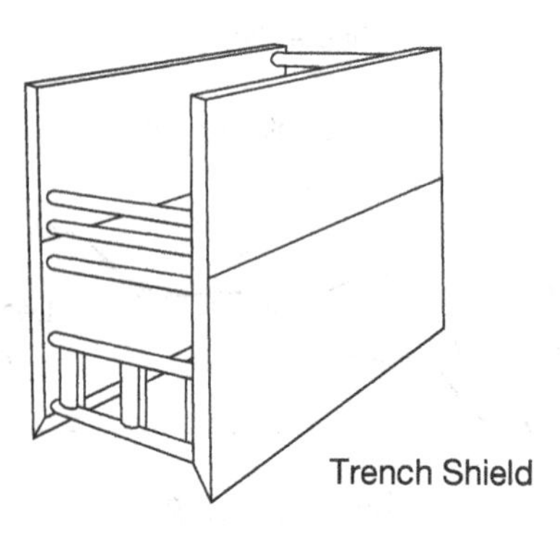

Trench Shield

RETAINING WALLS

Retaining walls must be stable enough to resist overturning, water uplift, sliding, and foundation pressures. Codes typically require a safety factor of 1.5 against lateral sliding and overturning. They are typically required for excavations deeper than 5 ft. Remember to check for buried utilities before digging. Also be prepared to install pumps to remove water accumulation.

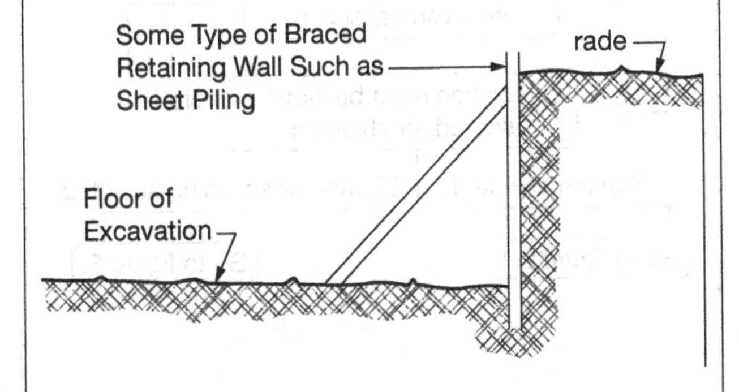

FIGURE 1 – OSHA EXCAVATION PRELIMINARY DECISION PROCESS OSHA 29CFR 1926

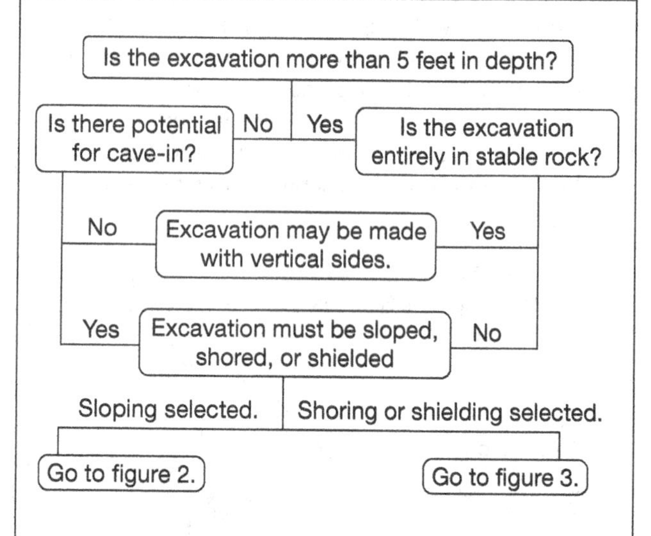

FIGURE 2 – OSHA EXCAVATION SLOPING OPTIONS OSHA 29CFR 1926

Sloping selected as the method of protection

Will soil classification be made in accordance with 1926.652(b)?

Yes

Excavation must comply with one of the following three options:

Option 1:
1926.652(b)(2), which requires Appendices A and B to be followed.

Option 2:
1926.652(b)(3), which requires other tabulated data (see definition) to be followed.

Option 3:
1926.652(b)(4), which requires the excavation to be designed by a registered profesional engineer.

No

Excavation must comply with 1926.652(b)(1) which requires a slope of $1\frac{1}{2}$ H: IV (34 degrees).

FIGURE 3 – OSHA EXCAVATION SHORING AND SHIELDING OPTIONS OSHA 29CFR 1926

Shoring or shielding selected as the method of protection

Soil classification is required when shoring or shielding is used. The excavation must comply with one of the following four options:

Option 1:
1926.652(c)(1), which requires Appendices A and C to be followed (e.g., timber shoring).

Option 2:
1926.652(c)(2), which requires manufacturers data to be followed (e.g., hydraulic shoring, trench jacks, air shores, shields).

Option 3:
1926.652(c)(3), which requires tabulated data (see definition) to be followed (e.g., any system as per the tabulated data).

Option 4:
1926.652(c)(4), which requires the excavation to be designed by a registered professional engineer (e.g., any designed system).

EXCAVATIONS WITH SLOPED WALLS

Instead of the walls of excavations or trenches being supported with some form of retaining wall, they can be sloped. Detailed requirements are given in Occupational Safety and Health Administration manual 29CFR 1926, Construction Industry. Recommended slope configurations for type A, B, and C soils and layered soils follow.

EXCAVATIONS IN TYPE A SOIL

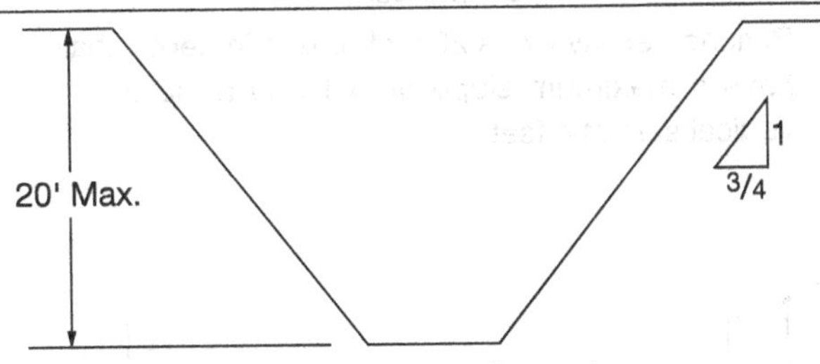

Simple Slope General

All simple slope excavations 20 feet or less in depth shall have a maximum allowable slope of ¾:1.

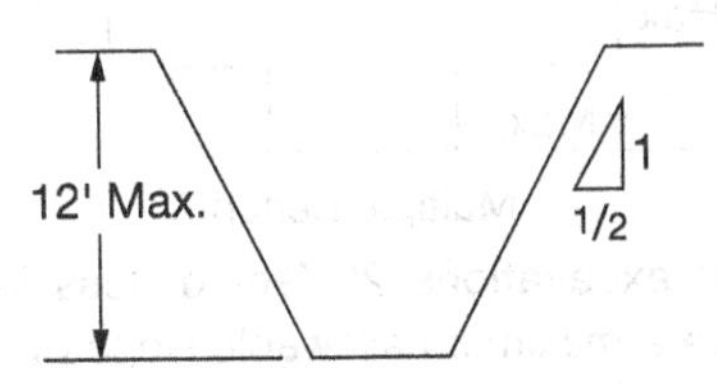

Simple Slope Short Term

Simple slope excavations which are open 24 hours or less (short term) and which are 12 feet or less in depth shall have a maximum allowable slope of ½ :1.

EXCAVATIONS IN TYPE A SOIL *(cont.)*

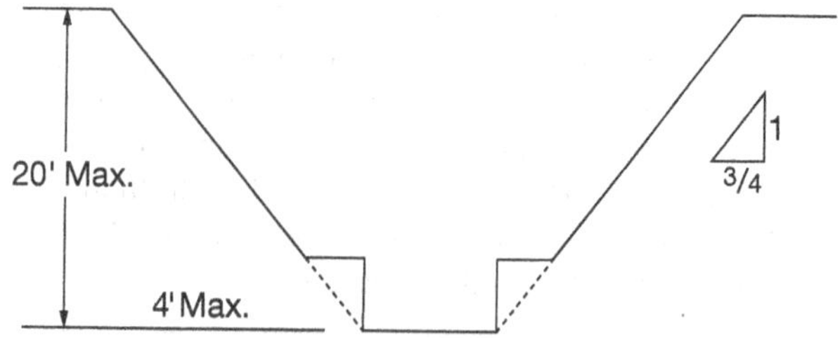

20' Max.

1

3/4

4' Max.

Simple Bench

Beached excavations 20 feet or less in depth shall have a maximum slope of ¾:1 and a maximum vertical side of 4 feet.

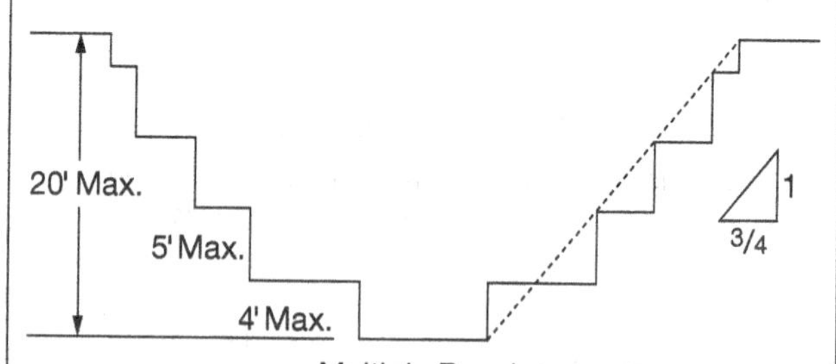

Multiple Bench

Benched excavations 20 feet or less in depth shall have a maximum allowable slope of ¾:1 and maximum allowable bench dimensions as shown in the above illustration.

EXCAVATIONS IN TYPE A SOIL *(cont.)*

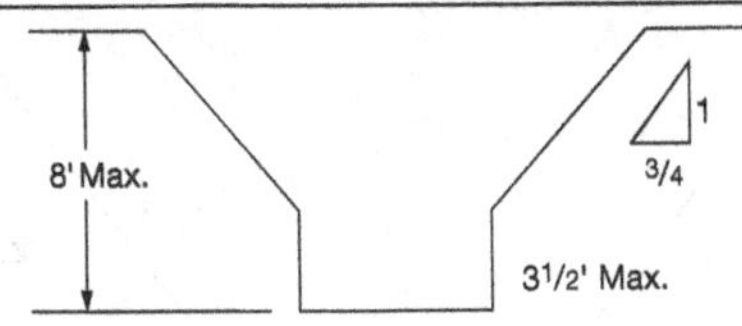

A simple bench with unsupported vertically sided lower portion in depths of 8 ft. and less shall have a maximum slope of ¾:1 and a maximum vertical side of 3½ ft.

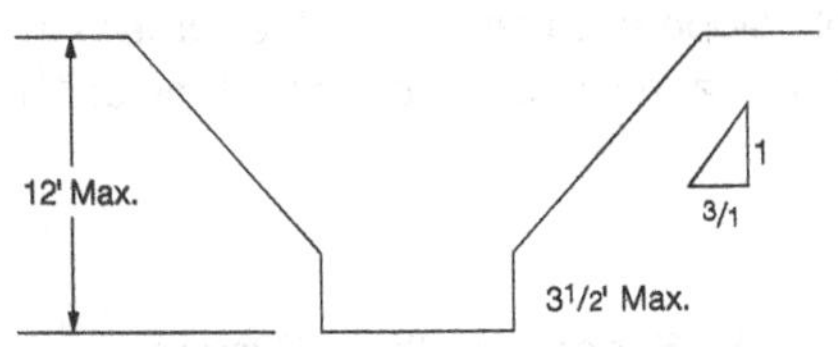

A simple bench with unsupported vertically sided lower portion in depths of 12 ft. or less shall have a maximum slope of 1:1 and a maximum vertical side of 3½ ft.

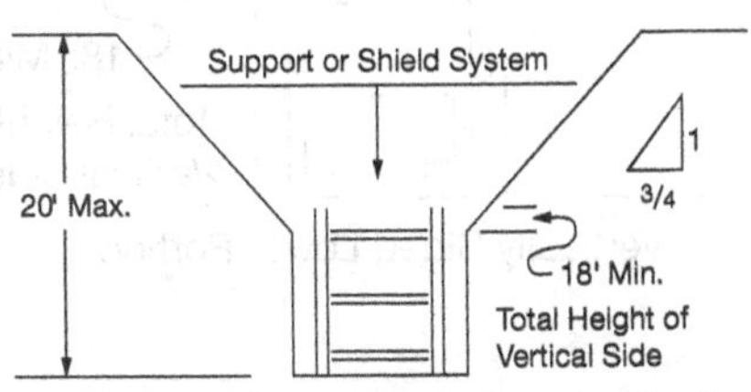

Simple bench with supported or vertically sided lower portion must have a maximum allowable slope of ¾:1. The support must extend 18 inches above the top of the support. Maximum depth 20 ft.

EXCAVATIONS IN TYPE B SOIL

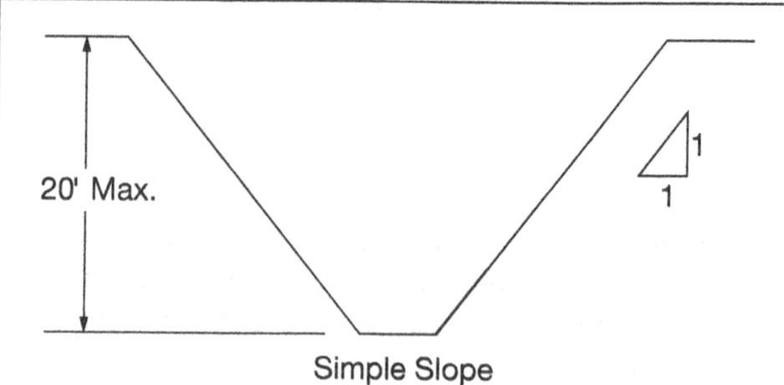

Simple Slope

Simple slope excavations 20 feet or less in depth shall have a maximum allowable slope of 1:1.

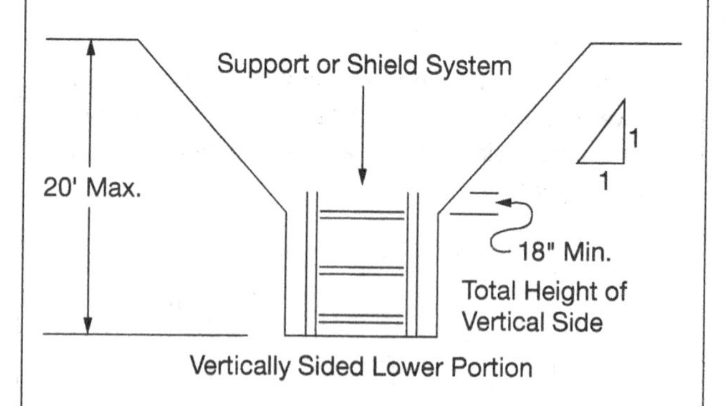

Vertically Sided Lower Portion

EXCAVATIONS IN TYPE B SOIL *(cont.)*

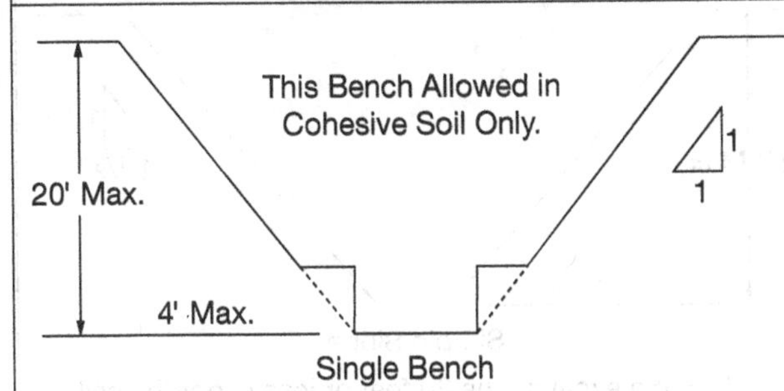

Single Bench

Benched excavations 20 feet or less in depth shall have a maximum allowable slope of 1:1 and a maximum vertical side of 4 feet.

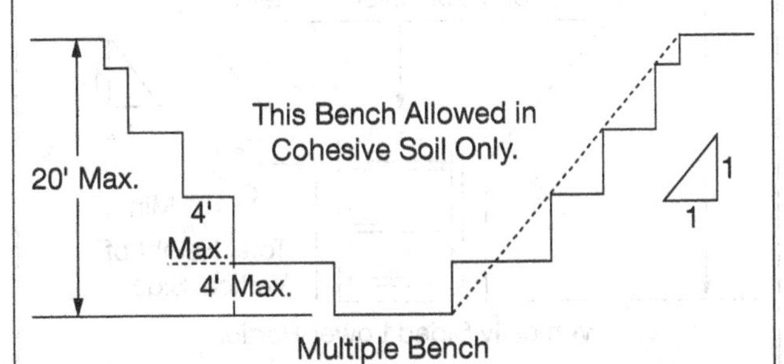

Multiple Bench

Benched excavations 20 feet or less in depth shall have a maximum allowable slope of 1:1 and maximum allowable bench dimension as shown in the above illustration.

EXCAVATIONS IN TYPE C SOIL

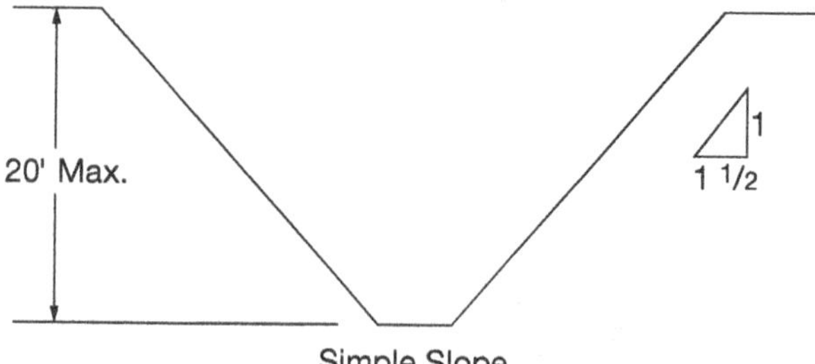

Simple Slope

All simple excavations 20 feet or less in depth shall have a maximum allowable slope of 1½:1.

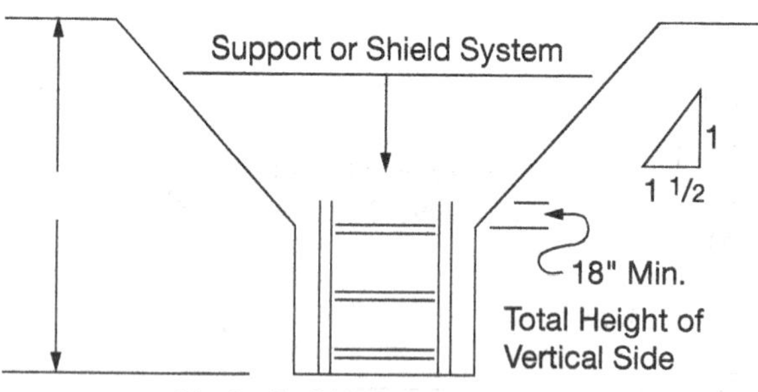

Vertically Sided Lower Portion

Excavations 20 feet or less in depth which have vertically sided lower portions shall be shielded or supported to a height of at least 18 inches above the vertical side. All excavations shall have a maximum allowable slope of 1½:1.

EXCAVATIONS MADE IN LAYERED SOIL

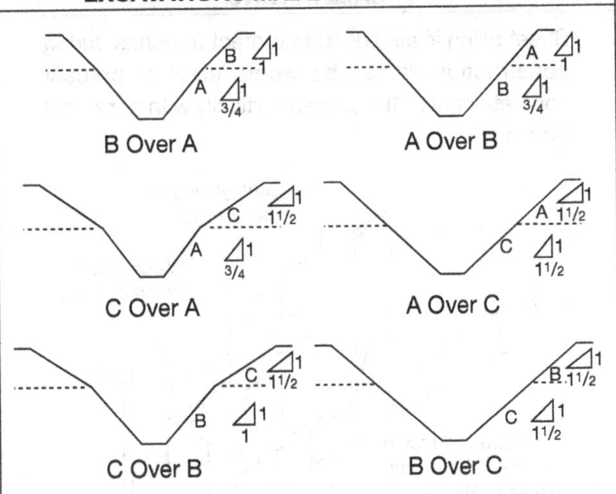

All excavations 20 feet or less in depth made in layered soils shall have a maximum allowable slope for each layer as set forth in the illustrations above.

SHEET PILING

Sheet piling is also used to protect trenches during construction. It can be wood, steel or precast concrete panels. It is placed vertically with horizontal bracing.

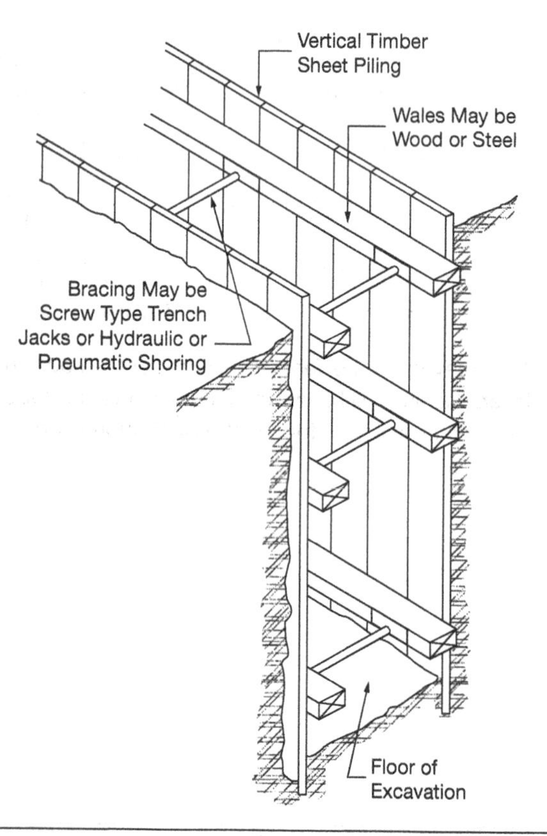

TYPICAL TRENCH DEPTH TO WIDTH RECOMMENDATIONS	
Trench Depths	**Minimum Trench Widths**
1 foot	16 inches
2 feet	17 inches
3 feet	18 inches
4 feet	20 inches
5* feet	22 inches

*Depths over 5 feet require shoring or a sheath.

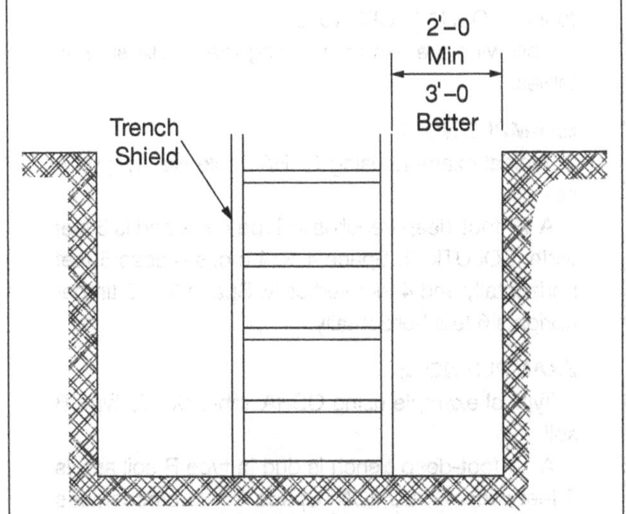

Excavated materials must be at least 2 feet back from the edge of the trench. OSHA.

TIMBER SHORING FOR TRENCHES

Typically, shoring is required on any excavation deeper than 6 feet; however, the sides can be cut on a slope, eliminating the need for shoring. Excavations on hillsides typically require shoring if they are deeper than 5 feet.

Excavations near areas where traffic will cause vibrations must be shored. On-site traffic or quantities of construction materials must be kept away from the edges of excavations.

The following shoring tables use soil classifications found in OSHA 29CRF 1926.

Following are examples using the OSHA shoring tables.

EXAMPLE NO. 1

Typical example using OSHA Table No. 1, type A soil:

A 13-foot-deep trench is in Type A soil and is 5 feet wide. SOLUTION: Space 4 × 4 cross braces 6 feet horizontally and 4 feet vertically. Space 3 × 8 timber uprights 6 feet horizontally.

EXAMPLE NO. 2

Typical example using OSHA Table No. 2, Type B soil:

A 13-foot-deep trench is dug in type B soil and is 5 feet wide. SOLUTION: Space 6 × 6 crossbraces 6 feet horizontally and 4 feet vertically. Space 8 × 8 wales at 5 feet vertically. Place 2 × 6 uprights at 2 feet horizontally.

EXAMPLE NO. 3

Typical example using OSHA Table No. 3, Type C soil:

A 13-foot-deep trench is in type C soil and is 5 feet wide. SOLUTION: Space 8 × 8 cross braces 6 feet horizontally and 5 feet vertically. Space 10 × 12 wales at 5 feet vertically. Place 2 × 6 uprights as close together as possible.

TIMBER TRENCH SHORING — MINIMUM TIMBER REQUIREMENTS FOR SOIL TYPE A

Pa = 25 × H + 72 psf (2 ft. Surcharge)

	Size (S4S) and spacing of members**													
	Cross braces							Wales		Uprights				
	Width of trench (feet)									Maximum allowable horizontal spacing (feet)				
Depth of trench (feet)	Horiz. spacing (feet)	Up to 4	Up to 6	Up to 9	Up to 12	Up to 15	Vertical spacing (feet)	Size (in.)	Vertical spacing (feet)	Close	4	5	6	8
5 to 10	Up to 6	4 × 4	4 × 4	4 × 4	4 × 4	4 × 6	4	Not req'd	Not req'd				4 × 6	
	Up to 8	4 × 4	4 × 4	4 × 4	4 × 6	4 × 6	4	Not req'd	Not req'd					4 × 8
	Up to 10	4 × 6	4 × 6	4 × 6	6 × 6	6 × 6	4	8 × 8	4			4 × 6		
	Up to 12	4 × 6	4 × 6	4 × 6	6 × 6	6 × 6	4	8 × 8	4				4 × 6	
10 to 15	Up to 6	4 × 4	4 × 4	4 × 4	6 × 6	6 × 6	4	Not req'd	Not req'd				4 × 10	
	Up to 8	4 × 6	4 × 6	4 × 6	6 × 6	6 × 6	4	6 × 8	4		4 × 6			
	Up to 10	6 × 6	6 × 6	6 × 6	6 × 6	6 × 6	4	8 × 8	4			4 × 8		
	Up to 12	6 × 6	6 × 6	6 × 6	6 × 6	6 × 6	4	8 × 10	4		4 × 6		4 × 10	
15 to 20	Up to 6	6 × 6	6 × 6	6 × 6	6 × 6	6 × 6	4	6 × 8	4	3 × 6				
	Up to 8	6 × 6	6 × 6	6 × 6	6 × 6	6 × 6	4	8 × 8	4	3 × 6	4 × 12			
	Up to 10	6 × 6	6 × 6	6 × 6	6 × 6	6 × 8	4	8 × 10	4	3 × 6				
	Up to 12	6 × 6	6 × 6	6 × 6	6 × 8	6 × 8	4	8 × 12	4	3 × 6	4 × 12			
Over 20	See Note 1													

*Douglas fir or equivalent with a bending strength not less than 1500 psi.

**Manufactured members of equivalent strength may be substituted for wood.

Occupational Safety and Health Administration, 29 CFR 1926, Construction Industry www.osha.gov

TIMBER TRENCH SHORING — MINIMUM TIMBER REQUIREMENTS FOR SOIL TYPE B

Pa = 45 × H + 72 psf (2 ft. Surcharge)

	Size (S4S) and spacing of members**												
	Cross braces						Wales		Uprights				
	Width of trench (feet)								Maximum allowable horizontal spacing (feet)				
Depth of trench (feet)	Horiz. spacing (feet)	Up to 4	Up to 6	Up to 9	Up to 12	Up to 15	Vertical spacing (feet)	Size (in.)	Vertical spacing (feet)	Close	2	3	4	5
5 to 10	Up to 6	4 × 6	4 × 6	4 × 6	6 × 6	6 × 6	5	6 × 8	5			3 × 12 4 × 8		4 × 12
	Up to 8	4 × 6	4 × 6	6 × 6	6 × 6	6 × 6	5	8 × 8	5		3 × 8		4 × 8	
	Up to 10	4 × 6	4 × 6	6 × 6	6 × 6	6 × 8	5	8 × 10	5			4 × 8		
	See Note 1													
10 to 15	Up to 6	6 × 6	6 × 6	6 × 6	6 × 8	6 × 8	5	8 × 8	5	3 × 6	4 × 10			
	Up to 8	6 × 8	6 × 8	6 × 8	8 × 8	8 × 8	5	10 × 10	5	3 × 6	4 × 10			
	Up to 10	6 × 8	6 × 8	8 × 8	8 × 8	8 × 8	5	10 × 12	5	3 × 6	4 × 10			
	See Note 1													
15 to 20	Up to 6	6 × 8	6 × 8	6 × 8	6 × 8	8 × 8	5	8 × 10	5	4 × 6				
	Up to 8	6 × 8	6 × 8	6 × 8	8 × 8	8 × 8	5	10 × 12	5	4 × 6				
	Up to 10	8 × 8	8 × 8	8 × 8	8 × 8	6 × 8	5	12 × 12	5	4 × 6				
	See Note 1													
Over 20	See Note 1													

*Douglas fir or equivalent with a bending strength not less than 1500 psi.
**Manufactured members of equivalent strength may be substituted for wood.
Occupational Safety and Health Administration, 29 CFR 1926, Construction Industry www.osha.gov

TIMBER TRENCH SHORING — MINIMUM TIMBER REQUIREMENTS FOR SOIL TYPE C

$$Pa = 80 \times H + 72 \text{ psf (2 ft. Surcharge)}$$

	Size (S4S) and spacing of members**													
	Cross braces							Wales		Uprights				
	Width of trench (feet)									Maximum allowable horizontal spacing (feet)				
Depth of trench (feet)	Horiz. spacing (feet)	Up to 4	Up to 6	Up to 9	Up to 12	Up to 15	Vertical spacing (feet)	Size (in.)	Vertical spacing (feet)	Close				
5 to 10	Up to 6	6 × 6	6 × 6	6 × 6	6 × 6	8 × 8	5	8 × 8	5	3 × 6				
	Up to 8	6 × 6	6 × 6	6 × 6	8 × 8	8 × 8	5	10 × 10	5	3 × 6				
	Up to 10	6 × 6	6 × 6	8 × 8	8 × 8	8 × 8	5	10 × 12	5	3 × 6				
	See Note 1													
10 to 15	Up to 6	6 × 8	6 × 8	6 × 8	8 × 8	8 × 8	5	10 × 10	5	4 × 6				
	Up to 8	8 × 8	8 × 8	8 × 8	8 × 8	8 × 8	5	12 × 12	5	4 × 6				
	See Note 1													
	See Note 1													
15 to 20	Up to 6	8 × 8	8 × 8	8 × 8	8 × 10	8 × 10	5	10 × 12	5	4 × 6				
	See Note 1													
	See Note 1													
	See Note 1													
Over 20	See Note 1													

*Douglas fir or equivalent with a bending strength not less than 1500 psi.
**Manufactured members of equivalent strength may be substituted for wood.
Occupational Safety and Health Administration, 29 CFR 1926, Construction Industry www.osha.gov

TIMBER SHORING EXAMPLE NO. 1

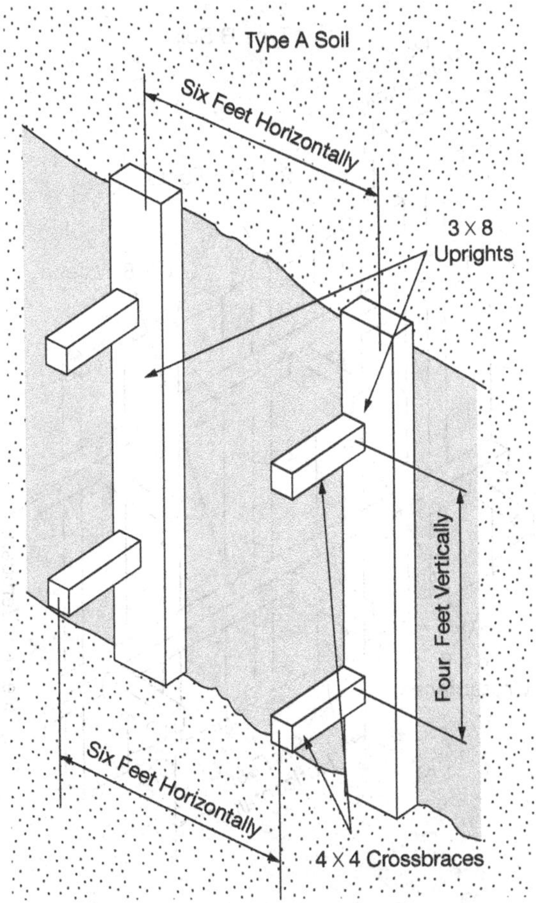

Type A Soil

Six Feet Horizontally

3 X 8
Uprights

Four Feet Vertically

Six Feet Horizontally

4 X 4 Crossbraces

TIMBER SHORING EXAMPLE NO. 2

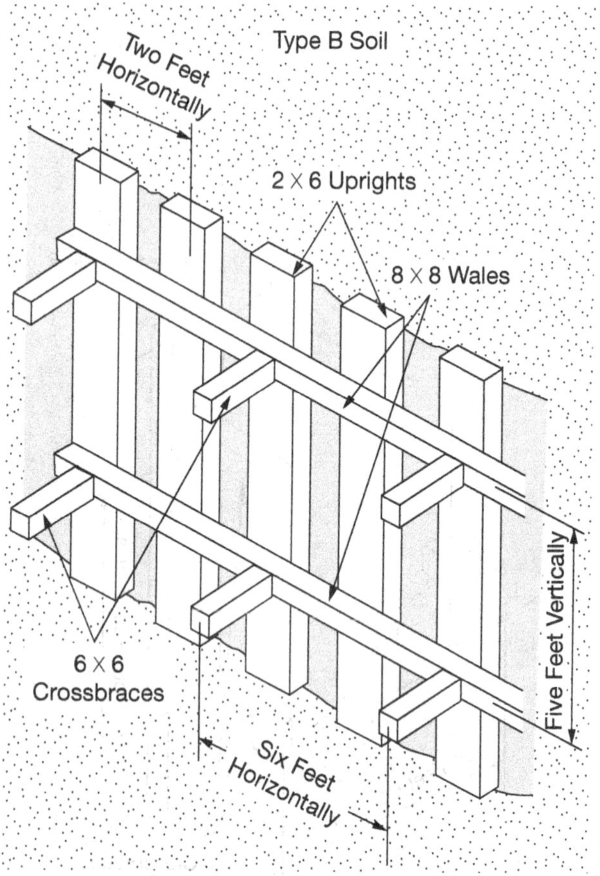

TIMBER SHORING EXAMPLE NO. 3

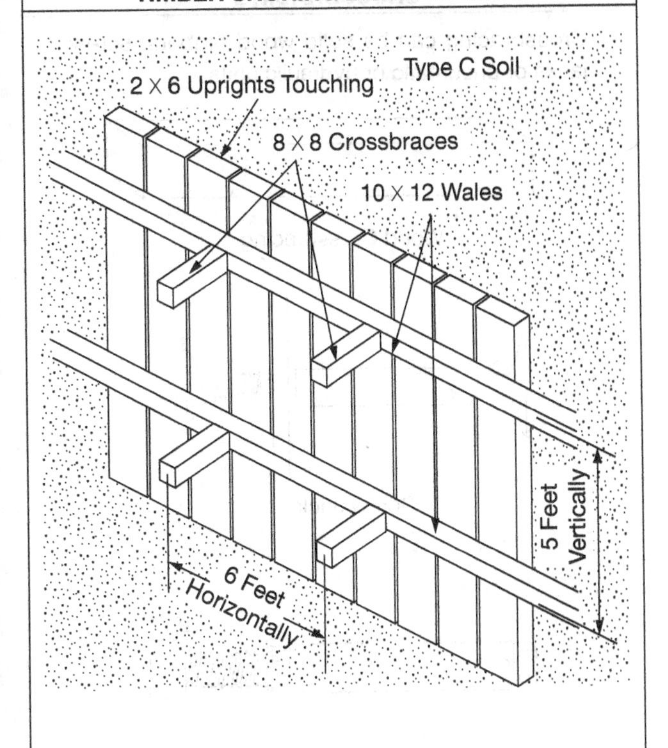

CROSSBRACING

Crossbracing can be solid wood timbers, screw jacks or pneumatic or hydraulic units.

Wood Crossbracing

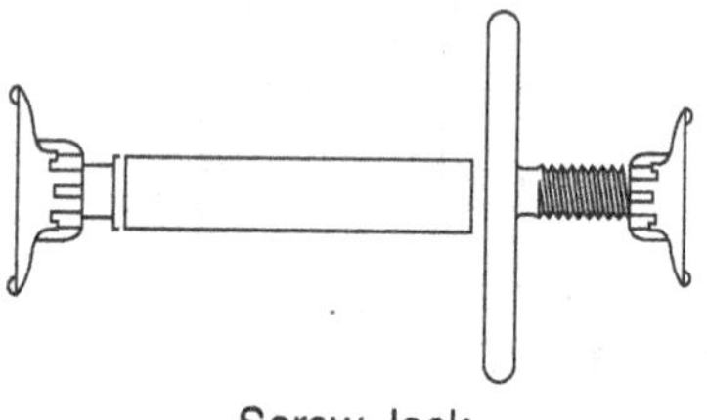

Screw Jack

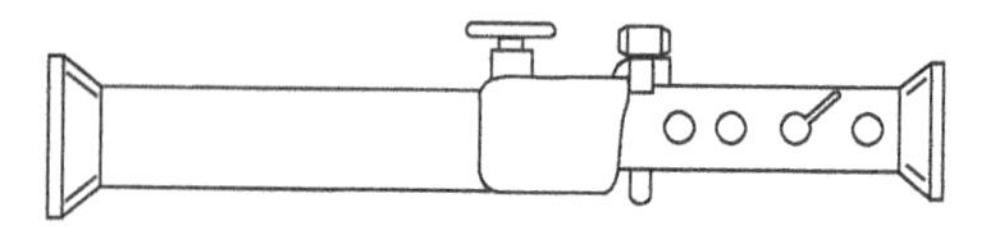

Pneumatic and Hydraulic Shoring

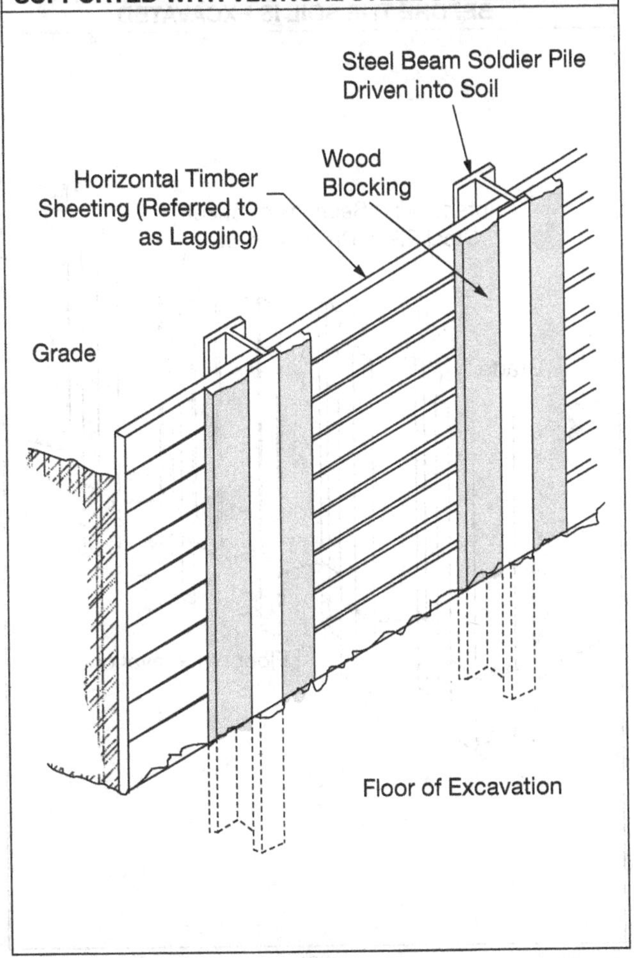
EXCAVATION WALL HELD WITH TIMBER LAGGING SUPPORTED WITH VERTICAL STEEL SOLDIER PILES
Steel Beam Soldier Pile Driven into Soil
Wood Blocking
Horizontal Timber Sheeting (Referred to as Lagging)
Grade
Floor of Excavation

EXCAVATION WALLS CAN BE HELD WITH STEEL SHEET PILING DRIVEN INTO THE EARTH BEFORE THE SOIL IS EXCAVATED

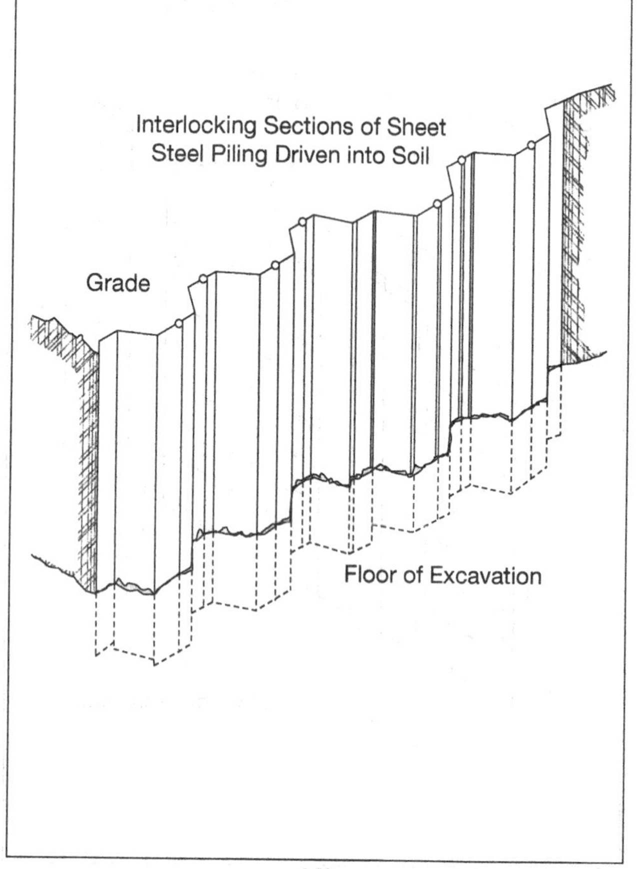

SHALLOW GRADES AT FOUNDATION

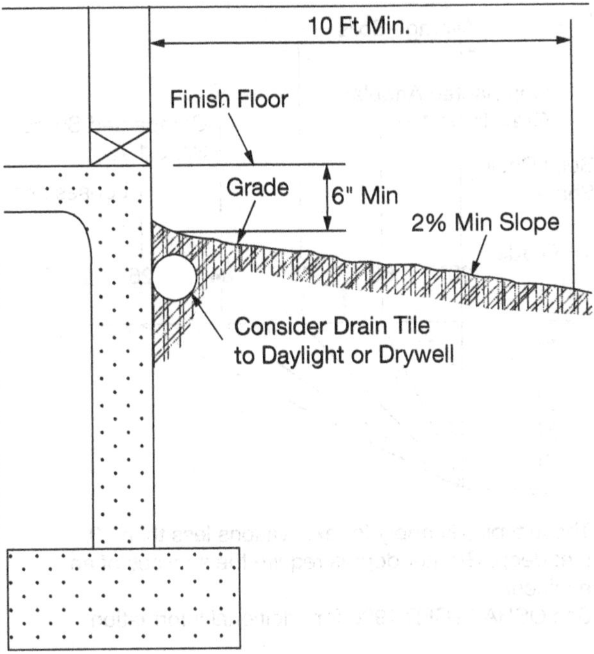

10 Ft Min.

Finish Floor

Grade

6" Min

2% Min Slope

Consider Drain Tile
to Daylight or Drywell

ANGLE OF REPOSE FOR COMMON SOIL TYPES

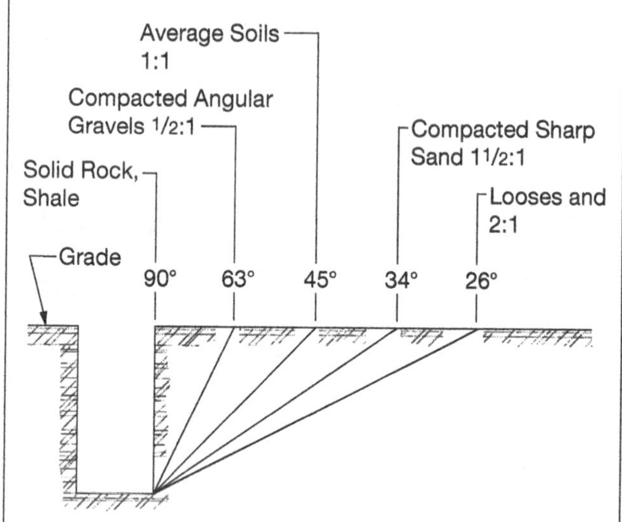

These typically apply for excavations less than 20 feet deep. Greater depths require the services of an engineer.

See OSHA 29CFR 1926 for additional information

GRADING TERMS

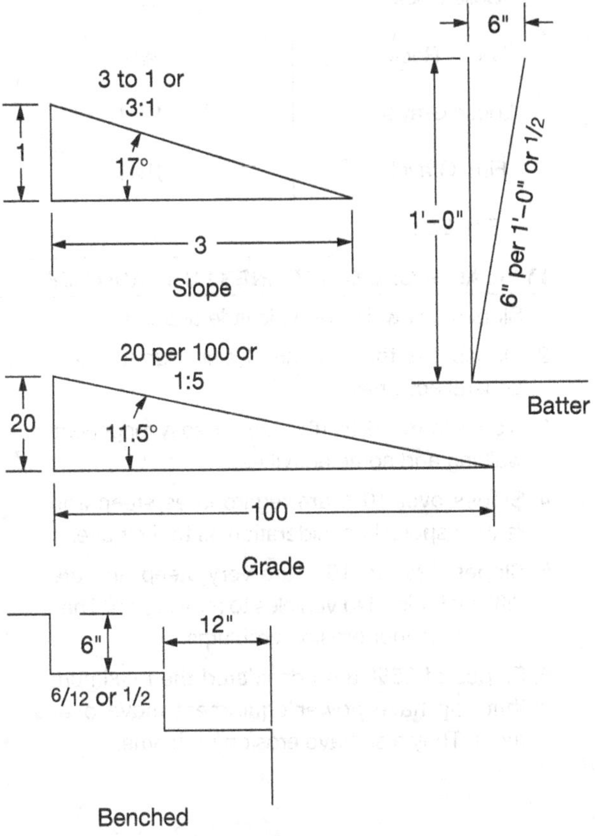

3 to 1 or
3:1

1

17°

3

Slope

6"

1'–0"

6" per 1'–0" or 1/2

Batter

20 per 100 or
1:5

20

11.5°

100

Grade

6"

12"

6/12 or 1/2

Benched

TYPICAL RECOMMENDED MAXIMUM FINISHED SLOPES	
Solid Rock	¼:1
Loose Rock	½:1
Loose Gravel	1½:1
Firm Gravel	1½:1
Soft Earth	2:1

TYPICAL SLOPE GUIDELINES FOR LAND USE

1. Slopes below 1% provide little drainage.
2. Slopes less than 5% are flat enough for most general activities.
3. Slopes from 5% to 10% are still easy for general walking and some activities.
4. Slopes over 10% are regarded as steep and require special consideration as to their use.
5. Slopes 12% to 15% are very steep and are difficult for loaded vehicles to move upon. They will have minor erosion problems.
6. Slopes of 25% are considered the maximum that can have power equipment move over them. They also have erosion problems.

METHODS FOR EXPRESSING SLOPES AND GRADIENTS

Percentage is rise divided by run multiplied by 100. Ratio is length of run to one unit of rise. Example, 1 ft rise to a 4 ft run is 4:1. Degrees are the actual angle between the sloping plane and the horizontal.

Percentage	Gradient Ratios	Degree
2%	50:1	
3%	33⅓:1	
4%	25:1	
5%	20:1	3° 00'
6%	17:1	
7%	14.3:1	
8%	12.5:1	
9%	11.1:1	
10%	10:1	6° 00'
15%	6.7:1	8° 30'
20%	5.:1	11° 30'
25%	4:1	14° 00'
30%	3.3:1	17° 00'
35%	2.9:1	19° 15'
40%	2.5:1	21° 30'
45%	2.2:1	24° 00'
50%	2:1	26° 30'
55%	1.8:1	28° 30'
60%	1.7:1	31° 00'
65%	1.5:1	33° 00'
70%	1.4:1	35° 00'
75%	1.3:1	36° 45'
100%	1:1	45° 00'

Gradient is the degree of slope or inclination of a surface that can be expressed as a ratio or a percentage.

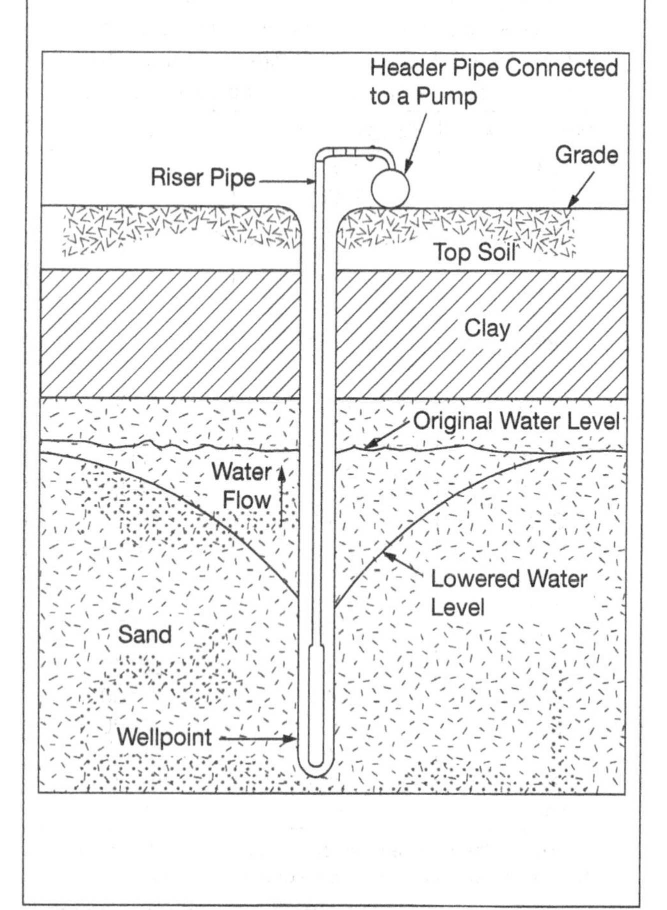
SINGLE WELLPOINT LOWERS THE WATER LEVEL
OF THE AREA AROUND IT
Header Pipe Connected
to a Pump
Grade
Riser Pipe
Top Soil
Clay
Original Water Level
Water
Flow
Lowered Water
Level
Sand
Wellpoint

MULTIPLE WELLPOINTS LOWER THE WATER LEVEL IN A LARGE AREA

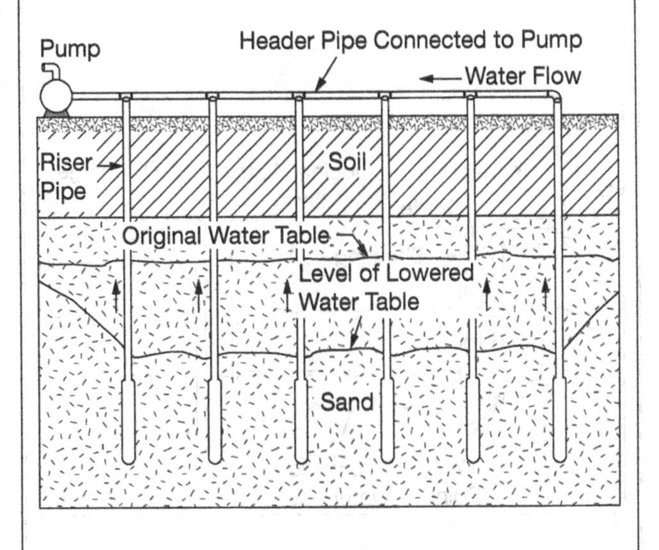

DEWATERING AN EXCAVATION DURING CONSTRUCTION

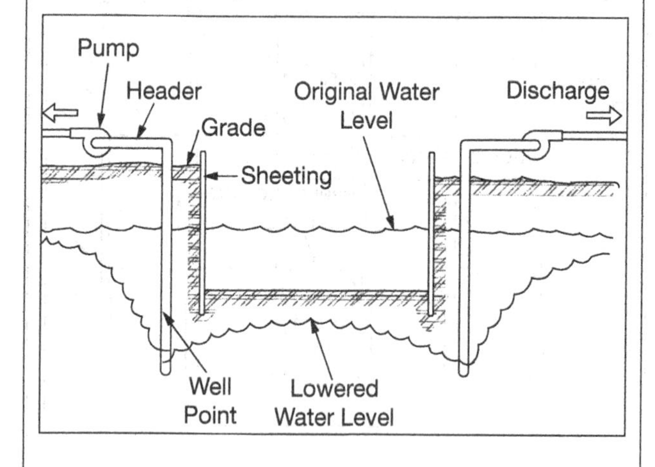

CHAPTER 5
Rigging

CRANES

The use of cranes, derricks, hoists, and related hardware presents situations that are hazardous to the construction worker. In addition to selecting the appropriate equipment for a job, the operating personnel must be thoroughly trained in the safe operation of the equipment and the handling of the loads.

Extensive information is available in the ASTM B30 series safety standards, CABLEWAYS, CRANES, DERRICKS, HOISTS, HOOKS, JACKS, and SLINGS. The publications are available from the American Society of Mechanical Engineers, www.asme.org. Additional information is available from the Crane Manufacturers Association of America, www.mhia.org/cmaa and the Materials Handling Institute, www.mhia.org.

TYPES OF CRANES

COMMERCIAL TRUCK-MOUNTED CRANE — NONTELESCOPING BOOM

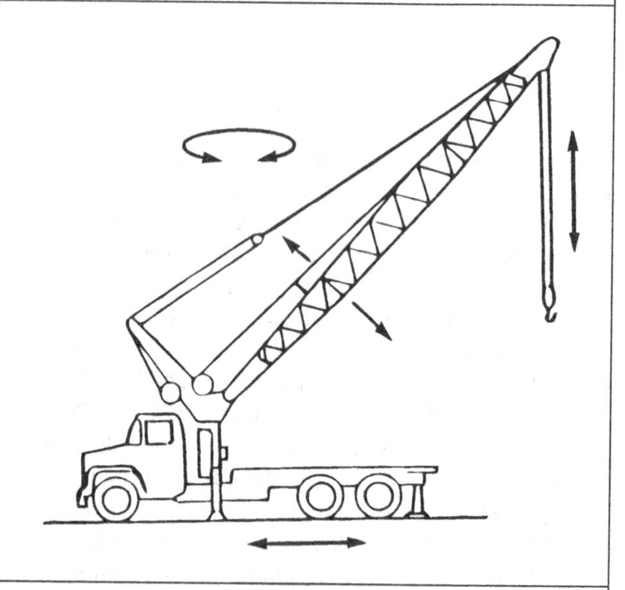

The boom may have a base boom structure of upper and lower sections between or beyond which additional sections may be added to increase its length, or it may consist of a base boom from which one or more boom extensions are telescoped for additional length.

TYPES OF CRANES *(cont.)*

COMMERCIAL TRUCK-MOUNTED CRANE — TELESCOPING BOOM

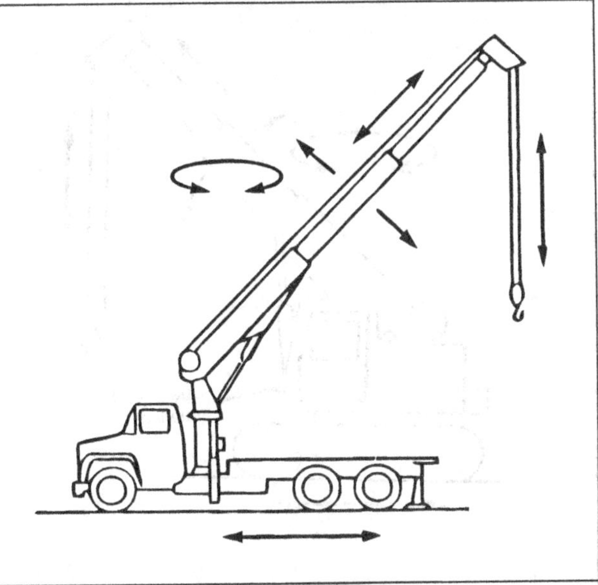

Reprinted from ASME B30.10-2005, by permission of the American Society of Mechanical Engineers.

The boom may have a base boom structure of upper and lower sections between or beyond which additional sections may be added to increase its length, or it may consist of a base boom from which one or more boom extensions are telescoped for additional length.

TYPES OF CRANES *(cont.)*

CRAWLER CRANE — TELESCOPING BOOM

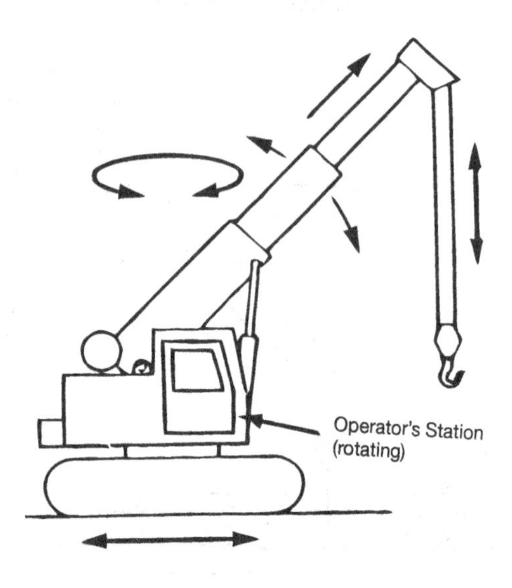

The boom may have a base boom structure of upper and lower sections between or beyond which additional sections may be added to increase its length, or it may consist of a base boom from which one or more boom extensions are telescoped for additional length.

TYPES OF CRANES *(cont.)*

CRAWLER CRANE

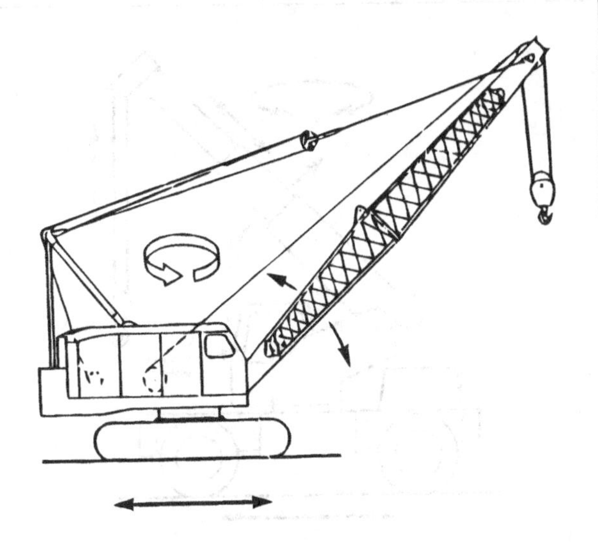

TYPICAL LOADS ON A STRUCTURE

WHEEL-MOUNTED CRANE — TELESCOPING BOOM (SINGLE CONTROL STATION, ROTATING)

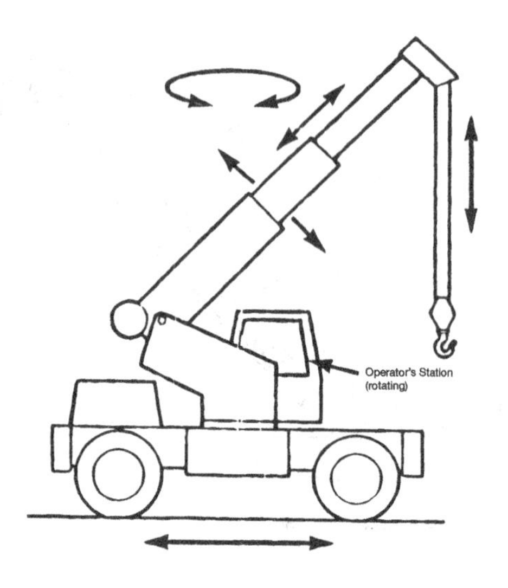

Reprinted from ASME B30.10-2005, by permission of the American Society of Mechanical Engineers.

The boom may have a base boom structure of upper and lower sections between or beyond which additional sections may be added to increase its length, or it may consist of a base boom from which one or more boom extensions are telescoped for additional length.

CRANE WORK AREA

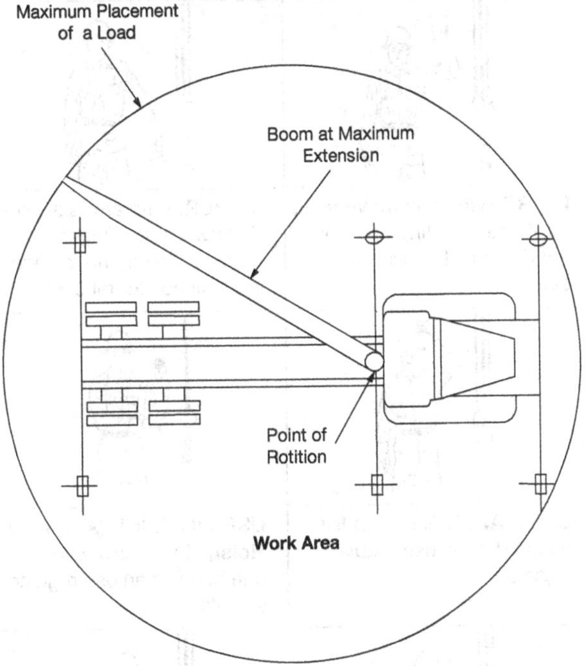

Maximum Placement of a Load

Boom at Maximum Extension

Point of Rotition

Work Area

The work area and the maximum placement of a load on all sides of the crane are limited by the boom length and boom angle established by the specified operating conditions of the crane.

STANDARD HAND SIGNALS FOR CONTROLLING CRANE OPERATIONS

HOIST. With forearm vertical, forefinger pointing up, move hand in small horizontal circle.	**LOWER.** With arm extended downward, forefinger pointing down, move hand in small horizontal circle.

USE MAIN HOIST. Tap fist on head; then use regular signals.	**USE WHIPLINE (Auxilliary Hoist).** Tap elbow with one hand; then use regular signals.

RAISE BOOM. Arm extended, fingers closed, thumb pointing upward.	**LOWER BOOM.** Arm extended, fingers cosed, thumb poingting downward.

STANDARD HAND SIGNALS FOR CONTROLLING CRANE OPERATIONS *(cont.)*

MOVE SLOWLY. Use one hand to give any motion signal and place other hand motionless in front of hand giving the motion signal slowly.	**RAISE BOOM AND LOWER LOAD.** With arm extended, thumb pointing up, flex fingers in and out as long as load movement is desired.

LOWER BOOM AND RAISE LOAD. With arm extended, thumb pointing down, flex fingers in and out as long as load movement is desired.	**EMERGENCY STOP.** Both arms extended, palms down, move arms back and forth horizontally.

STOP. Arm extended, palm down, move arm back and forth horizontally.	**SWING.** Arm extended, point with finger in direction of swing of boom.

STANDARD HAND SIGNALS FOR CONTROLLING CRANE OPERATIONS *(cont.)*

TRAVEL. Arm extended forward, hand open and slightly raised, make pushing motion in direction of travel.	**DOG EVERYTHING.** Clasp hands in front of body.

TRAVEL (Both Tracks). Use both fists in front of body, making a circular motion about each other, indicating direction of travel, forward or backward. (Land cranes only)	**TRAVEL (One Track).** Lock the track on side indicated by raised fist. Travel opposite track in direction indicated by circular motion of other fist, rotated vertically in front of body. (Land cranes only)

EXTEND BOOM (Telescoping Booms). Both fists in front of body with thumbs pointing outward.	**RETRACT BOOMS (Telescoping Booms).** Both fists in front of body with thumbs pointing toward each other.

STANDARD HAND SIGNALS FOR CONTROLLING CRANE OPERATIONS *(cont.)*

EXTEND BOOM (Telescoping Boom). One hand signal; one fist in front of chest with thumb tapping chest.

RETRACT BOOM (Telescoping Boom). One hand signal; one fist in front of chest, thumb pointing outward and heel of fist tapping chest.

HELICOPTER HAND SIGNALS

MOVE RIGHT. Left arm extended horizontally; right arm sweeps upward to position over head.

MOVE LEFT. Right arm extended horizontally; left arm sweeps upward to position over head.

HELICOPTER HAND SIGNALS *(cont.)*	

MOVE FORWARD. Combination of arm and hand movement in a collecting motion pulling toward body.	**MOVE REARWARD.** Hands above arm, palms out using a noticeable shovling motion.

RELEASE SLING LOAD. Left arm held down away from body. Right arm cuts across left arm in a slashing movement from above.	**HOLD HOVER.** The signal "Hold" is executed by placing arms over head with clenched fists.

TAKEOFF. Right hand behind back; left hand pointing up.	**LAND.** Arms crossed in front of body and pointing downward.

HELICOPTER HAND SIGNALS *(cont.)*

MOVE UPWARD. Arms extended, palms up; arms sweeping up.	**MOVE DOWNWARD.** Arms extended, palms down; arms sweeping down.

RIGGING HARDWARE

The selection and use of rigging hardware is critical for the safe operation of cranes, derricks, and hoists. ASME B30.26-2004, RIGGING HARDWARE, has provisions that apply to the construction, installation, inspection, and maintenance of detachable rigging hardware used for lifting purposes. This hardware includes shackles, links, rings, swivels, turnbuckles, eyebolts, hoist rings, wire rope clips, wedge sockets, and rigging blocks. These are shown in the following illustrations. They have been reproduced from ASTM B30.26-2004, RIGGING HARDWARE, by permission of the American Society of Mechanical Engineers. www.asme.org.

SHACKLES

ANGLE OF LOADING

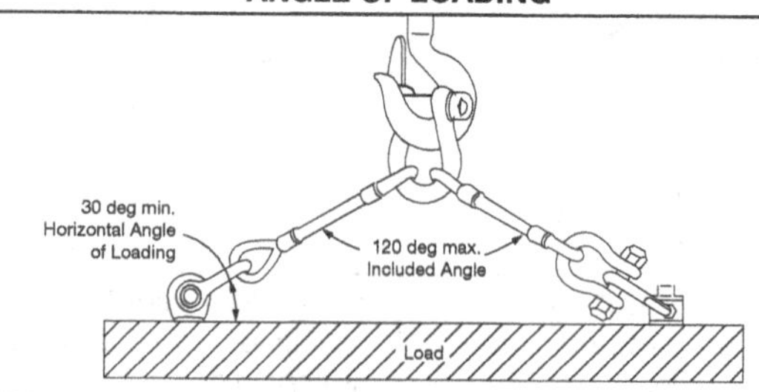

Horizontal Angle, Degree	Stress Multiplier
90	1,000
60	1,155
45	1,414
30	2,000

TYPICAL COMPONENTS

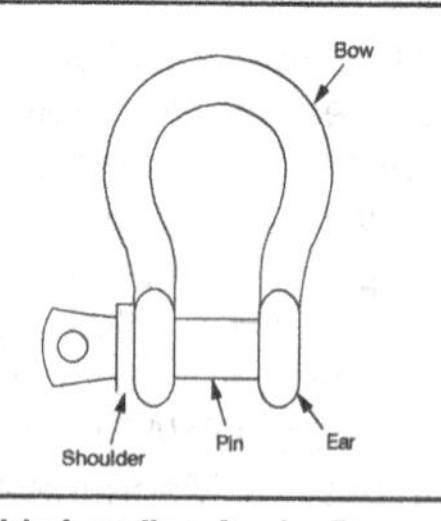

SIDE LOADING

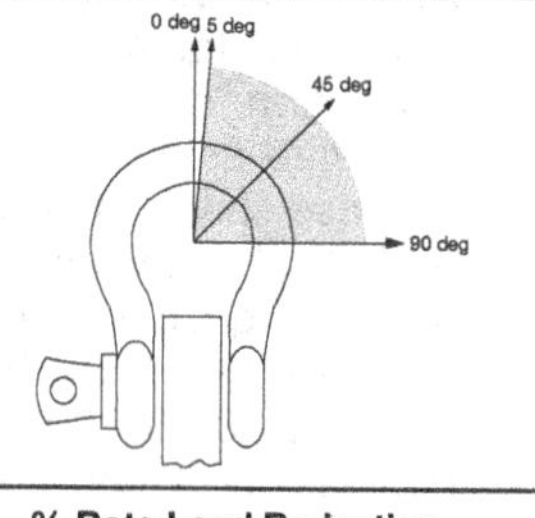

Side Loading Angle, Degree	% Rate Load Reduction
In-line (0) to 5	None
6 to 45	30%
46 to 90	50%
Over 90	Not recommended to load in this condition. Consult manufacturer or qualified person.

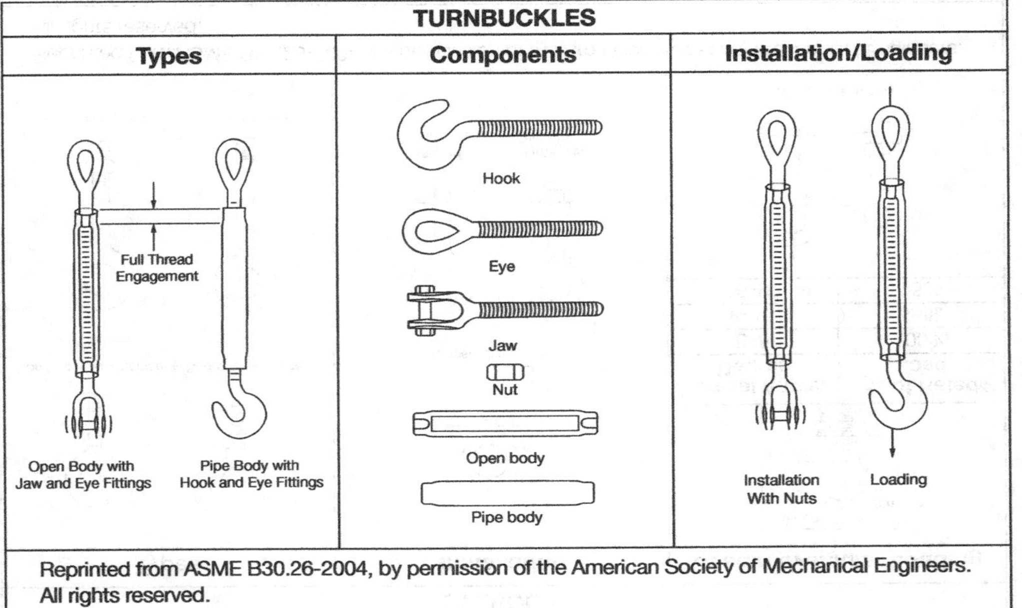
TURNBUCKLES
Types
Components
Installation/Loading
Full Thread
Engagement
Open Body with
Jaw and Eye Fittings
Pipe Body with
Hook and Eye Fittings
Hook
Eye
Jaw
Nut
Open body
Pipe body
Installation
With Nuts
Loading
Reprinted from ASME B30.26-2004, by permission of the American Society of Mechanical Engineers.
All rights reserved.

EYEBOLTS

Types

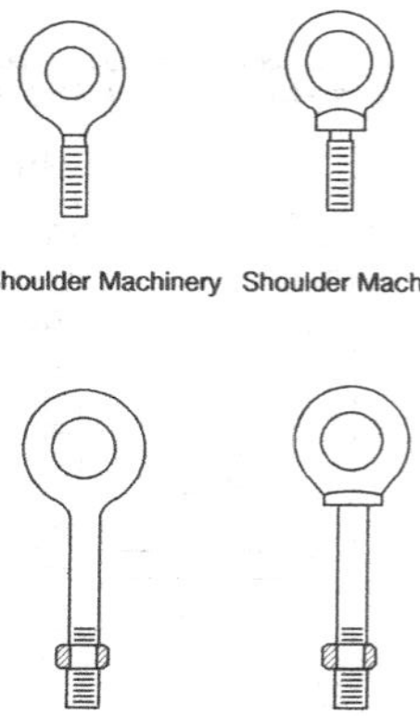

Non Shoulder Machinery

Shoulder Machinery

Non Shoulder Nut

Shoulder Nut

Installation

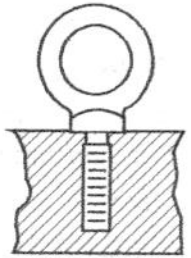

Tapped Blind Hole

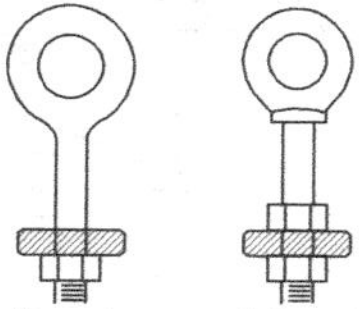

Tapped Through Hole

Untapped Through Hole

Loading/Angular Loading

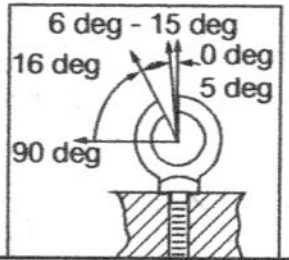

Vertical Angle, Degree	% of Rated Load
0–5	100%
6–15	55%
16–90	25%

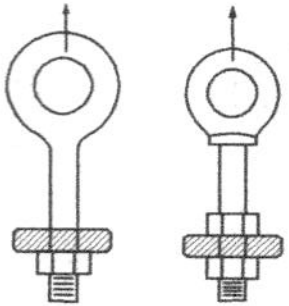

In-line loading only

Reprinted from ASME B30.26-2004, by permission of the American Society of Mechanical Engineers.

EYE NUTS

Types	Installation	Loading
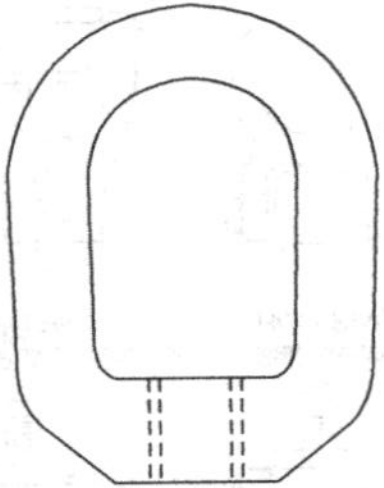 Typical	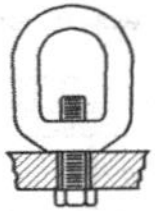 Through Hole No Nut 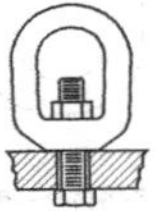 Through Hole Top Nut 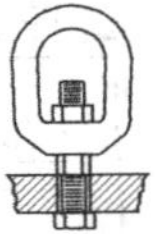 Through Hole Bottom Nut	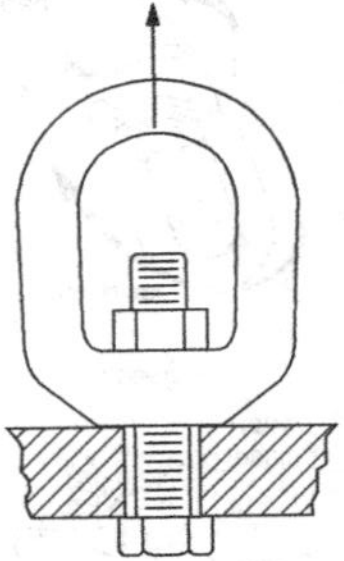 In-line Loading only

Reprinted from ASME B30.26-2004, by permission of the American Society of Mechanical Engineers.

SWIVEL HOIST RINGS

Types

Side Pull Swivel Hoist Ring

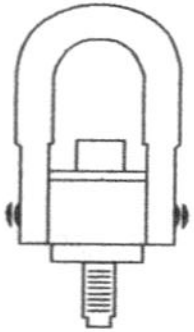

Bail Swivel Hoist Ring

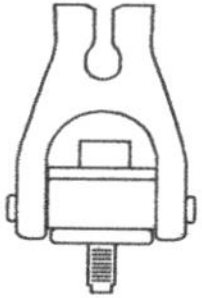

Chain Swivel Hoist Ring

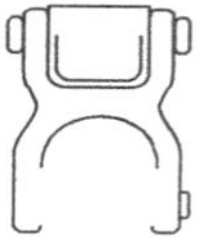

Webbing Swivel Hoist Ring

Installation

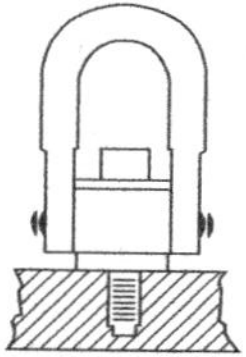

Tapped Hole

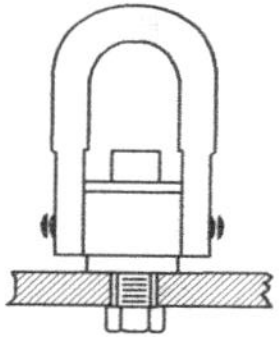

Through Hole

Components

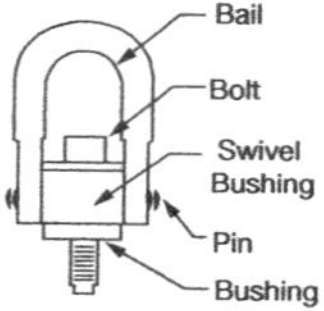

Bail

Bolt

Swivel Bushing

Pin

Bushing Range

Loading

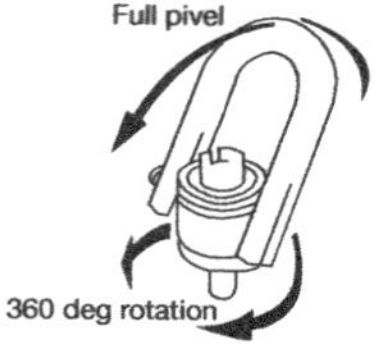

Full pivel

360 deg rotation

100% Loading at any Direction or Angle In-line With Ball

Reprinted from ASME B30.26-2004, by permission of the American Society of Mechanical Engineers.

ANGLE OF LOADING — ADJUSTABLE HARDWARE

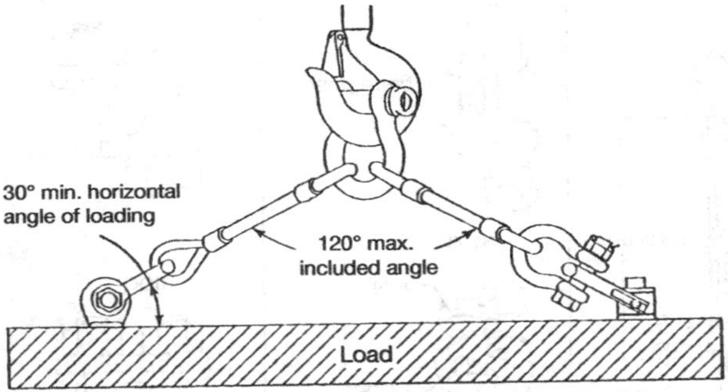

30° min. horizontal angle of loading

120° max. included angle

Load

Horizontal Angle, Degree	Stress Multiplier
90	1,000
60	1,155
45	1,414
30	2,000

Reprinted from ASME B30.26-2004, by permission of the American Society of Mechanical Engineers.

WIRE ROPE CLIPS

Types	Components	Installation and Loading	Notes
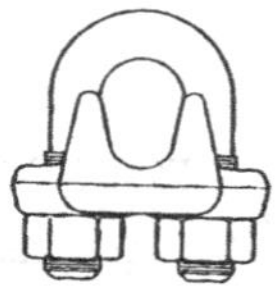 U-bolt Double Saddle	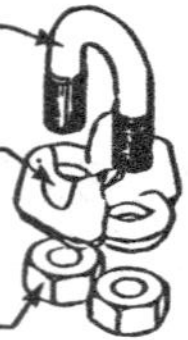 U-bolt Saddle Nut U-bolt Nut Saddle/leg	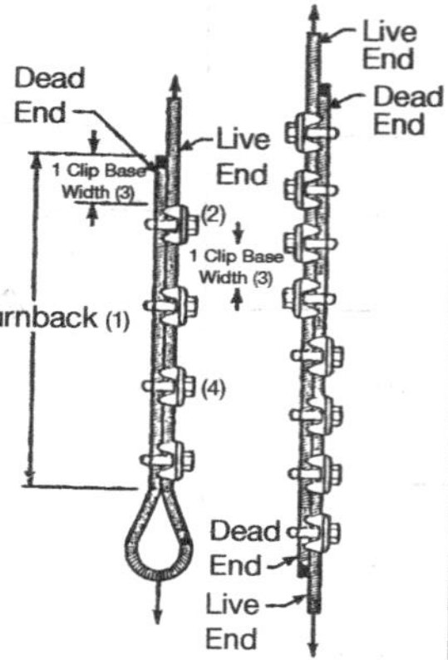 Dead End 1 Clip Base Width (3) Live End (2) 1 Clip Base Width (3) Turnback (1) (4) Live End Dead End Dead End Live End	Correct number of clips for wire rope size shall be used. 1) Correct turnback length should be used. 2) Correct orientation of saddle on live end shall be observed. 3) Correct spacing of clips should be used. 4) Correct torque on nuts shall be applied.

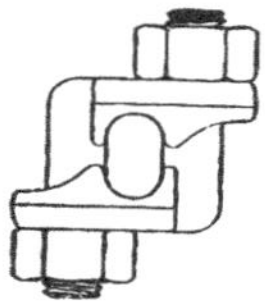

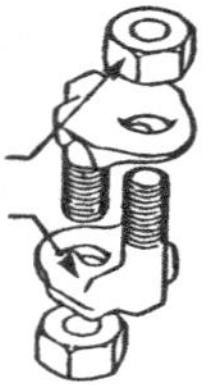

Reprinted from ASME B30.26-2004, by permission of the American Society of Mechanical Engineers.

WEDGE SOCKETS		
Components	**Correct Installation**	**Incorrect Installation**
Wedge Socket Body Pin Cotter	Live End Dead End	

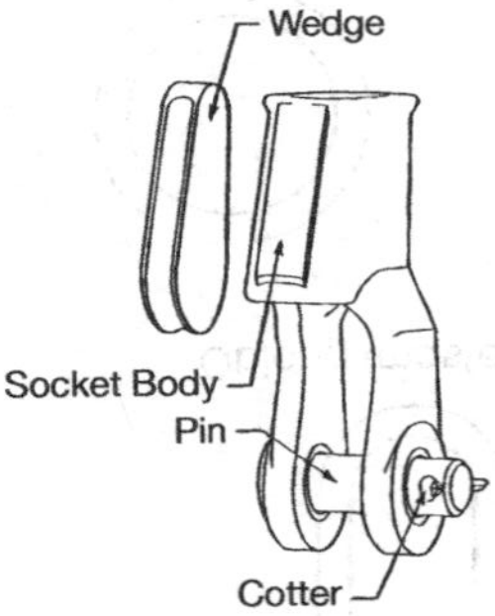

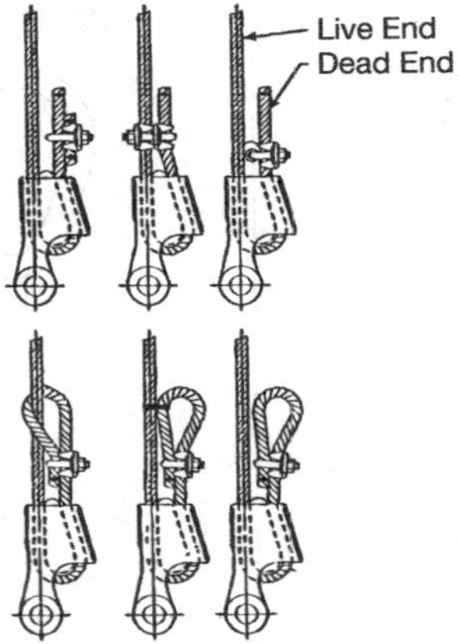

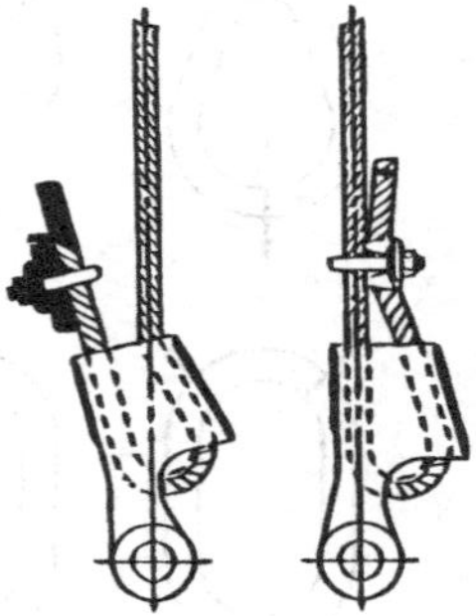

LINKS AND RINGS

Types	Loading
Oblong (master link) Round Pear shaped	

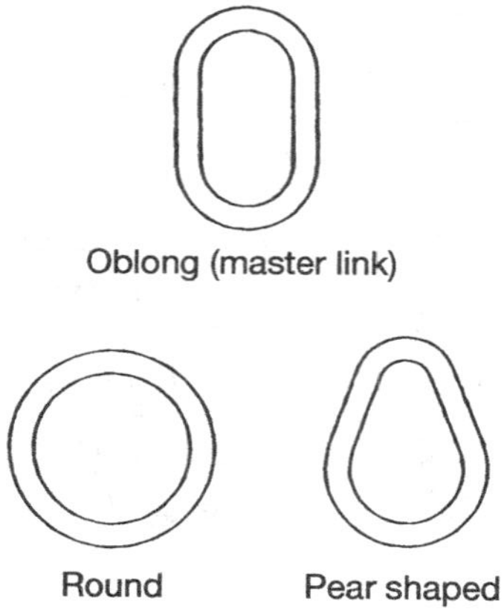

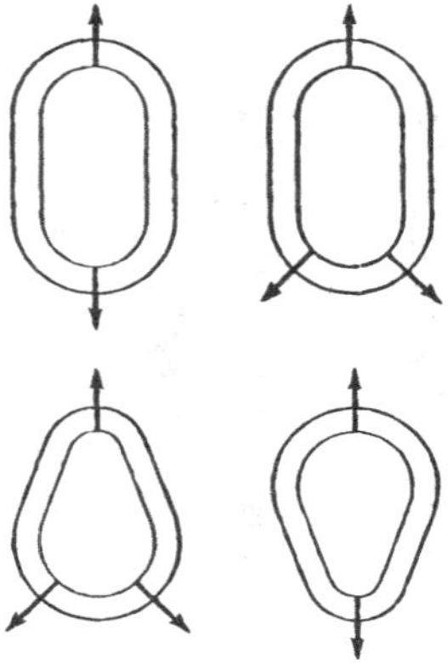

Reprinted from ASME B30.26-2004, by permission of the American Society of Mechanical Engineers.

SWIVELS

Types	Loading
Regular Chain Jaw End	In-line Loading Only

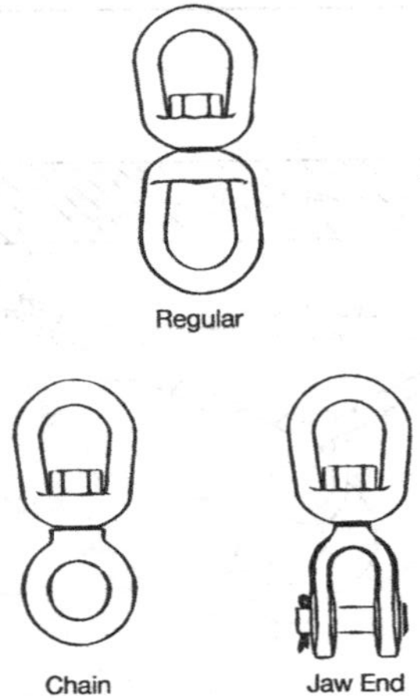

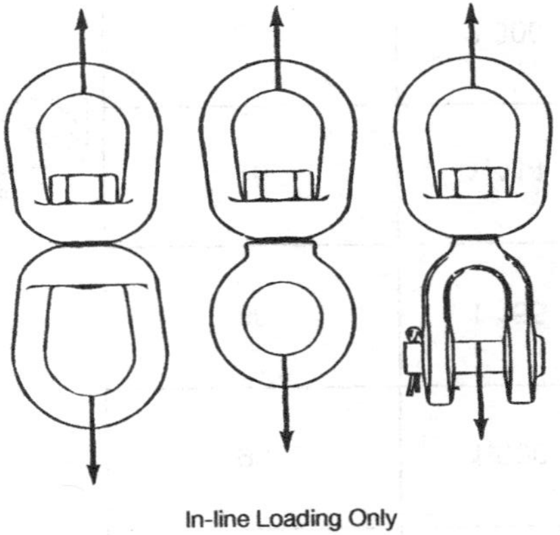

ANGLE OF LOADING — LINKS, RINGS, AND SWIVELS

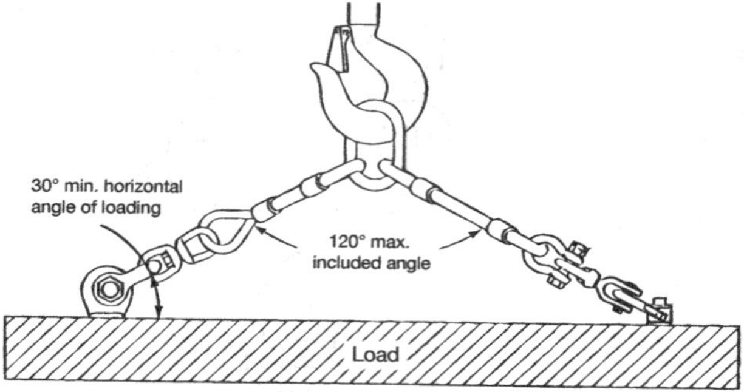

Horizontal Angle, Degree	Stress Multiplier
90	1,000
60	1,155
45	1,414
30	2,000

RIGGING BLOCK TYPES

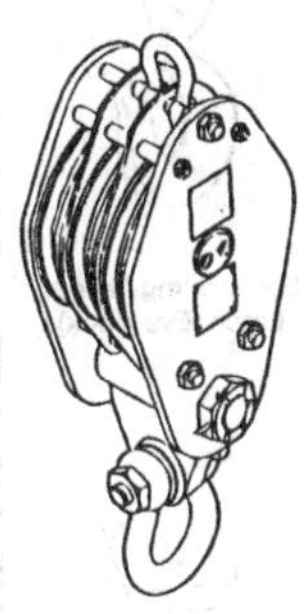

Tackle Block

Rolling Block

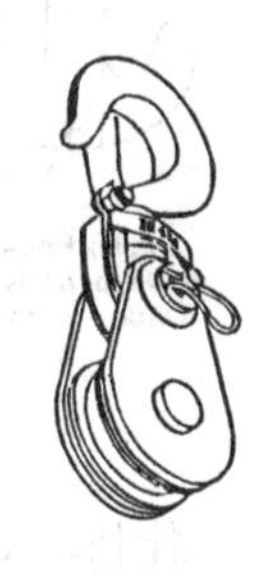

Snatch Block

Utility Block

Snatch Block

COMMON USE HOOKS

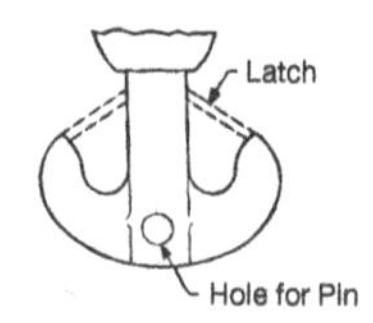

Duplex Hook (Sister)
(Hole for pin is optional)
(Latch—when required)

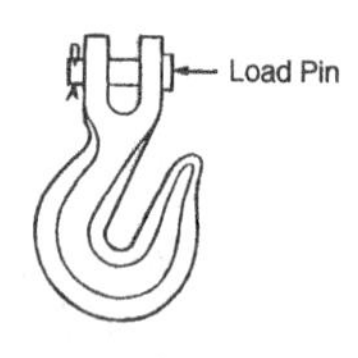

Clevis Grab Hook

Self-closing Flipper Latch (Eye Hook)

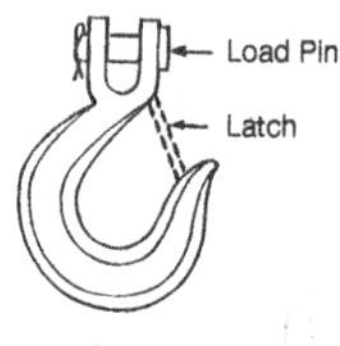

Clevis Hook
(Latch—when required)

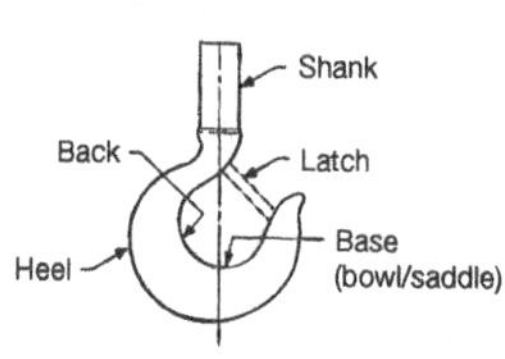

Shank Hook
(Latch—when required)

Choker Hook

Sorting Hook

Reprinted from ASME B30.26-2004, by permission of the American Society of Mechanical Engineers.

SLINGS

Selection of the proper sling and the correct use are important to safe movement of materials by a crane. The following tables and illustrations give basic information for several classifications. Additional information about sizes, loads, and safe use is available in the standard ASME B30.0-2006 SLINGS, which is available from The American Society of Mechanical Engineers, www.asme.org.

ALLOY STEEL CHAIN SLINGS — CONFIGURATIONS AND HITCHES

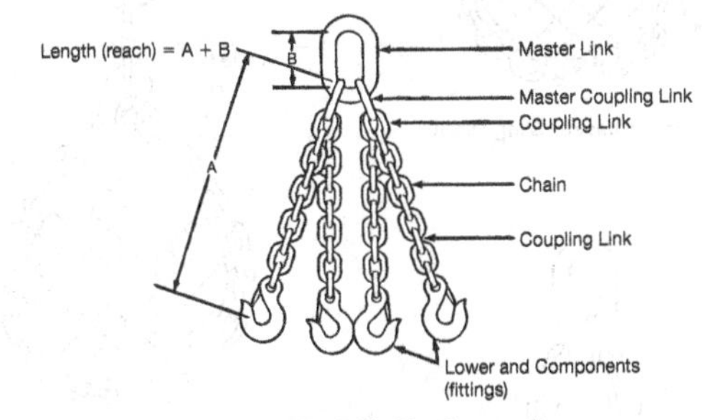

(a) Quadruple-leg Bridle Sling Components

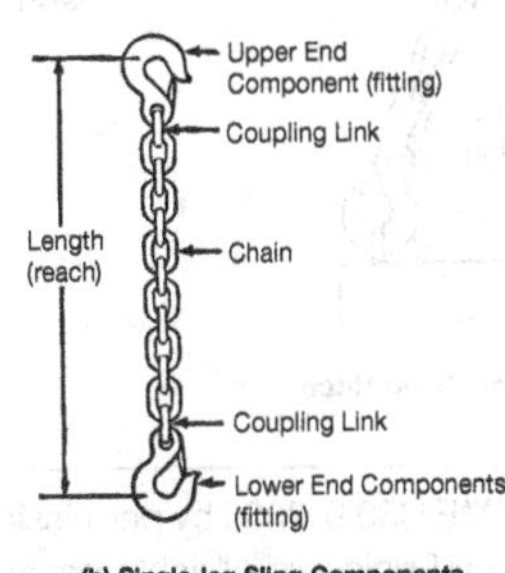

(b) Single-leg Sling Components

ALLOY STEEL CHAIN SLINGS — CONFIGURATIONS AND HITCHES *(cont.)*

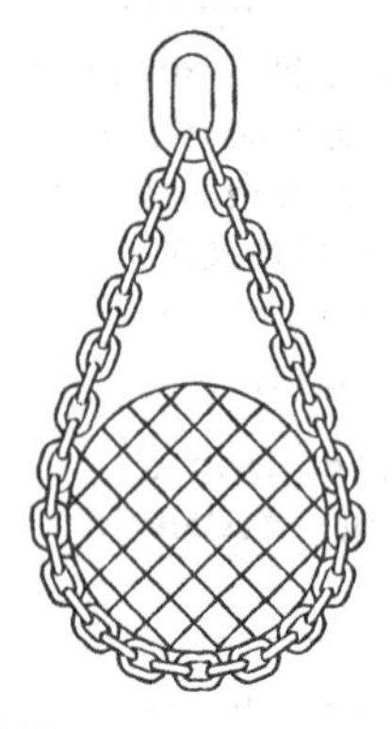

(c) Single-basket Sling Hitch

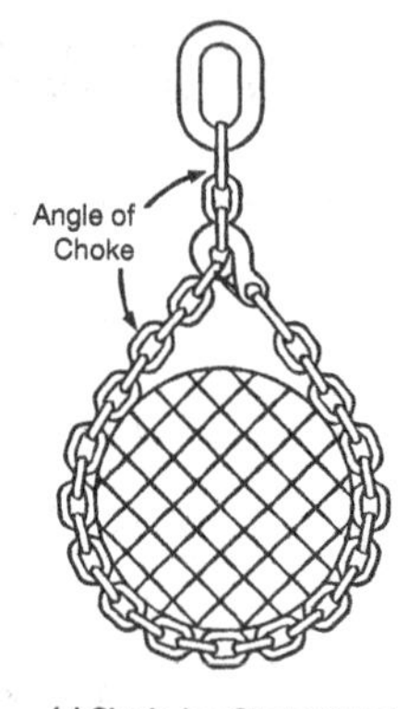

(e) Single-leg Choker Hitch

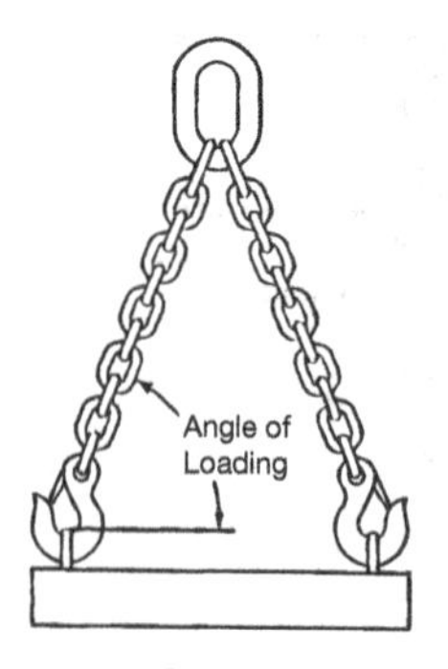

(d) Multiple-leg Bridle Sling Hitch

SYNTHETIC FIBER ROPE SLINGS

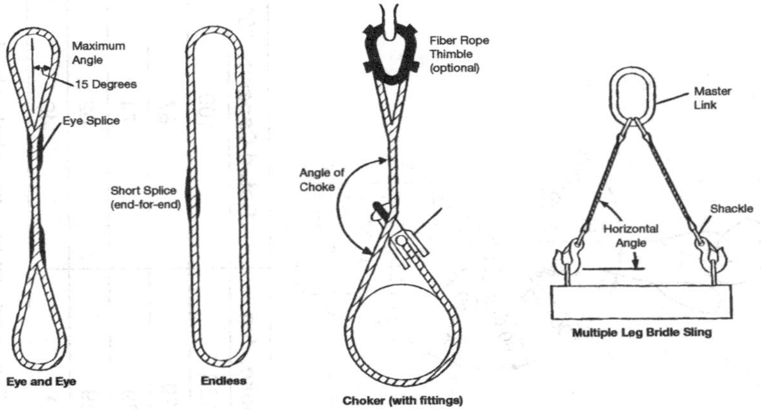

Maximum Angle
15 Degrees
Eye Splice
Eye and Eye

Short Splice (end-for-end)
Endless

Fiber Rope Thimble (optional)
Angle of Choke
Choker (with fittings)

Master Link
Shackle
Horizontal Angle
Multiple Leg Bridle Sling

General Note: Fittings designed for synthetic slings should be used.

ANGLE OF CHOKE

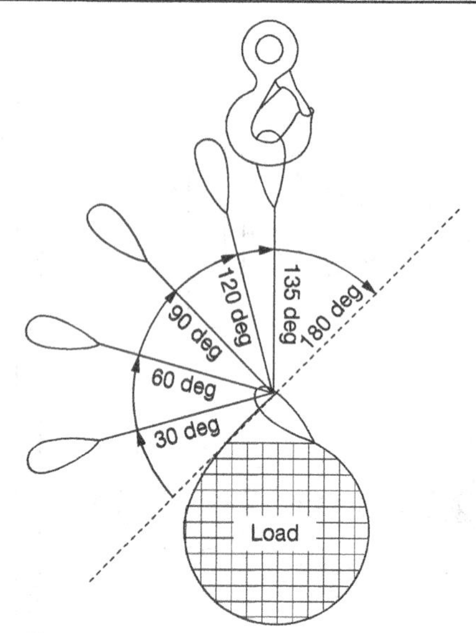

Angle of Choke, deg.	Rated Capacity, %*
Over 120	100
90–120	87
60–89	74
30–59	62
0–29	49

* Percent of sling rated capacity in a choker hitch.

D/d RATION

When *D* is 25 times the cmponent rope diameter (*d*) the *D/d* ratio is expressed as 25/1.

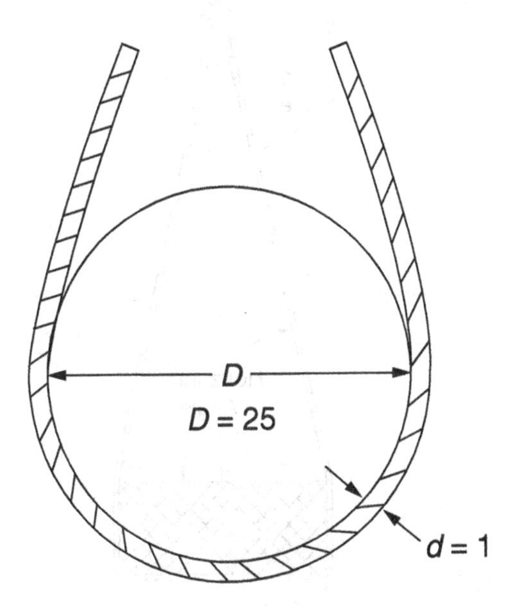

ANGLE OF LOADING

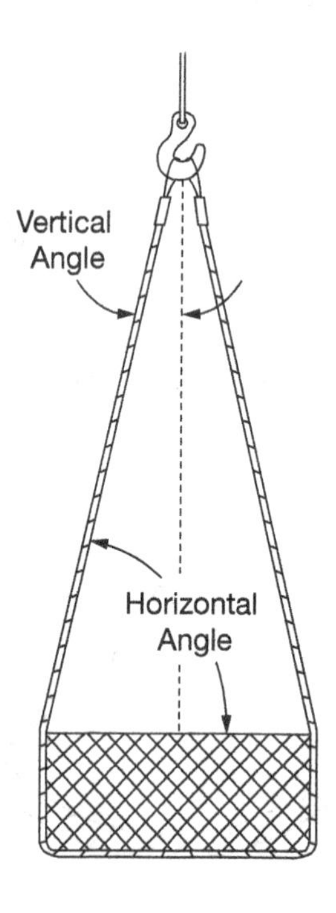

RATED LOAD FOR GRADE 100 ALLOY STEEL CHAIN SLINGS

Nominal Chain Size		Single-leg Vertical Choker Hitch [Note (1)]	Single-leg (Vertical Hitch)	Double-leg Bridle Sling Single Basket Slings [Note (1)]			Triple- and Quadruple-leg Bridle Slings Double Basket Slings [Note (1)]		
		Horizontal Angle, Degree [Note (2)]							
		90	90	60	45	30	60	45	30
in.	mm	lb.	lb.	lb.	lb.	lb.	lb.	lb.	lb.
7/32	5.5	2,100	2,700	4,700	3,800	2,700	7,000	5,700	4,000
9/32	7	3,500	4,300	7,400	6,100	4,300	11,200	9,100	6,400
5/16	8	4,500	5,700	9,900	8,100	5,700	14,800	12,100	8,500
3/8	10	7,100	8,800	15,200	12,400	8,800	22,900	18,700	13,200
1/2	13	12,000	15,000	26,000	21,200	15,000	39,000	31,800	22,500
5/8	16	18,100	22,600	9,100	32,000	22,600	58,700	47,900	33,900
3/4	20	28,300	35,300	61,100	49,900	35,300	91,700	74,900	53,000
7/8	22	34,200	42,700	74,000	60,400	42,700	110,900	90,600	64,000

Notes:

(1) For choker hitch, the angle at choke shall be 120° or greater.

(2) The horizontal angle is the angle formed between the inclined leg and the horiztonal plant of the load.

NYLON ROPE SLINGS — EYE-AND-EYE

Based on Design Factor = 5 and Rated Loads Expressed in Pounds (lb.)

Rope Dia. In.	Eye-and-Eye Sling					
			Two-leg Bridle or Basket			
			Horizontal Angle, Degree			
	Vertical	Choker	90	60	45	30
1/2	1,100	830	2,200	1,900	1,600	1,100
9/16	1,400	1,100	2,800	2,400	2,000	1,400
5/8	1,800	1,400	3,600	3,100	2,500	1,800
3/4	2,600	2,000	5,200	4,500	3,700	2,600
7/8	3,500	2,600	7,000	6,100	4,900	3,500
1	4,400	3,300	8,800	7,600	6,200	4,400
1 1/8	5,700	4,300	11,400	9,900	8,100	5,700
1 1/4	7,000	5,300	14,000	12,100	9,900	7,000
1 5/16	7,700	5,800	15,400	13,300	10,900	7,700
1 1/2	9,700	7,300	19,400	16,800	13,700	9,700
1 5/8	11,500	8,600	23,000	19,900	16,300	11,500
1 3/4	13,200	9,900	26,400	22,900	18,700	13,200
2	16,900	12,700	33,800	29,300	23,900	16,900
2 1/8	19,100	14,300	38,200	33,100	27,000	19,100
2 1/4	21,400	16,100	42,800	37,100	30,300	21,400
2 1/2	26,300	19,700	52,600	45,600	37,200	26,300
2 5/8	28,800	21,600	57,600	49,900	40,700	28,800
3	37,100	27,800	74,200	64,300	52,500	37,100

NYLON ROPE SLINGS — ENDLESS

Based on Design Factor = 5 and Rated Loads Expressed in Pounds (lb.)

Rope Dia. In.	Endless Sling					
			Basket			
			Horizontal Angle, Degree			
	Vertical	Choker	90	60	45	30
1/2	2,000	1,500	4,000	3,500	2,800	2,000
9/16	2,600	2,000	5,200	4,500	3,700	2,600
5/8	3,200	2,400	6,400	5,500	4,500	3,200
3/4	4,600	3,500	9,200	8,000	6,500	4,600
7/8	6,200	4,700	12,400	10,700	8,800	6,200
1	7,900	5,900	15,800	13,700	11,200	7,900
1 1/8	10,100	7,600	20,200	17,500	14,300	10,100
1 1/4	12,400	9,300	24,800	21,500	17,500	12,400
1 5/16	13,700	10,300	27,400	23,700	19,400	13,700
1 1/2	17,400	13,100	34,800	30,100	24,600	17,400
1 5/8	20,500	15,400	41,000	35,500	29,000	20,500
1 3/4	23,600	17,700	47,200	40,900	33,400	23,600
2	30,200	22,700	60,400	52,300	42,700	30,200
2 1/8	34,100	25,600	68,200	59,100	48,200	34,100
2 1/4	38,300	28,700	76,600	66,300	54,200	38,300
2 1/2	46,900	35,200	93,800	81,200	66,300	46,900
2 5/8	51,400	38,600	102,800	89,000	72,700	51,400
3	66,200	49,700	132,400	114,700	93,600	66,200

POLYESTER ROPE SLINGS — EYE-AND-EYE						
Based on Design Factor = 5 and Rated Loads Expressed in Pounds (lb.)						
	Eye-and-Eye Sling					
			Basket			
			Horizontal Angle, Degree			
Rope Dia. In.	Vertical	Choker	90	60	45	30
1/2	1,000	750	2,000	1,700	1,400	1,000
9/16	1,300	980	2,600	2,300	1,800	1,300
5/8	1,600	1,200	3,200	2,800	2,300	1,600
3/4	2,200	1,700	4,400	3,800	3,100	2,200
7/8	3,000	2,300	6,000	5,200	4,200	3,000
1	4,000	3,000	8,000	6,900	5,700	4,000
1 1/8	5,000	3,800	10,000	8,700	7,100	5,000
1 1/4	6,000	4,500	12,000	10,400	8,500	6,000
1 5/16	6,500	4,900	13,000	11,300	9,200	6,500
1 1/2	8,400	6,300	16,800	14,500	11,900	8,400
1 5/8	9,900	7,400	19,800	17,100	14,000	9,900
1 3/4	11,400	8,600	22,800	19,700	16,100	11,400
2	14,400	10,800	28,800	24,900	20,400	14,400
2 1/8	16,200	12,200	32,400	28,100	22,900	16,200
2 1/4	18,100	13,600	36,200	31,300	25,600	18,100
2 1/2	22,000	16,500	44,000	38,100	31,100	22,000
2 5/8	24,200	18,200	48,400	41,900	34,200	24,200
3	31,200	23,400	62,400	54,000	44,100	31,200

POLYESTER ROPE SLINGS — ENDLESS

Based on Design Factor = 5 and Rated Loads Expressed in Pounds (lb.)

Rope Dia. In.	Endless Sling					
			Basket			
			Horizontal Angle, Degree			
	Vertical	Choker	90	60	45	30
1/2	1,800	1,400	3,600	3,100	2,500	1,800
9/16	2,300	1,700	4,600	4,000	3,300	2,300
5/8	2,800	2,100	5,600	4,800	4,000	2,800
3/4	4,000	3,000	8,000	6,900	5,700	4,000
7/8	5,400	4,100	10,800	9,400	7,600	5,400
1	7,100	5,300	14,200	12,300	10,000	7,100
1 1/8	8,900	6,700	17,800	15,400	12,600	8,900
1 1/4	10,600	8,000	21,200	18,400	15,000	10,600
1 5/16	11,600	8,700	23,200	20,100	16,400	11,600
1 1/2	15,100	11,300	30,200	26,200	21,400	15,100
1 5/8	17,600	13,200	35,200	30,500	24,900	17,600
1 3/4	20,400	15,300	40,800	35,300	28,800	20,400
2	25,700	19,300	51,400	44,500	36,300	25,700
2 1/8	28,900	21,700	57,800	50,100	40,900	28,900
2 1/4	32,300	24,200	64,600	55,900	45,700	32,300
2 1/2	39,300	29,500	78,600	68,100	55,600	39,300
2 5/8	43,200	32,400	86,400	74,800	61,100	43,200
3	55,700	41,800	111,400	96,500	78,800	55,700

POLYPROPYLENE ROPE SLINGS — EYE-AND-EYE

Based on Design Factor = 5 and Rated Loads Expressed in Pounds (lb.)

Rope Dia. In.	Eye-and-Eye Sling					
	Vertical	Choker	Basket — Horizontal Angle, Degree			
			90	60	45	30
1/2	760	570	1,400	1,300	1,100	750
9/16	920	690	1,600	1,600	1,300	900
5/8	1,100	830	2,200	1,900	1,600	1,100
3/4	1,500	1,100	3,000	2,600	2,100	1,500
7/8	2,100	1,600	4,200	3,600	3,000	2,100
1	2,600	2,000	5,200	4,500	3,700	2,600
1 1/8	3,200	2,400	6,400	5,500	4,500	3,200
1 1/4	3,900	2,900	7,800	6,800	5,500	3,900
1 5/16	4,200	3,200	8,400	7,300	5,900	4,200
1 1/2	5,500	4,100	11,000	9,500	7,800	3,500
1 5/8	6,400	4,800	13,300	11,100	9,000	6,400
1 3/4	7,400	5,600	14,300	12,800	10,500	7,400
2	9,400	7,100	15,300	16,300	13,300	9,400
2 1/8	10,500	7,900	23,000	18,200	14,800	10,500
2 1/4	11,900	8,900	23,800	20,600	16,800	11,900
2 1/2	14,400	18,800	23,300	24,900	20,400	14,400
2 5/8	16,100	12,100	32,200	27,900	22,800	16,100
3	20,500	15,400	41,000	35,500	29,000	20,500

POLYPROPYLENE ROPE SLINGS — ENDLESS

Based on Design Factor = 5 and Rated Loads Expressed in Pounds (lb.)

Rope Dia. In.	Endless Sling					
			Basket			
			Horizontal Angle, Degree			
	Vertical	Choker	90	60	45	30
1/2	1,400	1,100	2,800	2,400	2,000	1,400
9/16	1,600	1,200	3,200	2,800	2,300	1,600
5/8	2,000	1,500	4,000	3,500	2,800	2,000
3/4	2,700	2,000	5,400	4,700	3,800	2,700
7/8	3,700	2,800	7,400	6,400	5,200	3,700
1	4,600	3,500	9,200	8,000	6,500	4,600
1 1/8	5,700	4,300	11,400	9,900	8,100	5,700
1 1/4	6,900	5,200	13,800	12,000	9,800	6,900
1 5/16	7,600	5,700	15,200	13,200	10,700	7,600
1 1/2	9,800	7,400	19,600	17,000	13,900	9,800
1 5/8	11,400	8,600	22,900	19,700	16,100	11,400
1 3/4	13,200	9,900	26,400	22,900	18,700	13,200
2	16,700	12,500	33,400	28,900	23,600	16,700
2 1/8	18,800	14,100	37,600	32,600	26,600	18,800
2 1/4	21,200	15,900	42,400	36,700	30,000	21,200
2 1/2	25,700	19,300	51,400	44,500	36,300	25,700
2 5/8	28,800	21,600	57,600	49,900	40,700	28,800
3	36,600	27,500	73,200	63,400	51,800	36,600

RATED LOAD FOR SINGLE- AND TWO-LEG SLINGS 6 × 19 OR 6 × 36 CLASSIFICATION

EXTRA IMPROVED PLOW STEEL (EIPS) GRADE FIBER CORE (FC) WIRE ROPE

Based on Design Factor = 5 and Rated Loads Expressed in Tons (2,000 lb.)

Rope Dia. In.	Single-Leg						
	Vertical			Choker	Vertical Basket		
	HT	MS	S	HT&MS	HT	MS	S
1/4	0.54	0.56	0.60	0.42	1.1	1.1	1.1
5/16	0.83	0.87	0.94	0.66	1.7	1.7	1.7
3/8	1.2	1.2	1.3	0.94	2.4	2.5	2.5
7/16	1.6	1.7	1.8	1.3	3.2	3.4	3.4
1/2	2.0	2.2	2.4	1.6	4.0	4.4	4.4
9/16	2.5	2.7	3.0	2.1	5.0	5.5	5.5
5/8	3.1	3.4	3.7	2.6	6.2	6.8	6.8
3/4	4.3	4.8	5.2	3.7	8.6	9.7	9.7
7/8	5.7	6.6	7.1	5.0	11	13	13
1	7.4	8.3	9.2	6.4	15	17	17
1 1/8	9.3	10	12	8.1	19	21	21
1 1/4	11	13	14	9.9	23	26	26

GENERAL NOTES:
(a) HT = hand-tucked splice
(b) MS = mechanical splice
(c) S = swaged or poured socket
(d) Rated loads based on min. *D/d* ratio of 25/1
(e) Rated load based on pin dia. no larger than natural eye width or less than the nominal sling dia.
(f) For choker hitch, the angle of choke shall be 120 deg. or greater

RATED LOAD FOR SINGLE- AND TWO-LEG SLINGS 6 × 19 OR 6 × 36 CLASSIFICATION *(cont.)*

EXTRA IMPROVED PLOW STEEL (EIPS) GRADE FIBER CORE (FC) WIRE ROPE

Based on Design Factor = 5 and Rated Loads Expressed in Tons (2,000 lb.)

Two-Leg Bridle or Basket

Rope Dia. In.	Vertical			60°		45°	
	HT	MS	S	HT	MS	HT	MS
1/4	1.1	1.1	1.1	0.94	0.97	0.77	0.79
5/16	1.7	1.7	1.7	1.4	1.5	1.2	1.2
3/8	2.4	2.5	2.5	2.0	2.2	1.7	1.8
7/16	3.2	3.4	3.4	2.7	2.9	2.2	2.4
1/2	4.0	4.49	4.4	3.5	3.8	2.9	3.1
9/16	5.0	5.5	5.5	4.4	4.8	3.6	3.9
5/8	6.2	6.8	6.8	5.3	5.9	4.4	4.8
3/4	8.6	9.7	9.7	7.4	8.4	6.1	6.8
7/8	11	13	13	9.8	11	8.0	9.3
1	15	17	17	13	14	10	12
1 1/8	19	21	21	16	18	13	15
1 1/4	23	26	26	20	22	16	18

GENERAL NOTES:
(a) HT = hand-tucked splice
(b) MS = mechanical splice
(c) S = swaged or poured socket
(d) Rated loads based on min. *D/d* ratio of 25/1
(e) Rated load based on pin dia. no larger than natural eye width or less than the nominal sling dia.
(f) For choker hitch, the angle of choke shall be 120 deg. or greater

RATED LOAD FOR SINGLE- AND TWO-LEG SLINGS 6 × 19 OR 6 × 36 CLASSIFICATION *(cont.)*

EXTRA IMPROVED PLOW STEEL (EIPS) GRADE FIBER CORE (FC) WIRE ROPE

Based on Design Factor = 5 and Rated Loads Expressed in Tons (2,000 lb.)

Rope Dia. In.	Two-Leg Bridle or Basket *(cont.)*		Horizontal Angle	
	30°		60°	30°
	HT	MS	S	HT
1/4	0.54	0.56	0.73	0.42
5/16	0.83	0.87	1.1	0.66
3/8	1.2	1.2	1.6	0.94
7/16	1.6	1.7	2.2	1.3
1/2	2.0	2.2	2.9	1.6
9/16	2.5	2.7	3.6	2.1
5/8	3.1	3.4	4.5	2.6
3/4	4.3	4.8	6.3	3.7
7/8	5.7	6.6	8.6	5.0
1	7.4	8.3	11	6.4
1 1/8	9.3	10	14	8.1
1 1/4	11	13	17	9.9

GENERAL NOTES:

(a) HT = hand-tucked splice

(b) MS = mechanical splice

(c) S = swaged or poured socket

(d) Rated loads based on min. *D/d* ratio of 25/1

(e) Rated load based on pin dia. no larger than natural eye width or less than the nominal sling dia.

(f) For choker hitch, the angle of choke shall be 120 deg. or greater

CHAPTER 6
Fasteners

There are an extensive variety of fasteners available. Some are standardized, while manufacturers also produce specialized products. Consult manufacturers' catalogs for details. Following are some of the most frequently used fasteners and related technical information.

STANDARD TAP DRILL SIZES FOR NATIONAL FINE THREADS

DATA GIVES 80% OF FULL THREAD DEPTH

Thread Dia.	Threads per Inch	Drill Size	Drill Decimal Equivalent
1½	12	1 27/64	1.4219
1⅜	12	1 19/64	1.2969
1¼	12	1 11/64	1.1719
1⅛	12	1 3/16	1.0469
1	14	15/16	.9375
⅞	14	13/16	.8125
¾	16	11/16	.6875
⅝	18	37/64	.5781
9/16	18	33/64	.5156
½	20	29/64	.4531
7/16	20	25/64	.3906
⅜	24	Q	.332
5/16	24	1	.272
¼	28	#3	.213

STANDARD TAP DRILL SIZES FOR NATIONAL COARSE THREADS			
DATA GIVES 80% OF FULL THREAD DEPTH			
Thread Dia. Inch	**Threads per Inch**	**Drill Size**	**Drill Decimal Equivalent**
2	4.5	1	1.7813
1¾	2	1	1.5469
1½	6	1	1.3281
1⅜	6	1	1.2188
1¼	7	1	1.1094
1⅛	7	63/64	.9844
1	8	⅞	.8750
⅞	9	49/64	.7656
¾	10	21/32	.6563
⅝	11	17/32	.5313
9/16	12	31/64	.4844
½	13	27/64	.4219
7/16	14	U	.368
⅜	16	5/16	.3125
5/16	18	F	.2570
¼	20	#8	.201

INCH TAP DRILL SIZES FOR AMERICAN STANDARD THREADS

Diam. of Thread	Threads per Inch	Drill*	Decimal Equiv.
No. 0-.060	80 NF**	3/64	.0469
1-.073	64 NC**	1.5 MM	.0591
	72 NF	53	.0595
2-.086	56 NC	50	.0700
	64 NF	50	.0700
3-.099	48 NC	5/64	.0781
	56 NF	45	.0820
4-.112	40 NC	43	.0890
	48 NF	42	.0935
5-.125	40 NC	38	.1015
	44 NF	37	.1040
6-.138	32 NC	36	.1065
	40 NF	33	.1130
8-.164	32 NC	29	.1360
	36 NF	29	.1360
10-.190	24 NC	25	.1495
	32 NF	21	.1590
12-.216	24 NC	16	.1770
	28 NF	14	.1820
1/4	20 NC	7	.2010
	28 NF	3	.2130
	32 NEF*	7/32	.2188
5/16	18 NC	F	.2570
	24 NF	1	.2720
	32 NEF	9/32	.2812
3/8	16 NC	5/16	.3125
	24 NF	Q	.3320
	32 NEF	11/32	.3438

*Drill diameter produces approximately 75% full thread.

**NF – National Fine NC – National Coarse

NEF – National Extra Fine

TAP DRILL SIZES FOR AMERICAN STANDARD THREADS *(cont.)*

Diam. of Thread	Threads per Inch	Drill*	Decimal Equiv.
7/16	14 NC	U	.3680
	20 NF	25/64	.3906
	28 NEF	Y	.4040
1/2	12 N	27/64	.4219
	13 NC	27/64	.4219
	20 NF	29/64	.4531
	28 NEF	15/32	.4687
9/16	12 NC	31/64	.4844
	18 NF	33/64	.5156
	24 NEF	33/64	.5156
5/8	11 NC	17/32	.5312
	12 N	25/64	.5469
	18 NF	14.5 MM	.5709
	24 NEF	37/64	.5781
11/16	12 N	39/64	.6094
	24 NEF	16.5 MM	.6496
	10 NC	16.5 MM	.6496
3/4	12 N	17 MM	.6693
	16 NF	17.5 MM	.6890
	20 NEF	45/64	.7031
13/16	12 N	18.5 MM	.7283
	16 N	3/4	.7500
	20 NEF	49/64	.7656
7/8	9 NC	49/64	.7656
	12 N	20 MM	.7874
	14 NF	25.5 MM	.8071
	16 N	13/16	.8125
	20 NEF	21 MM	.8268
15/16	12 N	35/64	.8594
	16 N	7/8	.8750
	20 NEF	22.5 MM	.8858
1	8 NC	7/8	.8750
	12 N	59/64	.9219
	14 NF	23.5 MM	.9252
	15 N	15/16	.9375
	20 NEF	61/64	.9531

METRIC THREAD DIMENSIONS AND TAP DRILL SIZES

Nominal Size Mm	COARSE SERIES			FINE SERIES	
	Approximately 75% Thread				
	Pitch mm	Tap Drill mm	Clearance Drill mm	Pitch mm	Tap Drill mm
1.4	0.3	1.1	1.55		
1.6	0.35	1.25	1.8		
2	0.4	1.6	2.2		
2.5	0.45	2.05	2.6		
3	0.5	2.5	3.2		
4	0.7	3.3	4.2		
5	0.8	4.2	5.2		
6	1.0	5.0	6.2		
8	1.25	6.75	8.2	1	7.0
10	1.5	8.5	10.2	1.25	8.75
12	1.75	10.25	12.2	1.25	10.50
14	2	12.00	14.2	1.5	12.50
16	2	14.00	16.45	1.5	14.50
18	2.5	15.50	18.20	1.5	16.50
20	2.5	17.50	20.50	1.5	18.50
22	2.5	19.50	22.80	1.5	20.50
24	3	21.00	24.60	2	22.00
27	3	24.00	27.95	2	25.00

Metric Thread Note:

M20 × 2.50

- 2.50 — Pitch
- 20 — Major Diameter in Millimeters
- M — Indicates Metric Thread

FRACTIONAL TWIST DRILLS INCHES	
Fractional Drill Sizes	**Decimal Equivalent**
1/64	.0156
1/32	.0312
3/64	.0469
1/16	.0625
5/64	.0781
3/32	.0937
7/64	.1094
1/8	.1250
9/64	.1406
5/32	.1562
11/64	.1719
3/16	.1875
13/64	.2031
7/32	.2187
15/64	.2344
1/4	.2500
17/64	.2656
9/32	.2812
19/64	.2969
5/16	.3125
21/64	.3281
11/32	.3437
23/64	.3594
3/8	.3750
25/64	.3906

FRACTIONAL TWIST DRILLS INCHES *(cont.)*	
Fractional Drill Sizes	**Decimal Equivalent**
13/32	.4062
. . .	.4130
27/64	.4219
7/16	.4375
29/64	.4531
15/32	.4687
31/64	.4844
1/2	.5000
33/64	.5156
17/32	.5312
35/64	.5469
9/16	.5625
37/64	.5781
19/32	.5937
39/64	.6094
5/8	.6250

FRACTIONAL TWIST DRILLS INCHES *(cont.)*	
Fractional Drill Sizes	**Decimal Equivalent**
41/64	.6406
21/32	.6562
43/64	.6719
11/16	.6875
45/64	.7031
23/32	.7187
47/64	.7344
3/4	.7500
49/64	.7656
25/32	.7812
51/64	.7969
13/16	.8125
53/64	.8281
27/32	.8437
55/64	.8594
7/8	.8750
57/64	.8906
29/32	.9062
59/64	.9219
15/16	.9375
61/64	.9531
31/32	.9687
63/64	.9844
1	1.0000

NUMBERED TWIST DRILLS INCHES			
Drill Number	**Decimal Equivalents**	**Drill Number**	**Decimal Equivalents**
80	.0135	40	.0980
79	.0145	39	.0995
78	.0160	38	.1015
77	.0180	37	.1040
76	.0200	36	.1065
75	.0210	35	.1100
74	.0225	34	.1110
73	.0240	33	.1130
72	.0250	32	.1160
71	.0260	31	.1200
70	.0280	30	.1285
69	.0292	29	.1360
68	.0310	28	.1405
67	.0320	27	.1440
66	.0330	26	.1470
65	.0350	25	.1495
64	.0360	24	.1520
63	.0370	23	.1540
62	.0380	22	.1570
61	.0390	21	.1590
60	.0400	20	.1610
59	.0410	19	.1660
58	.0420	18	.1695
57	.0430	17	.1720
56	.0465	16	.1770
55	.0520	15	.1800
54	.0550	14	.1820
53	.0595	13	.1850
52	.0635	12	.1890
51	.0670	11	.1910
50	.0700	10	.1935
49	.0730	9	.1960
48	.0760	8	.1990
47	.0785	7	.2010
46	.0810	6	.2040
45	.0820	5	.2055
44	.0860	4	.2090
43	.0890	3	.2130
42	.0935	2	.2210
41	.0960	1	.2280

TYPICAL METRIC DRILL SIZES	
DIA mm	Decimal Equiv Inch
6.50	0.2559
6.60	0.2598
6.70	0.2638
6.80	0.2677
6.90	0.2717
7.00	0.2756
7.10	0.2795
7.20	0.2835
7.30	0.2874
7.40	0.2913
7.50	0.2953
7.60	0.2992
7.70	0.3031
7.80	0.3071
7.90	0.3110
8.00	0.3150
8.10	0.3189
8.20	0.3228
8.30	0.3268
8.40	0.3307
8.50	0.3346
8.60	0.3386
8.70	0.3425
8.75	0.3445
8.80	0.3465
8.90	0.3504
9.00	0.3543
9.10	0.3583
9.20	0.3622
9.30	0.3661
9.40	0.3701
9.50	0.3740
9.60	0.3780
9.70	0.3819
9.80	0.3858
9.90	0.3898
10.00	0.3937

TYPICAL METRIC DRILL SIZES *(cont.)*

DIA mm	Decimal Equiv Inch
3.10	0.1220
3.20	0.1260
3.30	0.1299
3.40	0.1339
3.50	0.1378
3.60	0.1417
3.70	0.1457
3.80	0.1496
3.90	0.1535
4.00	0.1575
4.10	0.1614
4.20	0.1654
4.30	0.1693
4.40	0.1732
4.50	0.1772
4.60	0.1811
4.70	0.1850
4.80	0.1890
4.90	0.1929
5.00	0.1969
5.10	0.2008
5.20	0.2047
5.30	0.2087
5.40	0.2126
5.50	0.2165
5.60	0.2205
5.70	0.2244
5.75	0.2264
5.80	0.2283
5.90	0.2323
6.00	0.2362
6.10	0.2402
6.20	0.2441
6.30	0.2480
6.40	0.2520

TYPICAL METRIC DRILL SIZES *(cont.)*

DIA mm	Decimal Equiv Inch
10.2	.4016
10.3	.4055
10.4	.4134
10.8	.4252
11.0	.4331
11.2	.4409
11.5	.4528
11.8	.4646
12.0	.4724
12.2	.4803
12.5	.4921
13.0	.5118
13.5	.5315
14.0	.5512
14.5	.5709
15.0	.5906
15.5	.6102
16.0	.6299
16.5	.6496
17.0	.6693
17.5	.6890
18.0	.7087
18.5	.7283
19.0	.7480
19.5	.7677
20.0	.7874
20.5	.8071
21.0	.8268
21.5	.8465
22.0	.8661
22.5	.8858
23.0	.9055
23.5	.9252
24.0	.9449
24.5	.9846
25.0	.9843

SUGGESTED BIT SPEEDS FOR METAL USING HIGH SPEED STEEL DRILLS			
Drill Bit Size (inches/mm)	**Low-Carbon Steel, Cast Iron (soft), Malleable Iron**	**Medium-Carbon Steel, Cast Iron (hard)**	**Aluminum and Aluminum Alloys, Ordinary Brass**
⅛ (3.2)	2500 to 3000	2000 to 2500	6000 to 7000
¼ (6.35)	1200 to 1500	1000 to 1300	3000 to 4500
⅜ (9.39)	800 to 1100	700 to 800	2500 to 3000
½ (12.7)	600 to 800	500 to 600	1500 to 2300

SUGGESTED TWIST DRILL SPEEDS FOR BORING IN WOOD*	
Bit Size (inches)	**Spindle Speeds RPM**
Up to ¼	3600 to 3800
¼ to ½	2800 to 3100
½ to ¾	2000 to 2300
¾ to 1	1800 to 2000
Over 1	600 to 700

*The revolutions per minute to be used are affected by the characteristics of the wood and the rate at which the drill is moved into the wood. Bore a test hole in the stock and adjust the speed up or down as appears to be needed.

FASTENER LUBRICATIONS

Black fasteners are lubricated with a water-soluble oily lubricant applied by the manufacturer to nuts, washers, and bolts. They must feel oily to the touch. If the oily feeling is gone, reapply an oil-based lubricant, beeswax, stick wax, a liquid wax, or a spray lubricant.

Galvanized fasteners only require that the nut be lubricated. It is a wax-based material applied by the manufacturer that is dry to the touch. If relubrication is needed, apply a wax-based lubricant.

TORQUE TENSIONING

To achieve secure assemblies, fasteners should be tensioned to produce a clamping force greater than the external force tending to separate the joint. Torque is that convenient, measurable means of developing desired tension. Additional technical information is available from Fastener Technology International, 1867 W. Market Street, Akron, Ohio. www.fastenertech.com

TORQUE TENSIONING *(cont.)*

Suggested assembly torques to produce minimum specified tension in A325 structural bolts.

Bolt Dia (in)	Minimum Tension (lb)	Minimum Tension Plus 5% (lb)	Torque for Clean & Oiled Condition (ft-lb) (K = 0.15)	Torque for Nonclean Condition (ft-lb) (K = 0.20)
½	12,000	12,600	79	105
⅝	19,000	19,950	155	210
¾	28,000	29,400	275	365
⅞	39,000	40,950	448	595
1	51,000	53,550	670	890
1⅛	56,000	58,800	825	1100
1¼	71,000	74,550	1165	1550
1⅜	85,000	89,250	1535	2045
1½	103,000	108,150	2025	2700

Courtesy Fastener Technology International, www.fastenertech.com

TORQUE TENSIONING *(cont.)*

Suggested assembly torques to produce minimum specified tension in A490 structural bolts.

Bolt Dia (in)	Minimum Tension (lb)	Minimum Tension Plus 5% (lb)	Torque for Clean & Oiled Condition (ft-lb) (K = 0.15)	Torque for Nonclean Condition (ft-lb) (K = 0.20)
½	15,000	15,750	98	130
⅝	24,000	25,200	195	265
¾	35,000	36,750	345	460
⅞	49,000	51,450	565	750
1	64,000	67,200	840	1120
1⅛	80,000	84,000	1180	1575
1¼	102,000	107,150	1675	2230
1⅜	121,000	127,050	2185	2910
1½	148,000	155,400	2915	3885

Note: The calculated torque values may give varying results in bolt tension, depending on the condition of fastener components. To assure that adequate bolt tension has been attained, determine exact torque values with a bolt-tension-indicating device such as outlined in Section 8(d)(2) of *Specification for Structural Joints Using ASTM A325* or *A490 Bolts.*

Courtesy Fastener Technology International, www.fastenertech.com

SUGGESTED ASSEMBLY TORQUES FOR METRIC FASTENERS
(in metric units for metric calibrated wrenches and instruments)

Major Diameter and Thread Pitch	Stress Area mm^2	CLASS 4.6			CLASS 4.8			CLASS 5.8			CLASS 8.8		
		Clamp Load kN	Torque		Clamp Load kN	Torque		Clamp Load kN	Torque		Clamp Load kN	Torque	
			Dry K = .2 N·m	Lub. K = .15 N·m		Dry K = .2 N·m	Lub. K = .15 N·m		Dry K = .2 N·m	Lub. K = .15 N·m		Dry K = .2 N·m	Lub. K = .15 N·m
1.6 × 0.35	1.27				0.29	0.09	0.07						
2 × 0.4	2.07				0.48	0.19	0.14						
2.5 × 0.45	3.39				0.79	0.39	0.3						
3 × 0.5	5.03				1.17	0.7	0.5						
3.5 × 0.6	6.78				1.58	1.1	0.8						
4 × 0.7	8.78				2.05	1.6	1.2						
5 × 0.8	14.2	2.40	2.4	1.8	3.3	3.3	2.5	4.05	4.0	3.0			
6 × 1	20.1	3.39	4.0	3.0	4.67	5.6	4.2	5.73	6.9	5.2			
8 × 1.25	36.6	6.18	9.9	7.4	8.48	13.5	10.2	10.4	16.5	12.5			
10 × 1.5	58.0	9.83	19.7	14.7	13.5	27	20	16.5	33	25			

SUGGESTED ASSEMBLY TORQUES FOR METRIC FASTENERS
(in metric units for metric calibrated wrenches and instruments) *(cont.)*

Major Diameter and Thread Pitch	Stress Area mm²	CLASS 4.6			CLASS 4.8			CLASS 5.8			CLASS 8.8		
		Clamp Load kN	Torque		Clamp Load kN	Torque		Clamp Load kN	Torque		Clamp Load kN	Torque	
			Dry K = .2 N·m	Lub. K = .15 N·m		Dry K = .2 N·m	Lub. K = .15 N·m		Dry K = .2 N·m	Lub. K = .15 N·m		Dry K = .2 N·m	Lub. K = .15 N·m
12 × 1.75	84.3	14.3	34.3	25.7	19.6	47	35	24.0	58	43			
14 × 2	115	19.4	54.3	40.7	26.8	75	56	32.8	92	69			
16 × 2	157	26.5	84.8	63.6	36.5	117	88	44.8	143	108	70.7	226	170
20 × 2.5	245	41.3	165.2	124				69.8	279	209	110	440	330
24 × 3	353	60	288	216				100.5	482	362	159	763	572
30 × 3.5	561	94.5	567	425							253	1518	1139
36 × 4	817	138	994	745							368	2650	1987
Tensile Strength		400 MPa			420 MPa			520 MPa			830 MPa		
Proof Load		225 MPa			310 MPa			380 MPa			600 MPa		

Courtesy Fastener Technology International, www.fastenertech.com

SUGGESTED TIGHTENING TORQUES FOR 1960 INCH SERIES SOCKET-HEAD CAP SCREWS

Nominal Size	Basic Screw Diameter (inches)	Torque to Tighten Screws to Yield		Tension Induced in Screws Torqued to Yield (lb)		Tightening Torque*	
		UNRC	UNRF	UNRC	UNRF	UNRC	UNRF
		(in-lb)				(in-lb)	
0	0.0600	–	3.5	–	250	–	2.6
1	0.0730	6	6.5	370	390	4.5	4.8
2	0.0860	10	11	520	550	7.5	8
3	0.0990	15	16	680	730	11	12
4	0.1120	22	24	840	920	16	18
5	0.1250	32	33	1110	1160	24	24
6	0.1380	40	45	1270	1420	30	34
8	0.1640	74	78	1960	2060	55	58
10	0.1900	105	120	2450	2800	79	90

SUGGESTED TIGHTENING TORQUES FOR 1960 INCH SERIES SOCKET-HEAD CAP SCREWS *(cont.)*

Nominal Size	Basic Screw Diameter (inches)	Torque to Tighten Screws to Yield		Tension Induced in Screws Torqued to Yield (lb)		Tightening Torque*	
		UNRC	UNRF	UNRC	UNRF	UNRC	UNRF
		(ft-lb)				(ft-lb)	
¼	0.2500	22	25	4710	5390	17	19
5/16	0.3125	46	51	7760	8595	35	38
⅜	0.3750	83	93	11,485	13,020	62	70
7/16	0.4375	132	146	15,775	17,575	100	109
½	0.5000	203	230	21,015	23,715	150	172
⅝	0.6250	380	430	31,650	35,850	283	317
¾	0.7500	667	750	46,750	52,200	500	562
⅞	0.8750	917	1020	55,450	61,100	688	767
1	1.0000	1390	1520	72,700	79,550	1040	1080
1¼	1.2500	2780	3080	116,300	128,800	2080	2310
1½	1.5000	4850	5450	168,600	189,700	3625	4080

Courtesy Fastener Technology International, www.fastenertech.com

RECOMMENDED HOLE SIZES FOR VARIOUS BOLT DIAMETERS*									
Hole Diameters (inches)									
Bolt	½	⅝	¾	⅞	1	1⅛	1¼	1⅜	1½
Hole Type									
STD standard	9/16	11/16	13/16	15/16	1⅙	1 3/16	1 5/16	1 7/16	1 9/16
OVS oversized	⅝	13/16	15/16	1 1/16	1¼	1 7/16	1 9/16	1 11/16	1 13/16

*Hole sizes larger than standard must be approved by an engineer.

U, HOOK, AND EYE BOLTS

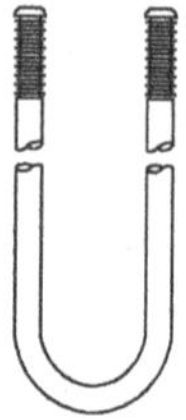

U-Bolt Round Bend

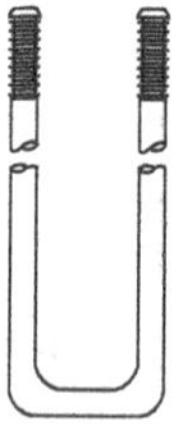

U-Bolt Square Bend

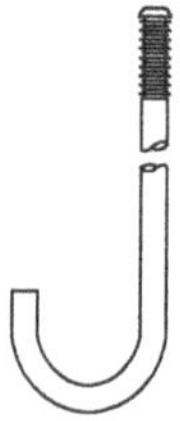

Hook Bolt Round Bend

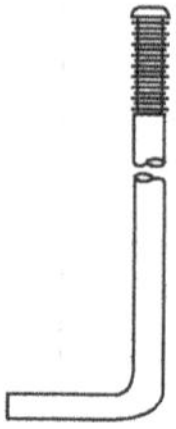

Right Angle Bend

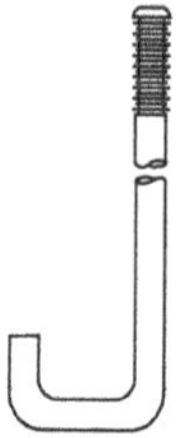

Square Bend

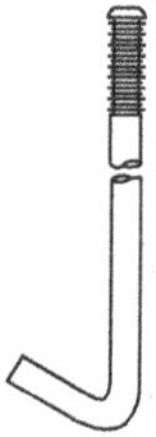

Special Bend

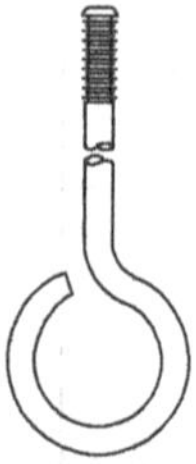

Eye Bolt (Closed)

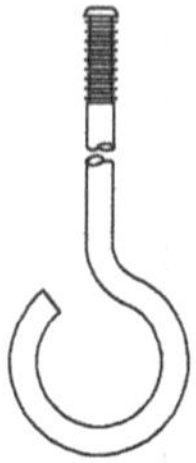

Eye Bolt (Open)

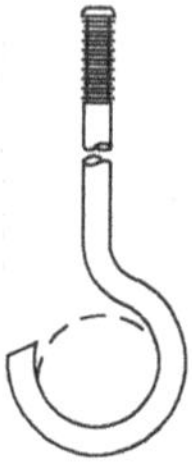

J-Bolt

SILL PLATE ANCHORS

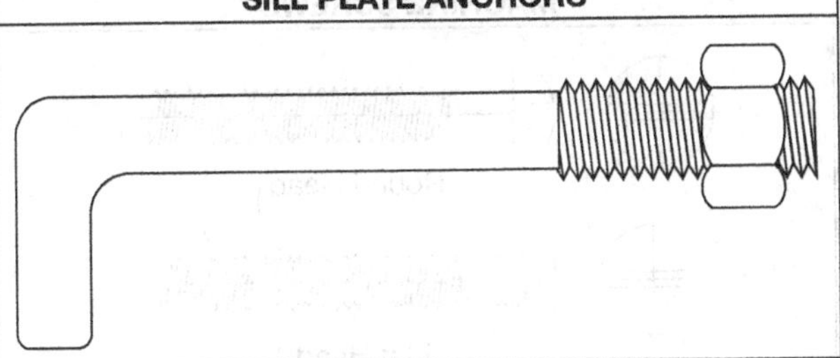

Diameter Inches	Length Inches
½	6, 8, 10, 12, 14
⅝	6, 8, 10, 12, 14, 16, 18, 20
¾	8, 9, 10, 12, 14, 16, 18, 20, 24
⅞ and 1	6, 8, 9, 10, 12, 14, 16, 18, 20, 24
1⅛ and 1¼	18, 20, 24

CARRIAGE BOLT

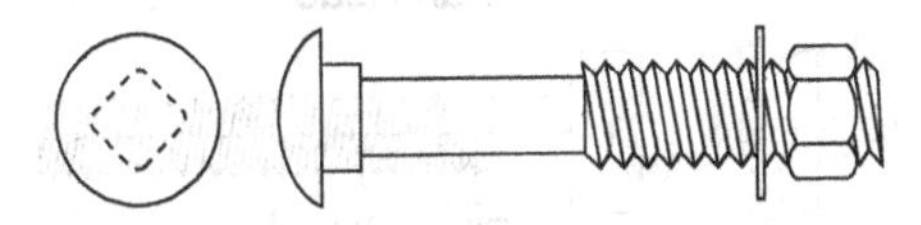

COMMONLY AVAILABLE SIZES OF CARRIAGE BOLTS

Diameter Inches	Lengths Inches
¼	½ to 8
5/16	⅝ to 12
⅜	¾ to 12
7/16	¾ to 8
½	1 to 12
⅝	1½ to 12
¾	2 to 12

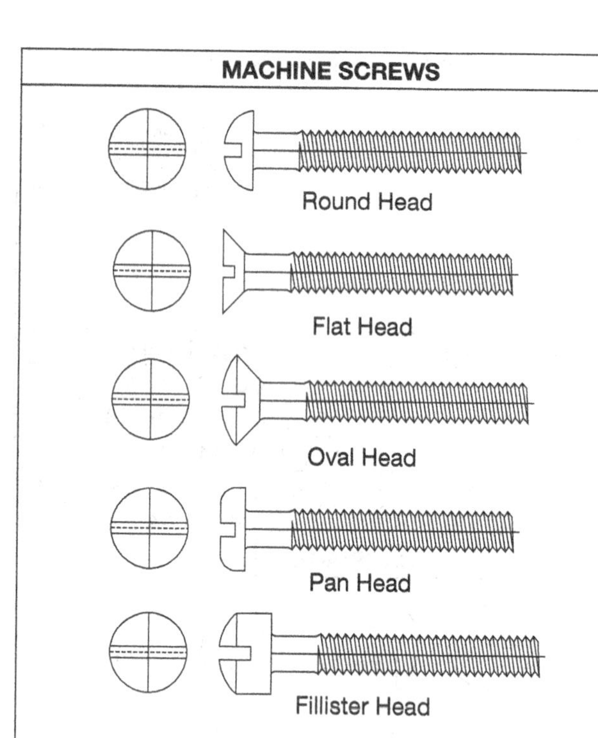

COMMONLY AVAILABLE SIZES OF MACHINE SCREWS

Diameter Inches	Length Inches		
	Round and Flat Head	Oval and Fillister Head	Pan Head
¼	5⁄16 to 6	5⁄16 to 3	¼ to 6
5⁄16	⅜ to 6	⅜ to 3	½ to 6
⅜	½ to 6	½ to 3	½ to 6

Length intervals: 1⁄16 in. up to ½", ⅛ in. ⅝" to 1¼", ¼ in. 1½" to 3", ½ in. 3½" to 6".

COMMON TYPES AND SIZES OF BOLTS

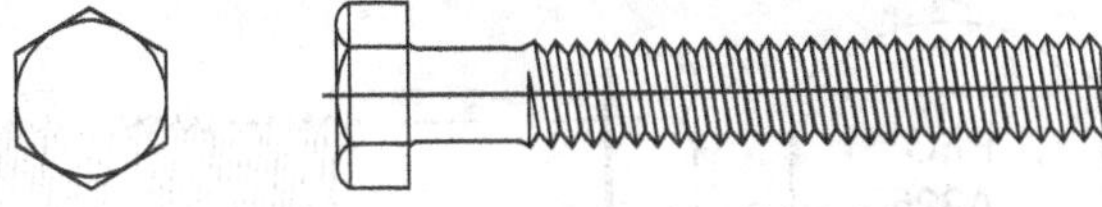

Regular Hex Head Bolt

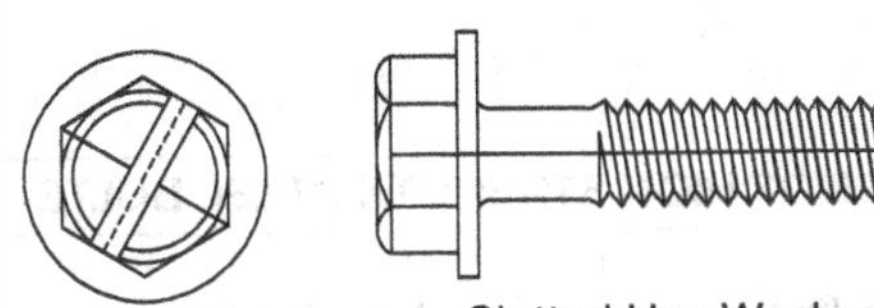

Slotted Hex Washer Head Bolt

COMMONLY AVAILABLE SIZES OF REGULAR STEEL BOLTS

Diameter, in.	Lengths, in.*
1/4	1/2 to 8
5/6	1/2 to 8 and 8 1/2 to 10
3/8	1/2 to 8 and 8 1/2 to 11 1/2
7/16	3/4 to 8 and 8 1/2 to 11 1/2
1/2	3/4 to 8 and 8 1/2 to 11 1/2
5/8	3/4 to 8 and 8 1/2 to 11 1/2
3/4	1 to 8 and 8 1/2 to 11 1/2
7/8	1 1/4 to 8 and 8 1/2 to 11 1/2
1	1 1/2 to 8 and 8 1/2 to 11 1/2
1 1/8	1 1/2 to 8 and 8 1/2 to 11 1/2
1 1/4	1 1/2 to 8 and 8 1/2 to 11 1/2

*Bolt lengths vary by 1/4 inch from 1/2-inch to 8-inch lengths and 1/2 inch from 8 1/2-inch to 11 1/2-inch lengths.

HIGH STRENGTH STEEL BOLT

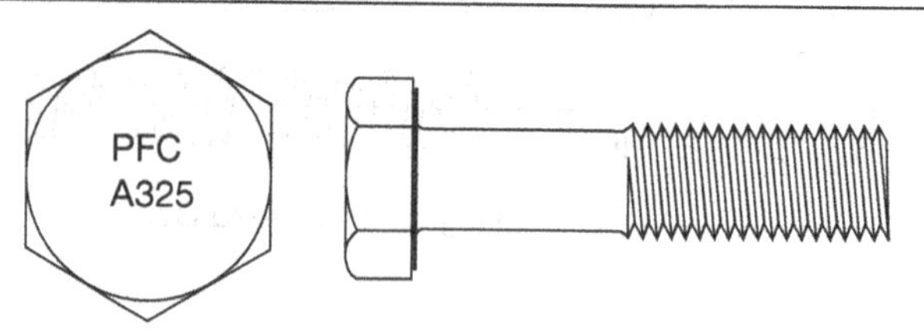

TYPICAL SIZES OF HIGH STRENGTH STEEL BOLTS

Diameter (in.)	Lengths (in.)
½	1¼ to 4
⅝	1½ to 8
¾	1½ to 9
⅞	2 to 8
1	2 to 8

Courtesy American Society for Testing and Materials, www.astm.org

ASTM A325 BOLT AND NUT IDENTIFICATION

Three Radial Lines at 60°

Three Radial Lines at 120°, Optional

A325 A325 A325

Type 3 Type 2 Type 1

Bolt Head Marking

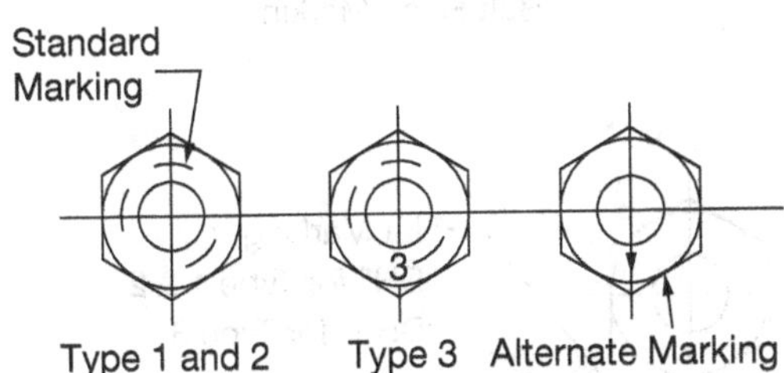

Nut Marking

Courtesy American Society for Testing and Materials, *www.astm.org*

ASTM A490 BOLT AND NUT IDENTIFICATION

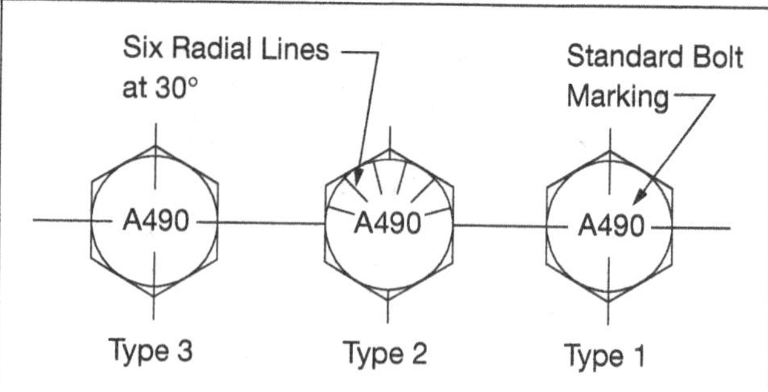

Bolt Head Marking

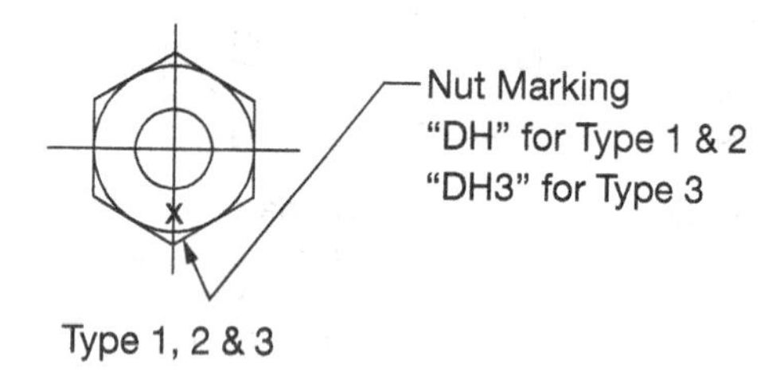

Nut Marking

Courtesy American Society for Testing and Materials, www.astm.org

SCREW TYPE FASTENERS

TYPES OF WOOD SCREW HEADS

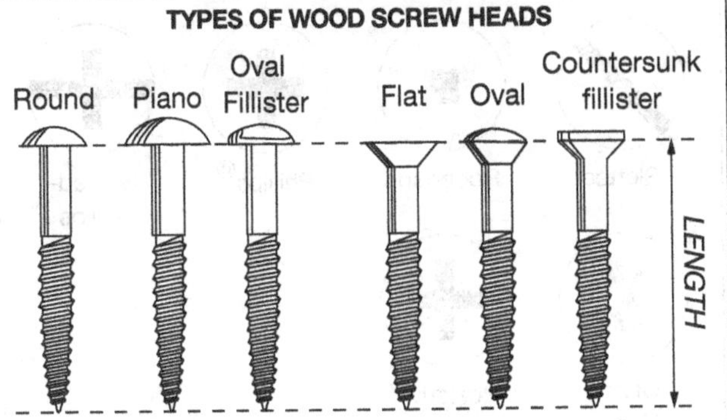

STANDARD WOOD SCREW LENGTHS AND GAUGES

Lengths Inches	Screw Gauge or Number
½	2 to 8
⅝	3 to 10
¾	4 to 11
⅞	6 to 12
1	6 to 14
1¼	7 to 17
1½	6 to 18
1¾	8 to 20
2	8 to 20
2¼	9 to 20
2½	12 to 20
2¾	14 to 20
3	16 to 20
3½	18 to 20
4	18 to 20

RECESSES AVAILABLE IN WOOD SCREW HEADS

Slotted

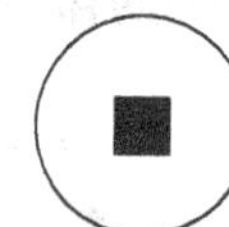

Robertson

Phillips®

Slotted–Phillips

Clutch Head

Pozidriv®

WOOD SCREW DIAMETERS IN INCHES AND BY GAUGE NUMBER

Wire Gauge Number	Decimal Size	Fractional Size
0	0.060	1/16
1	0.073	5/64
2	0.086	3/32
3	0.099	7/64
4	0.112	1/8
5	0.125	1/8
6	0.138	9/64
7	0.151	5/32
8	0.164	11/64
9	0.177	3/16
10	0.190	13/64
11	0.203	13/64
12	0.216	7/32
14	0.242	1/4
16	0.268	9/32
18	0.294	19/64
20	0.320	21/64

SCREW DIAMETERS AND DRILLS TO USE					
Screw Gauge	Screw Dia.	Flat or Oval Head Dia.	Shank Clearance Hold Dia.	Thread Pilot Hole Diameter	
				Softwood	Hardwood
2	.086	.164	3/32	1/32	3/64
3	.099	.190	7/64	3/64	1/16
4	.112	.216	7/64	3/64	1/16
5	.125	.242	1/8	1/16	5/64
6	.138	.268	9/64	1/16	5/64
7	.151	.294	5/32	1/16	3/32
8	.164	.320	11/64	5/64	3/32
9	.177	.346	3/16	5/64	7/64
10	.190	.371	3/16	3/32	7/64
11	.203	.398	13/64	3/32	1/8
12	.216	.424	7/32	7/64	1/8
13	.229	.450	15/64	7/64	1/8
14	.242	.476	1/4	7/64	9/64

COUNTERBORE DIAMETERS FOR STANDARD WOOD SCREWS

Screw gauge number	2	3	4	5	6	7	8	9	10	11	12	14	16	18	20
Counterbore diameter inches	3/16	1/4	1/4	1/4	5/16	5/16	3/8	3/8	3/8	7/16	7/16	1/2	9/16	5/8	11/16

SELECTED DRYWALL SCREWS

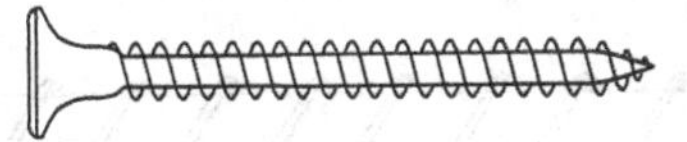

Bugle Head Type S
6 × 1", 6 × 1 1/4", 6 × 1 1/2", 6 × 1 5/8"

To Steel Framing

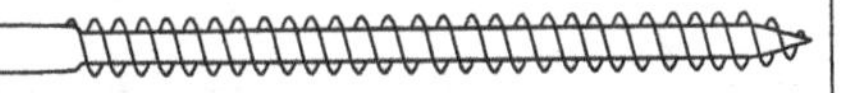

Bugle Head Type S
6 × 2", 6 × 2 1/4", 7 × 2 1/2", 8 × 3"

Multiple Layers to Steel Framing

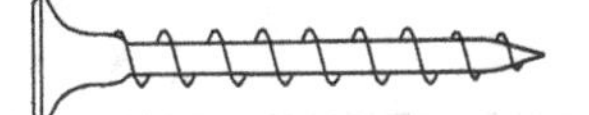

Bugle Head Type W
6 × 1 1/4"

To Wood Framing

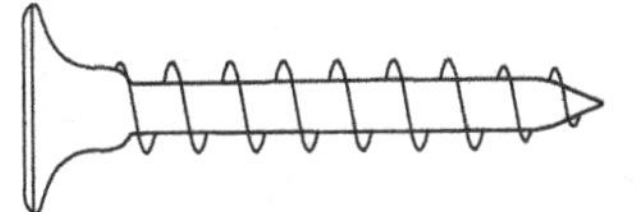

Bugle Head Laminating
13 × 1 1/2"

Laminating Gypsum to Gypsum

PARTICLEBOARD SCREWS

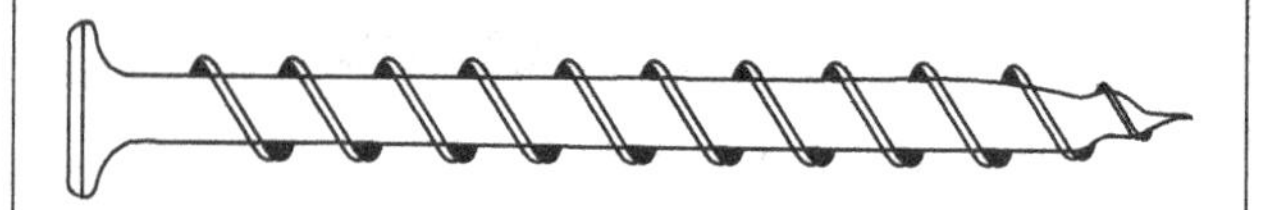

Lengths (inches) 1, 1¼, 1½, 1⅝, 2, 2¼, 2½, 3

CEMENT BOARD SCREWS

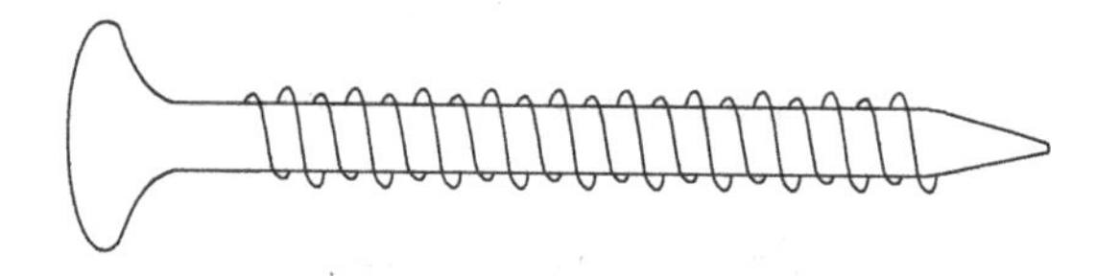

Diameter Inches	Lengths Inches
No. 8 and No. 9	1¼, 1⅝, 2¼

LAG SCREW

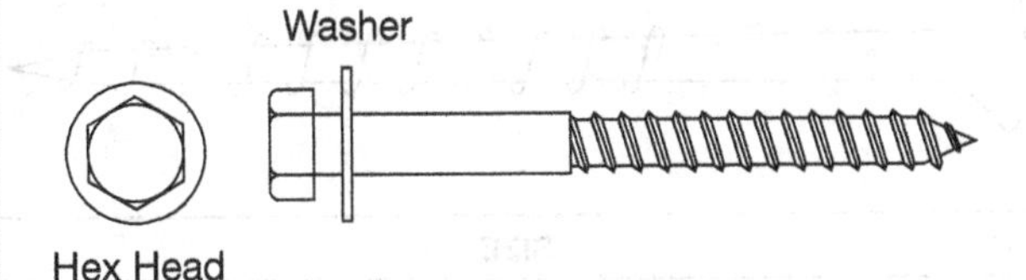

TYPICAL SIZES OF STANDARD LAG SCREWS	
Length* Inches	**Diameter Inches**
1 1½	¼ to ½**
2 2½ 3	¼ to ⅝**
3	¼ to ½** to 1¼***
4 5 6	¼ to ½** ⅝ to 1¼***

*Availiable in 1" to 16" lengths

**in 1/16" increments

***in ⅛" increments

WOOD DECKING SCREWS
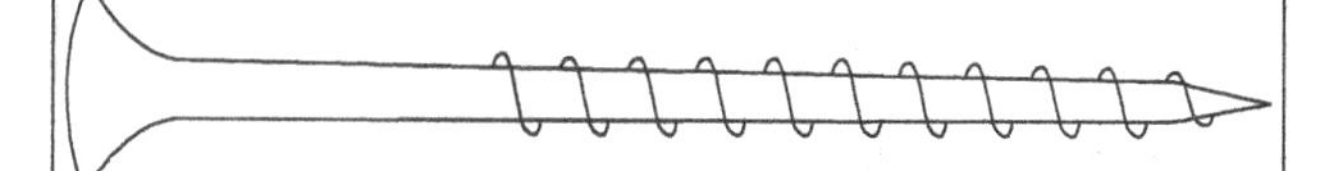
SIZE
6 × 1" 6 × 1¼" 6 × 1⅝" 6 × 2" 8 × 2½" 8 × 3" 10 × 3½" 10 × 4" 10 × 5" 10 × 6"

METALS USED IN SCREWS	
Zinc-plated steel	Most common in grades rated from 1 to 8, with grade 8 being the strongest.
Stainless steel	Better corrosion resistance.
Brass, bronze	Nonferrous metals.
Cadmium, chrome or nickel plated	For corrosion resistance and appearance.

COMMON SIZES OF SCREW EYES

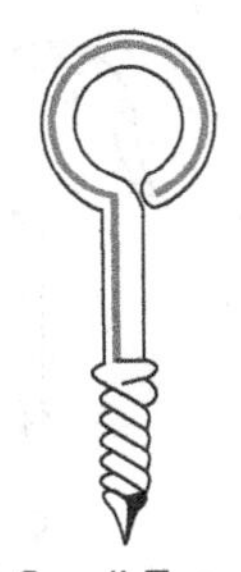

Medium Eye   Small Eye

SMALL SCREW EYE		MEDIUM SCREW EYE	
Eye Number	**Length Inches**	**Eye Number**	**Length Inches**
206	$1\frac{1}{2}$	0	$2\frac{7}{8}$
208	$1\frac{3}{8}$	4	$2\frac{3}{16}$
210	$1\frac{3}{16}$	6	$1\frac{15}{16}$
212	$\frac{15}{16}$	8	$1\frac{5}{8}$
214	$\frac{13}{16}$		
216	$\frac{11}{16}$		

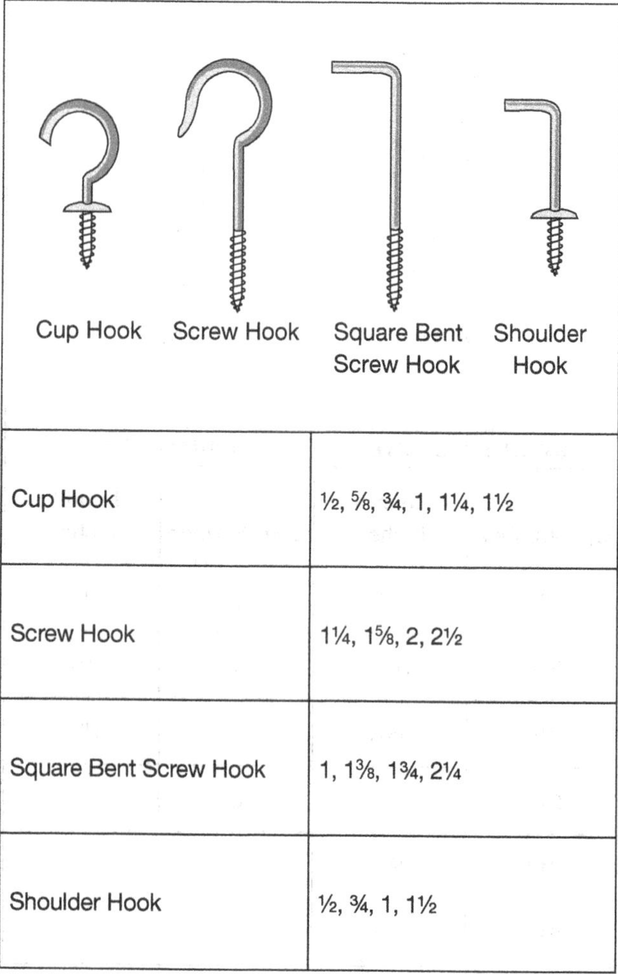

COMMON SIZES OF HOOK FASTENERS

Hook	Sizes
Cup Hook	½, ⅝, ¾, 1, 1¼, 1½
Screw Hook	1¼, 1⅝, 2, 2½
Square Bent Screw Hook	1, 1⅜, 1¾, 2¼
Shoulder Hook	½, ¾, 1, 1½

SHEET METAL SCREWS

Gimlet Point

Blunt Point

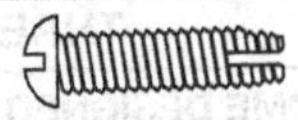

Cutting Slot Point

SHEET METAL SCREW SIZE AND INSTALLATION DATA

Screw Size Number	Screw Diameter (inches)	Inch Diameter of Drilled Hole	Number of Drill to Produce Hole	Diameter of Punched Hole (inches)	Gauge of Metal to be Secured
#4	.112	.084	#44	.086	28
			#44	.086	26
			#42	.093	24
			#42	.098	22
			#40	.100	20
#6	.137	.102	#39	.111	28
			#39	.111	26
			#39	.111	24
			#38	.111	22
			#36	.111	20
#7	.155	.116	#37	.121	28
			#37	.121	26
			#35	.121	24
			#33	.121	22
			#32	.121	20
			#31	–	18
#8	.164	.119	#33	.137	26
			#33	.137	24
			#32	.137	22
			#31	.137	20
			#30	–	18
#10	.188	.138	#30	.158	26
			#30	.158	24
			#30	.158	22
			#29	.158	20
			#25	.158	18
#12	.216	.161	#26	–	24
			#25	.185	22
			#24	.185	20
			#22	.185	18
#14	.246	.189	#15	–	24
			#12	.212	22
			#11	.212	20
			#9	.212	18

TYPICAL THREAD-FORMING AND THREAD-CUTTING SCREW SIZES	
SOME DESIGNED FOR USE IN WOOD AND OTHERS IN STEEL	
Size	**Drilling Thickness**
#6-20 × ⅜"	.036"–.100"
#6-20 × ½"	.036"–.100"
#6-20 × ¾"	.036"–.100"
#6-20 × 1"	.036"–.100"
#8-18 × ½"	.036"–.100"
#8-18 × ¾"	.036"–.100"
#8-18 × 1"	.036"–.100"
#8-18 × 1¼"	.036"–.100"
#8-18 × 1½"	.036"–.100"
#8-18 × 2"	.036"–.100"
#10-16 × ½"	.036"–.175"
#10-16 × ⅝"	.036"–.175"
#10-16 × ¾"	.036"–.175"
#10-16 × 1"	.036"–.175"
#10-16 × 1¼"	.036"–.175"
#10-16 × 1½"	.036"–.175"

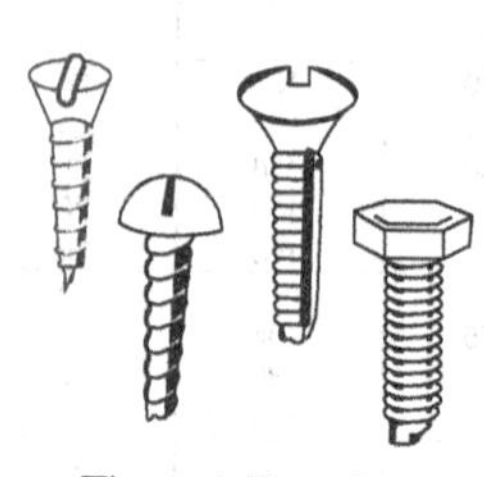

Thread-Forming Screws

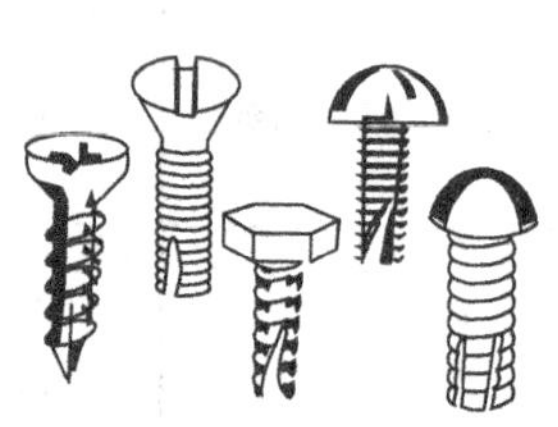

Thread-Cutting Screws

TYPICAL SCREWS USED IN LIGHT METAL BUILDING CONSTRUCTION

Other types and sizes are available. Consult the manufacturers.

SHEET METAL TO STEEL FRAMING (WITHOUT SEALING WASHER)

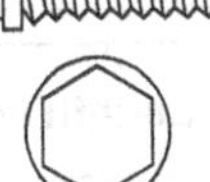

Hex Head

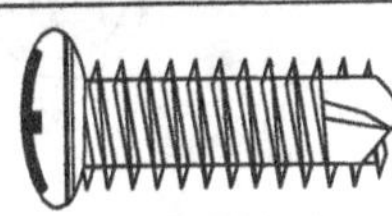

Pan Head

Total Steel Thickness (in.)	Typical Lengths (in.)
.036 to .100	½, ⅝, ¾, 1, 1½

SHEET METAL TO STEEL FRAMING (WITH SEALING WASHER)

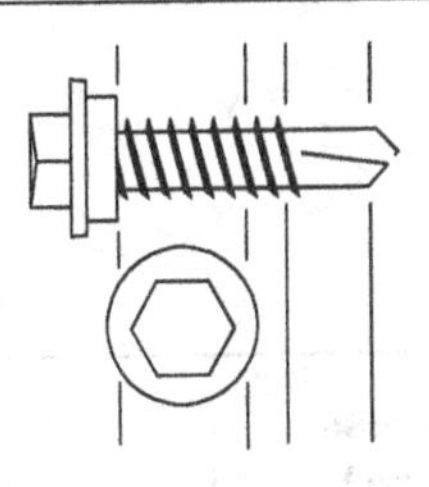

Hex Head

Total Steel Thickness (in.)	Typical Lengths (in.)
.036 to .210	1, 1¼

METAL TO WOOD WITH WASHER *(cont.)*	
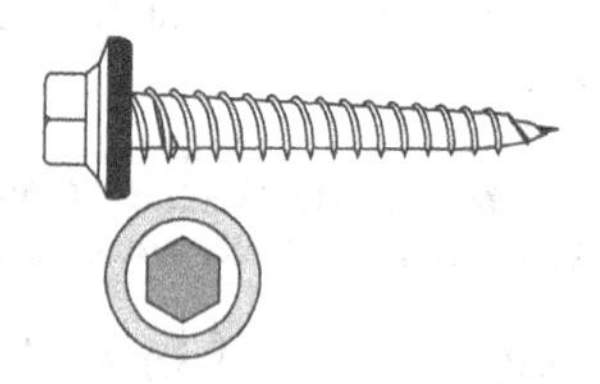	
Total Steel Thickness	**Length (in.)**
Up to 18 gauge	1, 1½, 2, 2½

WOOD TO METAL	
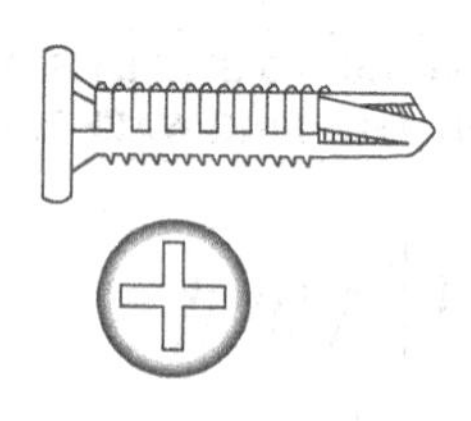	
Typical Thickness Allowances (in.)	**Length (in.)**
¼ to ⅝ wood to .036 to .210 thick steel	1

NAILS FOR EXTERIOR APPLICATIONS
[Hot-dipped zinc-coated twice in molten zino]

SIDING NAILS

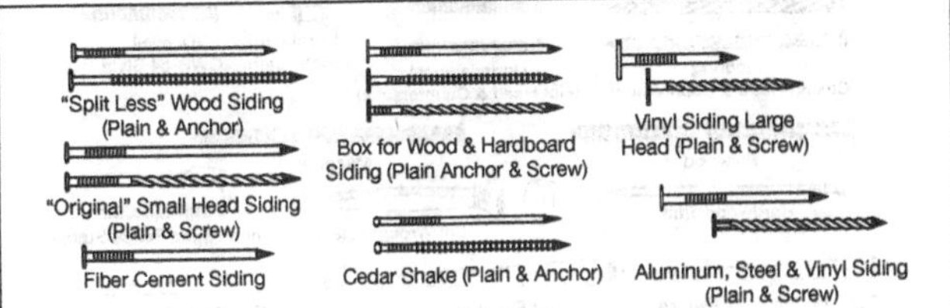

ROOFING, GUTTERING & SPIKES

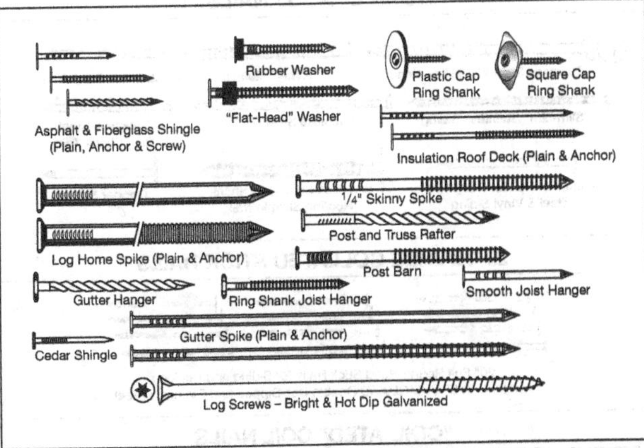

DECKING, TRIM NAILS & SCREWS

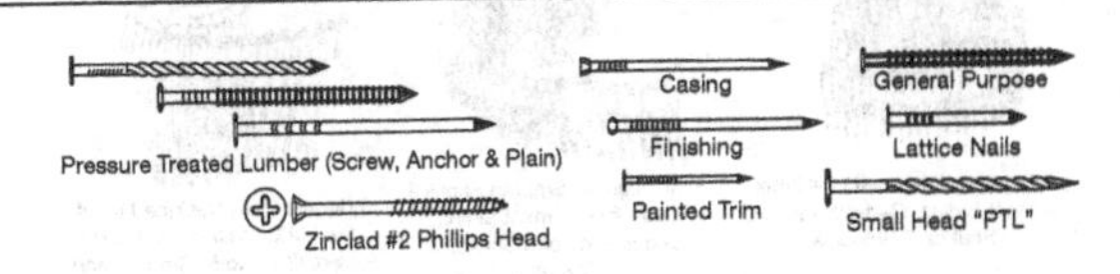

NAILS FOR EXTERIOR APPLICATIONS *(cont.)*

INTERIOR & OTHER NAILS

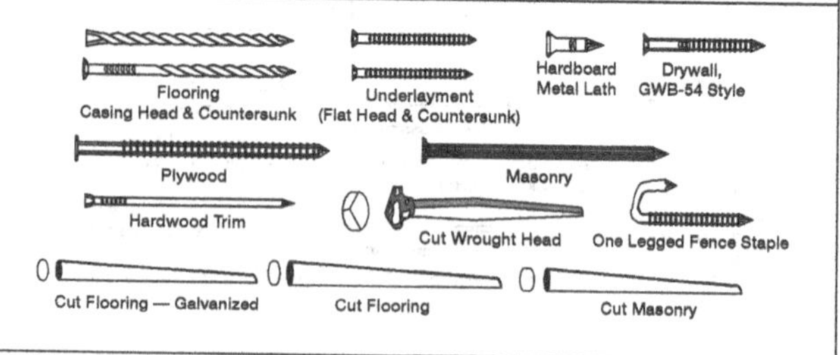

NON-FERROUS NAILS

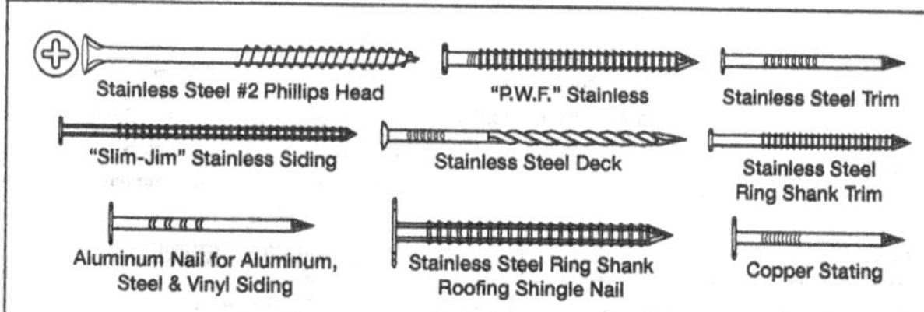

20 DEGREE COLLATED STICK NAILS

20" Full Round Head Stick Nails for Siding and Decking
Available in STORMGUARD Double Hot-Dipped and Stainless Steel

"COIL-ATED" COIL NAILS

0 Degree - Duo-Fast Coil Fiber Cement Siding, Cedar/Redwood Shakes & Shingles

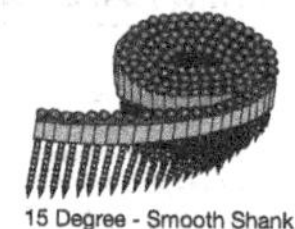

15 Degree - Smooth Shank Coil Fiber Cement Siding, Cedar/Redwood Shakes & Shingles

0 Degree Coils Designed to Fit the MAX. Multi-Impact Coil Nailers ($3^1/2$" to 5" Smooth and Ring Shanks)

Courtesy W.H. Maze Company, www.mazenails.com

TYPES OF NAIL SHANKS

Screw Thread

Ring Shank

Smooth Shank

COMMON NAILS

SIZES AND NUMBER OF NAILS PER POUND

Common Nail (General Construction)

Length	Penny	Gauge	Approx. Nails/lb
1	2	15	875
1¼	3	14	585
1½	4	12½	315
1¾	5	12½	270
2	6	11½	180
2¼	7	11½	160
2½	8	10¼	105
2¾	9	10¼	95
3	10	9	70
3¼	12	9	65
3½	16	8	50
4	20	6	30

BOX AND CASING NAILS

SIZES AND NUMBER OF NAILS PER POUND

Box Nail (Light Construction)

Casing Nail (Interior Trim)

Length	Penny	Gauge	Approx. Nails/lb
1	2	15½	1010
1¼	3	14½	635
1½	4	14	470
1¾	5	14	405
2	6	12½	235
2¼	7	12½	210
2½	8	11½	145
2¾	9	11½	130
3	10	10½	95
3¼	12	10½	86
3½	16	10	70
4	20	9	50

FINISHING NAILS

SIZE AND NUMBER OF NAILS PER POUND

Finishing Nail (Trim and Cabinets)

Length	Penny	Gauge	Approx. Nails/ lb
1	2	16½	135
1¼	3	15½	305
1½	4	15	545
1¾	5	15	500
2	6	13	310
2¼	7	12½	235
2½	8	12½	190
2¾	9	12½	170
3	10	11½	120
3¼	12	11½	110
3½	16	11	90
4	20	10	60

WIRE BRADS

SIZES AND GAUGES

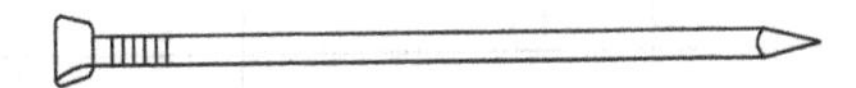

Wire Brads (Light Material)

Length (inches)	Gauge Number
½	20
¾	20
1	18
1¼	16
1½	14

SIZES AND GAUGES OF DUPLEX NAILS		
Length (inches)	**Penny**	**Gauge**
1¼	5	12½
2	6	11½
2¼	7	11½
2½	8	10¼
2¾	9	10¼
3	10	9
3¼	12	9
3½	16	8
4	20	6

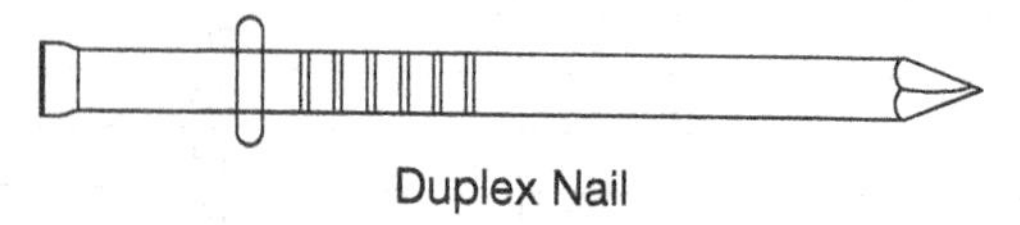

Duplex Nail

ANNULARLY AND HELICALLY THREADED NAILS

Sizes and Diameters of Helically and Annularly Threaded Nails.

Penny	Length Inches	Diameter Inches
6d	2	0.120
8d	2½	0.120
10d	3	0.135
12d	3¼	0.135
16d	3½	0.148
20d	4	0.177
30d	4½	0.177
40d	5	0.177
50d	5½	0.177
60d	6	0.177

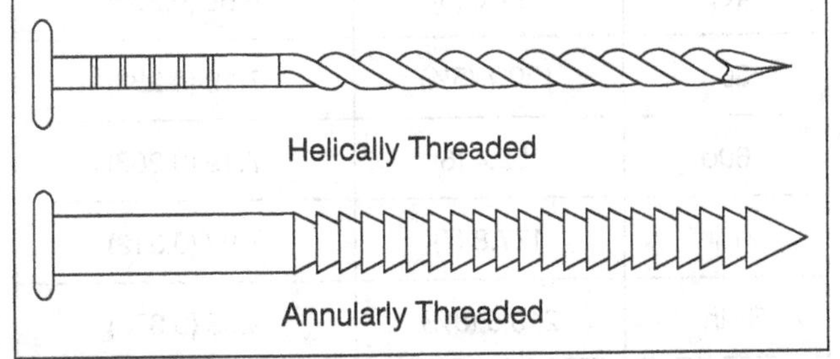

TYPICAL SPIKE SIZES

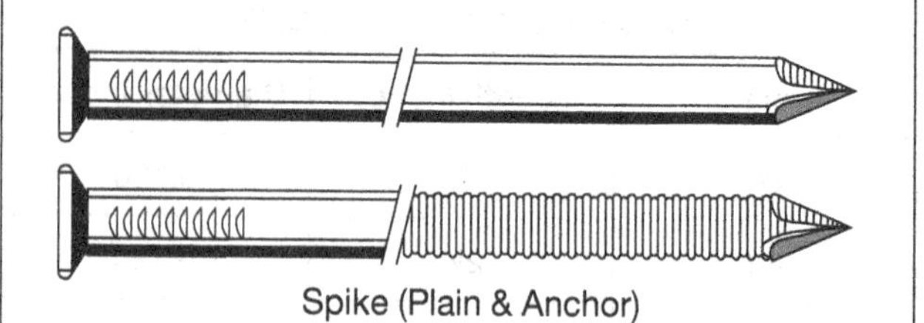

Spike (Plain & Anchor)

Size	Length (mm (in.))	Diameter (mm (in.))
10d	76.2 (3)	4.88 (0.192)
12d	82.6 (3¼)	4.88 (0.192)
16d	88.9 (3½)	5.26 (0.207)
20d	101.6 (4)	5.72 (0.225)
30d	114.3 (4½)	6.20 (0.244)
40d	127.0 (5)	6.68 (0.263)
50d	139.7 (5½)	7.19 (0.283)
60d	152.4 (6)	7.19 (0.283)
5⁄16 in.	177.8 (7)	7.92 (0.312)
⅜ in.	215.9 (8½)	9.53 (0.375)

NAILS COMMONLY USED TO SECURE GYPSUM PANELS TO WOOD FRAMING

Annular Ring

Cupped Head

Cooler

Color Pin (Predecorated Wallboard)

ANNULAR DRYWALL NAIL

Length (inch)	
1¼	Single-layer panel application requires 2000 nails per 1000 square ft.
1⅜	Double-layer panel application requires 1000 nails per 1000 square feet.
1½	
$1\frac{1}{58}$	

NAILS FOR JOIST HANGERS GALVANIZED RING SHANK	
Lengths (in.)	**Gauge**
1¼, 1½, 1¾, 2, 2½	10

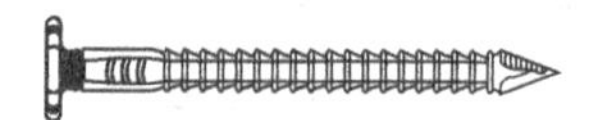

Courtesy W.H. Maze Company, www.mazenails.com

NAILS FOR ASPHALT AND FIBERGLASS SHINGLES SMOOTH AND RING SHAKE	
Lengths (in.)	**Gauge**
⅞, 1, 1¼, 1½, 1¾, 2, 2½	11

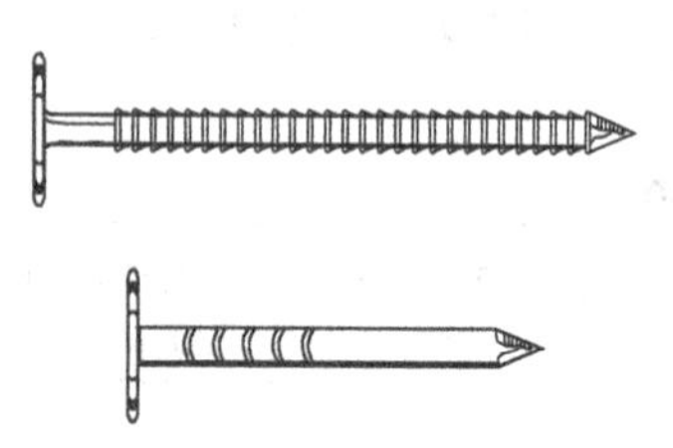

Courtesy W.H. Maze Company, www.mazenails.com

NAILS FOR FIBERGLASS AND METAL ROOFING RING SHANK SILICONE WASHER HEAD	
Lengths (in.)	**Gauge**
1½, 1¾, 2, 2½, 3, 3½	11
4	9

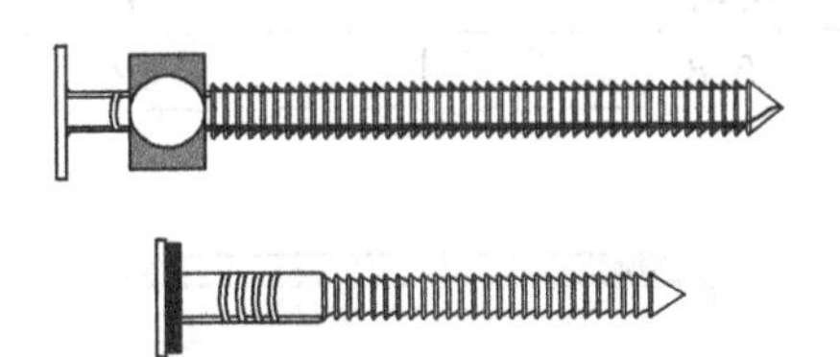

Courtesy W. H. Maze Company, www.mazenails.com

NAILS FOR DECKING SPIRAL SHANK STAINLESS STEEL	
Lengths (in.)	**Gauge**
2, 2½, 3	11½
3½	11

Courtesy W.H. Maze Company, www.mazenails.com

NAILS FOR NEW TREATED LUMBER SMOOTH, RING, AND SPIRAL SHANK STAINLESS STEEL	
Lengths (in.)	**Gauge**
2, 2½	11
3¼	10
3½	9
4, 4½	7
5, 6	5½

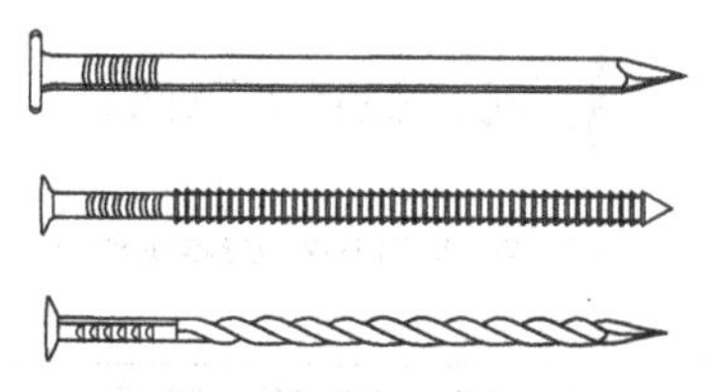

Courtesy W.H. Maze Company, www.mazenails.com

NAILS FOR ALUMINUM AND STEEL SIDING GALVANIZED	
Lengths (in.)	**Gauge**
1¼, 1½, 2	12

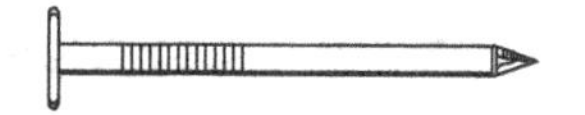

Courtesy W.H. Maze Company, www.mazenails.com

NAILS FOR FIBER CEMENT SIDING SMOOTH AND RING SHANK GALVANIZED	
Lengths (in.)	**Gauge**
2, 2½, 3	.113
3½	.120

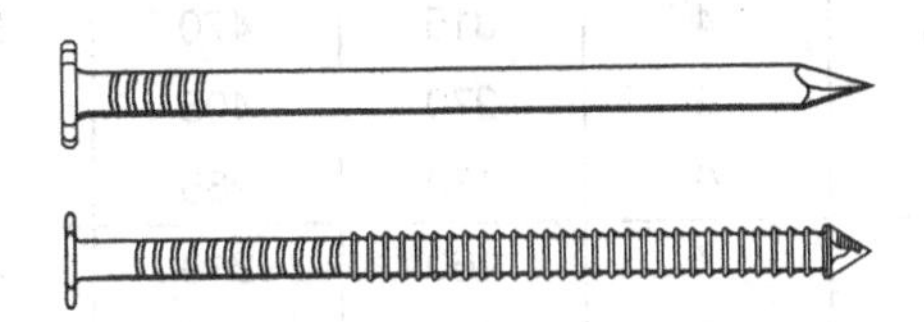

Courtesy W.H. Maze Company, www.mazenails.com

NAILS FOR WOOD SIDING SMOOTH AND RING SHANK GALVANIZED	
Lengths (in.)	**Gauge**
2, 2¼	14
2½, 2¾, 3	13
3¼, 3½, 4, 4½	12

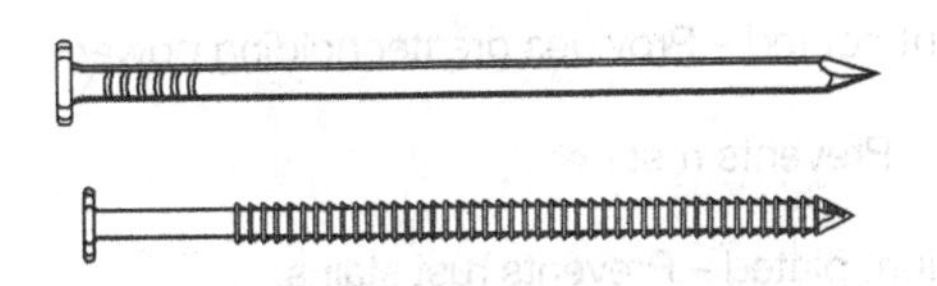

Courtesy W.H. Maze Company, www.mazenails.com

APPROXIMATE NUMBER OF NAILS PER POUND				
Length	Penny	Common Nails	Box and Casing Nails	Finishing Nails
1	2	875	1010	1135
1¼	3	585	635	850
1½	4	315	470	545
1¼	5	270	405	500
2	6	180	235	310
2¼	7	160	210	235
2½	8	105	145	190
2¾	9	95	130	170
3	10	70	95	120
3¼	12	65	86	110
3¼	16	50	70	90
4	20	30	50	60

NAIL FINISHES

Bright – No coating. Used when not exposed to moisture.

Galvanized – Zinc coated. Prevents rust.

Cement coated – Provides greater holding power.

Blued – Prevents rust.

Cadmium plated – Prevents rust stains.

Painted – Color matches material being nailed.

USES FOR NAILS MADE WITH VARIOUS METALS

Mild Steel – General framing

Aluminum – Securing gutters, roofing, aluminum siding

Copper – Areas where corrosion resistance is needed

Brass – Resist corrosion, decorative applications as hardware

Hardened steel – Securing wood flooring, masonry nails, other hard material

STAPLES AND CLEATS

FINE-WIRE STAPLES*

Crown (inches)	Gauge	Lengths (inches)
3⁄16	22	3⁄16, 1⁄4, 3⁄8, 1⁄2, 5⁄8
3⁄8	22	5⁄32, 3⁄16, 1⁄4, 5⁄16, 3⁄8, 7⁄16, 1⁄2, 5⁄8
1⁄2	22	1⁄4, 5⁄16, 3⁄8, 1⁄2

*Typical uses include furniture, interior trim, home insulation, roofing felt.

Other sizes available. Contact a manufacturer.

POWER-DRIVEN STAPLES*		
Crown	**Gauge**	**Lengths (inches)**
3/16	18	½, ⅝, ¾, ⅞, 1, 1⅛, 1¼
¼	18	⅜, 7/16, ½, ⅝, ¾, ⅞, 1, 1⅛, 1¼, 1½
⅜	18	⅜, ½, ⅝, ¾, ⅞, 1, 1¼, 1½
7/16	16	⅞, 1, 1⅛, 1, 1/14, 1⅜, 1½, 1⅝, 1¾, 2
7/16	15	1 ¼, 1 ½, 1¾, 2, 2¼, 2½
½	16	¾, 1, 1¼, 1⅜, 1½, 1¾, 2
5/16	16	⅝, ¾, ⅞, 1, 1⅛, 1¼, 1 13/8, 1½
1	16	⅝, ¾, ⅞, 1, 1⅛, 1¼, 1⅜, 1½

*Typical uses include installing underlayment, sheathing, paneling, molding, trim, soffits, gypsum, siding.

Other sizes available. Contact a manufacturer.

CLEATS AND STAPLES FOR HARDWOOD FLOORING

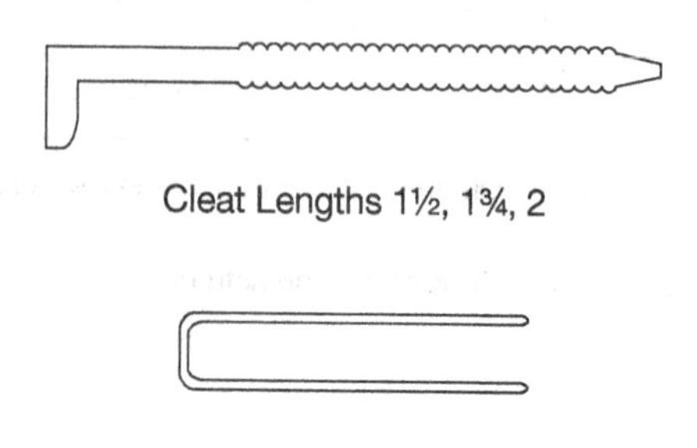

Cleat Lengths 1½, 1¾, 2

Staple Length (inches) 1, Crown 3/16, gauge 19

COMMON TYPES OF RIVETS

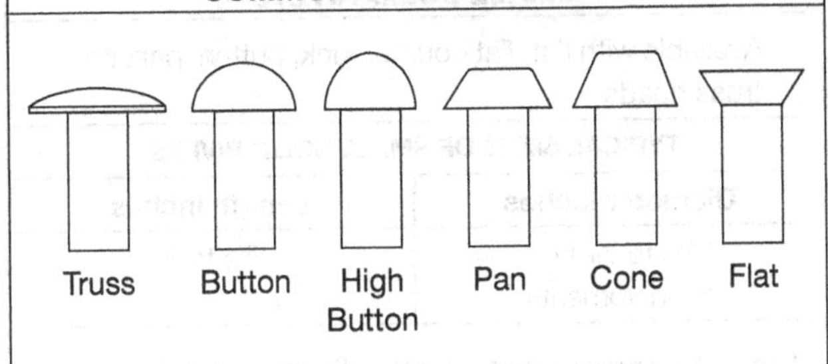

LARGE SOLID RIVETS

Available with button, high button, cone, flat countersunk, oval, countersunk, and pan heads

TYPICAL SIZES OF LARGE BUTTON HEAD SOLID RIVETS

Basic Diameter Inches	Lengths Inches
½	1⅝, 1¾, 1⅞, 2
⅝	1⅞, 2, 2⅛, 2¼
¾	1⅞, 2, 2⅛, 2¼
⅞	2, 2⅛, 2¼, 2⅜
1	2⅛, 2¼, 2⅜, 2½
1⅛	2¾, 2⅞, 3, 3⅛, 3⅜, 3½, 3¾, 3⅞
1¼	2⅞, 3, 3⅛, 3¼, 3½, 3¾, 3⅞

Other lengths and diameters are available. Rivets with other types of heads have similar diameters and lengths.

SMALL SOLID RIVETS	
Available with flat, flat countersunk, button, pan, and truss heads	
TYPICAL SIZES OF SMALL SOLID RIVETS	
Diameter Inches	**Length Inches**
1/16 to 7/16 in. 1/32 increments	3/16 to 3
See manufacturers' catalogs for specific size information.	

GRIP REQUIRED DETERMINES THE LENGTH OF THE RIVET USED

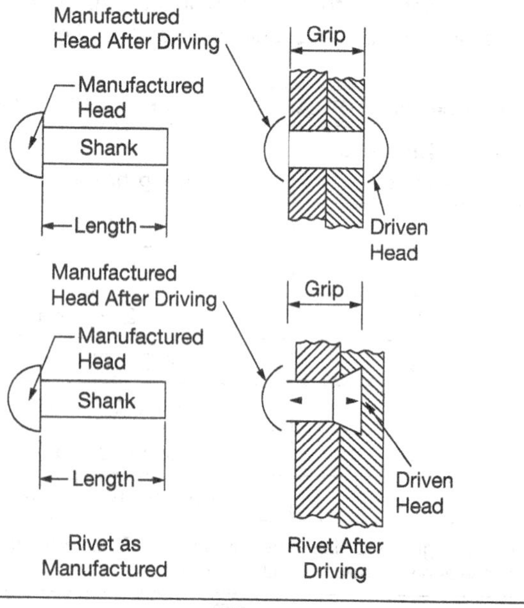

METAL DRIVE RIVET

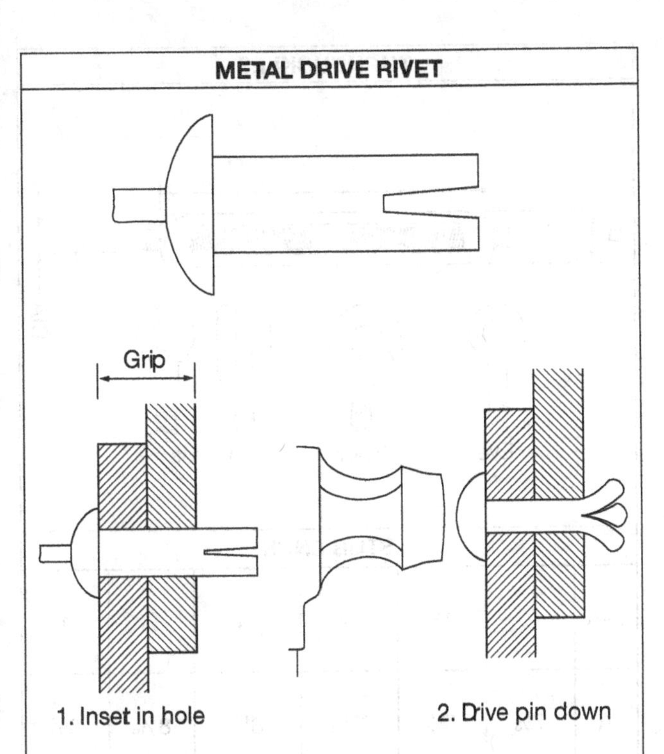

Rivet Size Inches	Grip Length	
	Min.	Max.
3/16 × 9/32	.076	.175
3/16 × 25/64	.176	.275
3/16 × 31/64	.276	.375
3/16 × 37/64	.376	.475
3/16 × 11/16	.476	.575
3/16 × 25/32	.576	.675
3/16 × 7/8	.676	.775

TURNBUCKLES

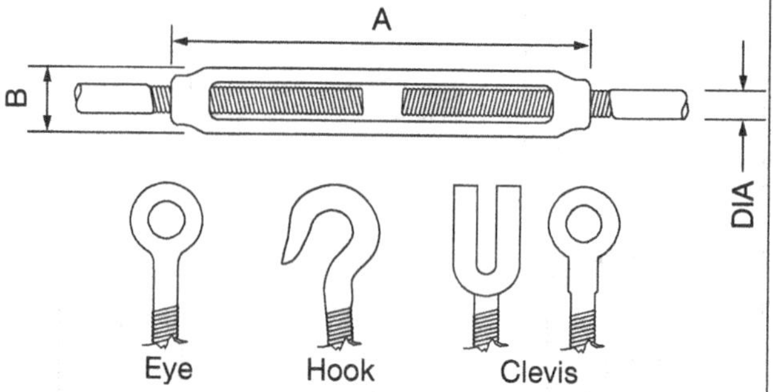

A

B

DIA

Eye Hook Clevis

STUB ENDS

DIA	3/8	1/2	5/8	3/4	7/8	1
A	7 1/8	7 1/2	7 13/16	8 1/8	8 7/16	8 3/4
B	1 1/32	1 5/16	1 1/2	1 23/32	1 7/8	2 1/32

Turnbuckles with diameters up to one inch. Larger diameter turnbuckles are available.

FLAT STEEL WASHERS

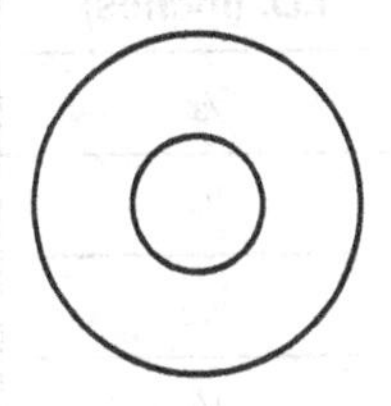

U.S. STANDARD FLAT STEEL WASHERS

Bolt Size	I.D. (inches)	O.D. (inches)	Bolt Size	I.D. (inches)	O.D. (inches)
3⁄16	1⁄4	9⁄16	1 1⁄8	1 1⁄4	2 3⁄4
1⁄4	5⁄16	3⁄4	1 1⁄4	1 3⁄8	3
5⁄16	3⁄8	7⁄8	1 3⁄8	1 1⁄2	3 1⁄4
3⁄8	7⁄16	1	1 1⁄2	1 5⁄8	3 1⁄2
7⁄16	1⁄2	1 1⁄4	1 5⁄8	1 3⁄4	3 3⁄4
1⁄2	9⁄16	1 3⁄8	1 3⁄4	1 7⁄8	4
9⁄16	5⁄8	1 1⁄2	1 7⁄8	2	4 1⁄4
5⁄8	11⁄16	1 3⁄4	2	2 1⁄8	4 1⁄2
3⁄4	13⁄16	2	2 1⁄4	2 3⁄8	4 3⁄4
7⁄8	15⁄16	2 1⁄4	2 1⁄2	2 5⁄8	5
1	1 1⁄16	2 1⁄2	3	3 1⁄8	5 1⁄2

S.A.E. FLAT STEEL WASHERS		
Bolt Size	**I.D. (inches)**	**O.D. (inches)**
6 1/8	5/32	3/8
8 5/32	3/16	7/16
10 3/16	7/32	1/2
12 7/32	1/4	9/16
1/4	9/32	5/8
5/16	11/32	11/16
3/8	13/32	13/16
7/16	15/32	59/64
1/2	17/32	1 1/16
9/16	19/32	1 3/16
5/8	21/32	1 5/16
3/4	13/16	1 1/2
7/8	15/16	1 3/4
1	1 1/6	2
1 1/8	1 3/16	2 1/4
1 1/4	1 5/16	2 1/2
1 3/8	1 7/16	2 3/4
1 1/2	1 9/16	3

FENDER FLAT STEEL WASHERS		
Bolt Size	**I.D. (inches)**	**O.D. (inches)**
3/16	7/32	1 1/8
3/16	7/32	1 1/4
1/4	9/32	1
1/4	17/64	1 1/8
1/4	9/32	1 1/4
1/4	5/16	1 1/4
1/4	9/32	1 1/2
5/16	3/8	1 1/4
5/16	11/32	1 1/4
5/16	11/32	1 1/2
3/8	25/64	1 1/2
3/8	13/32	1 1/2
1/2	17/32	1 1/2
1/2	17/32	2

SPLIT REGULAR LOCK WASHER

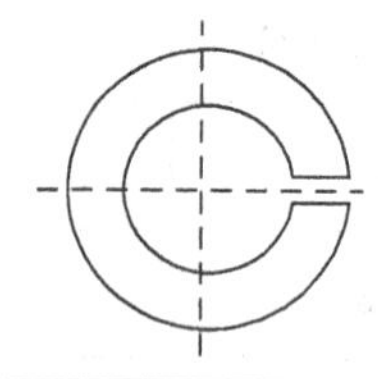

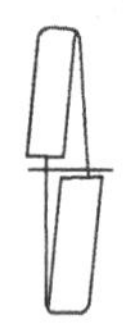

Nominal Washer Size		I.D. (inches)		O.D. (inches)
		Min.	Max.	Max.
	0.086	0.088	0.094	0.172
	0.099	0.101	0.107	0.195
	0.112	0.115	0.121	0.209
	0.125	0.128	0.134	0.236
	0.138	0.141	0.148	0.250
	0.164	0.168	0.175	0.293
	0.190	0.194	0.202	0.334
	0.216	0.221	0.229	0.377
1/4	0.250	0.255	0.263	0.489
5/16	0.312	0.318	0.328	0.586
3/8	0.375	0.382	0.393	0.683
7/16	0.438	0.446	0.459	0.779
1/2	0.500	0.509	0.523	0.873
9/16	0.562	0.572	0.587	0.971
5/8	0.625	0.636	0.653	1.079
11/16	0.688	0.700	0.718	1.176
3/4	0.750	0.763	0.783	1.271
13/16	0.812	0.826	0.847	1.367
7/8	0.875	0.890	0.912	1.464
15/16	0.938	0.954	0.978	1.560
	1.000	1.017	1.042	1.661
1 1/16	1.062	1.080	1.107	1.756
1 1/8	1.125	1.144	1.172	1.853
1 3/16	1.188	1.208	1/237	1.950
1 1/4	1.250	1.271	1.302	2.045
1 5/16	1.312	1.334	1.366	2.141
1 3/8	1.375	1.398	1.432	2.239
1 7/16	1.438	1.462	1.497	2.334
1 1/2	1.500	1.525	1.561	2.430

WELDING SYMBOLS USED ON ARCHITECTURAL DRAWINGS

				Or Butt Joints					
Back Weld	Fillet Weld 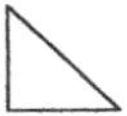	Plug or Slot 	Groove Square 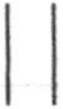	V 	Bevel 	U 	J 	Flare V 	Flare Bevel 

SUPPLEMENT WELD SYMBOLS

Backing	Spacer	Weld-All-Around	Field-Weld	Flush	Onvex
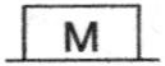 M	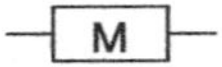 M				

MINIMUM LEG DIMENSIONS OF FILLET WELDS	
Base-Metal Thickness, in.	**Weld Size, in.**
To ¼ inch	⅛
Over ¼ to ½	3/16
Over ½ to ¾	¼
Over ¾ to 1½	5/16
Over 1½ to 2¼	⅜
Over 2¼ to 6	½
Over 6	⅝

BASIC TYPES OF WELDS

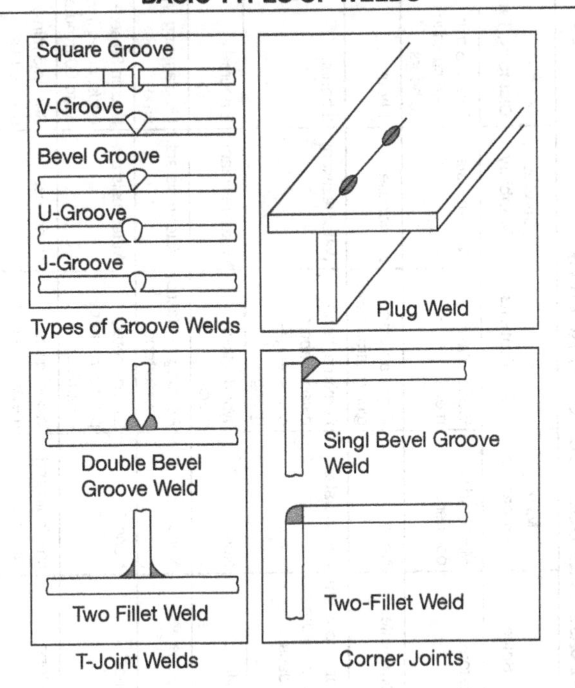

Types of Groove Welds

T-Joint Welds

Corner Joints

BONDING AGENTS

Various types of bonding agents are widely used in the construction industry. Glues are made from natural materials such as animal and vegetable products. Cements are rubber-based materials suspended in a liquid. Adhesives are synthetic materials.

FREQUENTLY USED BONDING AGENTS

Type	Form	Solvent	Mixing Procedure	Applications	Service Durability	Clamping Time
Acrylic	Liquid and powder	Acetone	Mix liquid w/ powder	Wood, glass, metal	Water-resistant	Sets in 5 minutes, cures in 12 hours
Aliphatic resin	Liquid (yellow glue)	Warm water	None	Edge- and face-gluing, laminating	Interior use	1 hour
Casein glue	Powder	Soap and warm water before it hardens	Mix w/water	Furniture, laminated timbers, doors, edge-gluing	Water-resistant	2 hours
Cellulose Cement	Clear liquid	Acetone	None	Wood, glass, metal	Water-resistant	2 hours
Contact cement	Liquid	Acetone	None	Bonding to plastic countertops	Water-resistant	No clamping time
Cyanoacrylate (Superglue)	Liquid or gel	None	None	Liquid-metal, plastic, rubber, ceramics	Water-resistant	Sets in 30 seconds, cures in 30 to 60 minutes
Epoxy	Liquid and a catalyst	Acetone	Mix liquid and catalyst	Wood, glass, metal, ceramics	Interior use	Sets in 5 minutes, cures in 2 hours

Hot-melts	Solid	Acetone	Melt w/ electric glue gun	Molding, overlays	Interior use	None
Liquid hide glue	Liquid	Warm water	None	Furniture, edge-gluing, laminating	Interior use	2 hours
Polyvinyl acetate (white glue)	Liquid	Soap and warm water	None	Edge- and face-gluing, laminating	Interior use	Sets in 8 hours, cures in 12 hours
Polyvinyl chloride	Liquid	Acetone	None	Wood, ceramics, glass, metal, plastic	Water-resistant	Sets in 5 minutes, cures in 12 hours
Resorcinal resin	Liquid and powder	Water before it hardens	Mix liquid w/ catalyst	Laminated timbers, sandwich panels, general bonding, boats	Waterproof	16 hours
Urea-Formaldehyde	Powder	Soap and warm water before it hardens	Mix w/ water	Lumber and hardwood plywood, assembly gluing	Water-resistant	16 hours
Urethane	Liquid	Alcohol before hardening	None	Gel-wood, porous materials, wood to wood, glass and metal	Interior use	1 hour

From THE HOME CARPENTER AND WOODWORKER'S REPAIR MANUAL, Courtesy Sterling Publishing Company, New York.

CERAMIC TILE BONDING AGENTS

Portland cement mortar. Mixture of Portland cement and sand. Used in wet area such as a shower.

Dry-set mortar. Mixture of Portland cement, sand, and additives. Use in wet areas.

Latex Portland cement mortar. Mixture of Portland cement, sand, and latex. Use in dry areas.

Modified epoxy emulsion mortar. Mixture of a resin and hardener added to Portland cement and sand. Binds well with little shrinkage.

Epoxy mortar. Mixture of a resin and a hardener with a silica filler. High bonding strength and impact resistance.

Organic adhesive. A mastic-type mixture. Dry interior use only.

Furan mortar. Mixture of furan resin and a hardener. Good chemical and high temperature resistance.

BONDING ANCHORING SYSTEMS

One type of bonding anchoring system has an acrylic or epoxy adhesive in a cartridge that is inserted in a power- or hand-operated dispenser that mixes the adhesive components in the proper ratio and places the adhesive at the bottom of the holes drilled in the concrete, into which steel reinforcing rods or threaded rods are inserted. Holes in grouted concrete block or masonry can also have rods bonded in holes in them. Manufacturers also have data on the allowable tension and shear loads on the connection.

BONDING ANCHORING SYSTEMS *(cont.)*

The air temperature greatly influences the time available for placing the adhesive. In hot weather, it must be placed in 4 or 5 minutes. In colder weather, it could give 20 to 40 minutes. In hot weather, the adhesive will cure in about one hour. In cold weather, it will take 24 hours or longer.

ANCHORS

A wide variety of anchors are available for securing materials to wood, concrete, and other products. Following are some that are widely used. Consult the manufacturer's catalog for additional types and sizes.

STUD TYPE ANCHOR

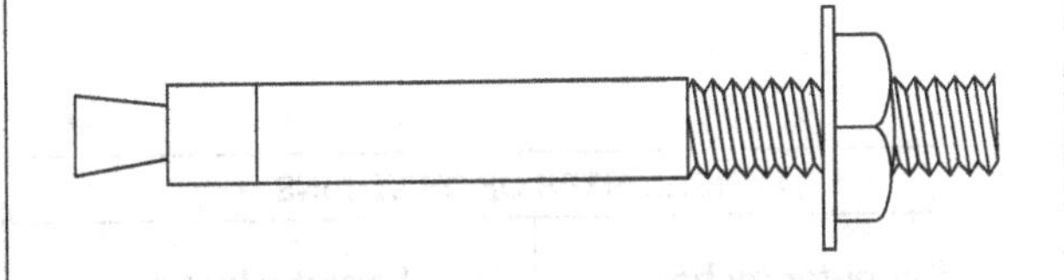

SIZES OF STUD TYPE ANCHORS

Anchor Dia.	Length Inches	Minimum Embedment
1/4	1 3/4, 2 1/4, 3 1/2	1 3/8
3/8	2 1/4, 3, 3 3/4	1 5/8
1/2	2 3/4, 4 1/4, 5 1/4	1 7/8
3/4	6 1/4	2 7/8

Sizes available vary with the manufacturer.

SEVERAL TYPES OF DRIVE PINS

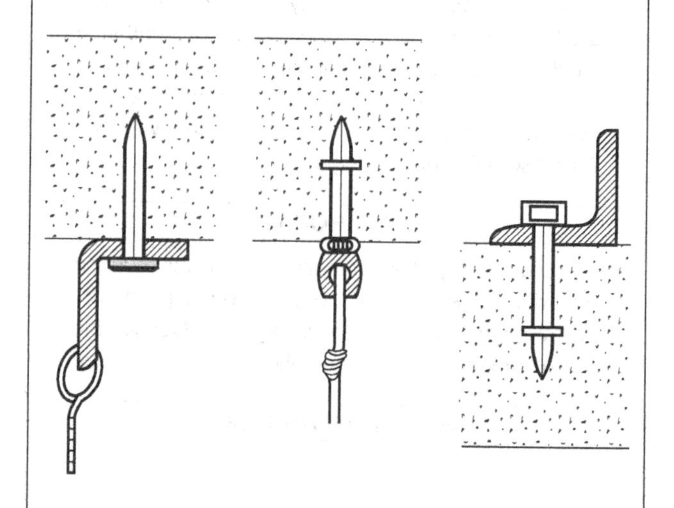

TYPICAL SIZES OF DRIVE PINS	
Diameter Inches	**Lengths Inches**
¼, ⅜	¾, 1⅛, 1⅝, 2, 2½, 3

Sizes vary with the manufacturer.

SELF-DRILLING ANCHOR

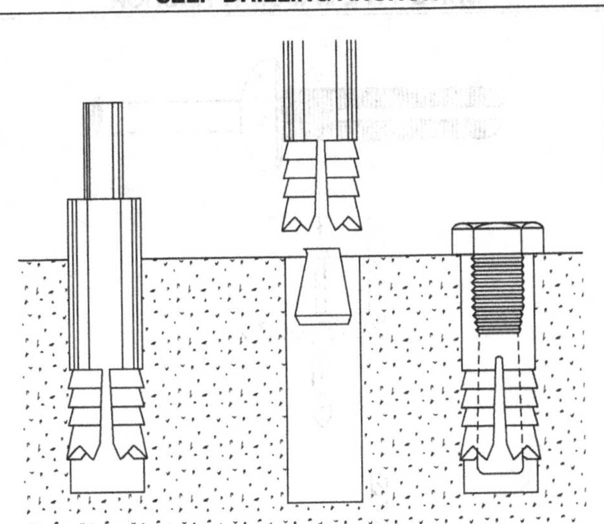

1. Drill the Hole With the Anchor

2. Insert the Expander Plug in the Hole and Drill the Anchor Back in the Hole

3. Snap off the Chucking End and Screw in the Bolt

SIZES OF SELF-DRILLING ANCHORS

Anchor Diameter	Anchor Length	Bolt Diameter
7/16	1¼	¼
9/16	1 7/16	⅜
11/16	1 15/16	½
27/32	2⅜	⅝
1	3	¾

Sizes vary with the manufacturer.

DRIVE-TYPE MASONRY ANCHOR

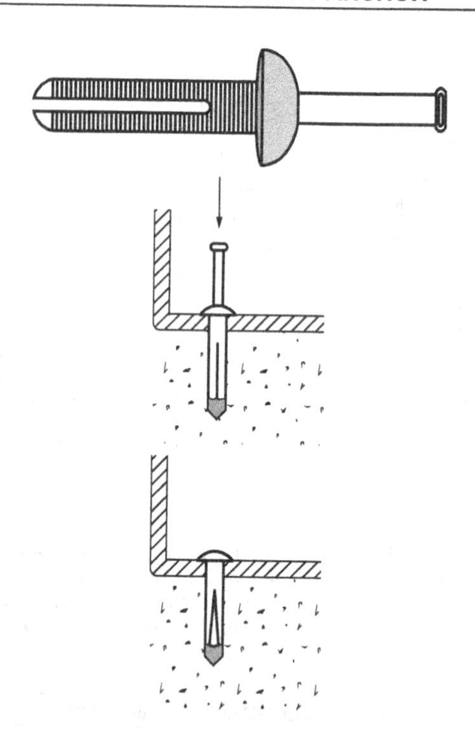

SIZES OF DRIVE-TYPE ANCHORS		
Anchor and Drill Size	**Length**	**Minimum Embedment**
3⁄16	7⁄8	½
¼	¾, 1, 1¼, 1½, 2	½

Sizes vary with the manufacturer.

PLASTIC SCREW ANCHOR

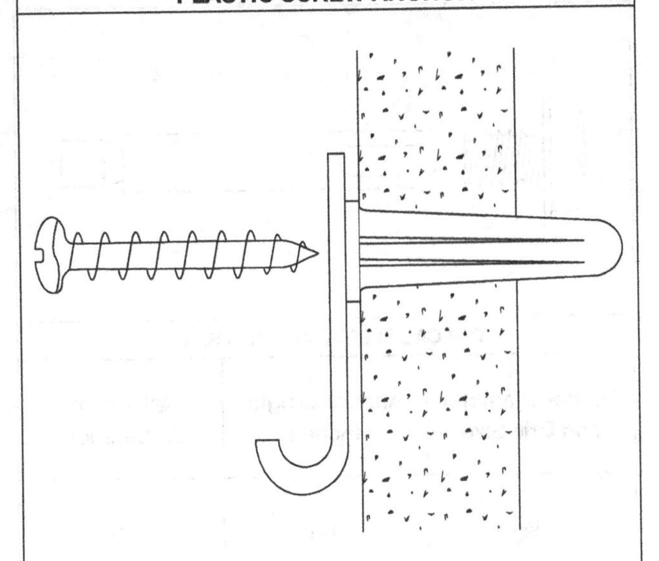

Install the anchor in a drilled hole. Place the object to be hung over it and install the screw.

Use in drywall, plaster, wood, some masonry.

TYPICAL SIZES OF PLASTIC SCREW ANCHORS

Diameter Inches	Length Inches	Hole Size Inches	Wood or Sheet Metal Screw Size
3/16	3/4	3/16	No. 8
3/16	7/8	3/16	No. 10
1/4	1	1/4	No. 12
5/16	1 3/8	5/16	No. 16

SLEEVE ANCHORS

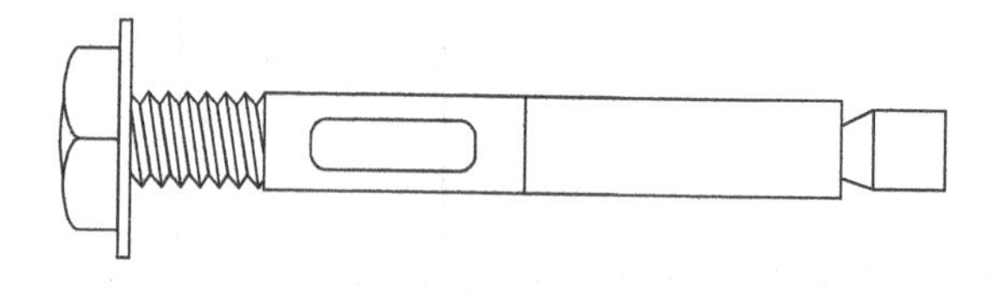

TYPICAL SLEEVE ANCHOR SIZES

Anchor Diameter and Drill Size	Anchor Length Inches	Minimum Embedment
¼	1⅜	1⅛
⅜	1⅞, 3	1½
½	2½, 3, 4	1⅞
⅝	4¼	2

Sizes available vary with the manufacturer.

HOLLOW WALL ANCHORS

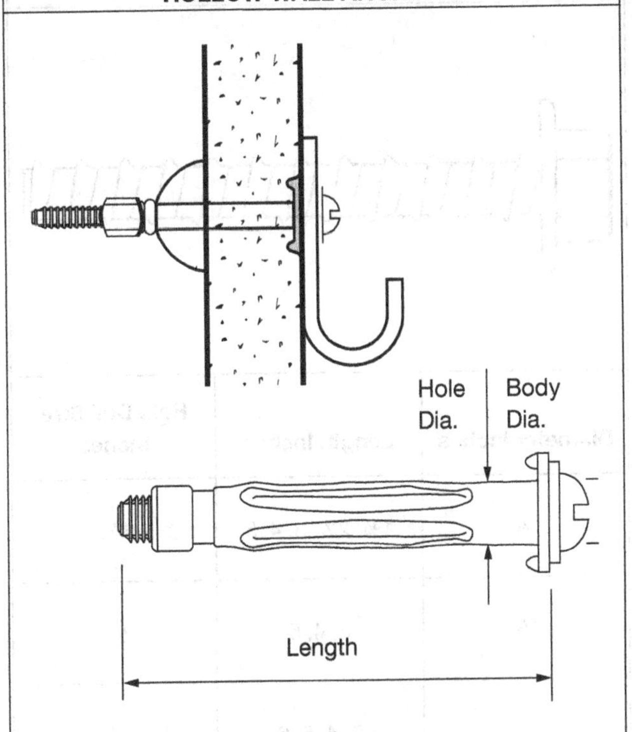

TYPICAL HOLLOW WALL ANCHOR SIZES		
Body Diameter	Anchor Length Inches	Maximum Expansion
1/4	1 13/16, 1 9/32, 1 23/32, 2 3/16	11/16
3/16	1 13/16, 2 5/16, 3 1/8	1 1/8
3/8	1 13/16, 2 5/16	1 1/8
7/16	3 1/8	1 1/8

SELF-THREADING CONCRETE ANCHOR

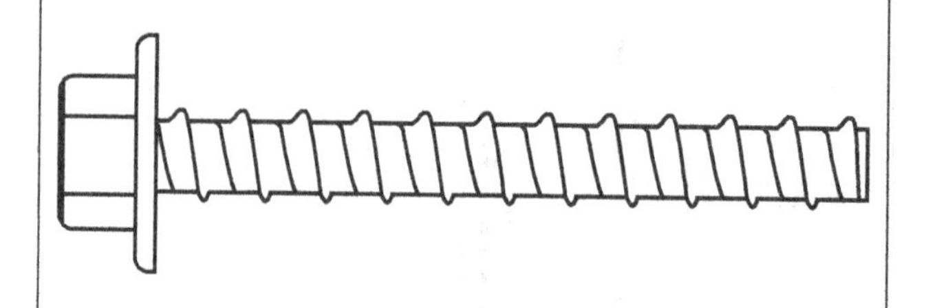

Diameter Inches	Length Inches	Hole Drill Size Inches
3/8	1¾, 2½, 3, 4, 5	5/16
½	3, 4, 5	7/16
5/8	3, 4, 5, 6	½
¾	4½, 5½, 6½	5/8

Drill the correct diameter hole in the concrete and screw the fastener in it.

TOGGLE BOLTS

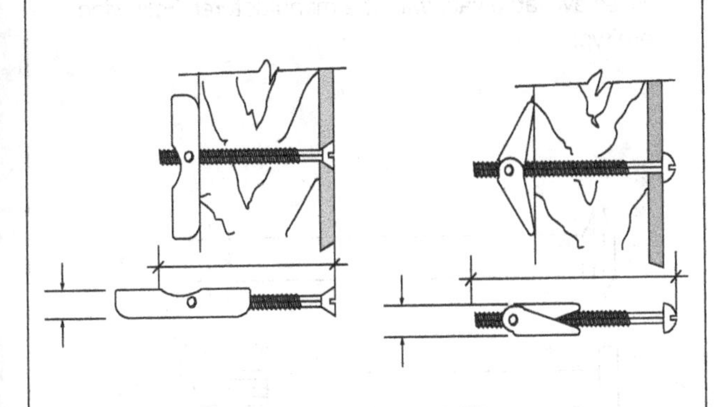

	TOGGLE BOLTS (INCHES)						
	Bolt diameter	⅛	5⁄32	3⁄16	¼	5⁄16	⅜
Tumble	Hole diameter	.375	.500	.500	.688	.875	.875
	Bolt length	2-4	2½-4	3-6	3-6	3-6	3-6
Spring Wing	Hole diameter	.375	.500	.500	.688	.875	1.000
	Bolt length	2-4	2½-4	2-6	2½-6	3-6	3-6

TYPICAL POWER DRIVE PINS

Available for use on concrete, steel, and plywood. Sizes available vary with the manufacturer. Following are typical.

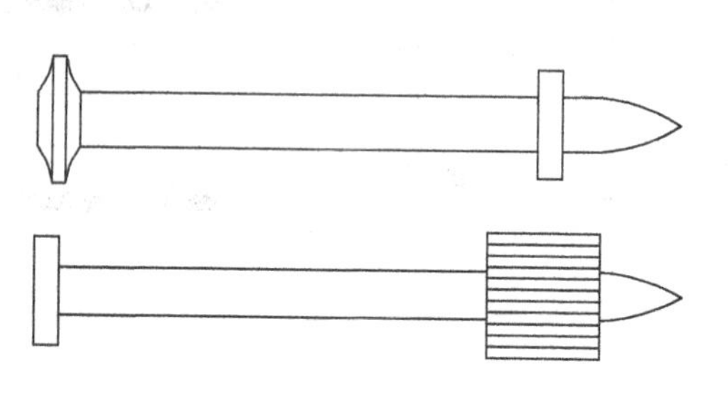

CONCRETE AND STEEL INSTALLATION	
Shank Diameter Inches	**Penetration Inches**
0.145	¾, 1, 1⅛, 1¼, 1½, 1¾, 2, 2⅜, 2½, 3

PLYWOOD INSTALLATION	
Shank Diameter Inches	**Penetration Inches**
0.145	Use on ½" plywood

LAG BOLT EXPANSION SHIELD

Concrete Anchor

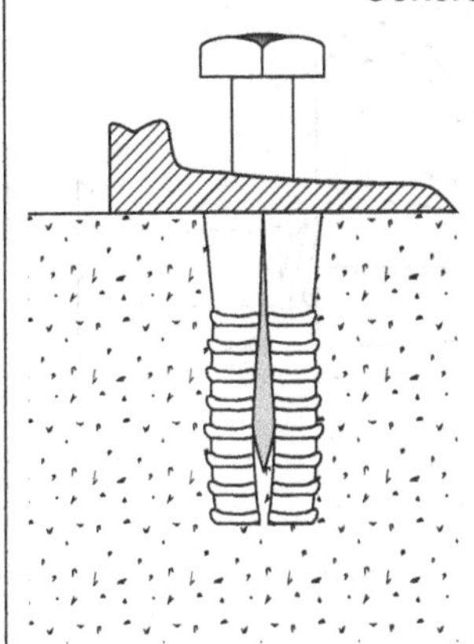

1. Insert the Expansion Shield in the Hole and Set in the Lag Bolt.

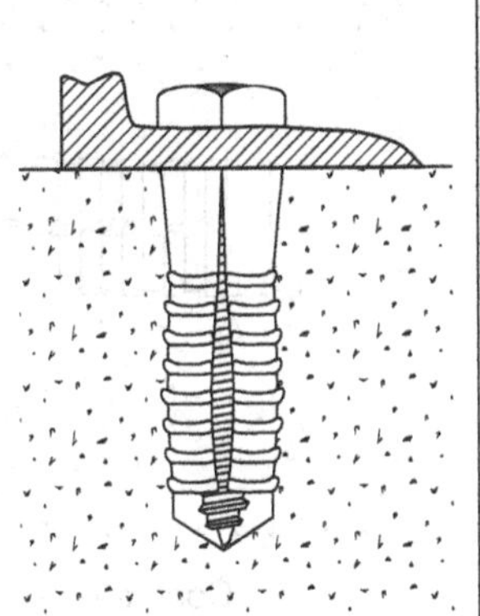

2. Screw the Lag Bolt into the Shield.

SHIELDS FOR LAG BOLTS			
Shield Dia. Inches	Lag Bolt Dia. Inches	Length Inches	
		Long	Short
½	¼	1½	1
½	5/16	1¾	1¼
5/8	3/8	2½	1¾
¾	½	3	2
7/8	5/8	3½	2
1	¾	3½	2

SHIELDS FOR WOOD SCREWS

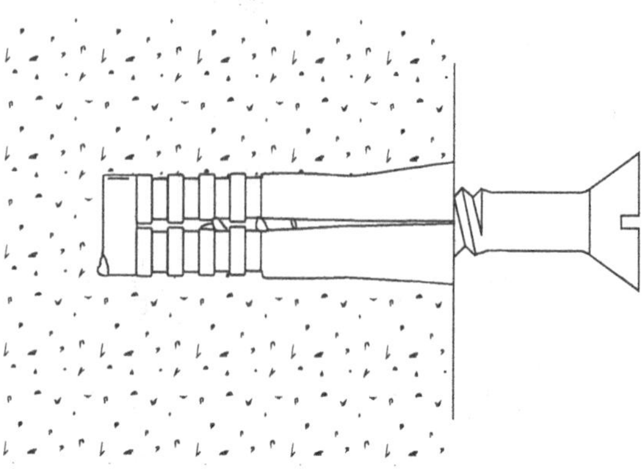

Concrete Anchor

Shield Dia. Inches	Shield Lengths Inches	Wood Screw No.
¼	¾ to 1½	6, 8
5/16	1 to 1½	10, 12, 14
⅜	1½	16, 18
7/16	1¾	20, 22

SLEEVE ANCHOR IN CONCRETE

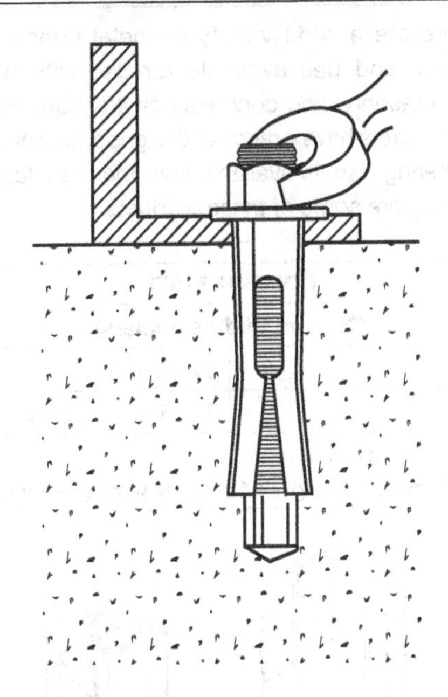

Tap the anchor into the hole. Insert the bolt in the anchor and tap it down into the anchor. Then tighten the nut.

TYPICAL SIZES OF SLEEVE ANCHORS		
Anchor Diameter Inches	**Anchor Length Inches**	**Bolt Diameter Inches**
3/8	2, 3	5/16
1/2	3¾, 4¾	3/8
5/8	4, 5, 6	1/2

CONNECTORS, ANCHORS, AND TIES

There are a wide variety of metal connectors, anchors, and ties available for use with wood, metal, masonry, and concrete construction. Various manufacturers have a range of designs and capacities. Engineering data is available from the manufacturer. Following are some of these products.

COLUMN BASES

Column Widths (inches)	
W1	3 9/16, 5/12, 7 1/8
W2	3 1/2, 3 9/16, 5 1/2, 5 7/16, 7

*Other sizes available.

Courtesy Simpson Strong-Tie Company. www.strongtie.com

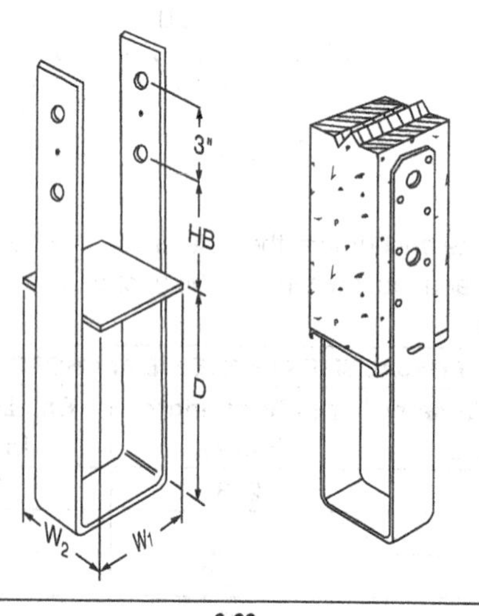

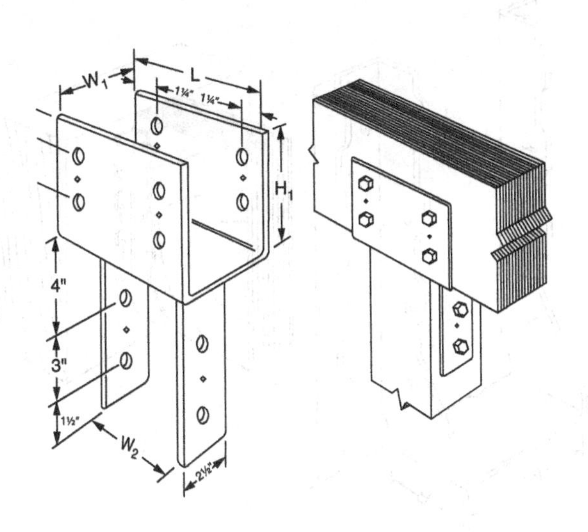

COLUMN CAPS*	
Cap Widths (inches)	
W1 Beam Widths	3⅝, 5½, 7⅛
W2 Post Widths	3⅝, 5½, 7⅛

*Other sizes available.

Courtesy Simpson Strong-Tie Company. www.strongtie.com

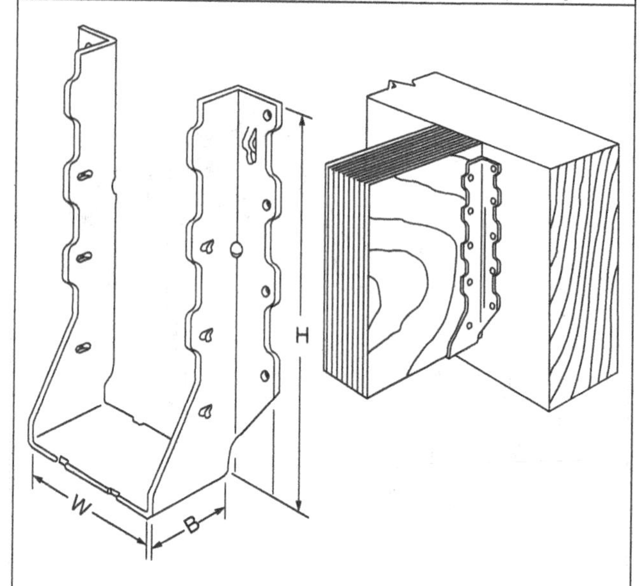

FACE MOUNT JOIST HANGER*	
Joist Sizes (inches)	
Width	**Height**
1½	9¼, 11¼, 14, 16, 20
2¼	9¼, 11¼, 14, 16, 18, 20

*Other designs and sizes available.

Courtesy Simpson Strong-Tie Company. www.strongtie.com

CONNECTORS, ANCHORS, AND TIES *(cont.)*

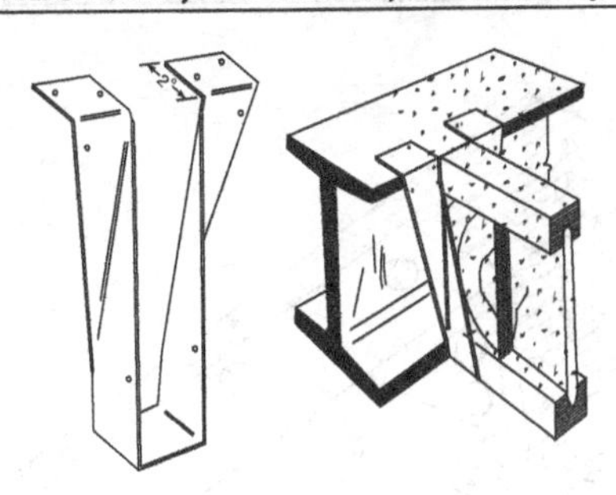

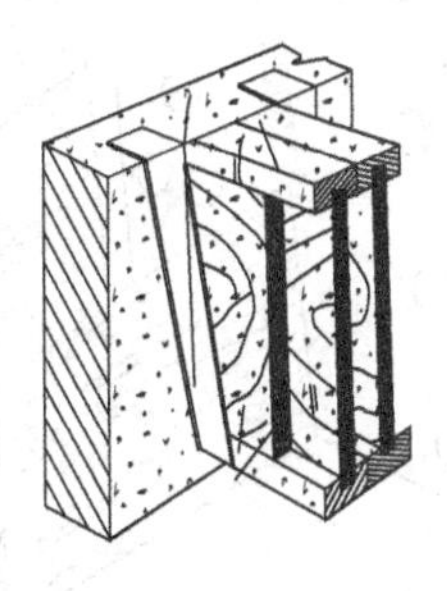

TOP HUNG I-JOIST HANGERS*	
Joist Sizes (inches)	
Width	**Height**
1¾	12, 14, 16
2¼	9½, 11¼, 12, 14, 16

*Other sizes available.

Courtesy Simpson Strong-Tie Company. www.strongtie.com

CONNECTORS, ANCHORS, AND TIES *(cont.)*

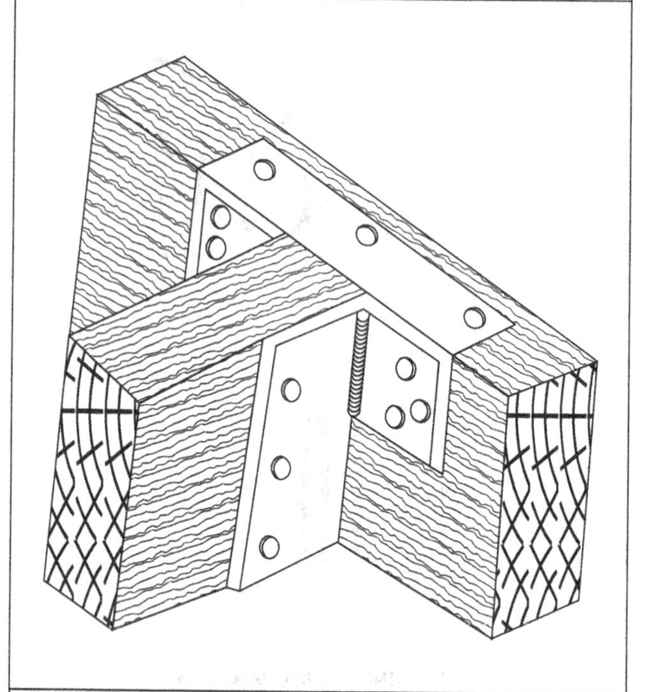

TOP HUNG BEAM HANGER		
Length	Saddle Width (inches)	Top Flange Length
9	3¼	2½
9, 11	5¼	2½
9, 11	6⅞	2½
Other sizes available.		

FLOOR-TO-FLOOR CONNECTIONS

UPLIFT FORCES IN POUNDS

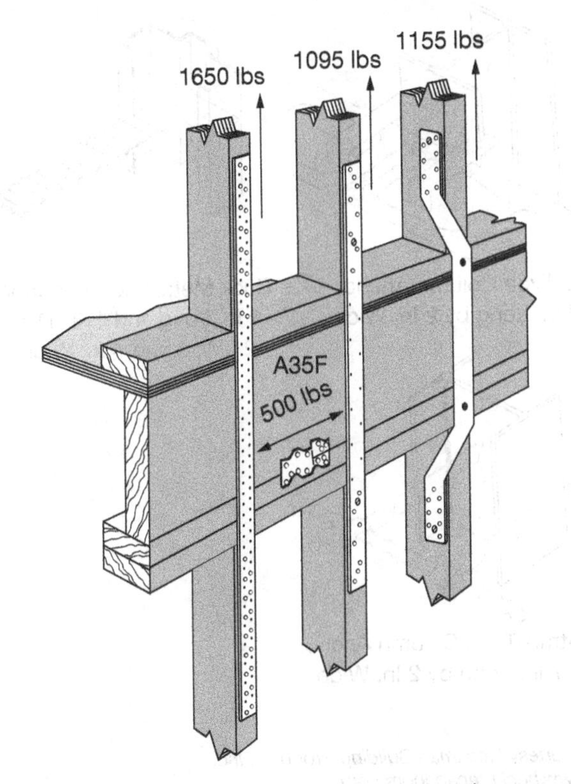

Courtesy Simpson Strong-Tie Company. www.strongtie.com

MASONRY ANCHORS FOR STRUCTURAL STEEL

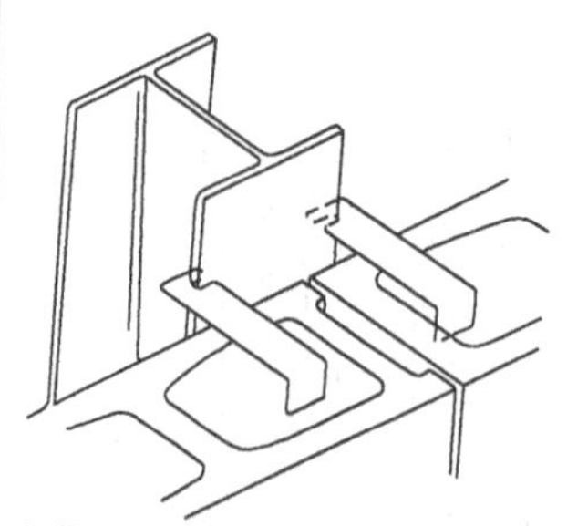

L-Type Column Anchor
7 In. Long by 2 In. Wide

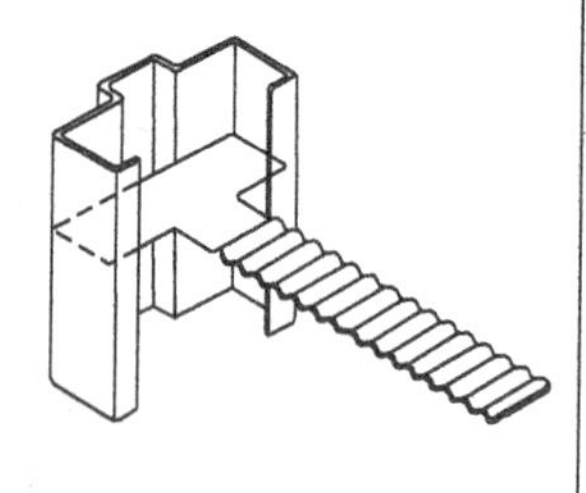

Buck Metal Frame Anchor
8 In. Long and 1 1/4, 1 1/2, 2 2 1/2, 3 and 4 In. Wide

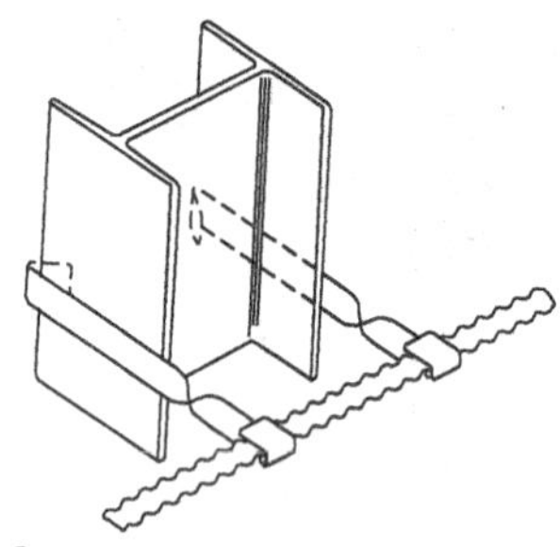

Strap-Type Column Anchor
7 In. Long by 2 In. Wide

Courtesy Heckman Building Products, Inc.
www.heckmananchors.com

BOLTED ANCHORS

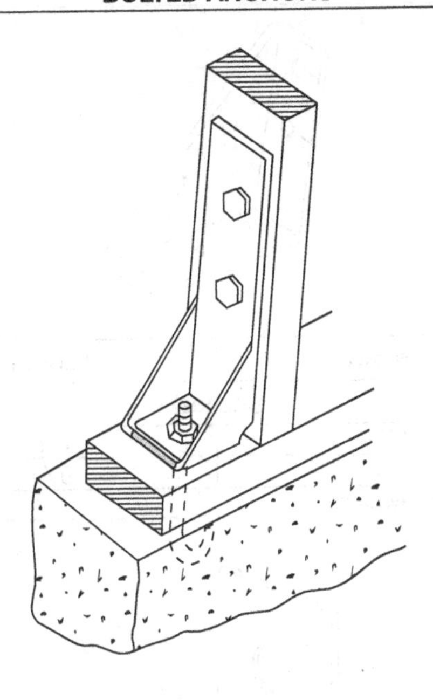

Length	Width
(inches)	
8	2⅝
9⅜	3
11	3½
15	3½
18	3½
20	4⅞

Other designs and sizes available.

ANGLE STIFFENERS

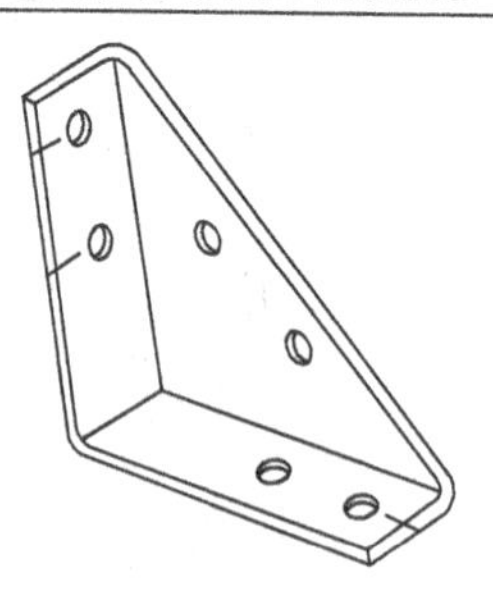

Width	Length
(inches)	
$2\frac{11}{16}$	$8\frac{1}{8}$
3	$9\frac{5}{16}$
Other sizes available.	

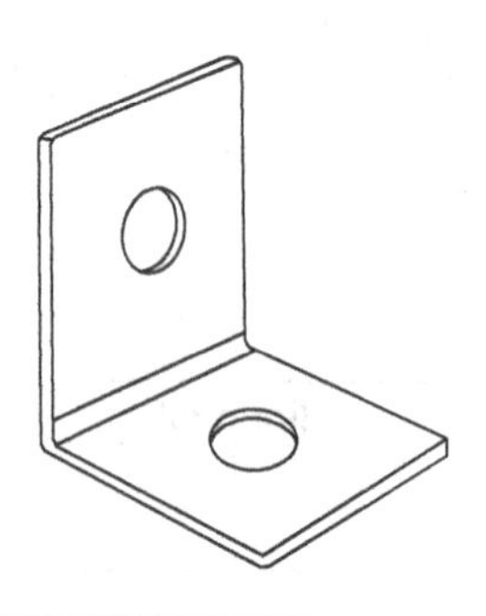

Width	Length
(inches)	
$3\frac{1}{4}$	$2\frac{1}{2}$, 5, $7\frac{1}{2}$
Other sizes available.	

ANGLE STIFFENERS *(cont.)*

Width	Length
(inches)	
5¾	5
Other sizes available.	

STRUCTURAL L-STRAP AND T-STRAP TIES

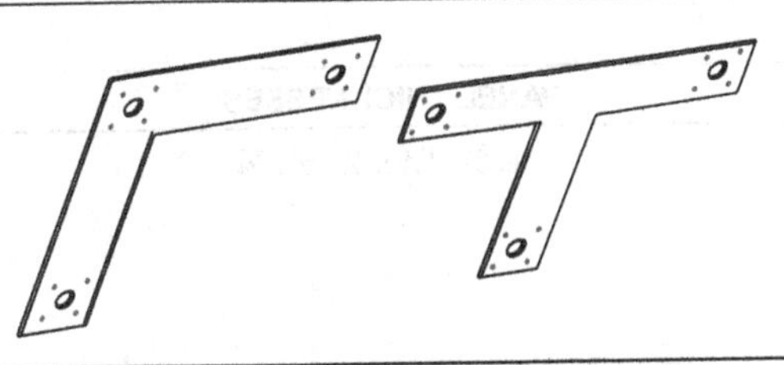

TYPICAL T-STRAP AND L-STRAP SIZES	
Width	Length
(inches)	
6	6
8	8
12	12
16	16
Other sizes available.	

PANEL SHEATHING CLUP

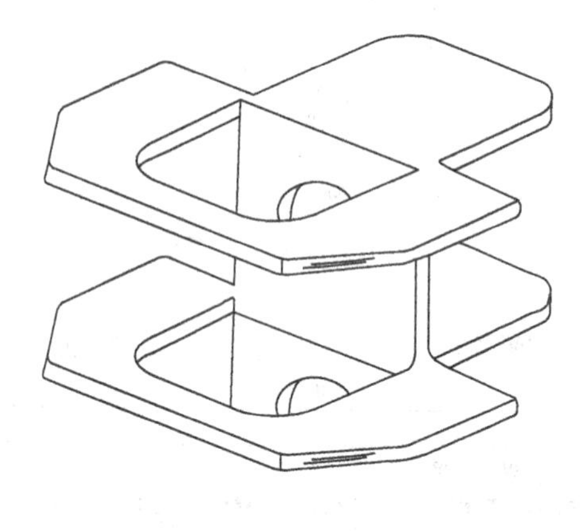

PANEL THICKNESSES

7/16, 3/8, 15/32, 1/2, 5/8, 3/4

WOOD TO CONCRETE STRAP ANCHOR

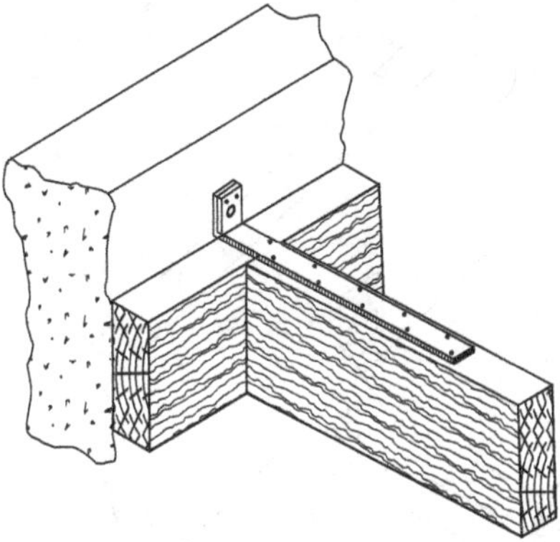

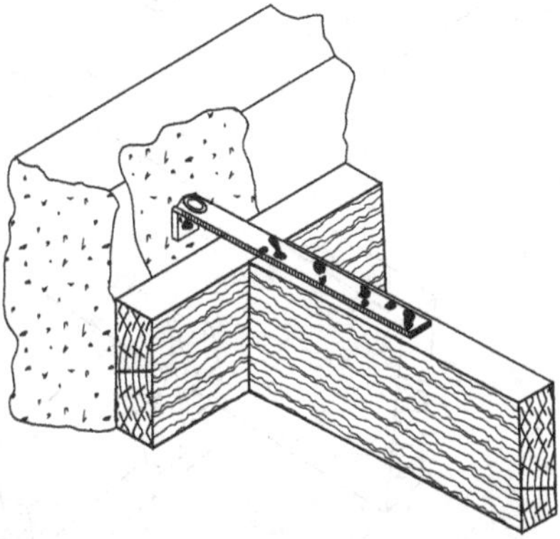

Width	Length
(inches)	
2¹⁄₁₆, 3¾	18 to 38

Other sizes available.

TRUSS AND RAFTER ANCHORS

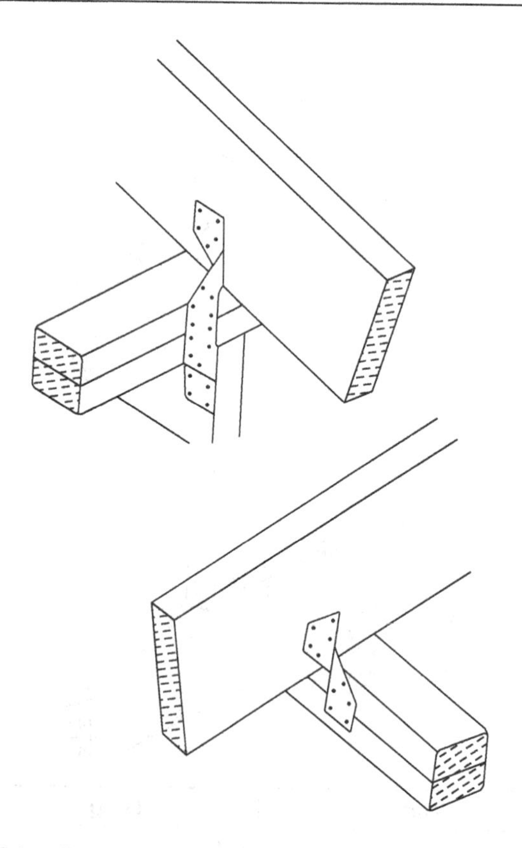

Other Designs Available.

CHAPTER 7
Concrete Construction

ASTM DESIGNATED TYPES OF PORTLAND CEMENT

Type I Normal – General use as in pavements, bridges, masonry building units.

Type IA Normal, Air-Entrained – General use, has entrained air for freezing conditions.

Type II Moderate – Protects against moderate sulfate attack, generates less heat, useful in large masses such as piers and retaining walls.

Type II A Moderate, Air-Entrained – Has entrained air, used in freezing conditions.

Type III High Early Strength – Provides high early strength faster than TYPE I and TYPE II. Reduces curing time.

Type IIIA, High Early Strength, Air-Entrained – Used where freezing conditions will be present.

Type IV, Low Heat of Hydration – Slower to develop strength, reduces rate and amount of heat developed. Used in large masses such as dams.

Type V, Sulfate Resisting – Used in concrete that is exposed to severe sulfate conditions generally found in ground water and some soils.

OTHER CEMENTS

White Portland Cement – Used where the finished product will be white, such as tile grout and precast curtain wall panels.

Blended Hydraulic Cements – These harden under water.

Masonry Cements – Used in the manufacture of masonry units.

Waterproof Portland Cement – Has calcium stearate or aluminum stearate added to Portland cement. Used where exposure to water is a factor.

Plastic Cements – They have a plasticizing agent added to TYPE I or TYPE II Portland cement. Used in stucco and plaster.

Expansive Cement – This is a hydraulic cement that expands during hardening, affecting drying shrinkage.

Regulated Set Cement – This is a hydraulic cement that can be compounded to set from 2 to 60 minutes.

TYPES OF FIBERS USED IN CONCRETE MIXES

The use of fiber reinforcement materials in concrete improves the strength, tensile strength, flexural strength, ductility and crack control.

Steel
AR-glass
Acrylic
Aramid
Carbon
Nylon
Polyester
Polyethylene
Polyproplyene
Wood cellulose
Sisal
Coconut
Bamboo
Jute
Elephant grass

SOME COMMONLY USED ADMIXTURES	
Types of Admixtures	**Effect on the Mix**
Accelerators	Accelerate the setting time and the development of early strength
Air detrainers	Reduce air content
Air-entraining admixtures	Improve workability and durability in freezing and thawing conditions
Bonding admixtures	Increase bond strength
Coloring admixtures	Add color to the mix
Dampproofing admixtures	Retard moisture penetration into the cured concrete
Pumping admixtures	Make pumped concrete flow easier
Retarders	Slow setting time to reduce the effect of hot weather
Superplasticers	Improve concrete flow, reduce water-cement ratio
Water reducers	Reduce water content in the mix
Additional types of admixtures and admixture standards are available from the American Society for Testing and Materials (ASTM) (www.astm.org) and The American Association of State Highway and Transportation Officials (AASHTO) (www.aashto.org).	

SOURCES OF CONCRETE SPECIFICATIONS

American Concrete Institute (ACI)j,
www.concrete.org

ACI MANUAL OF CONCRETE PRACTICE

American Society for Testing and Materials (ASTM),
www.astm.org

ASTM C94, STANDARD SPECIFICATION FOR READY MIXED CONCRETE

American Association of State Highway and Transportation Officials (AASHTO), www.aashto.org

AASHTO Standard Publications

Portland Cement Association (PSA),
www.cement.org

CONTROL OF CONCRETE MIXTURES

National Ready Mixed Concrete Association (NRMCA), www.nrmca.org

READY-MIXED CONCRETE

American Society of Concrete Contractors (ASCC),
www.ascconline.org

CONTRACTORS GUIDE TO QUALITY CONCRETE

STANDARDS FOR MATERIALS USED IN RESIDENTIAL CONCRETE

American Society for Testing and Materials, www.astm.org

Portland cement shall conform to ASTM C 150 or ASTM C 1157.

Blended hydraulic cement shall conform to ASTM C 595 or ASTM C 1157.

Fly ash and natural pozzolans shall conform to ASTM C 618.

Ground-granulated blast furnace slag shall conform to ASTM C 989.

Silica fume shall conform to ASTM C 1240.

Aggregates shall conform to ASTM C 33 OR ASTM C 330.

Water shall conform to ASTM C 94.

Air-entraining admixtures shall conform to ASTM C 260.

Water-reducing admixtures, retarding admixtures, accelerating admixtures, and water-reducing and accelerating admixtures shall conform to ASTM C 494 or ASTM 1017.

Calcium chloride shall conform to ASTM D 98.

Deformed steel reinforcing bars shall conform to ASTM A 615, ASTM A 706, or ASTM A 996.

Welded plain wire reinforcement shall conform to ASTM A 185.

Courtesy American Society for Testing and Materials, www.astm.org

FINE AND COARSE SIEVE SIZES	
Fine Sieve Sizes	
3/8" (9.5 mm)	
No. 4 (4.75 mm)	
No. 8 (2.36 mm)	
No. 16 (1.18 mm)	
No. 30 (600 μm)	
No. 50 (300 μm)	
No. 100 (150 μm)	
Coarse Sieve Sizes	
Size Number	**Nominal Size (Sieves with Square Openings)**
1	3½" to 1½" (90 to 37.5 mm)
2	2½" to 1½" (63 to 37.5 mm)
357	2" to No. 4 (50 to 4.75 mm)
467	1½" to No. 4 (37.5 to 4.75 mm)
57	1" to No. 4 (25.0 to 4.75 mm)
67	¾" to No. 4 (19.0 to 4.75 mm)
7	½" to No. 4 (12.5 to 4.75 mm)
8	3/8" to No. 8 (9.5 to 2.36 mm)
3	2" to 1" (50.0 to 25.0 mm)
4	1½" to ¾" (37.5 to 19.0 mm)

FIGURING BATCH WEIGHT USING MASS OR WEIGHT

The term MASS is replacing WEIGHT in ASTM standards. Mass is a measure of how much material or matter is contained in an object. Mass does not change with changes in gravitational force or altitude. Weight changes with the altitude.

Weight is a measure of force resulting from the effect of gravity on the mass of an object. The weight of a material will be greater at sea level than in areas of high elevation.

The instructions for determining the mass per unit volume (density) of concrete can be found in ASTM Test Method C 138/C 138M.

The volume of freshly mixed and unhardened concrete in a batch can be determined from the total mass of the batch divided by the mass per cubic foot of the concrete. The total mass of a batch is the sum of the masses of all material, including water, in the batch.

To find the batch yield using mass, calculate the number of cubic feet and divide this by 27 cubic feet, which equals one cubic yard. This step follows:

$$\text{Batch yield} = \frac{\text{Total mass of batch}}{\text{Mass per cubic foot of the concrete}} = \text{Cubic feet of concrete}$$

To find the batch yield using weight:

$$\text{Batch yield} = \frac{\text{Total weight in of batch in pounds}}{\text{Weight per cubic feet of the concrete}} = \text{Cubic feet of concrete}$$

FIGURING BATCH WEIGHT USING MASS OR WEIGHT *(cont.)*

To convert to cubic yards:

$$\text{Batch yield} = \frac{\text{Cubic feet of concrete}}{\text{27 cubic feet (equals one cubic yard)}} = \text{Cubic yards of concrete}$$

EXAMPLES OF SLUMPS TYPICALLY RECOMMENDED FOR VARIOUS INSTALLATIONS

Type of Concrete Work**	Maximum Slump* (inches)	Minimum Slump (inches)
Sidewalks, driveways, slabs on grade	4	2
Pavements	2	1
Reinforced slabs, walls, beams	5	2
Reinforced footings, foundations	4	2
Unreinforced footings, foundations	3	1
Mass concrete	2	1

*High number indicates wet mix, low number a dry mix.

**Slump responsibility transferred from the producer to the user occurs:

1. 30 min. after scheduled batch delivery time.
2. 30 min. after a slump adjustment has been made on the site.
3. 30 min. after a batch arrives on the site.

MINIMUM SPECIFIED COMPRESSIVE STRENGTH AT 28 DAYS AND MAXIMUM SLUMP OF CONCRETE

Type or Location of Concrete Construction	Weathering Probability			Maximum Slump, in.*
	Negligible psi	Moderate psi	Severe psi	
Type 1: Walls and foundations not exposed to weather. Interior slabs-on-ground, not including garage floor slabs	2500	2500	2500	6
Type 2: Walls, foundations, and other concrete work exposed to weather, except as noted in Type 3	2500	3000	3000	6
Type 3: Driveways, curbs, walk-ways, ramps, patios, porches, steps, and stairs exposed to weather and garage floors, slabs	2500	3500	4500	5

*Maximum slump refers to the characteristics of the specified mixture proportion based on *w/em* only. Mid-range and high-range water-reducing admixtures can be used to increase the slump beyond these maximums.

Courtesy American Concrete Institute, www.concrete.org

RELATIONSHIP BETWEEN WATER-CEMENT RATIO AND COMPRESSIVE STRENGTH OF CONCRETE		
Compressive Strength @ 28 Days (psi)	**Water-Cement Ratio, by Weight**	
	Non-Air-Entrained Concrete	**Air-Entrained Concrete**
6000	0.41	—
5000	0.48	0.41
4000	0.57	0.48
3000	0.68	0.59
2000	0.82	0.74

Courtesy American Concrete Institute, www.concrete.org

AIR CONTENT FOR TYPE 2 AND TYPE 3 CONCRETE UNDER MODERATE OR SEVERE WEATHER PROBABILITY		
Nominal Maximum Aggregate Size, in.	**Air Content (Tolerance ±0.015)**	
	Moderate	**Severe**
⅜	0.06	0.075
½	0.055	0.07
¾	0.05	0.07
1	0.045	0.06
1½	0.045	0.055

Courtesy American Concrete Institute, www.concrete.org

RECOMMENDED SLUMPS FOR VARIOUS TYPES OF CONSTRUCTION

Concrete Construction	Slump, mm (in.)	
	Maximum*	Minimum
Reinforced foundation walls and footings	75 (3)	25 (1)
Plain footings, caissons, and substructure walls	75 (3)	25 (1)
Beams and reinforced walls	100 (4)	25 (1)
Building columns	100 (4)	25 (1)
Pavements and slabs	75 (3)	25 (1)
Mass concrete	75 (3)	25 (1)

*May be increased 25 mm (1 in.) for consolidation by hand methods, such as rodding and spading.

Plasticizers can safely provide higher slumps.

Adapted from ACI 211.1.

Courtesy American Concrete Institute, www.concrete.org and Portland Cement Association, www.cement.org

PROPORTIONS BY MASS TO MAKE ONE TENTH CUBIC METER OF CONCRETE FOR SMALL JOBS – METRIC

Nominal Maximum Size Coarse Aggregate, mm	Air-Entrained Concrete				Non-Air-Entrained Concrete			
	Cement, kg	Wet Fine Aggregate, kg	Wet Coarse Aggregate, kg*	Water, kg	Cement, kg	Wet Fine Aggregate, kg	Wet Coarse Aggregate, kg	Water, kg
9.5	46	85	74	16	46	94	74	18
12.5	43	74	88	16	43	85	88	18
19.0	40	67	104	16	40	75	104	16
25.0	38	62	112	15	38	72	112	15
37.5	37	61	120	14	37	69	120	14

*If crushed stone is used, decrease coarse aggregate by 5 kg and increase fine aggregate by 5 kg.

PROPORTIONS BY MASS TO MAKE ONE CUBIC FOOT OF CONCRETE FOR SMALL JOBS – INCH-POUND

Nominal Maximum Size Coarse Aggregate, in.	Air-Entrained Concrete				Non-Air-Entrained Concrete			
	Cement, lb	Wet Fine Aggregate, lb	Wet Coarse Aggregate, lb*	Water, lb	Cement, lb	Wet Fine Aggregate, lb	Wet Coarse Aggregate, lb	Water, lb
⅜	29	53	46	10	29	59	46	11
½	27	46	55	10	27	53	55	11
¾	25	42	65	10	25	47	65	10
1	24	39	70	9	24	45	70	10
1½	23	38	75	9	23	43	75	9

*If crushed stone is used, decrease coarse aggregate by 3 lb and increase fine aggregate by 3 lb.

Courtesy American Concrete Institute, www.concrete.org *and Portland Cement Association,* www.cement.org

MAXIMUM WATER-CEMENTITIOUS MATERIAL RATIOS AND MINIMUM DESIGN STRENGTHS FOR VARIOUS EXPOSURE CONDITIONS

Exposure Condition	Maximum Water-Cementitious Material Ratio by Mass for Concrete	Minimum Design Compressive Strength, f'_c, MPa (psi)
Concrete protected from exposure to freezing and thawing, application of deicing chemicals, or aggressive substances	Select water-cementitious material ratio on basis of strength, workability, and finishing needs	Select strength based on structural requirements
Concrete intended to have low permeability when exposed to water	0.50	28 (4000)
Concrete exposed to freezing and thawing in a moist condition or deicers	0.45	31 (4500)
For corrosion protection for reinforced concrete exposed to chlorides from deicing salt water, brackish water, seawater, or spray from these sources	0.40	35 (5000)

Adapted from ACI 318 (2002).

Reproduced from Design and Control of Concrete Mixtures, Courtesy Portland Cement Association, www.cement.org

RECOMMENDED TOTAL TARGET AIR CONTENT FOR CONCRETE

Nominal Maximum Size Aggregate, mm (in.)	Air Content, Percent*		
	Severe Exposure**	Moderate Exposure†	Mild Exposure††
<9.5 (⅜)	9	7	5
9.5 (⅜)	7½	6	4½
12.5 (½)	7	5½	4
19.0 (¾)	6	5	3½
25.0 (1)	6	4½	3
37.5 (1½)	5½	4½	2½
50 (2)‡	5	4	2
75 (3)‡	4½	3½	1½

*Project specifications often allow the air content of the concrete to be within −1 to +2 percentage points of the table target values.

**Concrete exposed to wet-freeze-thaw conditions, deicers, or other aggressive agents.

†Concrete exposed to freezing but not continually moist, and not in contact with deicers or aggressive chemicals.

††Concrete not exposed to freezing conditions, deicers, or aggressive agents.

‡These air contents apply to the total mix, as for the preceding aggregate sizes. When testing these concretes, however, aggregate larger than 37.5 mm (1½ in.) is removed by handpicking or sieving and air content is determined on the minus 37.5 mm (1½ in.) fraction of mix. (Tolerance on air content as delivered applies to this value.)

Reproduced from Design and Control of Concrete Mixtures, Courtesy Portland Cement Association, www.cement.org

RELATIONSHIP BETWEEN WATER TO CEMENTITIOUS MATERIAL RATIO AND COMPRESSIVE STRENGTH OF CONCRETE – METRIC

Compressive Strength at 28 Days, MPa	Water-Cementitious Materials Ratio by Mass	
	Non-Air-Entrained Concrete	Air-Entrained Concrete
45	0.38	0.30
40	0.42	0.34
35	0.47	0.39
30	0.54	0.45
25	0.61	0.52
20	0.69	0.60
15	0.79	0.70

Strength is based on cylinders moist-cured 28 days in accordance with ASTM C 31 (AASHTO T 23). Relationship assumes nominal maximum size aggregate of about 19 to 25 mm.

Adapted from ACI 211.1 and ACI 211.3.

RELATIONSHIP BETWEEN WATER TO CEMENTITIOUS MATERIAL RATIO AND COMPRESSIVE STRENGTH OF CONCRETE – INCH-POUND UNITS

Compressive Strength at 28 Days, psi	Water-Cementitious Materials Ratio by Mass	
	Non-Air-Entrained Concrete	Air-Entrained Concrete
7000	0.33	–
6000	0.41	0.32
5000	0.48	0.40
4000	0.57	0.48
3000	0.68	0.59
2000	0.82	0.74

Strength is based on cylinders moist-cured 28 days in accordance with ASTM C 31 (AASHTO T 23). Relationship assumes nominal maximum size aggregate of about ¾ in. to 1 in.

Adapted from ACI 211.1 and ACI 211.3.

Reproduced from Design and Control of Concrete Mixtures, Courtesy Portland Cement Association, www.cement.org

PLACING CONCRETE IN COLD WEATHER

Cold weather procedures must be followed when the mean daily temperature is less than 40°F. Procedures are specified in ACI 301, Specifications for Structural Concrete. Typically the specifications of the job prohibit placing concrete outdoors when the temperature reaches 40°F and is dropping. It may be placed if the temperature is 40°F and rising. Rebar temperatures must be above freezing.

Heated water and aggregates may be used but might cause surface cracking and shrinkage. To avoid this, apply a curing compound to the surface and cover with insulating blankets. Insulating blankets made for this purpose may be made of fiberglass, sponge rubber, polyurethane foam, mineral wool, or cellulose fibers. Various board-type insulating products are also used. Expanded polystyrene and expanded polyurethane have high insulating properties per inch of thickness. Be certain to completely cover the edges and corners of the concrete area. The hydration of the cement creates heat, and this adds to the air temperature below the insulation.

Forms can be covered with insulating materials. A good way to do this is to place 1 inch or thicker polystyrene panels on the forms and cover them with ½-inch or ¾-inch plywood or particleboard. If the forms are wood, this adds additional insulation.

Concrete made with heated materials should have a temperature of 50 to 70°F when it is placed. Concrete should reach a strength of at least 500 psi before being exposed to freezing. Concrete is never placed on a frozen subgrade. If the subgrade appears frozen, thaw it by covering it with insulating blankets for several days. Once it has thawed, compact it before placing the concrete.

Consider adding accelerators such as Type 111 High Early Strength cement. Calcium chloride may be added to the cement at the rate of 1 to 2 percent of the weight of the cement; however, check the mix specifications to see if this is allowed.

All concrete to be exposed to freezing weather during placement or in permanent service should be air-entrained.

Heated enclosures can be used to control the air temperature. Aboveground installations will have enclosures on top and below. The heaters should be vented to the outside. Carbon dioxide created by burning fossil fuel interferes with the cement hydration and can cause the surface to dust. The carbon monoxide created is a major safety hazard, causing headaches and nausea, and it can cause death. Avoid rapid cooling of the concrete at the end of the heating time because it may cause cracking. Drop the temperature slowly over the first 24 hours.

MAXIMUM PERMISSIBLE WATER-CEMENT RATIOS FOR CONCRETE IN SEVERE EXPOSURES

Type of Structure	Structure Wet Continuously or Frequently Exposed to Freezing and Thawing	Structure Exposed to Sea Water or Sulfates
Thin sections (railings, curbs, sills, ledges, ornamental work) and sections less than 1 in. cover over steel.	0.45	0.40
All other structures	0.50	0.45

Courtesy American Concrete Institute, www.concrete.org

RECOMMENDED AIR CONTENTS FOR FROST-RESISTANT CONCRETE

Maximum Aggregate Size (in.)	Average Air Content (percent)*	
	Severe Exposure[1]	Moderate Exposure[2]
3/8	7½	6
½	7	5½
¾	6	5
1½	5½	4½
3	4½	3½
6	4	3

*A reasonable tolerance for air content in field construction is ±1½ percent.

[1]Outdoor exposure in a cold climate where the concrete may be in almost continuous contact with moisture prior to freezing, or where deicing salts are used. Examples are pavements, bridge decks, sidewalks, and water tanks.

[2]Outdoor exposure in a cold climate where concrete will be only occasionally exposed to moisture prior to freezing, and where no de-icing salts will be used. Examples are exterior walls, beams, girders, and slabs not in direct contact with soil.

Table 2.1 is adapted from ACI 201.2R, "Guide to Durable Concrete."

Courtesy American Concrete Institute, www.concrete.org

PLACING CONCRETE IN HOT WEATHER

Hot weather can lower the strength of the concrete and shorten the setting time, requiring faster action on finishing, and will dry the surface very rapidly, which affects the surface finish. Reducing concrete temperatures extends the setting time, reduces the rate of hydration, and reduces the rate of early strength gain, which increases later strength.

The temperature of the batch can be lowered by continuously sprinkling the coarse aggregate before it is introduced in the mixer. Crushed ice can be used in the mix instead of water. Use 50 pounds of ice for every 50 pounds of water. Water weighs 8.33 pounds per gallon. This will produce about a 10°F reduction in temperature.

Water-reducing retarders, fly ash, and slag added to the mix also provide cooling benefits. Sprinkle the ground before laying the concrete. Using a sprayed-on curing compound or covering the concrete with wet burlap or waterproof paper will reduce the temperatures subjected on the concrete. If forms are used, they can be sprayed with water and covered with wet burlap or waterproof paper.

RECOMMENDED CONCRETE TEMPERATURE FOR COLD-WEATHER CONSTRUCTION—AIR-ENTRAINED CONCRETE*

Line	Condition		Thickness of Sections, mm (in.)			
			Less than 300 (12)	300 to 900 (12 to 36)	900 to 1800 (36 to 72)	Over 1800 (72)
1	Minimum temperature of fresh concrete as *mixed* for weather indicated.	Above −1°C (30°F)	16°C (60°F)	13°C (55°F)	10°C (50°F)	7°C (45°F)
2		−18°C to −1°C (0°F to 30°F)	18°C (65°F)	16°C (60°F)	13°C (55°F)	10°C (50°F)
3		Below −18°C (0°F)	21°C (70°F)	18°C (65°F)	16°C (60°F)	13°C (55°F)
4	Minimum temperature of fresh concrete *as placed and maintained*.**		13°C (55°F)	10°C (50°F)	7°C (45°F)	5°C (40°F)

*Adapted from Table 3.1 of ACI 306R-88.

**Placement temperatures listed are for normal-weight concrete. Lower temperatures can be used for lightweight concrete if justified by tests.

For recommended duration of temperatures in Line 4, see Table 14-3.

Reproduced from Design and Control of Concrete Mixtures, Courtesy Portland Cement Association, www.cement.org

A. RECOMMENDED DURATION OF CONCRETE TEMPERATURE IN COLD WEATHER – AIR-ENTRAINED CONCRETE*				
Service Category	Protection from Early-Age Freezing		For Safe Stripping Strength	
	Conventional Concrete,** Days	High-Early Strength Concrete,† Days	Conventional Concrete,** Days	High-Early-Strength Concrete,† Days
No load, not exposed‡ favorable moist-curing	2	1	2	1
No load, exposed, but later has favorable moist-curing	3	2	3	2
Partial load, exposed			6	4
Fully stressed, exposed			See Table B below	

B. RECOMMENDED DURATION OF CONCRETE TEMPERATURE FOR FULLY STRESSED, EXPOSED, AIR-ENTRAINED CONCRETE

Required Percentage of Standard-Cured 28-Day Strength	Days at 10°C (50°F)			Days at 21°C (70°F)		
	Type of Portland Cement			Type of Portland Cement		
	I or GU	II or MS	III or HE	I or GU	II or MS	III or HE
50	6	9	3	4	6	3
65	11	14	5	8	10	4
85	21	28	16	16	18	12
95	29	35	26	23	24	20

*Adapted from Tables 5.1 and 5.3 of ACI 306. Cold weather is defined as that in which average daily temperature is less than 4°C (40°F) for 3 successive days except that if temperatures above 10°C (50°F) occur during at least 12 hours in any day, the concrete should no longer be regarded as winter concrete and normal curing practice should apply. For recommended concrete temperatures, see Table14-1. For concrete that is *not* air entrained, ACI Committee 306 states that protection for durability should be at least twice the number of days listed in Table A.

Part B was adapted from Table 6.8 of ACI 306R-88. The values shown are approximations and will vary according to the thickness of concrete, mix proportions, etc. They are intended to represent the ages at which supporting forms can be removed. For recommended concrete temperatures, see Table 14-1.

**Made with ASTM Type I, II, GU, or MS Portland cement.

†Made with ASTM Type III or HE cement, or an accelerator, or an extra 60 kg/m^3 (100 lb/yd^3) of cement.

‡"Exposed" means subject to freezing and thawing.

Reproduced from Design and Control of Concrete Mixtures, Courtesy Portland Cement Association, www.cement.org

CURING CONCRETE

Hydration of cement depends upon temperature, time, and the presence of moisture. The longer the concrete hardens, the stronger it becomes. Hydration stops when the concrete dries, so it must be kept from drying as long as possible. Warm concrete hardens faster than cold concrete. Keep concrete warm and prevent from freezing.

CONCRETE CURING METHODS

Method	Advantage	Considerations
Sprinkling with water	Excellent results	Must be kept wet
Straw	Provides insulation in winter	Can dry out and blow away
Ponding on flat surfaces	Excellent results, maintains uniform temperature	Requires considerable labor, not used in freezing weather
Moist earth	Low cost	Messy, stains concrete, can dry out
Curing compounds	Good results, easy to apply	Poor coverage will allow drying in spots, need a sprayer
Waterproof paper	Excellent protection	Expensive, hard to handle, is in rolls
Plastic film	Excellent results	Can tear, must be repaired, must be weighted down
Burlap and other absorbing fabrics	Excellent results, low cost	Must be kept wet

CHECKLIST OF COMMON FIELD PROBLEMS: CAUSES AND PREVENTION

Problem	Cause	Prevention or Correction
Fresh Concrete		
Excessive Bleeding	Insufficient fines in mix	Increase percent of fines—cement, fly ash, sand content. Introduce or increase air entrainment.
	Excess mix water	Reduce water content.
	Vapor retarder directly under slab	Compact 3 inches of sand or crusher-run gravel on the vapor retarder (also applicable to plastic shrinkage cracks and curling).
Segregation	High slump (excess water	Reduce water content. Use superplasticizer for desired slump.
	Overvibration	Don't vibrate concrete that is already flowable (superplasticized excepted).
	Inadequate vibration	Insert vibrator at closer intervals. Vibrate until concrete is flowable.
	Excessive drop in placing	Reduce free drop (use drop chutes).
	Lack of homogeneity in mix	Introduce air entrainment. Reduce coarse aggregate proportion in mix.
Sticky finish	High air content and/or over-sanded mix	Promote bleeding—reduce air, percent sand.
	Rapid drying of the surface	Dampen subgrade and forms. Apply fog spray.

CHECKLIST OF COMMON FIELD PROBLEMS: CAUSES AND PREVENTION *(cont.)*

Problem	Cause	Prevention or Correction
Fresh Concrete		
Rapid set (hot weather)	High concrete temperature High ambient temperature	Cool water, add ice or liquid nitrogen. Cool aggregate piles by sprinkling. Use maximum retarder dosage. Consider fly-ash mixes.
	High cement content	Introduces water-reducing retarder and/or fly ash into mix.
	Trucks waiting in sun	Schedule trucks for shortest wait, in shaded area if possible. Sprinkle outside of mixer drum.
Slow set (cool weather)	Lean mix—especially with fly ash or slag	Increase cement. Use accelerator. Heat aggregates and water.
	Cool or wet subgrade	Place plastic type (polyethylene) sheet on subgrade. Protect subgrade (cover with straw or mats).
Plastic shrinkage Cracking	Rapid evaporation of water from the surface primarily from wind and low humidity	Fog spray on surface at time of finishing. Induce water gain on surface (not to excess) by reducing sand and/or air entrainment. Reduce mixing water. Provide windbreaks. Reduce concrete temperature.

Courtesy American Concrete Institute, www.concrete.org

WELDED WIRE REINFORCEMENT

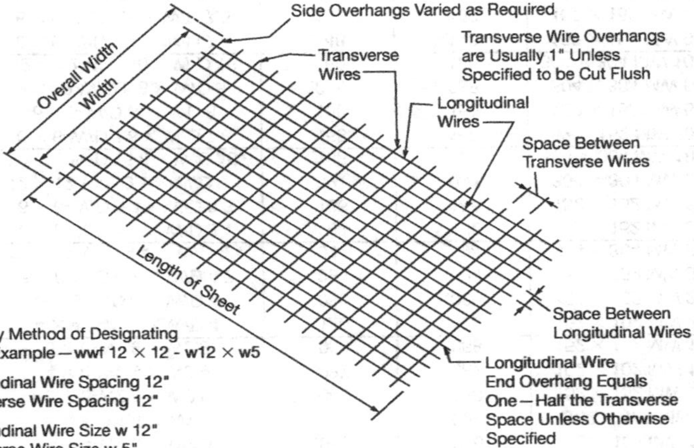

Side Overhangs Varied as Required

Transverse Wire Overhangs are Usually 1" Unless Specified to be Cut Flush

Transverse Wires

Longitudinal Wires

Space Between Transverse Wires

Overall Width

Width

Length of Sheet

Space Between Longitudinal Wires

Longitudinal Wire End Overhang Equals One—Half the Transverse Space Unless Otherwise Specified

Industry Method of Designating
Style: Example—wwf 12 × 12 - w12 × w5

Longitudinal Wire Spacing 12"
Transverse Wire Spacing 12"

Longitudinal Wire Size w 12"
Transverse Wire Size w 5"

Courtesy American Society for Testing and Materials, www.astm.org

COMMON SIZES OF WELDED WIRE REINFORCEMENT

Yield Strength (min) (fy in psi)	Style Designation (inches) (W = Plain, D = Deformed)	Steel Area (in.²/1'-0")		Metric (mm) Style Designation
		Longit.	Trans.	
65,000 (W only)	4 × 4-W1.4 × W1.4	.042	.042	102 × 102 MW 9.1 × MW 9.1
	4 × 4-W2.0 × W2.0	.060	.060	102 × 102 MW 13.3 × MW 13.3
	6 × 6-W1.4 × W1.4	.028	.028	152 × 152 MW 9.1 × MW 9.1
	6 × 6-W2.0 × W2.0	.040	.040	152 × 152 MW 13.3 × MW 13.3
	4 × 4-W2.9 × W2.9	.087	.087	102 × 102 MW 18.7 × MW 18.7
	6 × 6-W2.9 × W2.9	.058	.058	152 × 152 MW 18.7 × MW 18.7
70,000 (W & D)	4 × 4-W/D 4 × W/D 4	.120	.120	102 × 102 MW 25.8 × MW 25.8
	6 × 6-W/D 4 × W/D 4	.080	.080	152 × 152 MW 25.8 × MW 25.8
	6 × 6-W/D 4.7 × W/D 4.7	.094	.094	152 × 152 MW 30.3 × MW 30.3
	12 × 12-W/D 9.4 × W/D 9.4	.094	.094	304 × 304 MW 60.6 × MW 60.6
72,500 (W & D)	6 × 6-W/D 8.1 × W/D 8.1	.162	.162	152 × 152 MW 52.3 × MW 52.3
	6 × 6-W/D 8.3 × W/D 8.3	.166	.166	152 × 152 MW 53.5 × MW 53.5
	12 × 12-W/D 9.1 × W/D 9.1	.091	.091	304 × 304 MW 58.7 × MW 58.7
	12 × 12-W/D 16.6 × W/D 16.6	.166	.166	304 × 304 MW 107.1 × MW 107.1
75,000 (W & D)	6 × 6-W/D 7.8 × W/D 7.8	.156	.156	152 × 152 MW 100.6 × MW 100.6
	6 × 6-W/D 8 × W/D 8	.160	.160	152 × 152 MW 51.6 × MW 51.6
	12 × 12-W/D 8.8 × W/D 8.8	.088	.088	304 × 304 MW 56.8 × MW 56.8
	12 × 12-W/D 16 × W/D 16	.160	.160	304 × 304 MW 103.2 × MW 103.2
80,000 (W & D)	6 × 6-W/D 7.4 × W/D 7.4	.148	.148	152 × 152 MW 95.5 × MW 95.5
	6 × 6-W/D 7.5 × W/D 7.5	.150	.150	152 × 152 MW 48.4 × MW 48.4
	12 × 12 W/D 8.3 × W/D 8.3	.083	.083	304 × 304 MW 53.5 × MW 53.5

*These are soft conversions from the inch sizes

Courtesy Wire Reinforcement Institute, www.wirereinforcementinstitute.org

WELDED WIRE REINFORCEMENT WIRE SIZE EXAMPLES*

W = Plain Wire
D = Deformed Wire

in.-lb	Metric
W1.4 = 0.014 sq. in.	9 mm
W4 = 0.04 sq. in.	26 mm
W11 = 0.11 sq. in.	71 mm
D20 = 0.20 sq. in.	130 mm

*Wire size number = cross sectional area multiplied by 100.

Courtesy Wire Reinforcement Institute,
www.wirereinforcementinstitute.org

WELDED WIRE REINFORCEMENT WIRE SPACINGS

Inch 4, 6, 8, 12, 16, 18 (some up to 20, 24 and 48 available)

Metric (mm) 100, 150, 200, 300, 400 (some up to 600 and 1200 available)

Smaller sizes as low as 2" and 3" and 50 mm and 75 mm available.

Rolls 60" × 150' long and 96" × 100' long
(150 mm × 46 m long and 2400 mm × 30 m long)

Courtesy Wire Reinforcement Institute,
www.wirereinforcementinstitute.org

SOME TYPICAL REBAR BENDS

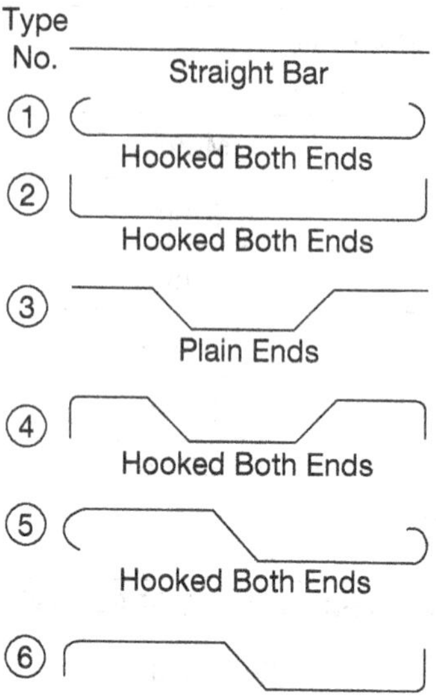

Type No.

Straight Bar

(1) Hooked Both Ends

(2) Hooked Both Ends

(3) Plain Ends

(4) Hooked Both Ends

(5) Hooked Both Ends

(6)

Some Reinforcing Both Bends

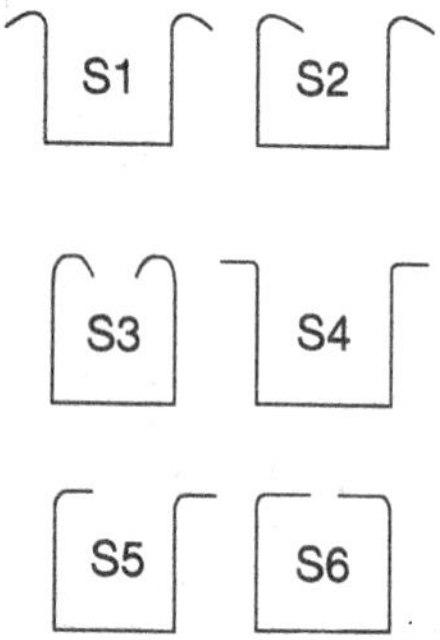

S1 S2

S3 S4

S5 S6

Some Typical Stirrup Bends

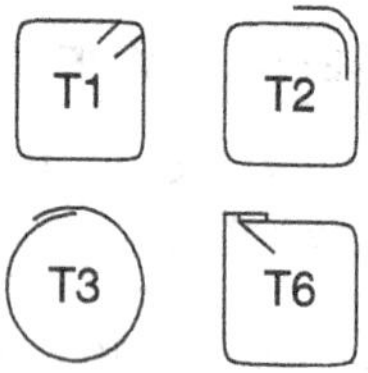

T1 T2

T3 T6

Some Typical Column Ties

Courtesy American Concrete Institute, www.concrete.org and American society of Concrete Contractors, www.ascconline.org

TYPES OF PORTLAND CEMENT		
Canadian Designation	**United States Designation**	**Type**
10	I	Normal
20	II	Moderate
30	III	High early strength
40	IV	Low heat of hydration
50	V	Sulfate resisting
Air-Entrained Types		
10A	IA	Normal air-entrained
20A	IIA	Moderate air-entrained
30A	IIIA	High early strength air-entrained

ASTM INCH-SIZE STEEL REINFORCING BARS

Bar Size Designation	Weight in Pounds Per Foot	Nominal Dimensions Diameters in Inches		Cross-Sectional Area in Square Inches
		Decimal	Fractions	
#3	0.376	0.375	3/8	0.11
#4	0.668	0.500	1/2	0.20
#5	1.043	0.625	5/8	0.31
#6	1.502	0.750	3/4	0.44
#7	2.044	0.875	7/8	0.60
#8	2.670	1.000	1.0	0.79
#9	3.400	1.128	1 1/8	1.00
#10	4.303	1.270	1 1/4	1.27
#11	5.313	1.410	1 3/8	1.56
#14	7.650	1.693	1 11/16	2.25
#18	13.60	2.257	2 1/4	4.00

ASTM METRIC-SIZE STEEL REINFORCING BARS

Bar Size Designation	Mass (kg/m)	Nominal Dimensions Diameters (mm)	Area (mm^2)
#10M	0.785	11.3	100
#15M	1.570	16.0	200
#20M	2.355	19.5	300
#25M	3.925	25.2	500
#30M	5.495	29.9	700
#35M	7.850	35.7	1000
#45M	11.775	43.7	1500
#55M	19.625	56.4	2500

Courtesy American Society for Testing and Materials, www.astm.org

IDENTIFICATION MARKS ASTM STANDARD REBARS

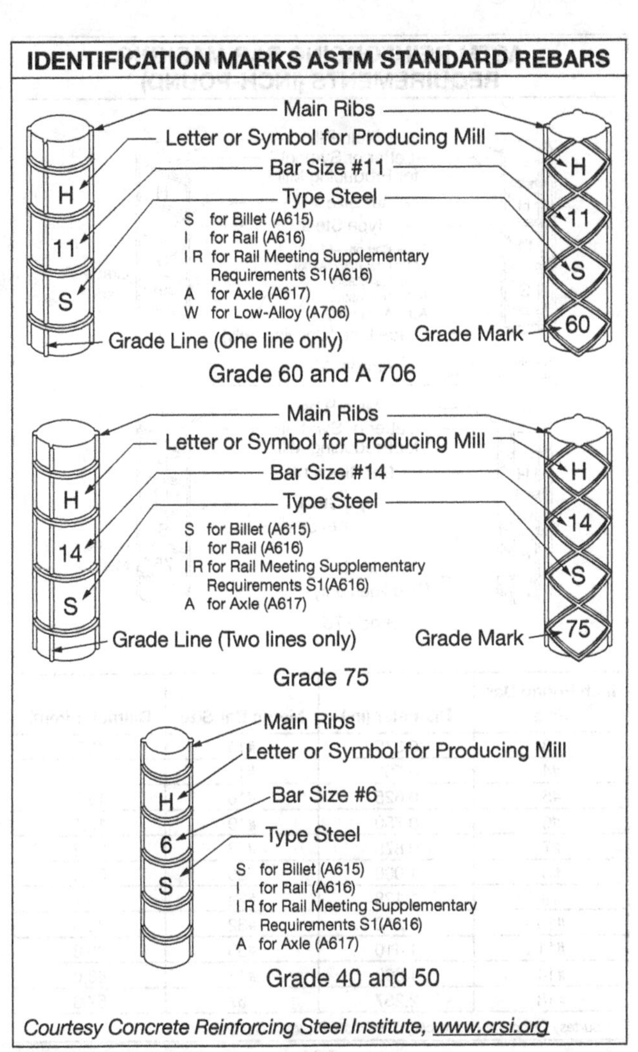

Courtesy Concrete Reinforcing Steel Institute, www.crsi.org

ASTM REINFORCING BAR MARKING REQUIREMENTS (INCH-POUND)

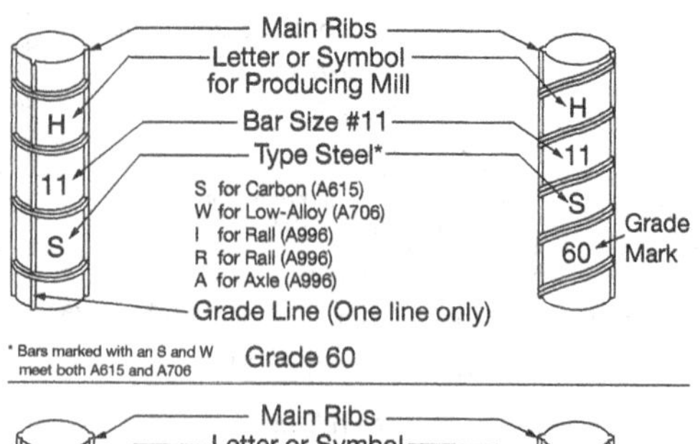

Grade 60

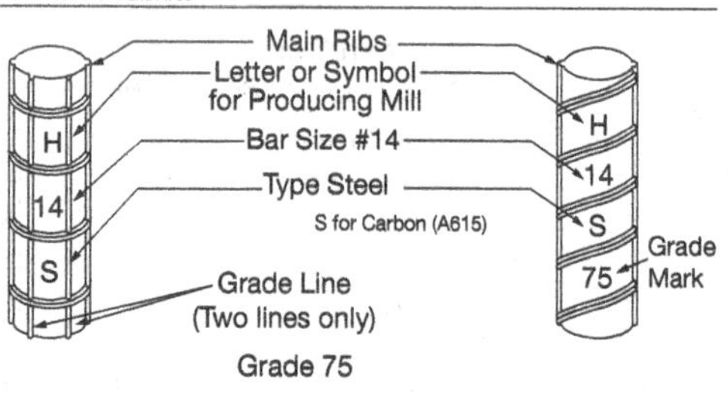

Grade 75

Inch Pound Bar Size	Diameter (in.)	Metric Bar Size	Diameter (mm)
#3	0.375	#10	9.5
#4	0.500	#13	12.7
#5	0.625	#16	15.9
#6	0.750	#19	19.1
#7	0.875	#22	22.2
#8	1.000	#25	25.4
#9	1.128	#29	28.7
#10	1.270	#32	32.3
#11	1.410	#36	35.8
#14	1.693	#43	43.0
#18	2.257	#57	57.3

Courtesy Concrete Reinforcing Steel Institute, www.crsi.org

MINIMUM SIZE AND REINFORCEMENT FOR ISOLATED FOOTINGS FOR RESIDENTIAL CONCRETE CONSTRUCTION*

	Allowable Soil-Bearing Capacity, lb/sq. ft.					
Tributary Area	**1500**	**2000**	**2500**	**3000**	**3500**	**4000**
Footing supporting roof load**	36 × 36 × 8 in. with 3 No. 4 each way	30 × 30 × 8 in. with 3 No. 4 each way	30 × 30 × 8 in. with 3 No. 4 each way	24 × 24 × 8 in. with 3 No. 4 each way	24 × 24 × 8 in. with 3 No. 4 each way	24 × 24 × 8 in. with 3 No. 4 each way
Footing supporting roof and one floor**	48 × 48 × 10 in. with 3 No. 4 each way	48 × 48 × 10 in. with 3 No. 4 each way	36 × 36 × 10 in. with 3 No. 4 each way	36 × 36 × 10 in. with 3 No. 4 each way	30 × 30 × 10 in. with 3 No. 4 each way	30 × 30 × 10 in. with 3 No. 4 each way
Footing supporting roof and two floors**	60 × 60 × 12 in. with 4 No. 5 each way	60 × 60 × 12 in. with 4 No. 5 each way	48 × 48 × 12 in. with 4 No. 5 each way	48 × 48 × 12 in. with 4 No. 5 each way	42 × 42 × 12 in. with 3 No. 5 each way	36 × 36 × 12 in. with 3 No. 5 each way

*Minimum concrete strength f shall be 2500 psi.

Minimum yield strength f shall be 40,000 psi.

**Maximum tributary area is 20 × 32 ft.

Courtesy American Concrete Institute www.concrete.org

READY-MIXED CONCRETE

Specifications for ready-mixed concrete are found in ASTM Designation: C94/C 94M – 04 Standard Specification for Ready-Mixed Concrete. This covers ready-mixed concrete manufactured and delivered to a purchaser in a freshly mixed and unhardened state. Additional information is available in the publication. Users Guide to ASTM Specification C94 on Ready-Mixed Concrete. It was developed by ASTM International (www.astm.org) and The National Ready-Mixed Concrete Association (www.nrmca.org).

READY-MIXED CONCRETE DELIVERY TIME

Maximum delivery time is the time elapsed between when the water is added to the cement in the batch to the time of discharge. Typically this is 90 minutes or 300 revolutions of the drum. In hot weather, typically 60 minutes is maximum. The delivery time may not be the same for various mixes and conditions. For example, consideration is given to the water content, types of cement, temperature of the material, ambient temperatures, and if fly ash or retarders have been added. ASTM C94 has provisions for waiving the time limit of the concrete under certain conditions.

THREE OPTIONS AVAILABLE TO THE CONTRACTOR WHEN ORDERING READY MIXED CONCRETE

Item		Option A	Option B	Option C
Maximum nominal aggregate size	By	Purchaser	Purchaser	Purchaser
Slump at point of delivery	By	Purchaser	Purchaser	Purchaser
Air content at point of delivery	By	Purchaser	Purchaser	Purchaser
Selection of ordering Option "A", "B," or "C"	By	Purchaser	Purchaser	Purchaser
Density (unit weight) for structural lightweight conc.	By	Purchaser	Purchaser	Purchaser
Mixture proportions	By	Manufacturer	Purchaser	Manufacturer
Cement content	By	Manufacturer	Purchaser	Manufacturer (above min.)
Compressive strength specified	By	Purchaser		Purchaser
Minimum cement content	By			Purchaser

THREE OPTIONS AVAILABLE TO THE CONTRACTOR WHEN ORDERING READY MIXED CONCRETE *(cont.)*

Item		Option A	Option B	Option C
List of ingredients and quantities of each	By	Manufacturer	Manufacturer	Manufacturer
Proof of mixture strength	By	Manufacturer		Manufacturer
Proof of material quality	By	Manufacturer		Manufacturer
Proof of other concrete qualities specified	By	Manufacturer		Manufacturer
Maximum water content	By		Purchaser	
Maximum water-cement ratio	By			
Required admixtures	By		Purchaser	Purchaser
Aggregate relative density (sp gr), gradation, sources	By		Manufacturer	

The purchaser may select any one of the three options and then furnish the specific information required.
Courtesy National Ready Mixed Concrete Association, www.nrmca.org

MINIMUM WIDTHS OF WALL FOOTINGS FOR RESIDENTIAL CONCRETE CONSTRUCTION*

	No. of Stories Above Grade**	Allowable Soil-Bearing Capacity, lb/sq. ft.					
		1500	2000	2500	3000	3500	4000
Convention wood frame construction (above grade)	One-story	16	12	10	8	7	6
	Two-story	19	15	12	10	8	7
	Three-story	22	17	14	11	10	9
4-in. brick veneer over wood frame: 8 in. hollow concrete masonry unit (above grade)	One-story	19	15	12	10	8	7
	Two-story	25	19	15	13	11	10
	Three-story	31	23	19	16	13	12
8-in. grouted concrete masonry unit	One-story	22	17	13	11	10	9
	Two-story	31	23	19	16	13	12
	Three-story	40	30	24	20	17	15

*Minimum concrete strength f_c' shall be 2500 psi.

Minimum footing widths that are greater than the wall thickness shall project a minimum of 2 in. on both sides of the wall. The footing width projection shall be measured from the face of the concrete to the edge of the footing.

**Table includes foundation (for example, a one-story includes the story above grade and a foundation).

Courtesy American Concrete Institute, www.concrete.org

APPROXIMATE COVERAGE PROVIDED BY ONE CUBIC YARD OF CONCRETE	
Slab Thickness Inches	Slab Area Square Feet
2	162
3	108
4	81
5	65
6	54
7	46
8	40

CUBIC YARDS OF CONCRETE IN SLABS						
Area in Square Feet (length × feet)	Slab Thickness (inches)					
	4	**5**	**6**	**8**	**10**	**12**
50	0.61	0.75	0.92	1.2	1.5	1.85
100	1.2	1.5	1.84	2.4	3.0	3.7
200	2.4	3.0	3.7	4.8	6.0	7.4

RECOMMENDED THICKNESSES FOR CONCRETE SLABS	
Sidewalks	4 to 6 inches
Driveways	6 to 8 inches
Porch floors	4 to 5 inches
Garage floors in Residences	4 to 5 inches
Slab on grade – residential floor	4 to 5 inches
Basement floors – residential	4 inches

CALCULATING CONCRETE VOLUME NEEDED

Multiply the length in feet by width in feet by thickness in feet. Example: a slab 40 feet long by 10 feet wide by 6 inches (.5 feet) thick equals 200 cubic feet. Concrete is sold by the cubic yard, so divide the 200 by 27 to get the required cubic yards, which is 7.4 cubic yards. There are 27 cubic feet in a cubic yard.

TYPICAL FOOTING SIZES USED FOR LIGHT FRAME CONSTRUCTION*

Number of Floors to be Supported	Width of Footing Inches	Thickness of Footing Inches
1	12	6
2	15	6
3	18	8

*Soil conditions can cause an increase in the size of the footing.

ALLOWABLE SERVICE LOAD ON EMBEDDED BOLTS (POUNDS)

Bolt Diameter (inches)	Minimum Embedment (inches)	Edge Distance (inches)	Spacing (inches)	Minimum Concrete Strength (psi)					
				$f'_c = 2500$		$f'_c = 3000$		$f'_c = 4000$	
				Tension	Shear	Tension	Shear	Tension	Shear
¼	2½	1½	3	200	500	200	500	200	500
⅜	3	2¼	4½	500	1100	500	1100	500	1100
½	4	3	6	950	1250	950	1250	950	1250
	4	5	5	1450	1600	1500	1650	1550	1750
⅝	4½	3¾	7½	1500	2750	1500	2750	1500	2750
	4½	6¼	7½	2125	2950	2200	3000	2400	3050
¾	5	4½	9	2250	3250	2250	3560	2250	3560
	5	7½	9	2825	4275	2950	4300	3200	4400
⅞	6	5¼	10½	2550	3700	2550	4050	2550	4050
1	7	6	12	3050	4125	3250	4500	3650	5300
1⅛	8	6¾	13½	3400	4750	3400	4750	3400	4750
1¼	9	7½	15	4000	5800	4000	5800	4000	5800

For Shear 1 inch = 25.4 mm, 1 pound per square inch = 0.00689 Mpa, 1 pound = 4.45N.

RELATIONSHIP BETWEEN WATER-CEMENT RATIO AND COMPRESSIVE STRENGTH OF CONCRETE		
Compressive Strength @ 28 Days (psi)	Water-Cement Ratio, by Weight	
	Non-Air-Entrained Concrete	Air-Entrained Concrete
6000	0.41	–
5000	0.48	0.40
4000	0.57	0.48
3000	0.68	0.59
2000	0.82	0.74

Courtesy American Concrete Institute, www.concrete.org and American Society of Concrete Contractors, www.ascconline.org

MAXIMUM PERMISSIBLE WATER-CEMENT RATIOS FOR CONCRETE IN SEVERE EXPOSURES		
Type of Structure	Structure Wet Continuously or Frequently Exposed to Freezing and Thawing	Structure Exposed to Sea Water or Sulfates
Thin sections (railings, curbs, sills, ledges, ornamental work) and sections less than 1 in. cover over steel	0.45	0.40
All other structures	0.50	0.45

Courtesy American Concrete Institute, www.concrete.org and American Society of Concrete Contractors, www.ascconline.org

RECOMMENDED AIR CONTENTS FOR FROST-RESISTANT CONCRETE

Maximum Aggregate Size (in.)	Average Air Content (Percent)*	
	Severe Exposure†	Moderate Exposure‡
3/8	7-1/2	6
1/2	7	5-1/2
3/4	6	5
1-1/2	5-1/2	4-1/2
3	4-1/2	3-1/2
6	4	3

*A reasonable tolerance for air content in field construction is ± 1-1/2 percent.

†Outdoor exposure in a cold climate where the concrete may be in almost continuous contact with moisture prior to freezing, or where deicing salts are used. Examples are pavements, bridge decks, sidewalks, and water tanks.

‡Outdoor exposure in a cold climate where concrete will be only occasionally exposed to moisture prior to freezing, and where no deicing salts will be used. Examples are exterior walls, beams, girders, and slabs not in direct contact with soil.

Table 2.1 is adapted from ACI 201.2R, "Guide to Durable Concrete."

Courtesy American Concrete Institute, www.concrete.org and American Society of Concrete Contractors, www.ascconline.org

EXAMPLE OF CONCRETE MIXES (NON-AIR-ENTRAINED)

		Mix #1		Mix #2		Mix #3	
	Specific Gravity	Absolute Volume, ft^3	Ingredients, lb/yd^3	Absolute Volume, ft^3	Ingredients, lb/yd^3	Absolute Volume, ft^3	Ingredients, lb/yd^3
Cement, lb	3.15	2.88	566	2.65	521	1.98	390
Fly ash, lb	2.40	0	0	0	0	1.00	150
Sand, SSD, lb	2.65	7.5	1240	8.12	1343	7.79	1288
Stone, SSD, lb	2.60	11.54	1872	11.54	1872	11.54	1872
Water, lb	1.00	4.81	300	4.42	276	4.42	276
Air (%)	—	0.27	—	0.27	—	0.27	—
Water-reducing admixture, oz.	—	—	0	—	31.3	—	32.4
Mix weight, lb	—	—	3978	—	4012	—	3976
Density, lb/ft^3	—	—	147.3	—	148.6	—	147.3
w/cm	—	—	0.53	—	0.53	—	0.51

Courtesy American Concrete Institute, www.concrete.org and American Society of Concrete Contractors, www.ascconline.org

EXAMPLE CONCRETE MIX #1 (NON-AIR-ENTRAINED), EFFECT OF FREE MOISTURE IN AGGREGATES (FMA) ON BATCH WEIGHT

	Specific Gravity	Absolute Volume, ft^3	Ingredients, Batch A. lb/yd^3	FMA Sand 0% Stone 0% (SSD)	Ingredients, Batch B, lb/yd^3	FMA Sand 0% Stone 0% (dry)	Ingredients, Batch C, lb/yd^3	FMA Sand 0% Stone 0% (normal)	Ingredients, Batch D, lb/yd^3	FMA Sand 0% Stone 0% (wet)
Cement, lb	3.15	2.88	566	—	566	—	566	—	566	—
Fly ash, lb	2.40	0	0		0	—	0	—	0	—
Sand, SSD, lb	2.65	7.5	1240	0	1278	38	1305	65	1333	93
Stone, SSD, lb	2.60	11.54	1872	0	1872	0	1872	0	1891	19
Water, lb	1.00	4.81	300	—	262	—	235	—	188	—
Air (%)	—	0.27	—	—	—	—	—	—	—	—
Water-reducing admixture, oz.	—	—	0	—	0	—	0	—	0	—
Mix weight, lb	—	—	3978	—	3978	—	3978	—	3978	—
Density, lb/ft^3	—	—	147.3	—	147.3	—	147.3	—	147.3	—
w/cm	—	—	0.53		0.53		0.53		0.53	

Courtesy American Concrete Institute, www.concrete.org and American society of Concrete Contractors, www.ascconline.org

CHAPTER 8
Masonry Construction

TYPES OF MORTAR
TYPE M – High-early strength, compressive strength of 2,500 psi (17 MPa). Use below grade, in contact with earth and areas of severe frost action. Use for reinforced masonry.
TYPE S – Medium-high strength, compressive strength 1,800 psi (12.5 MPa). Use in wall construction above grade and with reinforced and unreinforced masonry.
TYPE N – Medium strength, compressive strength 750 psi (5 MPa). Use for general above-grade construction where requirements for compressive and laterial strength are not high.
TYPE O – Medium-strength, compressive strength 350 psi (2.5 MPa). Use for general internal purposes where compressive strength does not exceed 100 lb per sq. in.

TYPICAL COMPRESSIVE STRENGTHS OF VARIOUS TYPES OF MORTAR	
Mortar	**Compressive Strength (PSI)**
Type M	2500
Type S	1800
Type N	750
Type O	350
Type K	75

MORTAR ADMIXTURES

Admixtures are ingredients added to mortar to alter the properties. They are used sparingly and only on the advice of a concrete specialist.

AIR ENTRAINMENT – Used in mortar that will be exposed to freeze-thaw cycles. It improves workability. Typical ingredients include neutralized vinsol resins, organic acid salts, fatty acids, and hydrocarbon derivaties.

ACCELERATORS – Used in cold weather construction to speed the hydration process. Typical ingredients include calcium chloride and various nonchloride products.

BOND MODIFIERS – Used to improve adhesion and surface density. Typical ingredients include acrylic polymer latex, polyvinyl acetate, styrene butadiene rubber, and metholcellulose.

MORTAR ADMIXTURES *(cont.)*

CORROSION INHIBITORS – Used to prevent steel from corroding and to offset reactions from chloride. Typically this is calcium nitrate.

COLOR PIGMENTS – Used to color the mortar. Typical ingredients used include carbon black, red iron oxide, cobalt, and chromium oxide.

EXTENDED-LIFE RETARDERS – Used to slow the reaction time of the Portland cement and extend the useful life typically from 12 to 24 hours. The typical ingredient is hydroxycarboxylic acids.

INTEGRAL WATER REPELLANTS – Used to reduce water absorption of mortar after it has hardened. Typical ingredients include polymer compounds or stearates.

RETARDERS – Used to slow the set of fresh mortar for several hours and helps control shrinkage. Typical ingredients include sodium gluconate, sodium lignosulfate, or sodium citrate.

MORTAR COLOR AND PIGMENTS

Mortar Color	Pigment
Black	Synthetic + carbon black
Bright red	Synthetic red iron oxide
Buff	Natural yellow + red
Blue	Cobalt
Chocolate	Natural red + yellow + carbon black
Red/pink	Natural red iron oxide
Green	Chromium oxide

PORTLAND CEMENT AND MASONRY MORTAR QUANTITIES

Portland cement-lime mortars are a mixture of Portland cement, sand, and lime.

Masonry cements are a mixture of a Portland cement and sand.

QUANTITIES TO MAKE A CUBIC YARD OF CEMENT-LIME MORTAR

Mortar Type	Sand Cubic Ft.	Portland Cement Cubic Ft.	Lime Cubic Ft.
M	27	7.5	2.0
S	27	6.0	3.0
N	27	4.5	4.5
O	27	3.0	6.0

QUANTITIES TO MAKE A CUBIC YARD OF MORTAR CEMENT

Mortar Type	Sand Cubic Ft.	Portland Cement Cubic Ft.
M	27	9.0
S	27	9.0
N	27	9.0
O	27	9.0

READY-MIXED MORTAR

The standard for ready-mixed mortar is ASTM C1142, Standard Specification for Ready-Mixed Mortar for Unit Masonry. It is mixed at a plant and delivered to the site in containers or ready-mix trucks. It must be stored in covered containers and protected from freezing. A set retarder keeps it plastic and workable for 24 to 36 hours, and in some cases longer. The mortar sets in the normal manner.

Types of ready-mixed mortar are RM, RS, RN and RO. They have the same water retention and compressive strength as the conventional mortars M, S, N and O.

PROPERTY REQUIREMENTS OF READY-MIXED MORTAR

Mortar Type	Minimum Compressive Strength[1] (psi)	Minimum Water Retention (%)	Maximum Air Content[2] (%)
RM	2500	75	18
RS	1800	75	18
RN	750	75	18
RO	350	75	18

[1]Average minimum compressive strength at 28 days.

[2]When structural reinforcement is incorporated into mortar, the maximum air content must be 12 percent, or bond strength test data must be provided to justify higher air content.

ASTM C1142, Standard Specification for Ready-Mixed Mortar for Unit Masonry, with permission American Society for Testing and Materials, www.astm.org.

RECOMMENDATIONS FOR PLACING MORTAR IN COLD WEATHER	
Ambient Temperature	**Cold Weather Procedures for Work in Progress**
Above 40°F (4.4°C)	No special requirements.
Below 40°F (4.4°C)	Do not lay glass unit masonry.
32°F to 40°F (0°C to 4.4°C)	Heat sand or mixing water to produce mortar temperature between 40°F and 120°F (4.4°C and 48.9°C) at the time of mixing. Heat materials for grout only if they are below 32°F (0°C).
25°F to 32°F (−3.9°C to 0°C)	Heat sand or mixing water to produce mortar temperature between 40°F and 120°F (4.4°C and 48.9°C) at the time of mixing. Keep mortar above freezing until used in masonry. Heat materials to produce grout temperature between 70°F and 120°F (21.1°C and 48.9°C) at the time of mixing. Keep grout temperature above 70°F (21.1°C) at the time of placement.
20°F to 25°F (−6.7°C to −3.9°C)	In addition to requirements for 25°F to 32°F (−3.9°C to 0°C), heat masonry surfaces under construction to 40°F (4.4°C) and use wind breaks or enclosures when the wind velocity exceeds 15 mph (24 km/h). Heat masonry to a minimum of 40°F (4.4°C) prior to grouting.

RECOMMENDATIONS FOR PLACING MORTAR IN COLD WEATHER *(cont.)*	
20°F (−6.7°C) and below	In addition to all of the above requirements, provide an enclosure and auxiliary heat to keep air temperature above 32°F (0°C) within the enclosure.
Ambient Temperature (Minimum for Grouted, Mean Daily for Ungrouted)	**Cold Weather Procedures for Newly Completed Masonry**
Above 40°F (4.4°C)	No special requirements.
25°F to 40°F (−3.9°C to 4.4°C)	Cover newly constructed masonry with a weather-resistive membrane for 24 hours after being completed.
20°F to 25°F (−6.7°C to −3.9°C)	Cover newly constructed masonry with weather-resistive insulating blankets (or equal protection) for 24 hours after being completed. Extend the time period to 48 hours for grouted masonry, unless the only cement used in the grout is ASTM C 150 Type III.
20°F (−6.7°C) and below	Keep newly constructed masonry above 32°F (0°C) for at least 24 hours after being completed. Use heated enclosures, electric heating blankets, infrared lamps, or other acceptable methods. Extend the time period to 48 hours for grouted masonry, unless the only cement used in the grout is ASTM C 150 Type III.
Courtesy National Concrete Masonry Association	

CONCRETE MASONRY UNITS

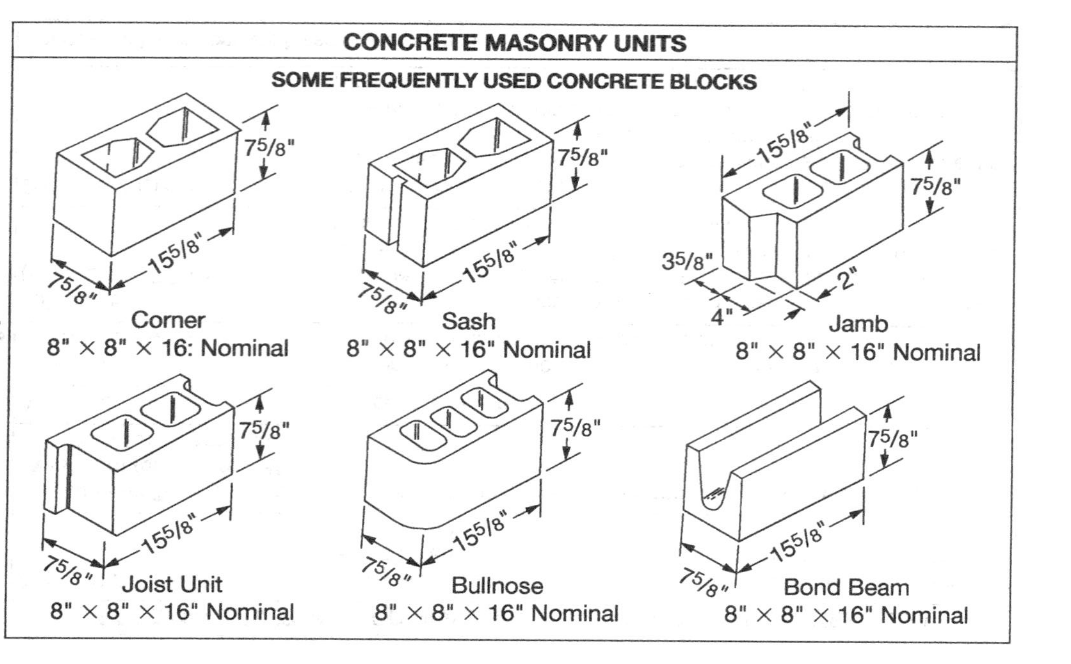

CONCRETE MASONRY UNITS *(cont.)*

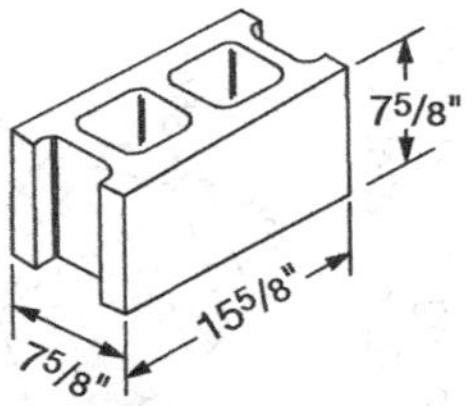

7 5/8"
15 5/8"
7 5/8"

Stretcher (Two Core)
8" × 8" × 16" Nominal

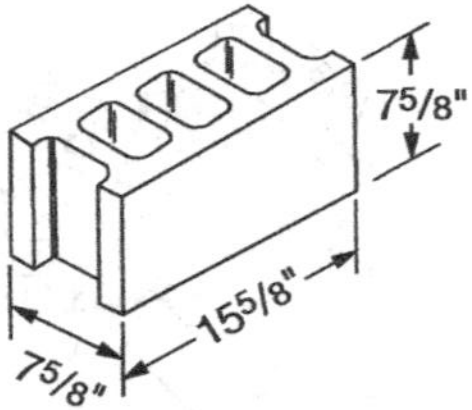

7 5/8"
15 5/8"
7 5/8"

Stretcher (Three Core)
8" × 8" × 16" Nominal

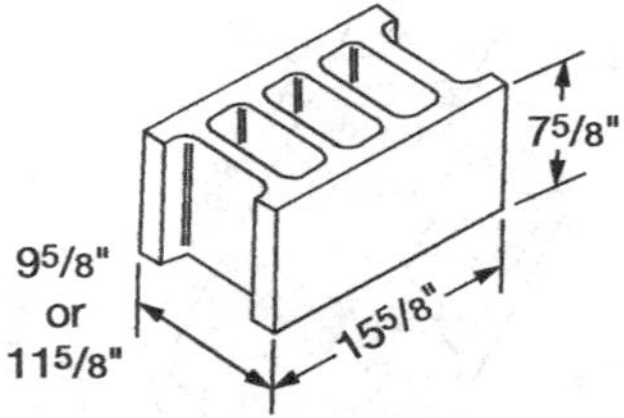

7 5/8"
9 5/8" or 11 5/8"
15 5/8"

10" and 12" Stretcher
10" × 8" × 16" and
12" × 8" × 16" Nominal

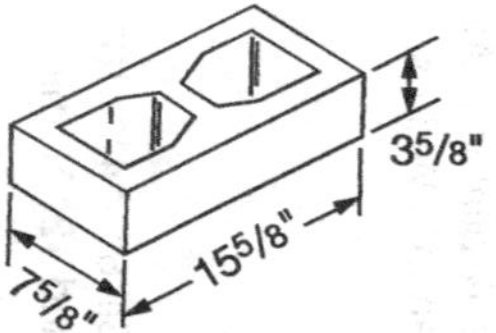

3 5/8"
15 5/8"
7 5/8"

Stretcher
8" × 4" × 16" Nominal

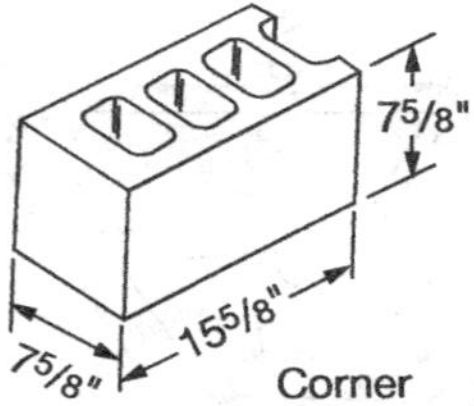

7 5/8"
15 5/8"
7 5/8"

Corner
8" × 8" × 16" Nominal

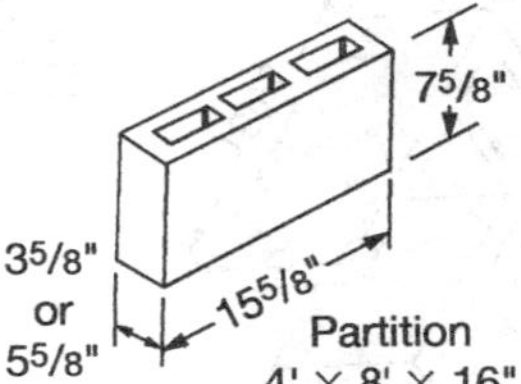

7 5/8"
3 5/8" or 5 5/8"
15 5/8"

Partition
4' × 8' × 16"
6' × 8" × 16" Nominal

CONCRETE MASONRY UNITS *(cont.)*

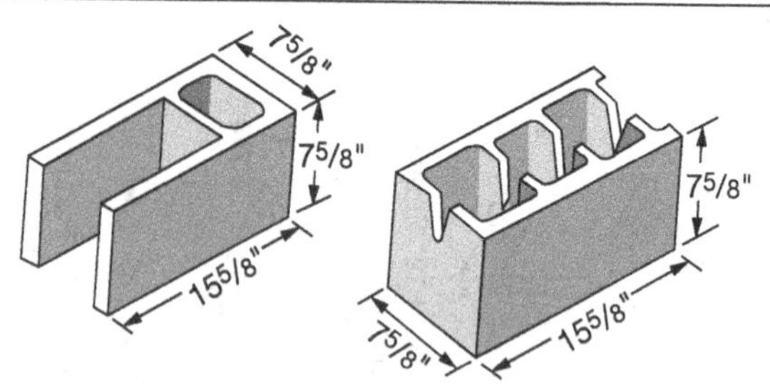

Plumbing or Conduit
8" × 8" × 16" Nominal

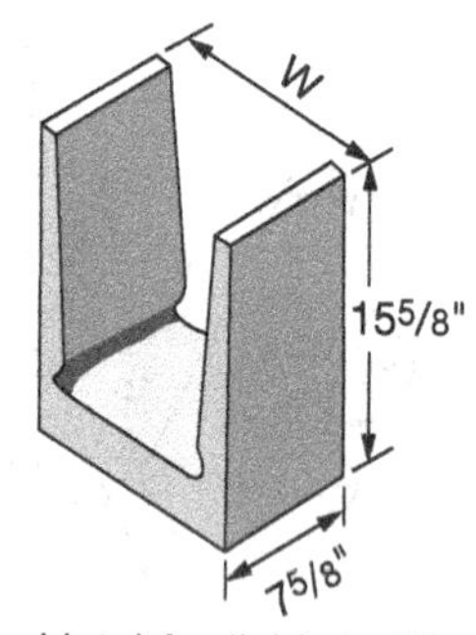

Lintel Available in All
Standard Widths

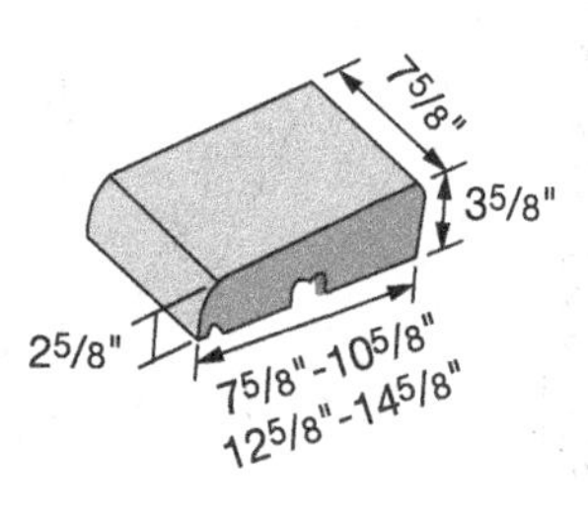

Coping 8" × 4" ×10"
Various Widths

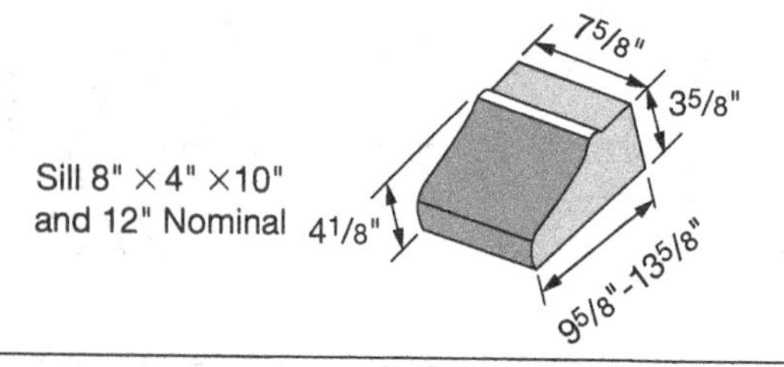

Sill 8" × 4" ×10"
and 12" Nominal

CONCRETE MASONRY UNITS *(cont.)*

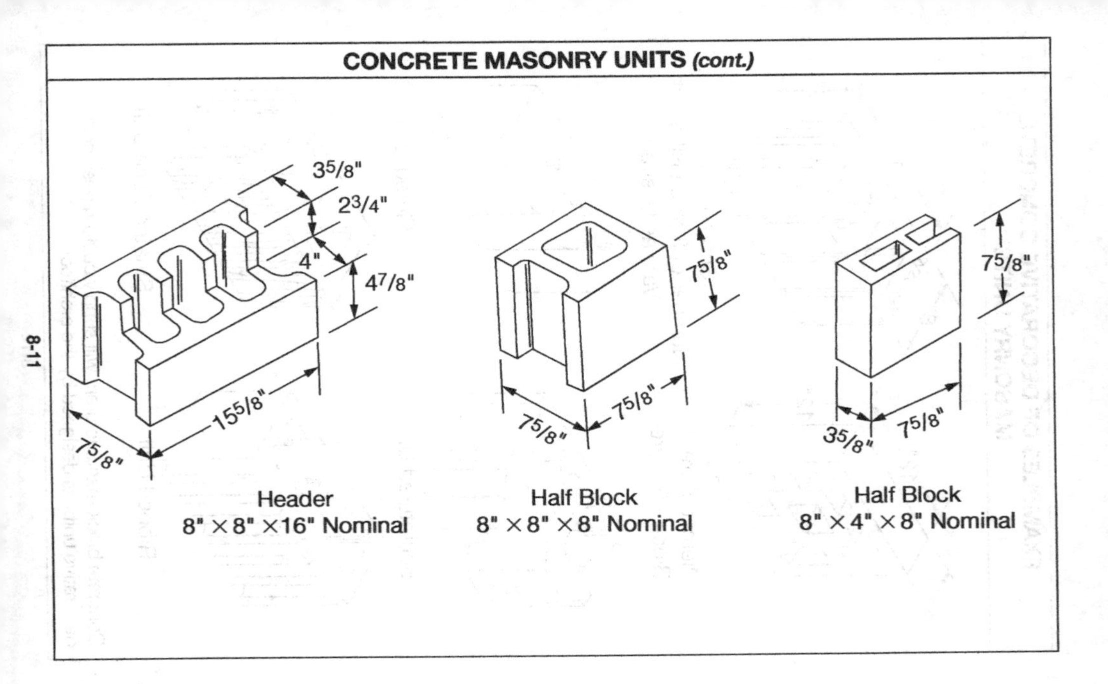

EXAMPLES OF DECORATIVE CONCRETE MASONRY UNITS

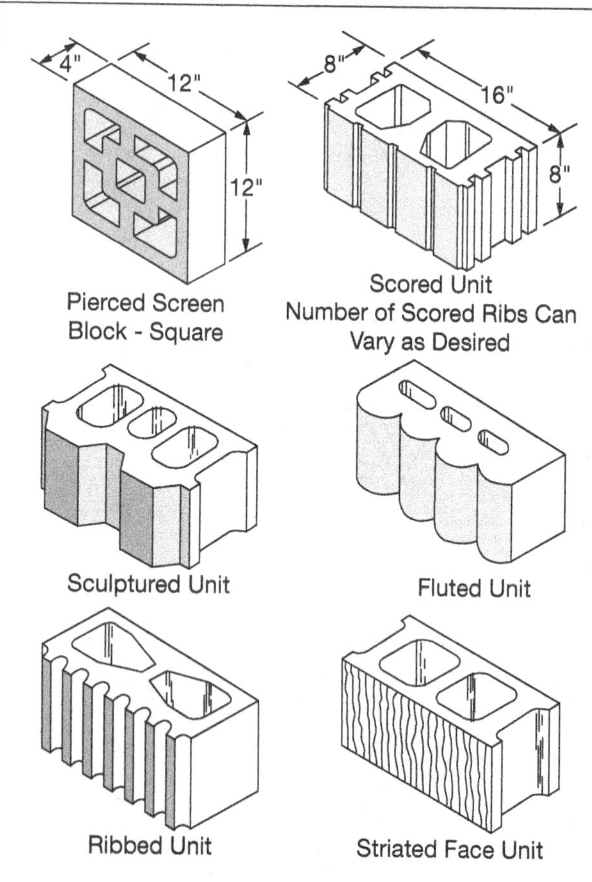

Pierced Screen Block - Square

Scored Unit Number of Scored Ribs Can Vary as Desired

Sculptured Unit

Fluted Unit

Ribbed Unit

Striated Face Unit

Concrete block manufacturers will supply custom designed decorative units as designed by the architect.

MATERIALS REQUIRED FOR 100 SQUARE FEET OF CONCRETE BLOCK WALL				
Type of Block	Wall Thickness	Number of 8 × 16 Inch Blocks	Cement	Sand
Standard	8 inches	112	1 sack	3 cu. ft.

TYPICAL QUANTITIES OF BLOCK AND MORTAR FOR CONCRETE BLOCK WALLS				
	For 100 Square Feet of Wall			Per 100 Blocks
Actual Block Sizes (width × height × length), Inches	Nominal Wall Thickness, Inches	Number of Blocks	Mortar, Cubic Feet	Mortar, Cubic Feet
3⅝ × 3⅝ × 15⅝	4	225	5	2.5
5⅝ × 3⅝ × 15⅝	6	225	5	2.5
7⅝ × 3⅝ × 15⅝	8	225	5	2.5
3¾ × 5 × 11¾	4	221	4	2.3
5¾ × 5 × 11¾	6	221	4	2.3
7¾ × 5 × 11¾	8	221	4	2.3
3⅝ × 7⅝ × 15⅝	4	112.5	3	3
5⅝ × 7⅝ × 15⅝	6	112.5	3	3
7⅝ × 7 ⅝ x 15⅝	8	112.5	3	3
11⅝ × 7⅝ × 15⅝	12	112.5	3	3
Based on ⅜-inch mortar joint				

ESTIMATING THE NUMBER OF 8 × 8 × 16-INCH CONCRETE BLOCKS NEEDED TO BUILD A WALL

To find the number of 16-inch concrete blocks in the wall, find the square feet (length × width) of the wall and multiply by 1.125. For example, a wall with 100 square feet will require 112.5 blocks. This is based on a standard 15⅝ × 7⅝-inch block and ⅜-inch mortar joints.

100 square feet of 8-inch wide concrete block wall requires one sack of cement and 3 cubic feet of sand.

ESTIMATING THE NUMBER OF 16-INCH CONCRETE BLOCKS PER COURSE

To find the number of 16-inch concrete blocks in a course, calculate the length of the wall in inches and divide by 16. For example, a 10-foot wall has 120 inches that, when divided by 16, indicates the course requires 7.5 blocks. This is based on a standard 16-inch block and a ⅜-inch head joint.

ESTIMATING THE NUMBER OF COURSES OF 8-INCH CONCRETE BLOCK FOR THE HEIGHT OF A WALL

To find the number of 8-inch concrete block courses, divide the height of the wall in inches by 8. For example, a wall 10 feet high has 120 inches, which when divided by 8 indicates that 15 courses are required. This is based on a standard 7⅝-inch block and ⅜-inch mortar joints.

TYPICAL CONCRETE BLOCK WALL CONSTRUCTIONS

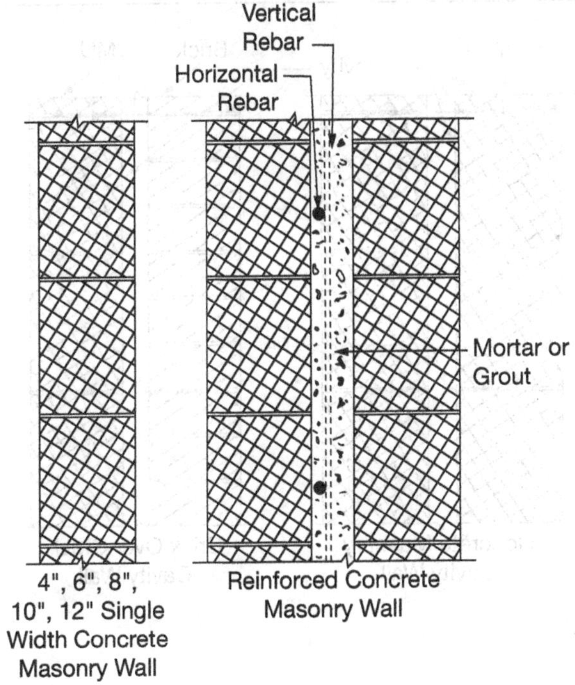

Vertical Rebar

Horizontal Rebar

Mortar or Grout

4", 6", 8", 10", 12" Single Width Concrete Masonry Wall

Reinforced Concrete Masonry Wall

TYPICAL CONCRETE BLOCK WALL CONSTRUCTIONS *(cont.)*

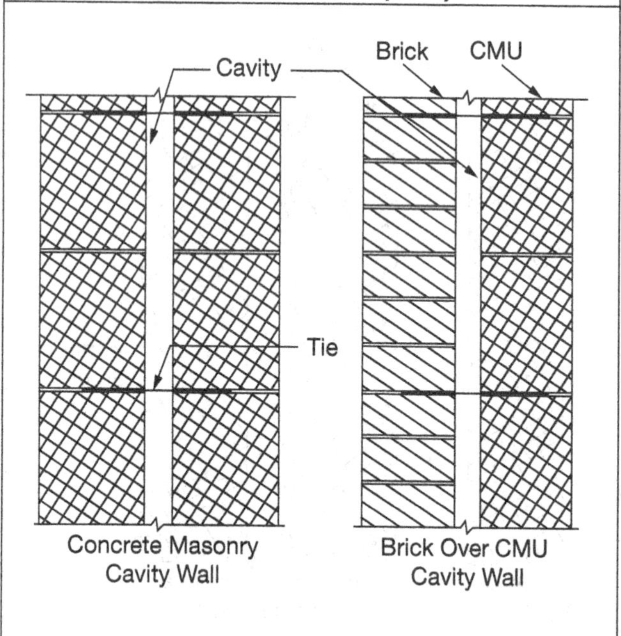

LINTELS USED IN MASONRY CONSTRUCTIONS

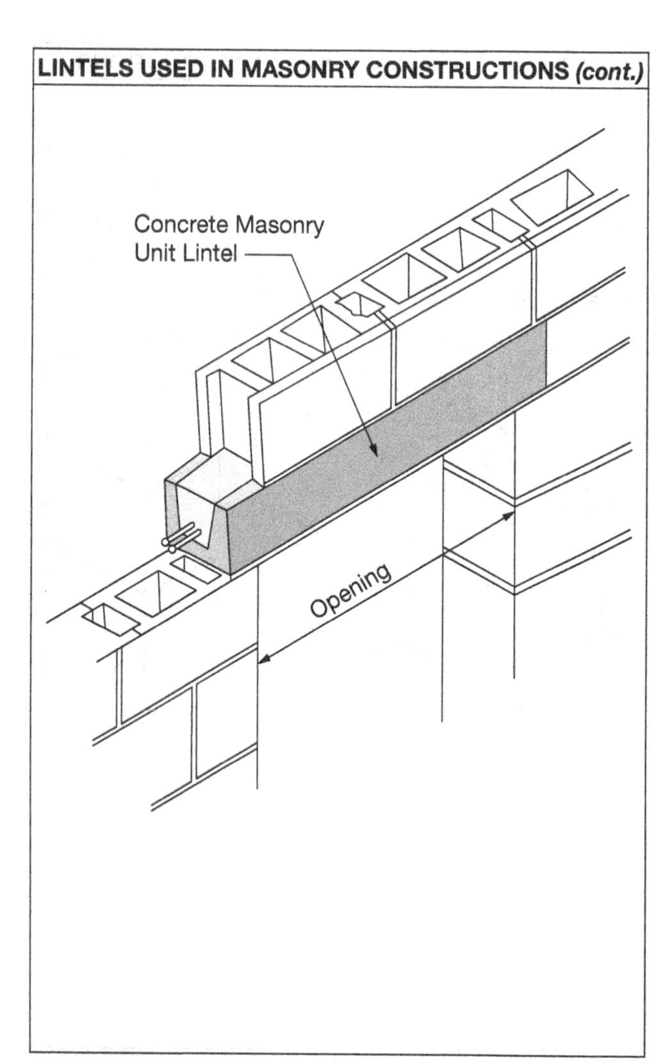
LINTELS USED IN MASONRY CONSTRUCTIONS (cont.)
Concrete Masonry
Unit Lintel
Opening

LINTELS USED IN MASONRY CONSTRUCTIONS *(cont.)*

Precast Concrete Lintel

Opening

MASONRY COLD WEATHER PROTECTION REQUIREMENTS

Mean Daily Air Temperature, C	Protection
0 to 4	Masonry shall be protected from rain or snow.
−4 to 0	Masonry shall be completely covered for 48 hours.
−7 to −4	Masonry shall be completed covered with insulating blankets for 48 hours.
−6 and below	The masonry temperature shall be maintained above 0°C for 48 hours by enclosure and supplementary heat.

The amount of insulation required to properly cure masonry in cold weather is determined on the basis of the expected air temperature and wind velocity and the size and shape of the structure.

With the permission of Canadian Standards Association, material is reproduced from CSA Standard, CAN/CSA A371-04, Masonry Construction for Buildings, copyrighted by Canadian Standards Association, 178 Rexdale Blvd., Toronto, Ontario, M9W 1R3. While use of this information has been authorized, CSA shall not be responsible for the manner in which the information is presented, nor for any interpretation thereof. For more information on CSA or to purchase standards, please visit our website at www.shopcsa.ca or call 1-800-463-6727.

BRICK CONSTRUCTION

NOMINAL BRICK SIZES

Unit Designation	Dimensions			
	Thickness	Height	Length	Modular Coursing
Standard modular	4"	2⅔"	8"	3C = 8"
Engineer	4"	3⅕"	8"	5C = 16"
Economy	4"	4"	8"	1C = 4"
Double	4"	5⅓"	8"	3C = 16"
Roman	4"	2"	12"	2C = 4"
Norman	4"	2⅔"	12"	3C = 8"
Norwegian	4"	3⅕"	12"	5C = 16"
Utility	4"	4"	12"	1C = 4"
Triple	4"	5⅓"	12"	3C = 16"
SCR brick	6"	2⅔"	12"	3C = 8"
6" Norwegian	6"	3⅕"	12"	5C = 16"
6" Jumbo	6"	4"	12"	1C = 4"
8" Jumbo	8"	4"	12"	1C = 4"

MATERIALS REQUIRED FOR 100 SQUARE FEET OF BRICK WALL

Type of Brick	Wall Thickness	Number of Bricks	Cement	Sand
Standard	4 inch	616	3 sacks	9 cubic feet
Standard	8 inch	1232	7 sacks	21 cubic feet

AMOUNT OF BRICK AND MORTAR NEEDED TO CONSTRUCT 4-INCH AND 8-INCH THICK BRICK WALLS

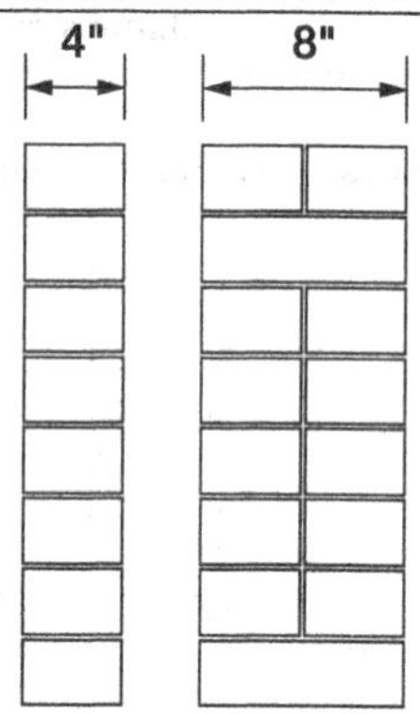

	4-Inch Wall		8-Inch Wall	
Area of Wall (sq. ft.)	No. of Brick	Cubic ft. of Mortar	No. of Bricks	Cubic ft. of Mortar
1	6.2	0.075	12.4	0.195
10	62	1	124	2
20	124	2	247	4
30	185	2½	370	6
40	247	3½	493	8
50	309	4	617	10
60	370	5	740	12
70	432	5½	863	14
80	493	6½	986	16
90	555	7	1109	18
100	617	8	1233	20
200	1,233	15	2465	39
300	1,849	23	3697	59
400	2,465	30	2929	78
500	3,081	38	6161	98

EXPOSURE GRADES OF BRICK		
Grade	**Compressive Strength**	**Exposure**
NW	1500 psi (10MPa)	Interior or backup location, dry, nonfreezing
MW	2500 psi (17MPa)	Exterior wall, moderate moisture, moderate freezing and thawing
SW	3000 psi (21MPa)	Exterior location, wet location, severe freezing and thawing

APPEARANCE GRADES OF BRICK	
FBX	Face Brick Extra
FBS	Face Brick Standard
HBX	Hollow Brick Extra
HBS	Hollow Brick Standard
TBX	Thin Brick Extra
TBS	Thin Brick Standard

TYPICAL BRICK VENEER WALL CONSTRUCTION

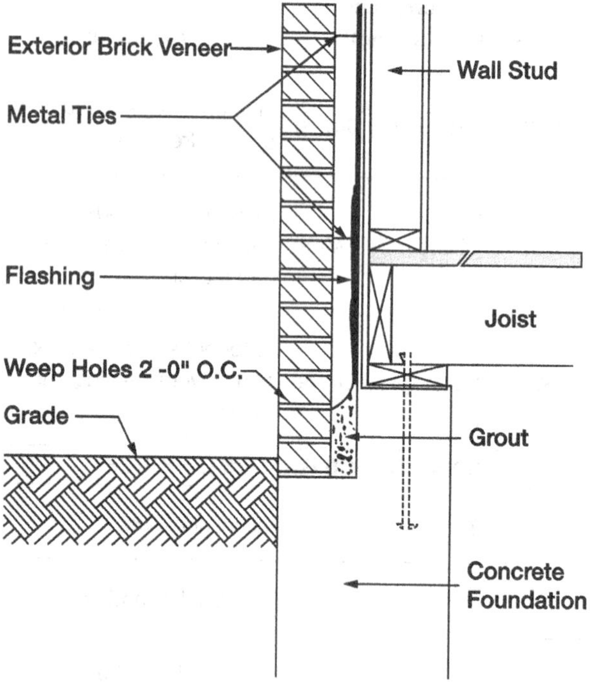

Exterior Brick Veneer
Wall Stud
Metal Ties
Flashing
Joist
Weep Holes 2 -0" O.C.
Grade
Grout
Concrete Foundation

Brick Veneer Over Frame Wall with Wood Floor Framing

TYPICAL BRICK VENEER WALL CONSTRUCTION *(cont.)*

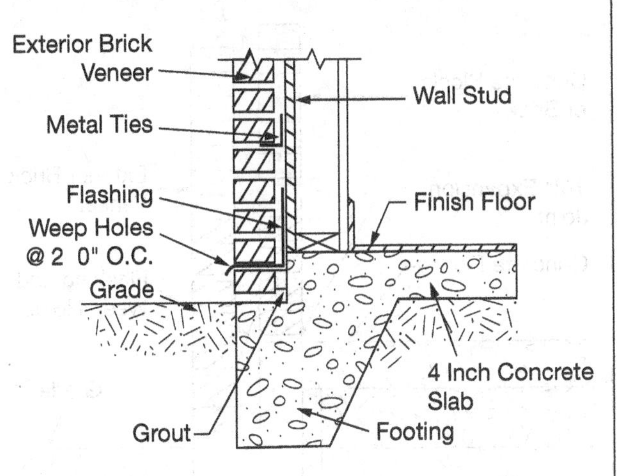

Brick Veneer Over Frame Wall with Concrete Floor

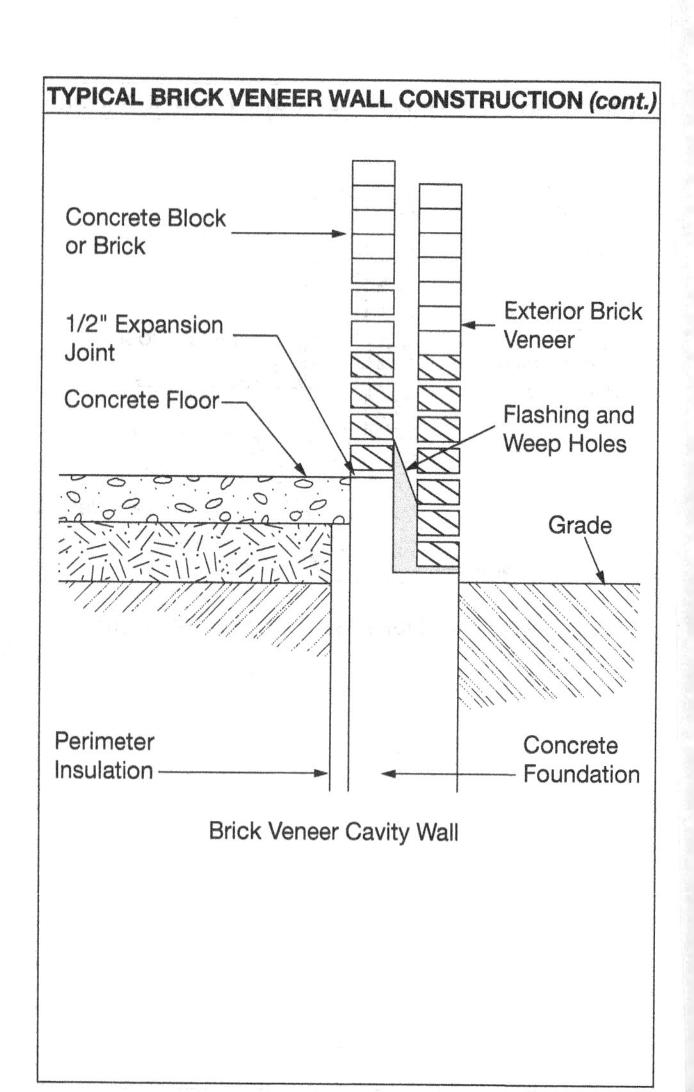

Brick Veneer Cavity Wall

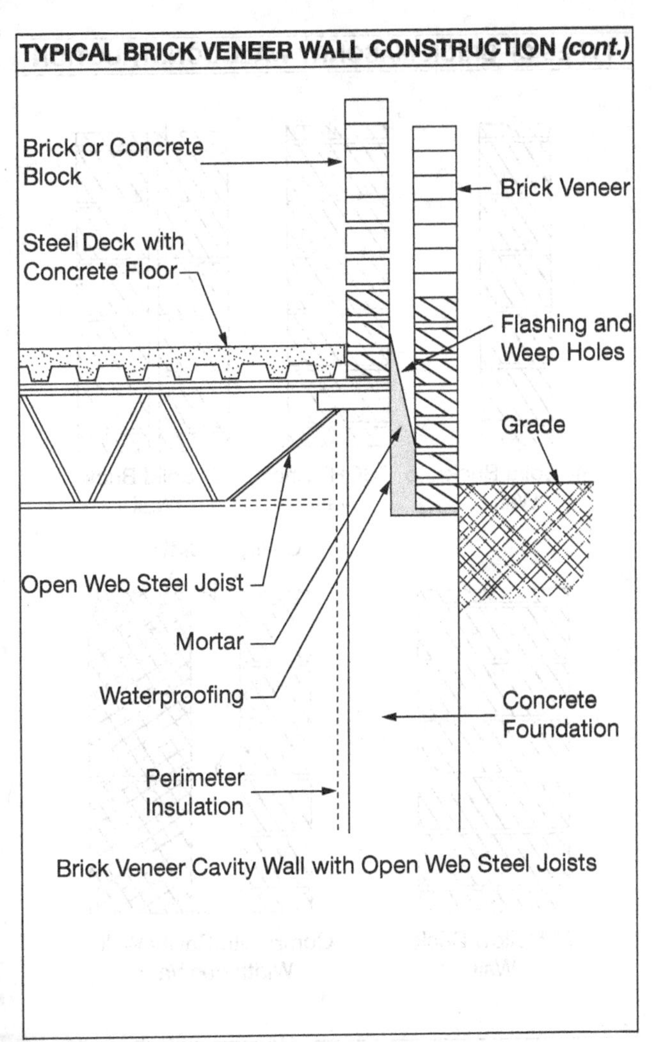

Brick Veneer Cavity Wall with Open Web Steel Joists

TYPES OF BRICK MASONRY WALL CONSTRUCTION

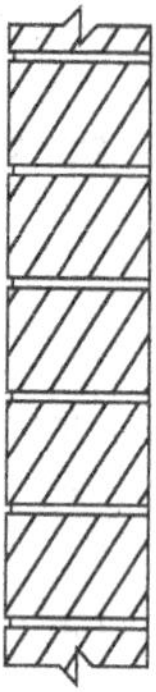

4" Solid Brick Wall

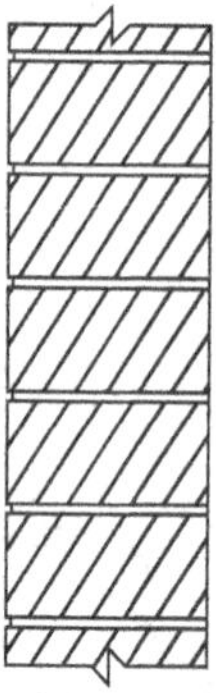

6" "SCR" Brick Wall

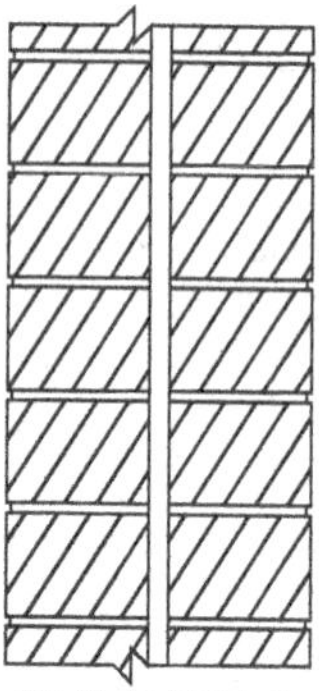

8" Solid Brick Wall

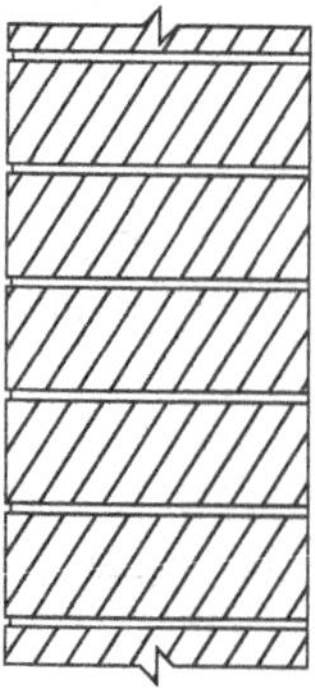

Cavity

CMU

Tie

8" Hollow Brick Wall

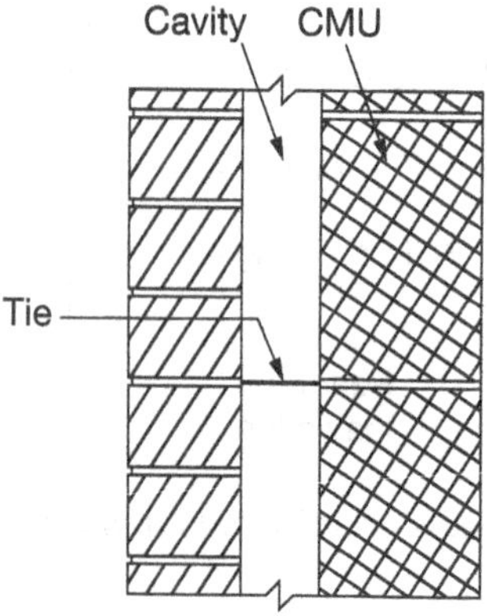

Composite Cavity Wall Width can Vary

TYPES OF BRICK MASONRY WALL CONSTRUCTION *(cont.)*

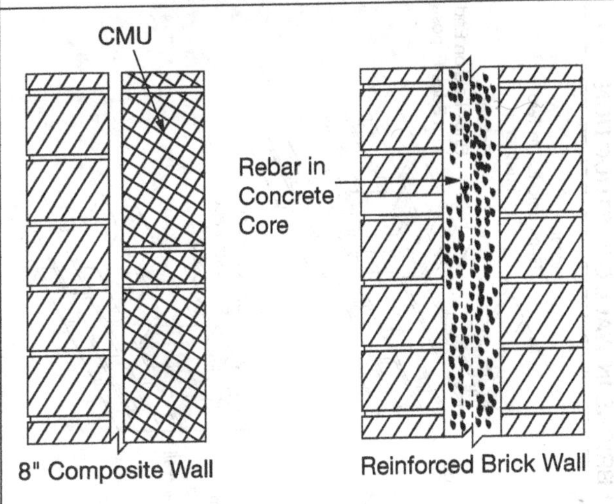

8" Composite Wall

Reinforced Brick Wall

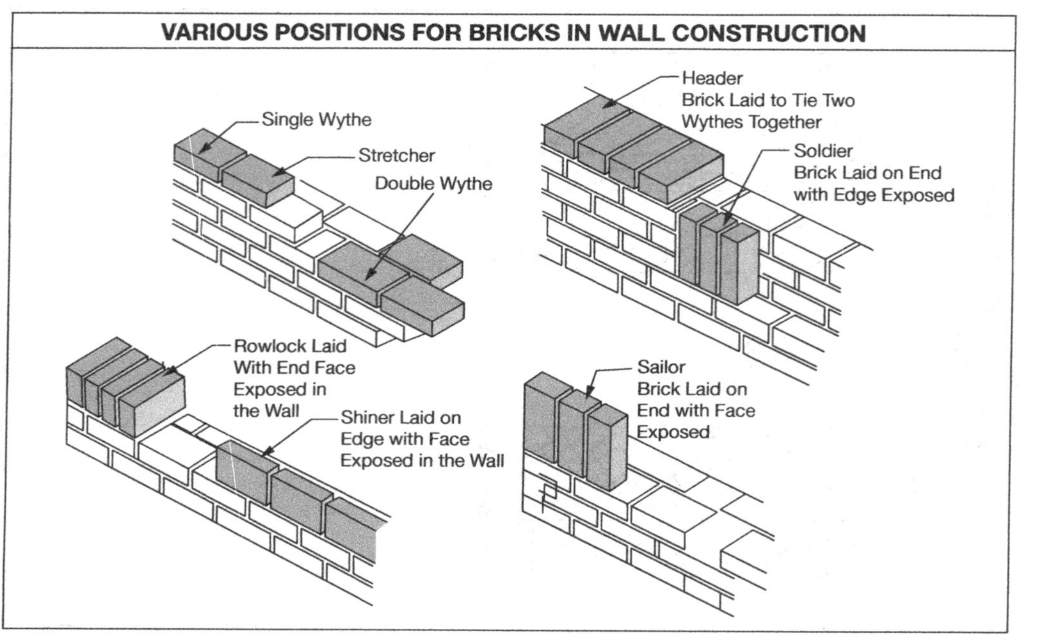
VARIOUS POSITIONS FOR BRICKS IN WALL CONSTRUCTION
Single Wythe
Stretcher
Double Wythe
Header
Brick Laid to Tie Two
Wythes Together
Soldier
Brick Laid on End
with Edge Exposed
Rowlock Laid
With End Face
Exposed in
the Wall
Shiner Laid on
Edge with Face
Exposed in the Wall
Sailor
Brick Laid on
End with Face
Exposed

WEATHERABILITY OF FREQUENTLY USED MORTAR JOINTS

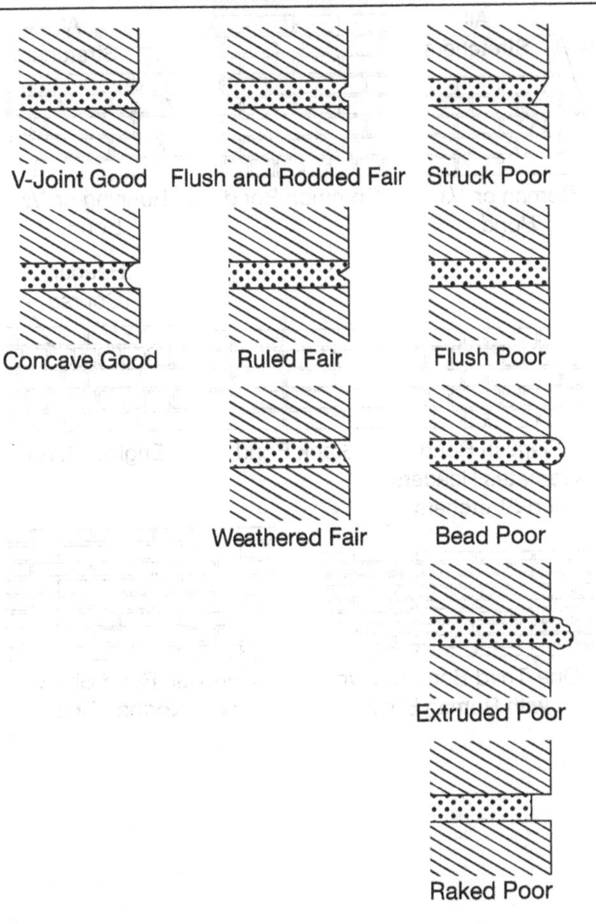

FREQUENTLY USED BRICK BOND PATTERNS

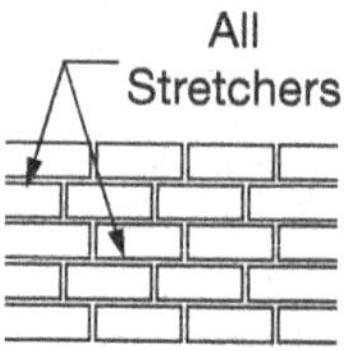

All Stretchers

Roman or 1/3 Bond

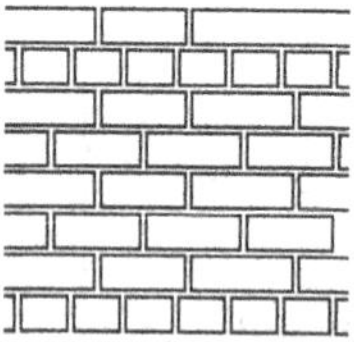

Common Bond

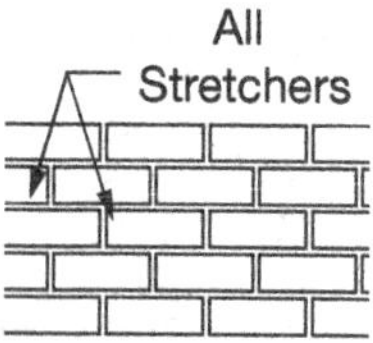

All Stretchers

Running or 1/2 Bond

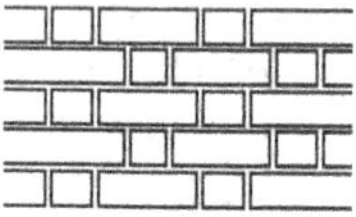

Header Course

Flemish Bond Alternates Headers and Stretchers

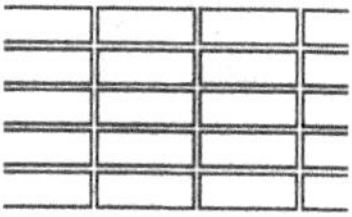

Stack Bond

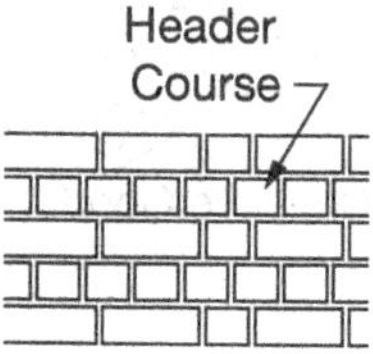

English Bond

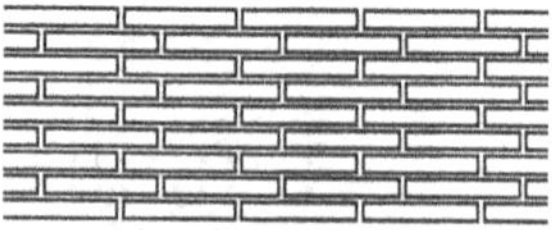

One-Third Bond Shown with Roman Brick

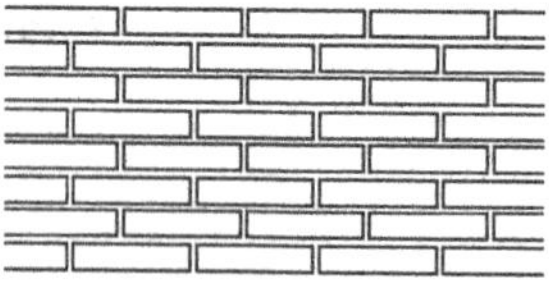

One-Half Bond Shown with Norman Brick

STRUCTURAL CLAY TILE CONSTRUCTION

Structural Clay Tile

Structural clay tiles are hollow burned-clay masonry units with parallel cells.

Two general types of structural clay tile are available. Side-construction tile are designed to receive the principle stress at right angles to the axis of the cells. The other, end-construction tile, are designed to receive the principal stress parallel to the axis of the cells.

Type	Grade
Load-bearing	LBX, LB
Non-load-bearing	NB

LBX – For general use in masonry construction and adapted for use in masonry exposed to weathering

LB – For general use in masonry not exposed to frost action. Used in exposed masonry protected with a facing of 3 inches or more of brick, stone, terra cotta, or other masonry.

NB – Non-load-bearing units made from surface clay, shale, or burned clay

STRUCTURAL CLAY TILE CONSTRUCTION *(cont.)*

Facing Tile

Type	Grade
Flat panels Standard Special duty Glazed units	FTX, FTS

FTX – General use in exposed exterior and interior masonry walls where low absorption is required

FTS – Smooth and rough textured face tile used in exposed exterior masonry walls where moderate absorption is required

Standard – General use in exterior and interior walls

Special Duty – General use in exterior and interior masonry walls requiring resistance to impact and moisture transmission

Glazed Units – Ceramic glazed structural clay tile

Floor Tile

Grade: FT1, FT2

Both suitable for flat or segmented panels

SIZES OF SOME OF THE COMMONLY USED STRUCTURAL CLAY TILE (INCHES)		
Width	**Length**	**Height**
3¾	11¾	5$\frac{1}{16}$
5¾	11¾	5$\frac{1}{16}$
1¾	11¾	5$\frac{1}{16}$
3¾	15¾	7¾
1¾	15¾	7¾
5¾	15¾	7¾

See manufacturers' catalogs for the many other sizes and shapes. These include stretchers, sills, caps, miters, bases, corners, and jamb tiles.

SOME OF THE MANY TYPES OF CLAY TILE MASONRY UNITS

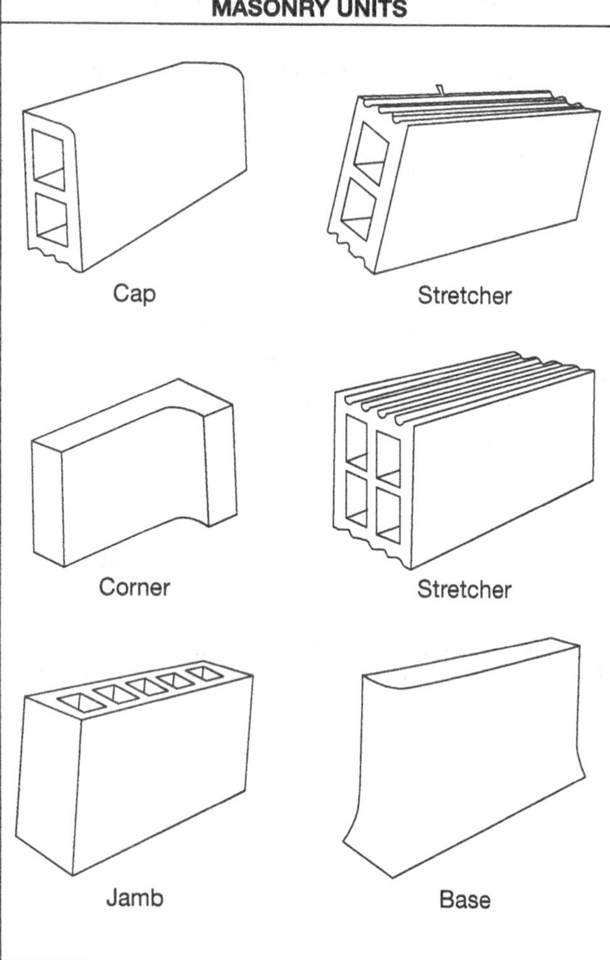

TYPICAL STRUCTURAL CLAY TILE WALL

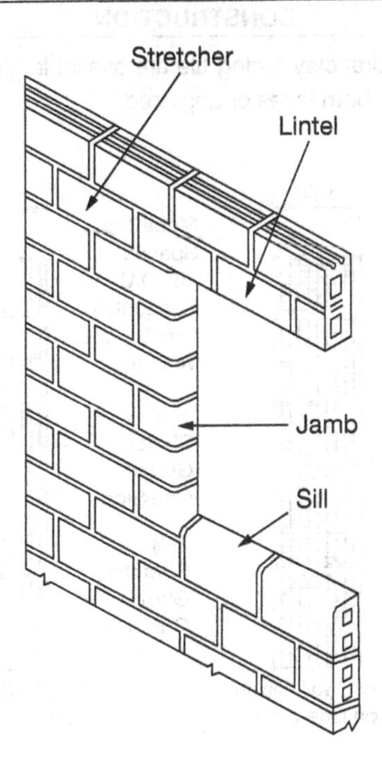

TYPICAL STRUCTURAL CLAY FACING TILE WALL CONSTRUCTION

Structural clay facing tile are available glazed on one face, both faces or unglazed

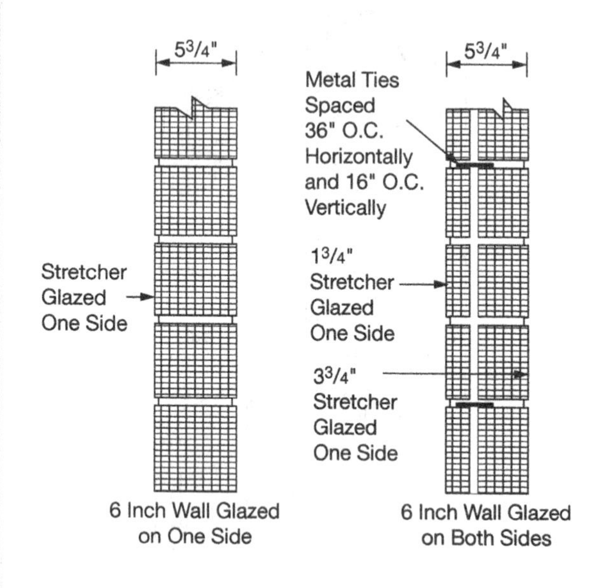

6 Inch Wall Glazed on One Side

6 Inch Wall Glazed on Both Sides

TYPICAL STRUCTURAL CLAY FACING TILE WALL CONSTRUCTION *(cont.)*

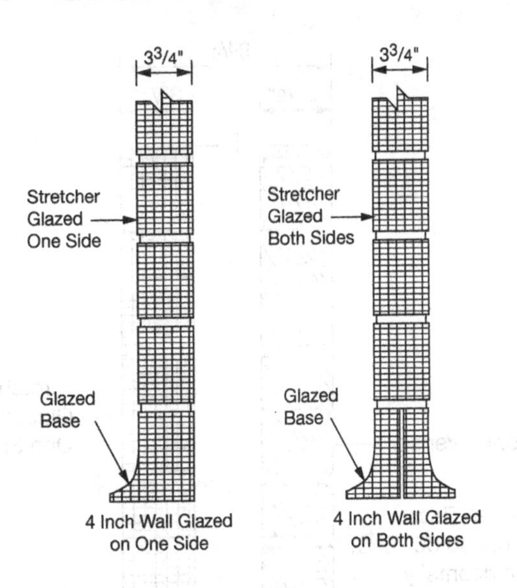

TYPICAL STRUCTURAL CLAY FACING TILE WALL CONSTRUCTION *(cont.)*

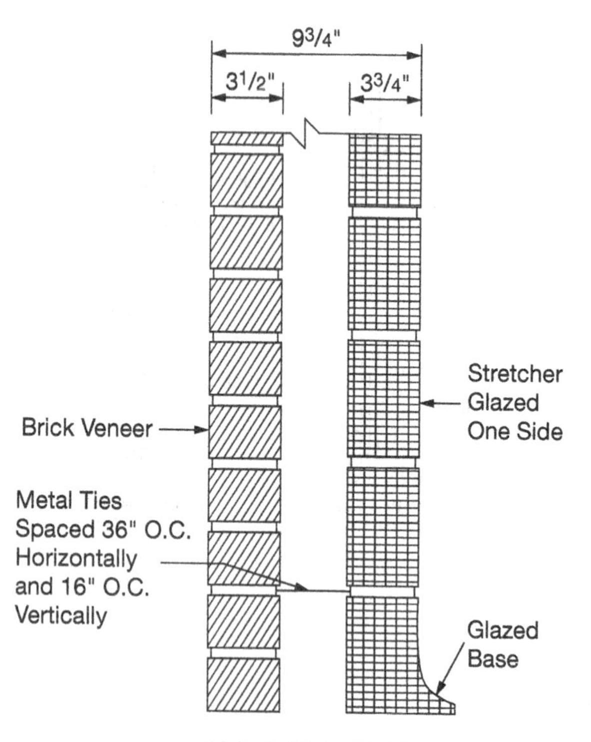

10 Inch Wall with Glazed Tile on One Side and Brick Veneer on the Other

STONE CONSTRUCTION

COMMONLY USED TYPES OF STONE IN BUILDING CONSTRUCTION

Name	Type of Stone	Color	Hardness	Typical Uses
Granite, dark	Igneous	Gray to black	Hard	Exterior veneer
Granite, light	Igneous	White to gray	Hard	Exterior veneer, paving blocks
Bluestone	Sedimentary	Blue-gray	Fairly hard	Flagstones, flooring, countertops
Limestone	Sedimentary	White, light gray	Soft	Exterior veneer, flagstones
Marble	Metamorphis	Varied, white blue-gray, black, pink	Fairly hard	Exterior and interior Flooring, furniture tops
Sandstone	Sedimentary	Tan to light brown, brick red	Fairly hard	Exterior and interior veneers, flagstones
Slate	Metamorphic	Gray-green, gray, blue, red tint	Medium hard	Shingles, flooring, counter-tops
Soapstone	Metamorphic	Green, gray, blue	Medium hard	Exterior Veneer, flagstone, flooring, furniture tops

COMMON PATTERNS FOR BUILDING STONE WALLS

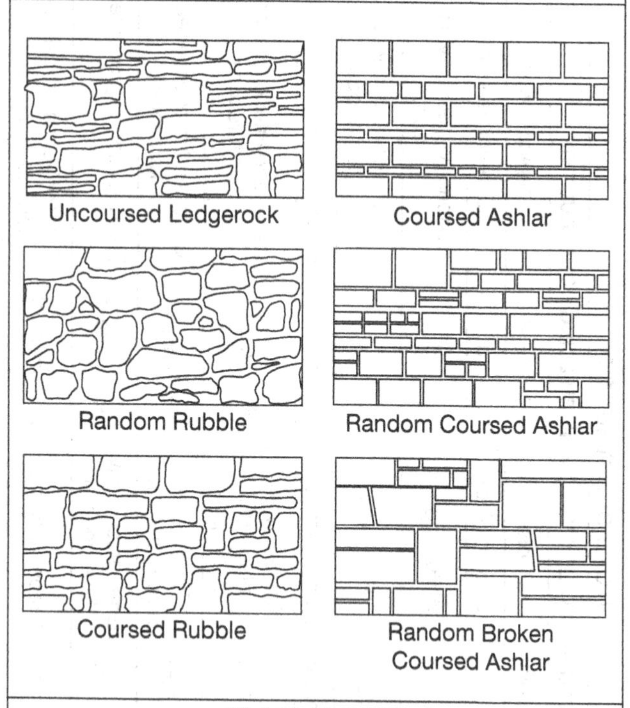

GLASS BLOCKS

STANDARD UNITS are 3½ inches (98 mm) thick.

THIN UNITS are 3⅛ inches (79 mm) thick for hollow units and 3 inches (76 mm) for solid units.

Exterior Standard-Unit Panels

Maximum panel area allowed is 144 sq. ft. (13.4 m^2) when design wind pressure is 20 psf (958 N/m^2).

Maximum panel dimensions between supports is 25 ft. (7620 mm) in width or 20 ft. (6096 mm) in height. Area may be adjusted for other wind pressures as allowed by codes.

Exterior Thin-Unit Panels

Maximum area allowed is 85 sq. ft. (7.9 m^2). Maximum distance between structural supports shall be 15 ft. (4572 mm) in width of 10 ft (3048 mm) in height. These units must not be used where wind design pressure exceeds 20 psf (958 N/m^2).

Interior Panels

Maximum area for a standard unit panel is 250 sq. ft. (23.2 m^2). Maximum area for thin unit panels shall be 150 sq. ft. (13.9 m^2). Maximum distance between structural supports shall be 25 ft. (97620 mm) in width or 20 ft. (6096 mm) in height.

Solid Units

The maximum area of solid units in exterior or interior panels shall be 100 sq. ft. (9.3 m^2).

TYPICAL SIZES OF GLASS BLOCKS	
Nominal Sizes	
3 in. Thick	**4 in. Thick**
3 × 6 4 × 8 4 × 12 6 × 6 6 × 8 8 × 8 12 × 12	4 × 8 6 × 6 6 × 8 8 × 8 12 × 12

3-inch nominal thickness is 3⅛ inches.
4-inch nominal thickness is 3⅞ inches.
Face dimensions are additional sizes and installation instructions.

Consult the International Building Code Standard Section 2110, Glass Masonry, for installation requirements. It provides limits on the panel dimensions between structural supports and areas for various types of units and other specific requirements.

MAXIMUM GLASS BLOCK PANEL DIMENSIONS

	Standard Unit (3⅞ in)			Thin Unit (3⅛ in)			Solid Unit (3 in)		
	Area sq. ft.	Width ft.	Height ft.	Area sq. ft.	Width ft.	Height ft.	Area sq. ft.	Width ft.	Height ft.
Exterior	144	25	20	85	15	10	100	10	10
Interior	250	25	20	150	25	20	100	10	10

TYPICAL MORTAR MIX FOR LAYING GLASS BLOCKS		
Portland Cement	**Sand**	**Lime**
1 part	0.5 part	3.4 parts

NUMBER OF BLOCK FOR 100 SQ. FT. PANEL					
Block Sizes (nominal)	6 in.	8 in.	12 in.	4 × 8 in.	6 × 8 in.
Number of Block	400	225	100	450	300

CHAPTER 9
Foundations

Building codes establish standards that set minimum requirements for size, strength, composition, reinforcement, and installation of concrete and concrete block walls and footings. Codes reference American Concrete Institute standards and publications of the Portland Cement Association.

The material used to construct foundations must resist compressive and horizontal forces. Compressive forces are due to the weight of the floors, building contents, and the roof. Horizontal forces are due to the wind, subsurface water, soil, and earthquake rise and fall. The type of soil upon which the foundation rests has a significant influence upon its design.

TYPICAL FOUNDATIONS USED FOR RESIDENTIAL AND LIGHT COMMERCIAL CONSTRUCTION

FORCES ON THE FOUNDATION

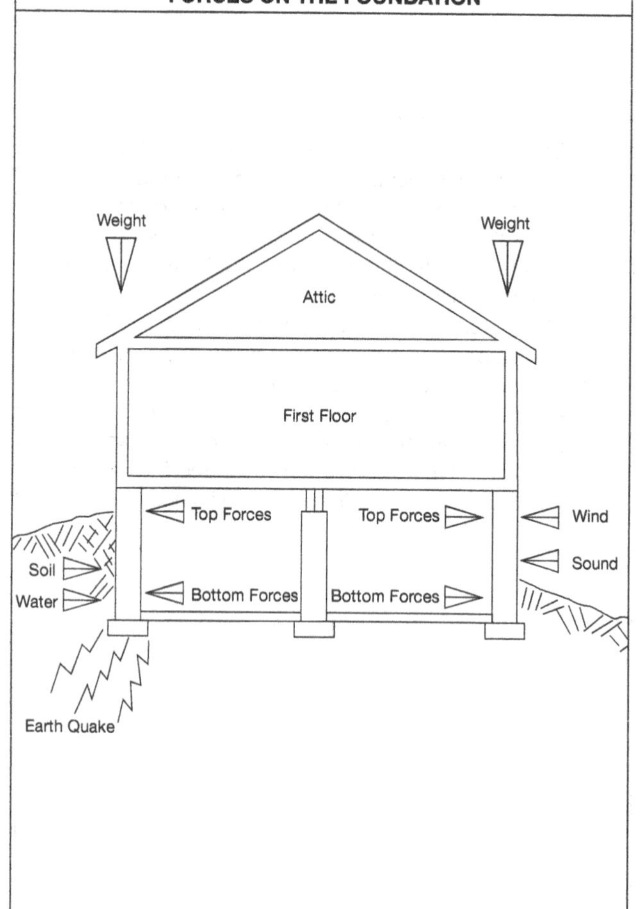

TYPICAL LOADS ON A STRUCTURE

Actual weights will vary depending upon materials and spans. See Chapter 3 for weights of materials and recommended design loads.

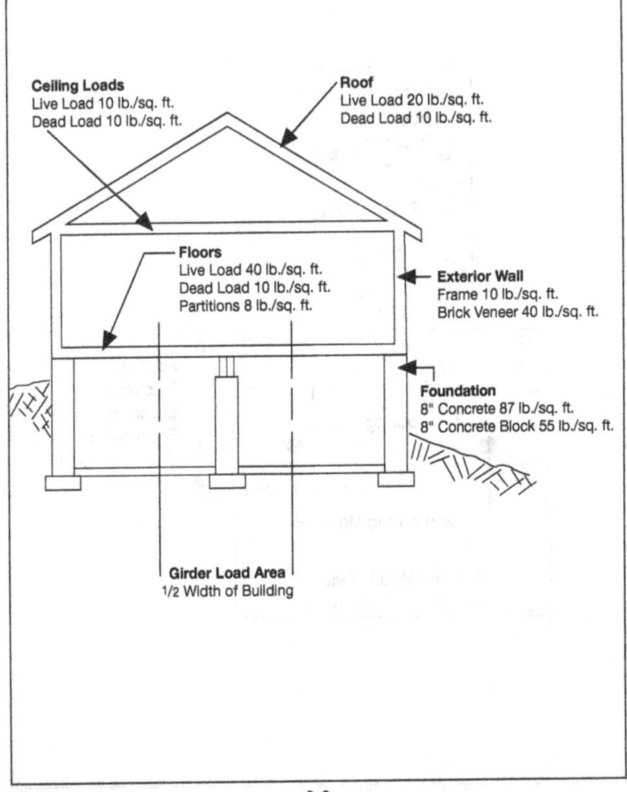

TYPICAL RATIO OF FOOTING SIZE IN RELATION TO THE FOUNDATION THICKNESS

Actual design solution will depend upon the forces on the foundation and the bearing capacity of the soil.

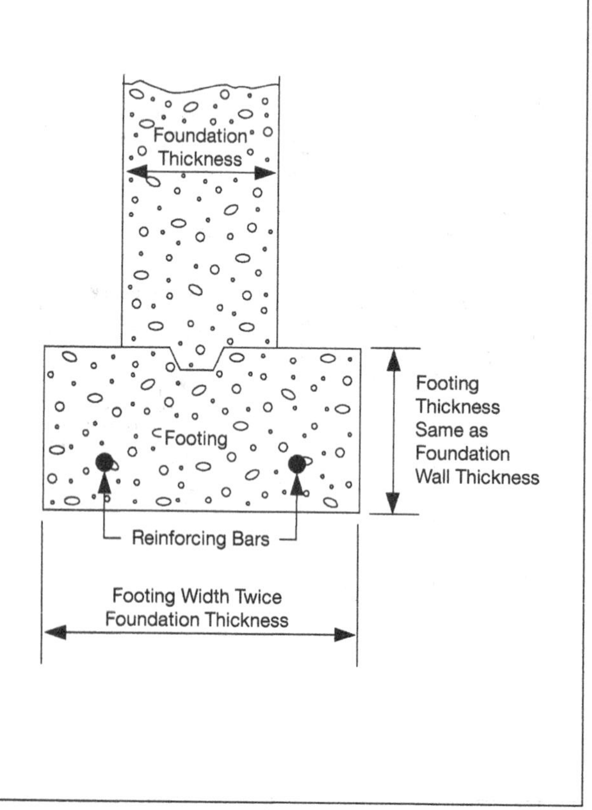

TYPICAL RATIO OF FOOTING SIZE IN RELATION TO THE FOUNDATION THICKNESS *(cont.)*

Actual design solution will depend upon the forces on the foundation and the bearing capacity of the soil.

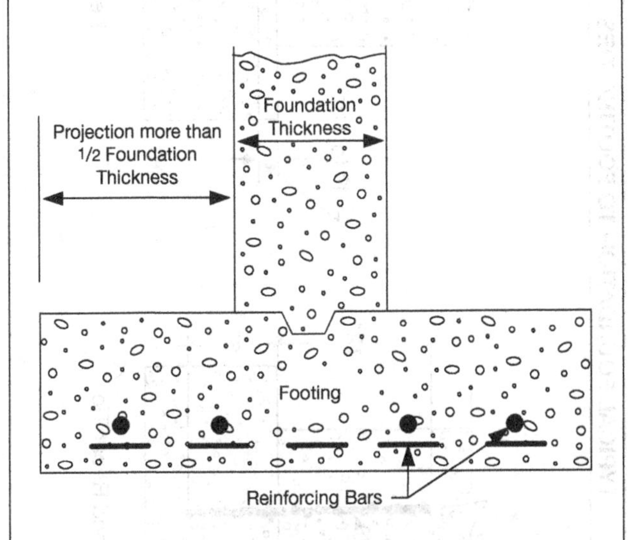

This condition subjects the footing to bending forces and requires additional reinforcing.

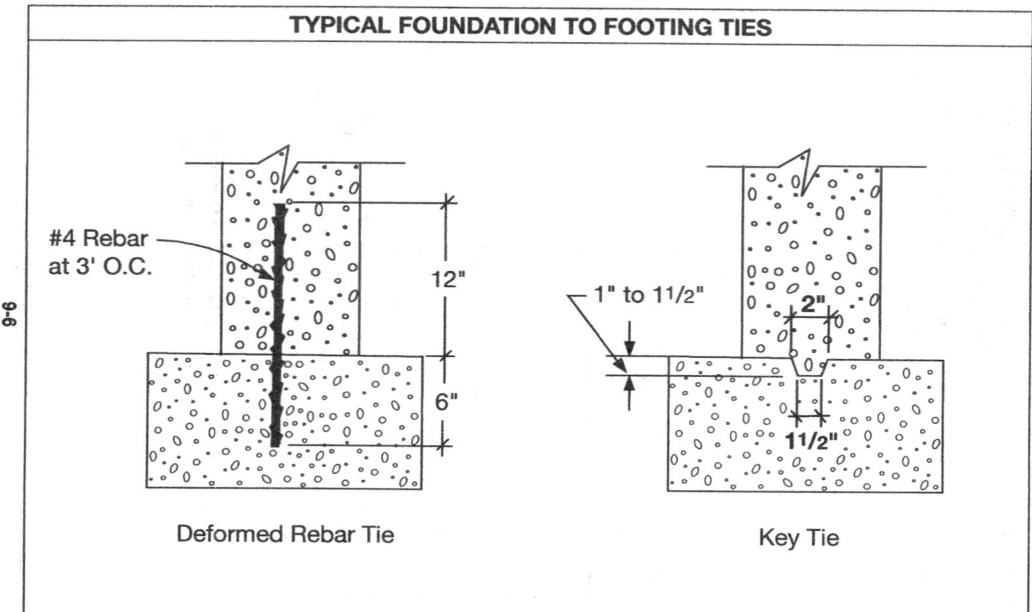

TYPICAL FOUNDATION TO FOOTING TIES
#4 Rebar
at 3' O.C.
12"
6"
Deformed Rebar Tie
1" to 1 1/2"
2"
1 1/2"
Key Tie

MINIMUM EXCAVATION REQUIREMENTS FOR BENCHED FOUNDATION EXCAVATION

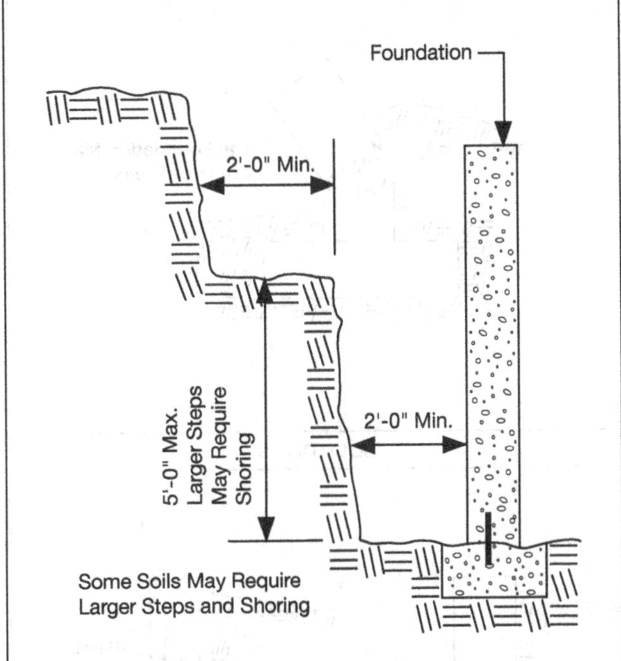

See Chapter 4 for additional excavation requirements.

See Chapter 18, Soils and Foundations, in the *2006 International Building Code* for additional information on foundations and safety regulations.

POSSIBLE RETAINING WALL FOUNDATION PROBLEMS

PROBLEM

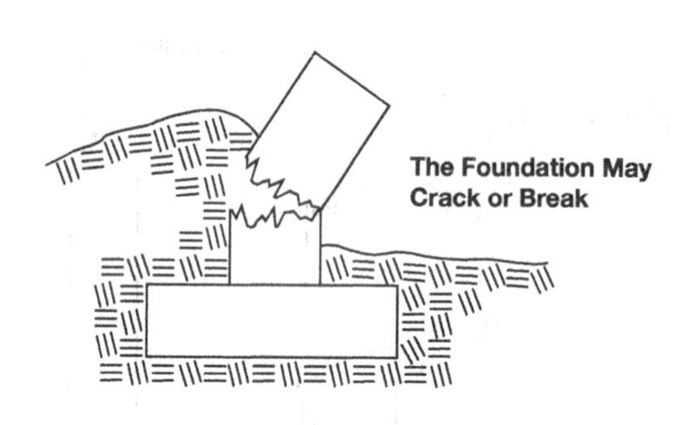

SOLUTIONS

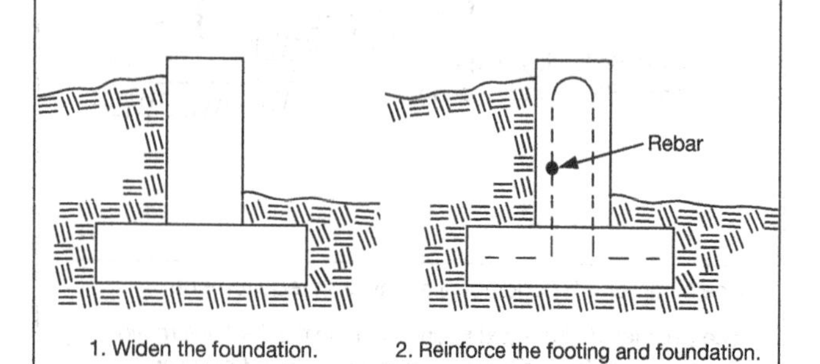

1. Widen the foundation.

2. Reinforce the footing and foundation.

POSSIBLE RETAINING WALL FOUNDATION PROBLEMS *(cont.)*

PROBLEM

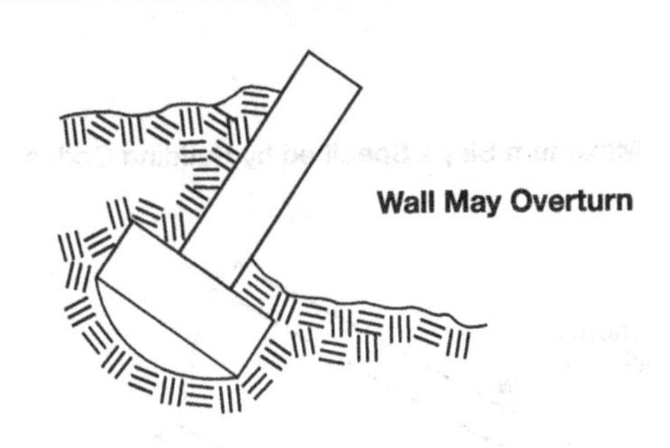

SOLUTIONS

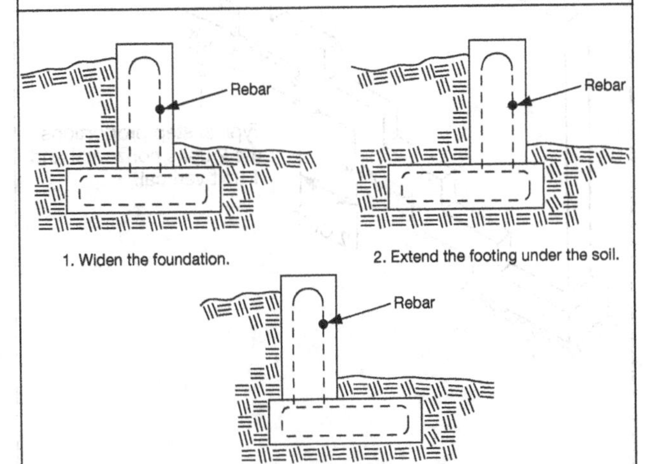

1. Widen the foundation.
2. Extend the footing under the soil.
3. Extend the footing to compensate for the soil pressure.

STEPPED FOUNDATION

Maximum Slope Specified by Building Codes

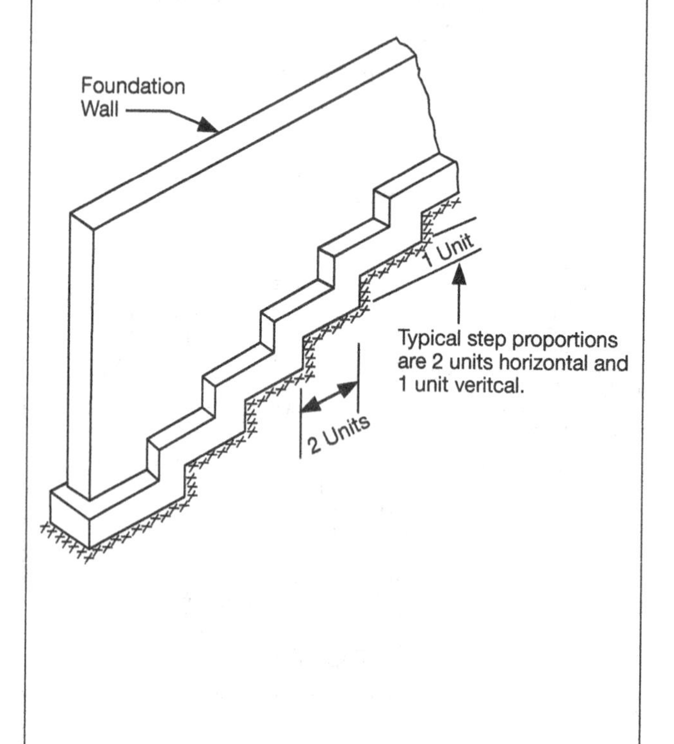

CONTINUOUS FOUNDATION

CONCRETE GRADE BEAM ON PIERS FOUNDATION WITH A WOOD FRAMED FLOOR

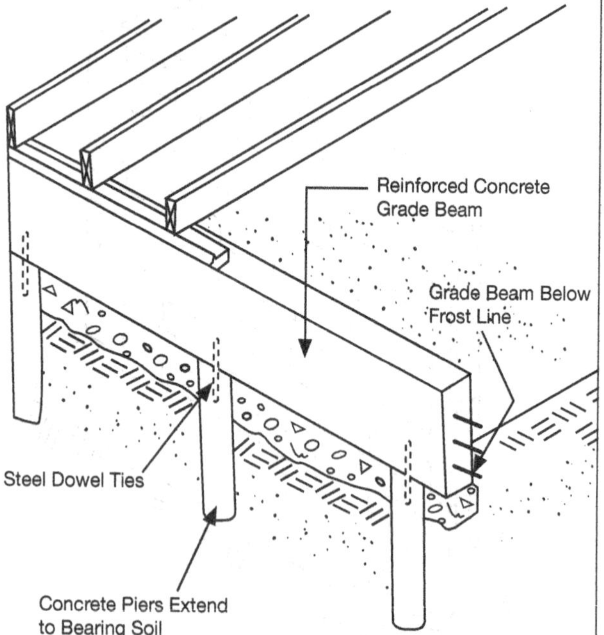

Reinforced Concrete Grade Beam

Grade Beam Below Frost Line

Steel Dowel Ties

Concrete Piers Extend to Bearing Soil

CONCRETE GRADE BEAM ON PIERS WITH A CONCRETE SLAB FLOOR

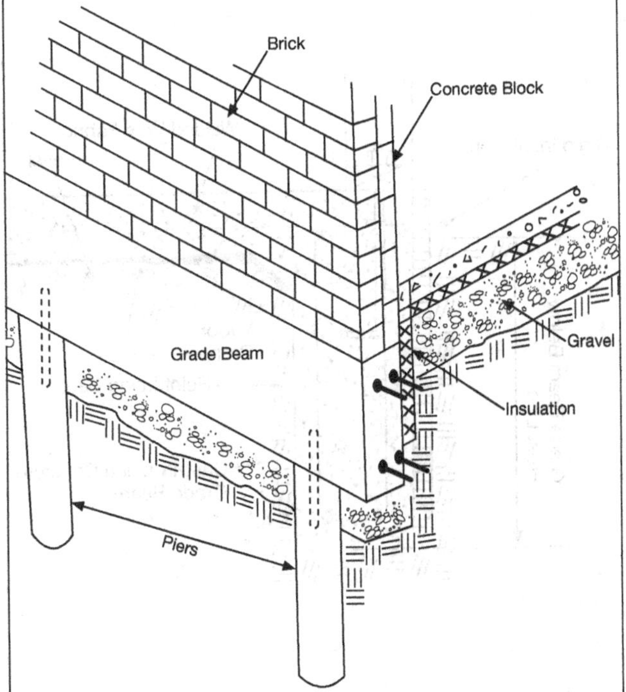

FOUNDATIONS WITH CONCRETE SLAB FLOORS

MONOLITHICALLY CAST GRADE BEAM AND SLAB

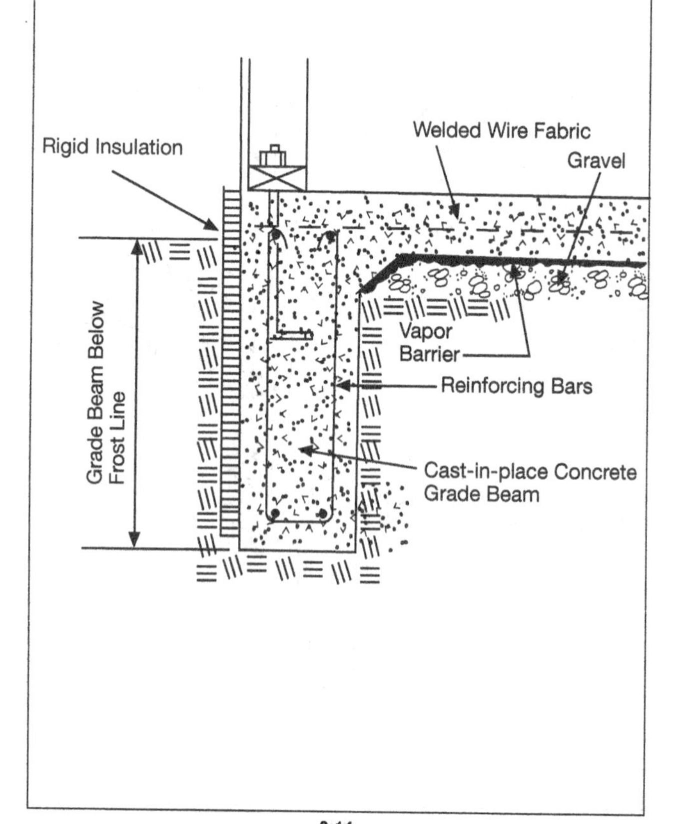

CONCRETE BLOCK MASONRY FOUNDATION WITH A GROUND SUPPORTED SLAB FLOOR

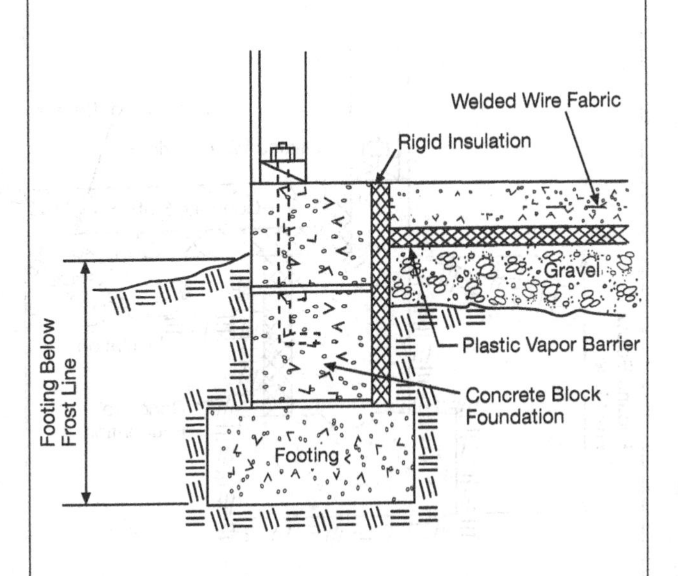

POURED CONCRETE FOUNDATIONS WITH EDGE SUPPORTED SLAB FLOOR

HEAVY INSULATION USED IN COLD CLIMATES

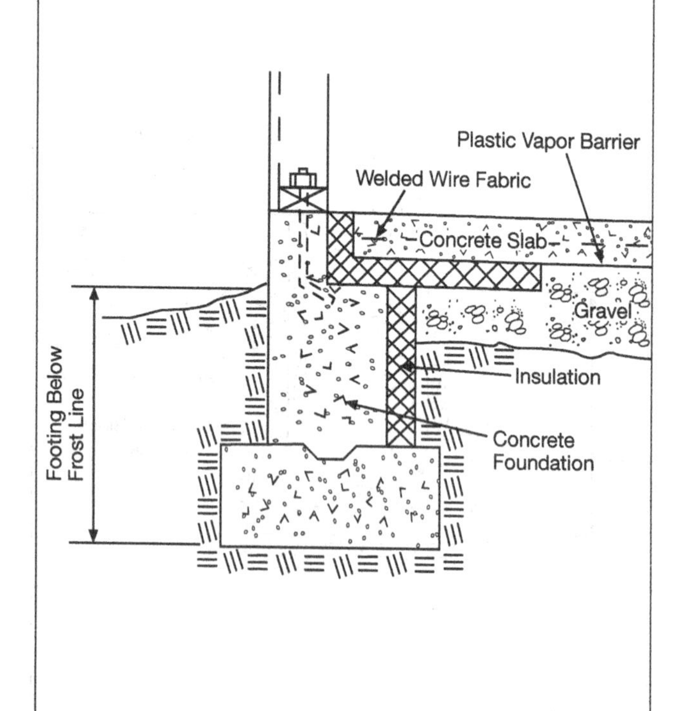

A MONOLITHICALLY CAST CONCRETE FOOTING AND FLOOR SLAB FOR INTERIOR LOAD BEARING WALLS

USED IN WARM CLIMATES

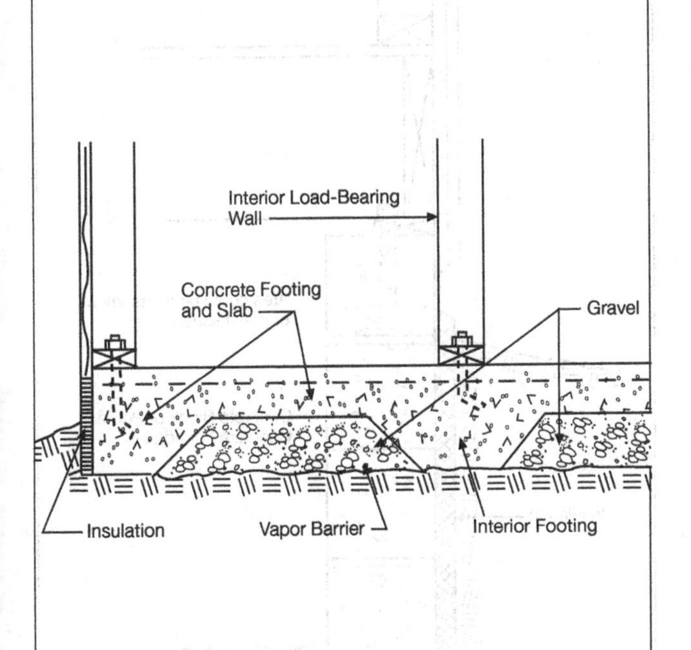

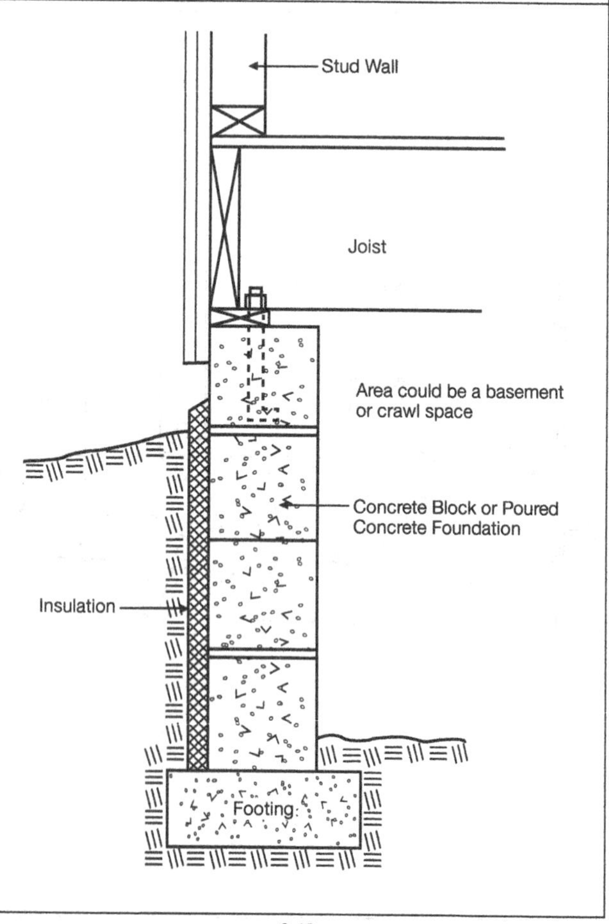
CONCRETE BLOCK OR POURED CONCRETE FOUNDATION WITH WOOD FRAME FLOOR AND FRAME EXTERIOR WALL
Stud Wall
Joist
Area could be a basement or crawl space
Concrete Block or Poured Concrete Foundation
Insulation
Footing

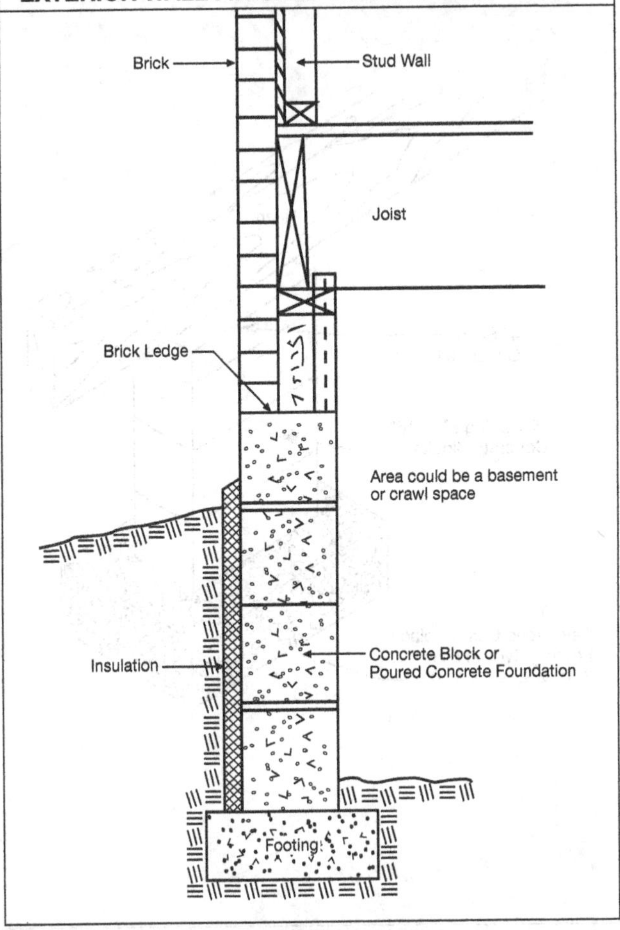
CONCRETE BLOCK OR POURED CONCRETE FOUNDATION WITH WOOD FRAMED FLOOR AND EXTERIOR WALL FINISHED WITH A BRICK VENEER
Brick
Stud Wall
Joist
Brick Ledge
Area could be a basement or crawl space
Insulation
Concrete Block or Poured Concrete Foundation
Footing

TYPICAL CONCRETE BLOCK PIER

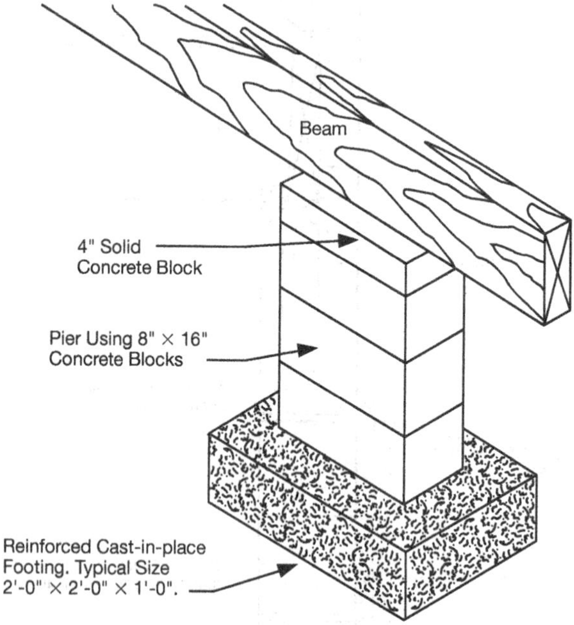

Beam

4" Solid
Concrete Block

Pier Using 8" × 16"
Concrete Blocks

Reinforced Cast-in-place
Footing. Typical Size
2'-0" × 2'-0" × 1'-0".

STEEL COLUMNS SUPPORT I-BEAMS AND WOOD BEAMS

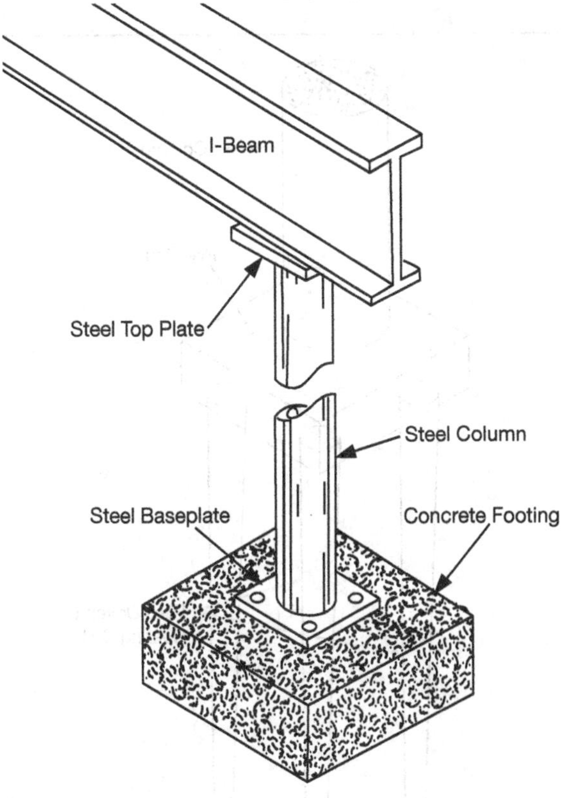

I-Beam

Steel Top Plate

Steel Column

Steel Baseplate

Concrete Footing

COLUMN SUPPORT

When conditions warrant, its columns can be supported by a pile cap which is supported by piles extended to adequate bearing soil.

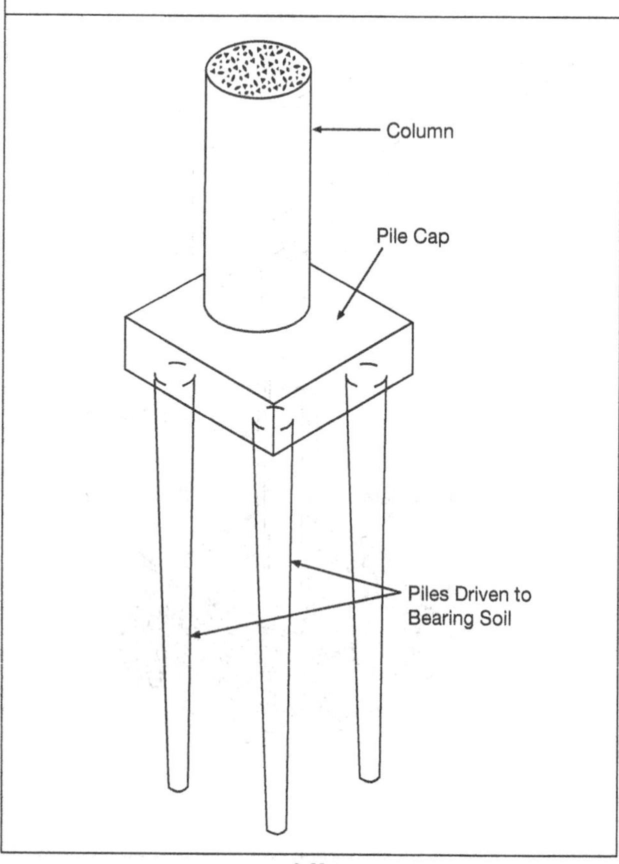

HEAVY STRUCTURAL LOADS

Heavy structural loads and/or pour surface soil requires piles driven to adequate bearing soil.

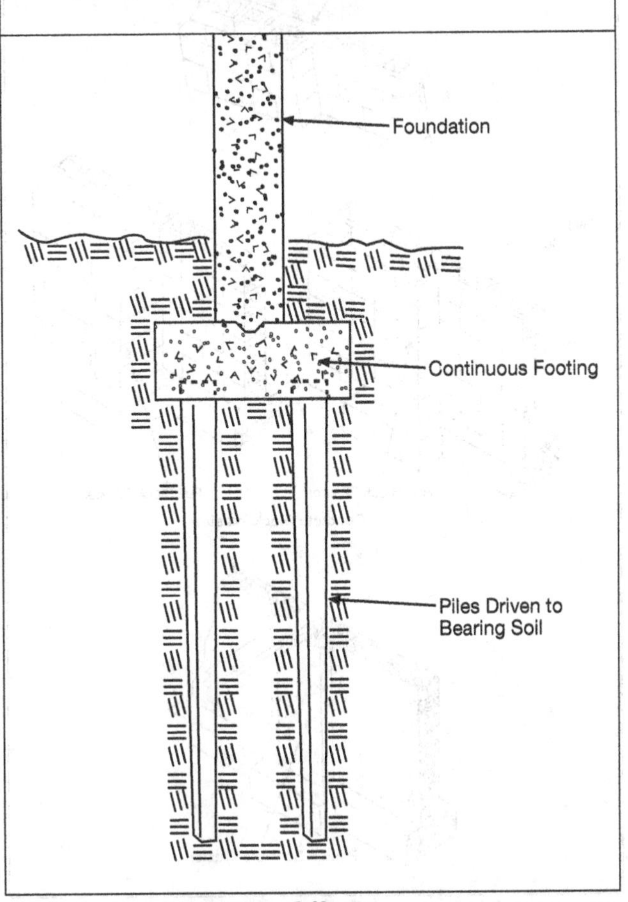

COMMONLY BUILT PILASTERS

Precast Concrete
Pilaster Blocks

Fill Pilaster Cores
with Concrete

Half Block Pilaster

Full Block Pilaster

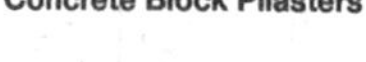

Concrete Block Pilasters

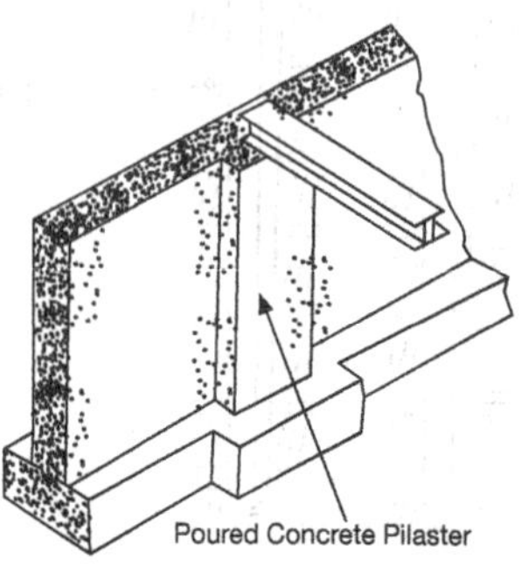

TYPICAL FORM FOR CONCRETE WALKS, DRIVEWAYS, AND SLABS

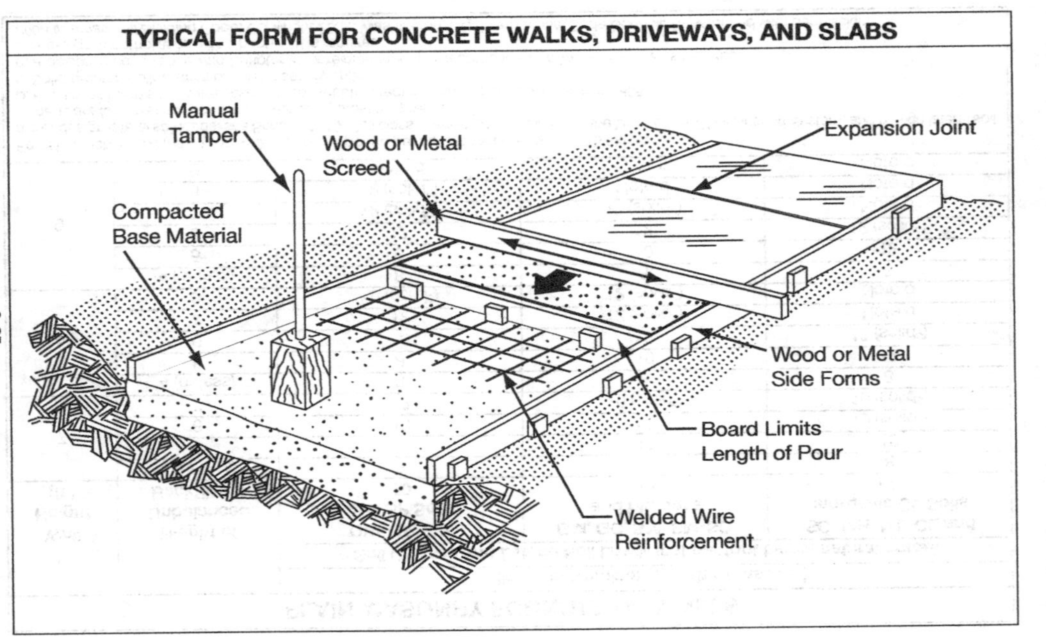

PLAIN MASONRY FOUNDATION WALLS

Wall Height (ft.)	Height of Unbalanced Backfill (ft.)	Minimum Nominal Wall Thickness (in.)		
		Soil Classes and Lateral Soil Load[a] (psf per foot below natural grade)		
		GW, GP, SW, and SP Soils 30	GM, GC, SM, SM-SC, and ML Soils 45	SC, MH, ML-CL and Inorganic CL Soils 60
7	4 (or less)	8	8	8
	5	8	10	10
	6	10	12	10 (solid[c])
	7	12	10 (solid[c])	10 (solid[c])
8	4 (or less)	8	8	8
	5	8	10	12
	6	10	12	12 (solid[c])
	7	12	12 (solid[c])	Note d
	8	10 (solid[c])	12 (solid[c])	Note d
9	4 (or less)	8	8	8
	5	8	10	12
	6	12	12	12 (solid[c])
	7	12 (solid[c])	12 (solid[c])	Note d
	8	12 (solid[c])	Note d	Note d
	9	Note d	Note d	Note d

For S1: 1 inch = 25.4 mm, 1 foot = 304.8 mm, 1 pound per square foot per foot = 0.157 kPa/m.

a. For design lateral soil loads, see Section 1610. Soil classes are in accordance with the Unified Soil Classification System, and design lateral soil loads are for moist soil conditions without hydrostatic pressure.

b. Provisions for this table are based on construction requirements specified in Section 1805.5.2.

c. Solid grouted hollow units or solid masonry units.

d. A design in compliance with Chapter 21 or reinforcement in accordance with Table 1805.5(2) is required.

e. A design in compliance with Chapter 19 is required.

PLAIN CONCRETE FOUNDATION WALLS

Wall Height (ft.)	Height of Unbalanced Backfill (ft.)	Minimum Nominal Wall Thickness (in.)		
		Soil Classes and Lateral Soil Load[a] (psf per foot below natural grade)		
		GW, GP, SW, and SP Soils 30	GM, GC, SM, SM-SC, and ML Soils 45	SC, MH, ML-CL and Inorganic CL Soils 60
7	4 (or less)	7 1/2	7 1/2	7 1/2
	5	7 1/2	7 1/2	7 1/2
	6	7 1/2	7 1/2	8
	7	7 1/2	8	10
8	4 (or less)	7 1/2	7 1/2	7 1/2
	5	7 1/2	7 1/2	7 1/2
	6	7 1/2	7 1/2	10
	7	7 1/2	10	10
	8	10	10	12
9	4 (or less)	7 1/2	7 1/2	7 1/2
	5	7 1/2	7 1/2	7 1/2
	6	7 1/2	7 1/2	10
	7	7 1/2	10	10
	8	10	10	12
	9	10	12	Note e

For S1: 1 inch = 25.4 mm, 1 foot = 304.8 mm, 1 pound per square foot per foot = 0.157 kPa/m.

a. For design lateral soil loads, see Section 1610. Soil classes are in accordance with the Unified Soil Classification System, and design lateral soil loads are for moist soil conditions without hydrostatic pressure.

b. Provisions for this table are based on construction requirements specified in Section 1805.5.2.

c. Solid grouted hollow units or solid masonry units.

d. A design in compliance with Chapter 21 or reinforcement in accordance with Table 1805.5(2) is required.

e. A design in compliance with Chapter 19 is required.

FOOTINGS SUPPORTING WALLS OF LIGHT-FRAME CONSTRUCTION

Number of Floors Supported by the Footing[f]	Width of Footing (in.)	Thickness of Footing (in.)
1	12	6
2	15	6
3	18	8[g]

For S1: 1 inch = 25.4 mm, 1 foot = 304.8 mm.

a. Depth of footings shall be in accordance with Section 1805.2.

b. The ground under the floor is permitted to be excavated to the elevation of the top of the footing.

c. Interior-stud-bearing walls are permitted to be supported by isolated footings. The footing width and length shall be twice the width shown in this table, and footings shall be spaced not more than 6' on center.

d. See Section 1910 for additional requirements for footings of structures assigned to Seismic Design Category C, D, E, or F.

e. For thickness of foundation walls, see Section 1805.5.

f. Footings are permitted to support a roof in addition to the stipulated number of floors. Footings supporting roof only shall be as required for supporting one floor.

g. Plain concrete footings for Group R=3 occupancies are permitted to be 6" thick.

TYPICAL THICKNESS OF RESIDENCE FOUNDATION WALLS			
Type of Foundation Wall Construction	Maximum Height of Unbalanced Fill	Minimum Wall Thickness (in.)	
		Frame	Masonry or Masonry Veneer
Hollow Masonry (i.e. Concrete Block)	3	8	8
	5	8	8
	7	12	10
Solid Masonry	3	6	8
	5	8	8
	7	10	8
Poured Concrete	3	6	8
	5	6	8
	7	6	8

TYPICAL SAFE CAPACITIES PER LINEAL FOOT OF FOOTING							
Width (in.) per Lineal Foot	Bearing Area (sq. ft.)	Safe Soil Bearing Capacity (lbs.)					
		1,000	**2,000**	**3,000**	**4,000**	**6,000**	**8,000**
8	0.66	670	1,340	2,000	2,670	4,000	5,340
10	0.83	835	1,665	2,500	3,335	5,000	6,665
12	1.00	1,000	2,000	3,000	4,000	6,000	8,000
14	1.16	1,165	2,335	3,500	4,665	7,000	9,335
16	1.33	1,330	2,670	4,000	5,330	8,000	10,670
18	1.5	1,500	3,000	4,500	6,000	9,000	12,000

1,000 – Clay, sandy clay, silty clay
2,000 – Sand, silty sand, silty gravel
3,000 – Sandy gravel, gravel
4,000 – Foliated rock, sedimentary rock
6,000 – Moderately dry clay, coarse sand and clay
8,000 – Hard dry clay, coarse firm sand

TYPICAL PIER AND PIER FOOTING SIZES FOR SINGLE FAMILY RESIDENCE ON AVERAGE SOIL

Pier Material	Minimum Pier Size (in.)	Minimum Pier Footing Size (in.)	Pier Spacing	
			Right Angle to Joists	Parallel to Joists
Solid or Grouted Masonry	8 × 12	12 × 24 × 8	8'-0" O.C.	12'-0" O.C.
Hollow Masonry, Interior Pier	8 × 16	16 × 24 × 8	8'-0" O.C.	12'-0" O.C.
Plain Concrete	10 x 10 or 12" diameter	20 × 20 × 8	8'-0" O.C.	12'-0" O.C.

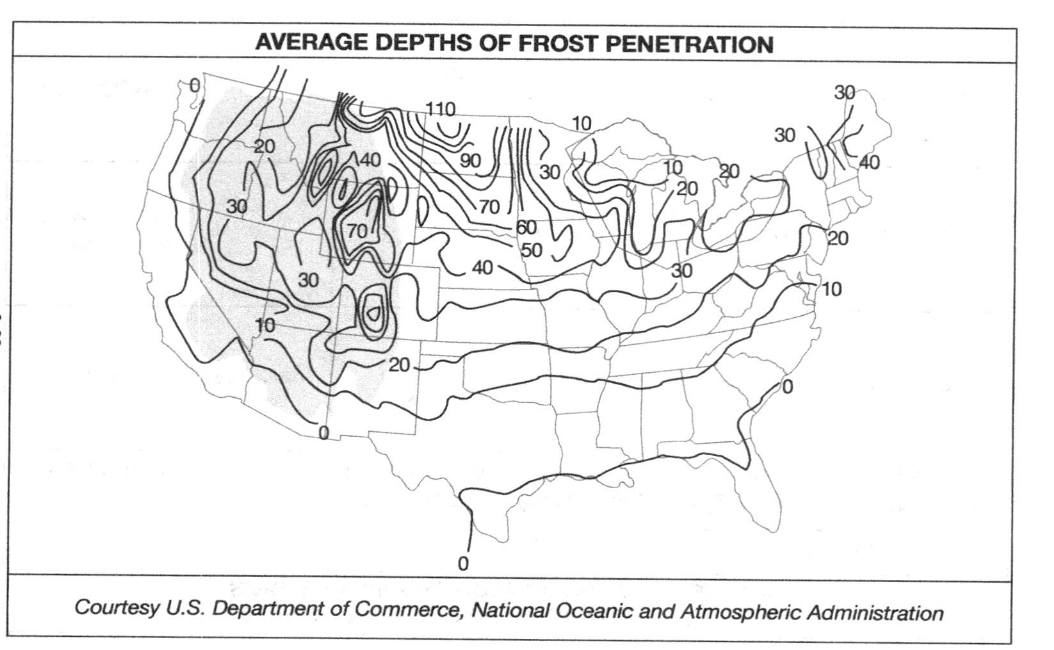

Courtesy U.S. Department of Commerce, National Oceanic and Atmospheric Administration

OSHA EXCAVATION SUPPORT SYSTEMS

OSHA requires that a support system such as shoring and bracing be provided when excavations will occur near the foundation of an existing building to ensure that the adjacent structure or pavement remains stable during excavation and construction. OSHA prohibits excavation below the base of any foundation or retaining wall unless the following recommendations are observed.

1. A support system is provided such as underpinning. Underpinning examples follow.
2. The excavation is in stable rock.
3. A registered professional engineer determines that the structure is far enough away from the excavation and that the excavation will not pose a hazard to employees.

A TYPICAL UNDERPINNING PROCEDURE

Underpinning is used to support buildings close to a proposed excavation and when the existing foundation needs extra support as provided by underpinning. This requires considerable engineering support, and special considerations are needed because each situation is usually different. Following is an example of a typical underpinning installation.

WATER IN EXCAVATIONS

OSHA standards prohibit employees from working without adequate protection in excavations where water has accumulated or is accumulating. Water removal equipment must be monitored by competent, trained personnel. Diversion ditches, dikes, and other suitable means to prevent surface water from entering an excavation and to provide drainage to an adjacent area are required. A competent person must inspect excavations when water inflow has occurred.

INGRESS TO AND EGRESS FROM EXCAVATIONS

Safe access into and egress from excavations 4 feet or deeper are required. This can include ladders, stairs, or ramps.

SOILS, TRENCHING, AND EXCAVATIONS

See Chapter 4 for information on soils, trenching, and excavations.

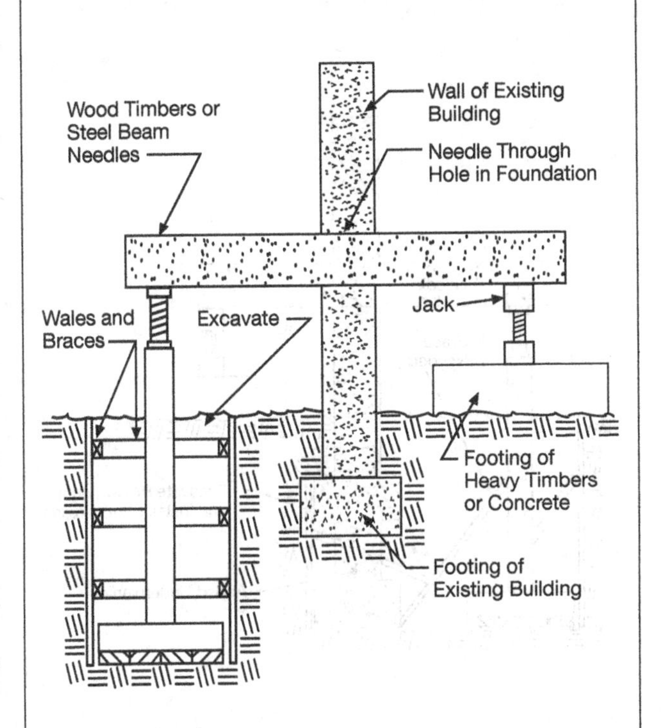

1. Excavate a small area for the needle jack. Brace the sides as needed. Install the needles through holes cut in the exterior wall.

FOUNDATIONS WITH UNDERPINNING INSTALLED BY SUPPORTING THE BUILDING WITH NEEDLES *(cont.)*

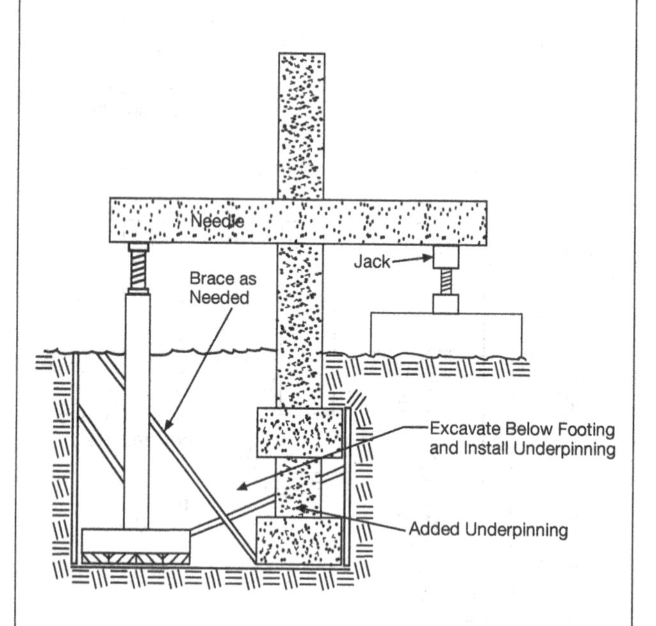

2. After the needles are in place excavate the area under the footing, brace the walls of the excavation with sheathing and form and pour the underpinning.

FOUNDATION CLEARANCES FROM SLOPES

Face of Footing

Top of Slope

Face of Structure

Toe of Slope

H

H/3—Need Not Exceed 40' Max.

H/2—Need Not Exceed 15' Max.

Alternate setbacks must be approved by local building officials.

TYPICAL SOIL BEARING CAPACITIES

GRAVELLY SOILS

Symbol	Description	lb./sq. ft.
GW	Well-graded Gravel and Sand-gravel Mixes	8,000
GP	Poorly Graded Gravel and Sand-gravel Mixes	8,000
GM	Silty Gravels, Sand-silt-gravel Mixes	4,000
GC	Clayey Gravels, Clay-sand-gravel Mixes	4,000

SANDY SOILS

Symbol	Description	lb./sq. ft.
SW	Well-graded Sand or Gravelly Sand	6,000
SP	Poorly Graded Sands, Gravelly Sands, Little Fines	5,000
SM	Silty Sands	4,000
SC	Sand-clay Mixes	4,000

CLAYS AND SILTS

Symbol	Description	lb./sq. ft.
ML	Inorganic Silts, Clayey Fine Sands	2,000
CL	Inorganic Clays, Silty Clays, Gravelly Clays	2,000
CH	Inorganic Clay, Medium to High Plasticity	2,000

ORGANIC SOILS

Peat and other organic materials. Not suitable for foundation construction.

Soil classification and investigation should be made by a registered design professional. Data is submitted to the local building official. Satisfactory data from investigations on adjacent areas can be accepted by the building official in place of a new soil investigation.

COLD WEATHER CONSTRUCTION REQUIREMENTS	
Ambient Temperature	**Cold Weather Procedures for Work in Progress**
Above 40°F (4.4°C)	No special requirements.
Below 40°F (4.4°C)	Do not lay glass unit masonry.
32°F to 40° F (0°C to 4.4°C)	Heat sand or mixing water to produce mortar temperature between 40°F and 120°F (4.4°C and 48.9°C) at the time of mixing. Heat materials for grout only if they are below 32F° (0°C).
25°F to 32°F (-3.9°C to 0°C)	Heat sand or mixing water to produce mortar temperature between 40°F and 120°F (4.4°C and 48.9°C) at the time of mixing. Keep mortar above freezing until used in masonry. Heat materials to produce grout temperature between 70°F and 120°F (21.1°C and 48.9°C) at the time of mixing. Keep group temperature above 70°F (21.1°C) at the time of placement.
20°F to 25°F and Below (-6.7°C to -3.9°C)	In addition to requirements for 25°F (-3.9°C to 0°C), heat masonry surfaces under construction to 40°F (4.4°C) and use wind breaks or enclosures when the wind velocity exceeds 15 mph (24 km/h). Heat masonry to a minimum of 40°F (4.4°C) prior to grouting.
20°F (-6.7°C) and below	In addition to all of the above requirements, provide an enclosure and auxiliary heat to keep air temperature above 32°F (0°C) within the enclosure.

COLD WEATHER CONSTRUCTION REQUIREMENTS *(cont.)*

Ambient Temperature (minimum for grouted; mean daily for ungrouted)	Cold Weather Procedures for Newly Completed Masonry
Above 40°F (4.4°C)	No special requirements.
25°F to 40°F (-3.9°C to 4.4°C)	Cover newly constructed masonry with a weather-resistant membrane for 24 hours after being completed.
20°F to 25°F (-6.7°C to -3.9°C)	Cover newly constructed masonry with weather-resistive insulating blankets or equal protection for 24 hours after being completed. Extend the time period to 48 hours for grouted masonry, unless the only cement used in the grout is ASTM C 150 Type III.
20°F (-6.7°C) and Below	Keep newly constructed masonry above 32°F (0°C) for at least 24 hours after being completed. Use heated enclosures, electric heating blankets, infrared lamps, or other acceptable methods. Extend the time period to 48 hours for grouted masonry, unless the only cement used in the grout is ASTM C 150 Type III.

Courtesy Portland Cement Association, www.cement.org

CHAPTER 10
Carpentry

The products and construction techniques related to wood construction are extensive. Often, technical data vary for a single type of product, and the manufacturer must be contacted to get the specifics related to their product. The types of fasteners available are also many and varied. See Chapter 6 for information on fasteners and Chapter 13 for wood flooring information.

SOFTWOOD LUMBER GRADES

The actual grade designations and the properties of the lumber will vary depending upon the grading agency. Some are rule-writing agencies that publish grading rule books. Others are independent rule-setting agencies. Most softwood species used in construction are stress graded under the American Softwood Lumber Standard PS 20 of the U.S. Department of Commerce.

Softwood lumber is graded in three major categories: Stress-Graded, Nonstress-Graded and Appearance.

SOFTWOOD LUMBER GRADES *(cont.)*

Stress-graded lumber has the properties listed in the National Grading Rule, which is part of the American Softwood Lumber Standard. Most softwood lumber used in construction is stress-graded. It can be graded visually or machine stress-rated (MSR).

Nonstress-graded lumber has the size of the pieces combined with visual grade requirements that provide a measure of its structural integrity. Grades for various species of wood will vary because of the differences in their properties. The grade names will vary depending upon the grading association. Some nonstress-graded lumber is used for subflooring and sheathing where appearance is not important but where minimum strength requirements are met.

Appearance lumber must meet appearance requirements as well as strength requirements. They may be graded using stress-graded or nonstress-graded rules.

DRY SOFTWOOD LUMBER SIZES*		
Nominal Size	Dressed	
Inches	Inches	mm
	THICKNESS	
3/8	5/16	8
1/2	7/16	11
5/8	9/16	14
3/4	5/8	16
1	3/4	19
1 1/4	1	25
1 1/2	1 1/4	32
2	1 1/2	38
2 1/2	2	51
3	2 1/2	64
3 1/2	3	76
4	3 1/2	89
4 1/2	4	102
6	5 1/2	140
8	7 1/2	190
	FACE WIDTH	
2	1 1/2	38
3	2 1/2	64
4	3 1/2	89
5	4 1/2	114
6	5 1/2	140
7	6 1/2	165
8	7 1/4	184
9	8 1/4	210
10	9 1/4	235
11	10 1/4	260
12	11 1/4	286
14	13 1/4	337
16	15 1/4	387

*Dry lumber contains less than 19% moisture.

TYPICAL GRADE CLASSIFICATIONS FOR VISUALLY STRESS-GRADED SOFTWOOD LUMBER		
Lumber Classification	**Grade Name**	**Strength Ratio***
		PCT
Light framing (2 to 4 in. thick, 2 to 4 in. wide)[2]	Construction Standard Utility	34 19 9
Structural light framing (2 to 4 in. thick, 2 to 4 in. wide)	Select Structural 1 2 3	67 55 45 26
Studs (2 to 4 in. thick, 2 to 6 in. wide, 10 ft. and shorter)	Stud	26
Structural joists and planks (2 to 4 in. thick, 6 in. and wider)	Select Structural 1 2 3	65 55 45 26
Appearance framing (2 to 4 in. thick, 2 in. and wider)	Appearance	55

*Strength ratio gives a comparison of the quality of the grade.
National Grading Rule PS-20, U.S. Department of Commerce.

TYPICAL NONSTRESS-GRADED SOFTWOOD LUMBER CLASSIFICATIONS	
No. 1	Best grade, strongest, some consideration is given to appearance.
No. 2	Larger knots and knot holes are permitted. Used for subfloors, wall and roof sheathing, concrete formwork.
No. 3	More knots and holes permitted. These influence the strength, appearance not important.
*Grade classification identification will vary because of differences in wood species and grading agency rules.	

TYPICAL APPEARANCE GRADE SOFTWOOD LUMBER CLASSIFICATIONS*	
B & Btr	Primarily clear, some pin holes permitted, can be stained.
C	Permits small surface checks, small tight knots, paint or natural finish.
C & Btr	Contains pieces of both B & Btr and C grades. Paint or natural finish.
D	Minor defects acceptable, economical grade for paint or natural finish.
*Classifications in use will vary due to differences in wood species and the grading agency rules.	

COMMON TYPES OF WORKED LUMBER

Worked lumber is that which in addition to being surfaced has machining operations performed upon it.

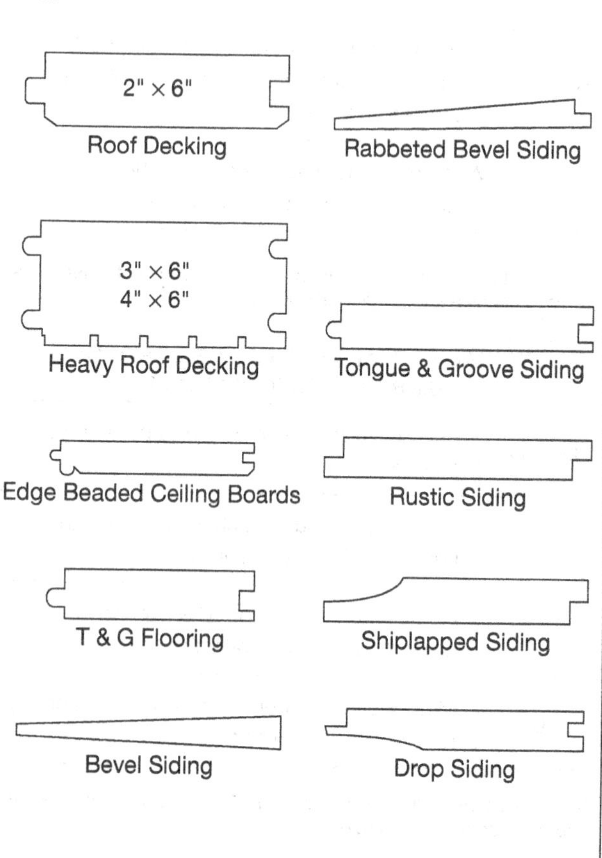

Roof Decking

Rabbeted Bevel Siding

Heavy Roof Decking

Tongue & Groove Siding

Edge Beaded Ceiling Boards

Rustic Siding

T & G Flooring

Shiplapped Siding

Bevel Siding

Drop Siding

PRESSURE-TREATED WOOD AND PLYWOOD

EPA-approved wood preservatives commonly used on wood products in general use in building construction include alkaline copper quat (ACQ), copper azole (CA), and bardac 22C50.

ACQ is an EPA-approved nonchromium, water-based preservative. It is used for full exposure to aboveground, ground contact, and fresh water applications. It can be painted and stained. ACQ-B is used to treat Western wood species. ACQ-C is used in all parts of the United States. ACQ-D is used everywhere except on the West Coast.

CA is a water-based preservative used to prevent decay from fungi and insects. It is used to treat species of softwoods in full exposure to aboveground and fresh water applications.

Bardac is used to treat framing lumber and plywood that are continuously exposed to the weather. It is not suitable for ground contact use. It protects against wood-destroying insects and fungi.

Other preservatives having special uses include creosote and methyl isothiocyanate. Creosote is used to treat industrial products exposed to the ground and water, such as utility poles. Methyl isothiocyanate is used on large timbers such as pilings. Its use is restricted.

GLUING PROPERTIES OF SELECTED WOODS	
Bond easily	Alder Aspen Fir: white, grand, noble, pacific Red cedar, western Redwood Spruce, Sitka
Bond well	Elm, American rock Maple, soft Sycamore Walnut, black Yellow poplar Douglas fir Larch, western Pine: sugar, ponderosa Red cedar, eastern Mahogany: African, American
Bond satisfactorily	Ash, white Birch: sweet, yellow Cherry Hickory: pecan, true Oak: red, white Pine, southern
Bond with Difficulty	Persimmon Teak Rosewood
Courtesy Forest Products Laboratory, USDA Forest Service, Madison, Wis.	

LUMBER SURFACING DESIGNATIONS

Sawed lumber may be surfaced in a number of ways to meet specific needs. Following are frequently used designations.

Rough Lumber – Sawed lumber cut to standard sizes. The surfaces are rough.

Surfaced Lumber – Lumber that has been dressed by planing. This smoothes the surfaces and reduces the pieces to their finished size.

Surface Lumber Designations

S1S – Surfaced on 1 side
S2E – Surfaced on one edge
S2S – Surfaces on two sides
S2E – Surfaced on two edges
S1S2E – Surfaced on one side and one edge
S1S2E – Surfaced on one side and two edges
S2S1E – Surfaced on two sides and one edge
S4S – Surfaced on four sides

STORING LUMBER ON THE SITE

Lumber should be stored so it is not subject to moisture, which will increase the moisture content and lead to swelling and warping. It should be stored flat off the ground and covered with a waterproof tarp or plastic sheeting. All sides and ends must be carefully covered, and taped if necessary to keep the wind from blowing off the cover. If it is to be stored outside for an extended period of time, a plastic ground cover below it would be recommended.

STORING LUMBER ON THE SITE *(cont.)*

Lumber stored indoors, if not covered, will pick up moisture from the air. This typically happens when the drywall is being taped. This could increase the moisture content beyond that specified. This is especially important when storing interior trim materials, wood flooring, and other unfinished products such as doors.

CALCULATING BOARD FEET

A board foot is a piece of wood 1 inch thick by 12 inches wide by 12 inches long. It contains 144 cubic inches. To calculate the number of board feet in a piece of wood, use the following formula: No. of Bd. Ft. = thickness (inches) × width (inches) × length (feet).

Example: A piece 1 in. × 8 in. × 10 ft. is figured

$$\frac{1 \times 8 \times 10}{12} = \frac{80}{12} = 6.7 \text{ bd. ft.}$$

Wood that is less than 1 inch thick is figured as if it were 1 inch thick.

Wood with a fractional thickness, as 1½ inch, has the total thickness changed to a fraction as $\frac{3}{2}$. The numerator (3) is placed above the line and the denominator (2) is placed below the line.

See the following example: A piece 1½ × 8 × 10 is shown as $\frac{3 \times 8 \times 10}{2 \times 12} = 10$ bd. ft.

BOARD FEET FOR SELECTED LUMBER SIZES									
Nominal Size (inches)	Length (feet)								
	8	10	12	14	16	18	20	22	24
1 × 2	1⅓	1⅔	2	2⅓	2⅔	3	3⅓	3⅔	4
1 × 3	2	2½	3	3½	4	4½	5	5½	6
1 × 4	2⅔	3⅓	4	4⅔	5⅓	6	6⅔	7⅓	8
1 × 5	3⅓	4⅙	5	5⅚	6⅔	7½	8⅓	9⅙	10
1 × 6	4	5	6	7	8	9	10	11	12
1 × 7	4⅔	5⅚	7	8⅙	9⅓	10½	11⅔	12⅚	14
1 × 8	5⅓	6⅔	8	9⅓	10⅔	12	13⅓	14⅔	16
1 × 10	6⅔	8⅓	10	11⅔	13⅓	15	16⅔	18⅓	20
1 × 12	8	10	12	14	16	18	20	22	24
1¼ × 4	3⅓	4⅙	5	5⅚	6⅔	7½	8⅓	9⅙	10
1¼ × 6	5	6¼	7½	8¾	10	11¼	12½	13¾	15
1¼ × 8	6⅔	8⅓	10	11⅔	13⅓	15	16⅔	18⅓	20
1¼ × 10	8⅓	10 5/12	12½	14 7/12	16⅔	18¾	20⅚	22 11/12	25
1¼ × 12	10	12½	15	17½	20	22½	25	27½	30
1½ × 4	4	5	6	7	8	9	10	11	12
1½ × 6	6	7½	9	10½	12	13½	15	16½	18
1½ × 8	8	10	12	14	16	18	20	22	24
1½ × 10	10	12½	15	17½	20	22½	25	27½	30
1½ × 12	12	15	18	21	24	27	30	33	36
2 × 2	2⅔	3⅓	4	4⅔	5⅓	6	6⅔	7⅓	8
2 × 4	5⅓	6⅔	8	9⅓	10⅔	12	13⅓	14⅔	16
2 × 6	8	10	12	14	16	18	20	22	24
2 × 8	10⅔	13⅓	16	18⅔	21⅓	24	26⅔	29⅓	32
2 × 10	13⅓	16⅔	20	23⅓	26⅔	30	33⅓	36⅔	40
2 × 12	16	20	24	28	32	36	40	44	48
3 × 3	6	7½	9	10½	12	13½	15	16½	18
3 × 6	12	15	18	21	24	27	30	33	36
3 × 8	16	20	24	28	32	36	40	44	48
3 × 10	20	25	30	35	40	45	50	55	60
3 × 12	24	30	36	42	48	54	60	66	72
4 × 4	10⅔	13⅓	16	18⅔	21⅓	24	26⅔	29⅓	32
4 × 6	16	20	24	28	32	36	40	44	48
4 × 8	21⅓	26⅔	32	37⅓	42⅔	48	53⅓	58⅔	64
4 × 10	26⅔	33⅓	40	46⅔	53⅓	60	66⅔	73⅓	80
4 × 12	32	40	48	56	64	72	80	88	96

NAILS AND STAPLES FASTENING SCHEDULE		
Connection	**Fastening[a,m]**	**Location**
1. Joist to sill or girder	3 – 8d common 3 - 3" × 0.131" nails 3 – 3" 14 gage staples	toenail
2. Bridging to joist	2 – 8d common 2 – 3" × 0.131" nails 2 – 3" 14 gage staples	toenail each end
3. 1" × 6" subfloor or less to each joist	2 – 8d common	face nail
4. Wider than 1" × 6" subfloor to each joist	3 – 8d common	face nail
5. 2" subfloor to joist or girder	2 – 16d common	blind and face nail
6. Sole plate to joist or blocking	16d at 16" o.c. 3" × 0.131" nails at 8" o.c. 3" 14 gage staples at 12" o.c.	typical face nail
Sole plate to joist or blocking at braced wall panel	3 – 16d at 16" 4 – 3" × 0.131" nails at 16" 4 – 3" 15 gage staples per 16"	braced wall panels
7. Top plate to stud	2 – 16d common 3 – 3" × 0.131" nails 3 – 3" 14 gage staples	end nail
8. Stud to sole plate	4 – 8d common 4 – 3" × 0.131" nails 3 – 3" 14 gage staples	toenail
	2 – 16d common 3 – 3" × 0.131" nails 3 – 3" 14 gage staples	end nail

NAILS AND STAPLES FASTENING SCHEDULE *(cont.)*		
9. Double studs	16d at 24" o.c. 3" × 0.131" nail at 8" o.c. 3" 14 gage staple at 8" o.c.	face nail
10. Double top plates	16d at 16" o.c. 3" × 0.131" nail at 12" o.c. 3" 14 gage staple at 12" o.c.	typical face nail
	8 – 16d common 12 – 3" × 0.131" nails 12 – 3" 14 gage staples	lap splice
11. Blocking between joists or rafters to top plate	3 – 8d common 3 – 3" × 0.131" nails 3 – 3" 14 gage staples	toenail
12. Rim joist to top plate	8d at 6" o.c. 3" × 0.131" nail at 6" o.c. 3" 14 gage staple at 6" o.c.	toenail
13. Top plates, laps and intersections	2 – 16d common 3 – 3" × 0.131" nails 3 – 3" 14 gage staples	face nail
14. Continuous header, two pieces	16d common	16" o.c. along edge
15. Ceiling joists to plate	3 – 8d common 5 – 3" × 0.131" nails 5 – 3" 14 gage staples	toenail
16. Continuous header to stud	4 – 8d common	toenail

NAILS AND STAPLES FASTENING SCHEDULE *(cont.)*		
17. Ceiling joists, laps over partitions (see Section 2308.10.4.1, Table 2308.10.4.1)	3 – 16d common minimum, Table 2308.10.4.1 4 – 3" × 0.131" nails 4 – 3" 14 gage staples	face nail
18. Ceiling joists to parallel rafters (see Section 2308.10.4.1, Table 2308.10.4.1)	3 – 16d common minimum, Table 2308.10.4.1 4 – 3" × 0.131" nails 4 – 3" 14 gage staples	face nail
19. Rafter to plate (see Section 2308.10.1, Table 2308.10.1)	3 – 8d common 3 – 3" × 0.131" nails 3 - 3" 14 gage staples	toenail
20. 1" diagonal brace to each stud and plate	2 – 8d common 2 – 3" × 0.131" nails 2 – 3" 14 gage staples	face nail
21. 1" × 8" sheathing to each bearing wall	2 – 8d common	face nail
22. Wider than 1" × 8" sheathing to each bearing	3 – 8d common	face nail
23. Built-up corner studs	16d common 3" × 0.131" nails 3" 14 gage staples	24" o.c. 16" o.c. 16" o.c.
24. Built-up girder and beams	20d common 32" o.c. 3" × 0.131" nail at 24" o.c. 3" 14 gage staple at 24" o.c.	face nail at top and staggered on opposite sides
	2 – 20d common 3 – 3" × 0.131" nails 3 – 3" 14 gage staples	face nail at ends and at each splice

NAILS AND STAPLES FASTENING SCHEDULE *(cont.)*		
25. 2" planks	16d common	at each bearing
26. Collar tie to rafter	3 – 10d common 4 – 3" × 0.131" nails 4 – 3" 14 gage staples	face nail
27. Jack rafter to hip	3 – 10d common 4 – 3" × 0.131" nails 4 – 3" 14 gage staples	toenail
	2 – 16d common 3 – 3" × 0.131" nails 3 – 3" 14 gage staples	face nail
28. Roof rafter to 2-by ridge beam	2 – 16d common 3 – 3" × 0.131" nails 3 – 3" 14 gage staples	toenail
	2 – 16d common 3 – 3" × 0.131" nails 3 – 3" 14 gage staples	face nail
29. Joist to band joist	3 – 16d common 5 – 3" × 0.131" nails 5 – 3" 14 gage staples	face nail
30. Ledger strips	3 – 16d common 4 – 3" × 0.131" nails 4 – 3" 14 gage staples	face nail

NAILS AND STAPLES FASTENING SCHEDULE *(cont.)*

31. Wood structural panels and particleboard:[b] Subfloor, roof and wall sheathing (to framing): Single Floor (combination subfloor-underlayment to framing):	½" and less ⅞" to 1" 1⅛" to 1¼" ¾" and less ⅞" to 1" 1⅛" to 1¼"	6d[c,1] 2⅜" × 0.113" nail[n] 1¾" 16 gage[o] 19/32" to ¾" 8d[d] or 6d[e] 2⅜" × 0.113" nail[p] 2" 16 gage[p] 8d[c] 10d[d] or 8d[e] 6d[e] 8d[e] 10d[d] or 8d[e]
32. Panel siding (to framing)	½" or less ⅝"	6d[f] 8d[f]
33. Fiberboard sheathing[g]	½" 25/32"	No. 11 gage roofing nail[h] 6d common nail No. 16 gage staple[i] No. 11 gage roofing nail[h] 8d common nail No. 16 gage staple[i]
34. Interior paneling	¼" ⅜"	4d[j] 6d[k]

For S1: 1 inch = 25.4 mm.

[a]Common or box nails are permitted to be used except where otherwise stated.

[b]Nails spaced at 6 inches on center at edges, 12 inches at intermediate supports except 6 inches at supports where spans are 48 inches or more. For nailing of wood structural panel and particleboard diaphragms and shear walls, refer to Section 2305. Nails for wall sheathing are permitted to be common, box, or casing.

NAILS AND STAPLES FASTENING SCHEDULE *(cont.)*

[c]Common or deformed shank.
[d]Common.
[e]Deformed shank.
[f]Corrosion-resistant siding or casing nail.
[g]Fasteners spaced 3 inches on center at exterior edges and 6 inches on center at intermediate supports.
[h]Corrosion-resistant roofing nails with 7⁄16-inch diameter head and 1½-inch length for ½-inch sheathing and 1¾-inch length for 25⁄32-inch sheathing.
[i]Corrosion-resistant staples with nominal 7⁄16-inch crown and 1⅛-inch length for ½-inch sheathing and 1½-inch length for 25-32-inch sheathing. Panel supports at 16 inches (20 inches if strength axis in the long direction of the panel, unless otherwise marked).
[j]Casing or finish nails spaced 6 inches on panel edges, 12 inches at intermediate supports.
[k]Panel supports at 24 inches. Casing or finish nails spaced 6 inches on panel edges, 12 inches at intermediate supports.
[l]For roof sheathing applications, 8d nails are the minimum required for wood structural panels.
[m]Staples shall have a minimum crown width of 7⁄16 inch.
[n]For roof sheathing applications, fasteners spaced 4 inches on center at edges, 8 inches at intermediate supports.
[o]Fasteners spaced 4 inches on center at edges, 8 inches at intermediate supports for subfloor and wall sheathing and 3 inches on center at edges, 6 inches at intermediate supports for roof sheathing.
[p]Fasteners spaced 4 inches on center at edges, 8 inches at intermediate supports.

HARDWOOD LUMBER

Hardwood lumber is the product of deciduous trees, which have broad leaves and lose them in the winter. Commonly available hardwoods include oak, walnut, beech, birch, basswood, cherry, chestnut, gum, elm, hickory, mahogany, maple, and yellow poplar.

HARDWOOD LUMBER THICKNESSES

Rough Size Inches	Dressed Two Sides Inches
1	13⁄16
1¼	1 1⁄16
1½	1 5⁄16
1¾	1½
2	1¾
2½	2¼
3	2¾
3½	3¼
4	3¾

HARDWOOD LUMBER GRADES	
First and Seconds	The best grade. Has 83.3% clear wood on the poorest side
Select – No. One Common	Has 66.6% clear on poorest side
Select – No. 2 Common	Has 50% clear on poorest side
Select – No. 3 Common	Has 50% clear on poorest side

WOOD MOISTURE CONTENT

The moisture content of wood affects its strength and size. When dried, the strength increases and the size decreases.

MAXIMUM CRUSHING STRENGTH OF WOOD PARALLEL WITH THE GRAIN WHEN COMPARED TO THE WOOD CONTAINING 2% MOISTURE

% Moisture	Red Spruce	Longleaf Pine	Douglas Fir
2	1.000	1.000	1.000
4	0.926	0.894	0.929
6	0.841	0.790	0.850
8	0.756	0.702	0.774
10	0.681	0.623	0.714
12	0.617	0.552	0.643
14	0.554	0.488	0.589
16	0.505	0.431	0.535
18	0.463	0.377	0.494
20	0.426	0.328	0.458

U.S. Forest Service Circular 108

REQUIRED MOISTURE CONTENT LEVELS IN WOOD PRODUCTS BEFORE INSTALLATION

Uses	Geographic Areas		
	Most of the United States	Dry Western States such as Utah, Nevada, Arizona, New Mexico	Warm Moist Coastal Areas – Texas through Virginia
Exterior applications: framing, siding, sheathing, trim, laminated timbers	12%	9%	12%
Interior uses: flooring, trim, exposed structural members	8%	6%	11%

FIRE-RETARDANT TREATMENTS

Wood is treated with water-borne or organic solvent-borne fire retardant chemicals. It is treated with inorganic salts or complex fire-retardant chemicals. Some are pressure applied while others are applied to the surface of the wood. Most treatments do not resist exposure to the weather. Leach-resistant types can be used on products that are exposed to the weather, such as wood shingles.

Woods with crystal salt treatments have an abrasive effect on cutting tools. Use carbide-tipped cutting tools. Since the chemicals corrode fasteners, those known to resist corrosion must be used.

TYPICAL SPANS FOR SOLID AND BUILT-UP WOOD BEAMS

Spans Based on Allowable Fiber Stress of 1500 Psi

Width of Structure	Girder Size (in.) (Built-up or Solid)	Maximum Span When Supporting a Load-Bearing Interior Partition	
		1 Story	1½ or 2 Story
Up to 26 ft.	6 × 8	7'-0"	6'-0"
	6 × 10	9'-0"	7'-6"
	6 × 12	10'-6"	9'-0"
26 ft. to 32 ft.	6 × 8	6'-6"	5'-6"
	6 × 10	8'-0"	7'-0"
	6 × 12	10'-0"	8'-0"

Federal Housing Administration Minimum Property Standards

TYPICAL GLUED LAMINATED TIMBER SIZES*

(1½ inch lamination)

Width Inches	Depth Inches
2½	9 to 27
3⅛	6 to 24
5⅛	6 to 36
6¾	7.5 to 48
8¾	9 to 36
10¾	10.5 to 81

*Consult the manufacturer for specific load capacities for lumber species, spans, and applications.

TYPICAL LOADS* FOR GLULAM ROOF BEAMS**

Span in Feet	3" 3 6"	3" 3 9"	3" 3 12"	3" 3 15"
10	272	745	1104	1380
14	–	335	676	986
16	–	224	518	789
20	–	115	272	505

*Loads shown are pounds per lineal foot including weight of beam.

**Applicable for straight, simply supported beams. (Consult the manufacturer for specific load data.)

TYPICAL LOADS* FOR GLULAM FLOOR BEAMS**

Span in Feet	3" 3 6"	3" 3 9"	3" 3 12"	3" 3 15"
10	266	896	1200	1500
14	–	167	397	774
16	–	112	266	519
20	–	–	136	266

*Loads shown are pounds per lineal foot including weight of beam.

**Applicable for straight, simply supported beams. (Consult the manufacturer for specific load data.)

TWO STORY FLOOR LOADS

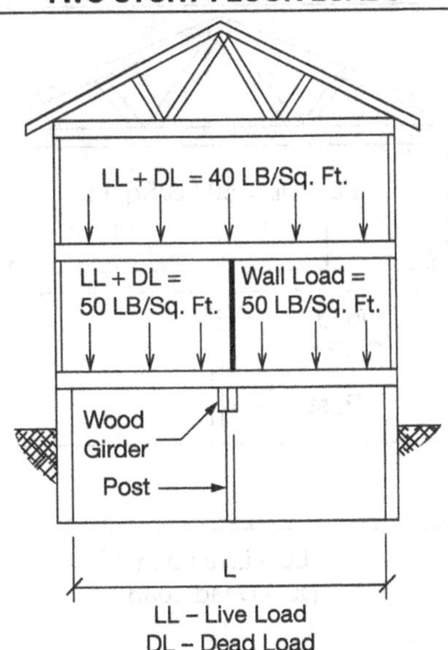

Maximum Span Designs for Girders Supporting Two-Story Floor Loads Using Lumber Having an Allowable Bending Stress Not Less Than 1000 psi*

Nominal Lumber Sizes	Girder Spans 5 B (ft)*								
	House Widths 5 L (ft)								
	20	22	24	26	28	30	32	34	36
3–2 × 8	4'-7"	4'-2"	–	–	–	–	–	–	–
2–2 × 12	4'-9"	4'-4"	4'–0"	–	–	–	–	–	–
3–2 × 10	5'–10"	5'–4"	4'–11"	4'–7"	4'–3"	4'–0"	–	–	–
3–2 × 12	7'-1"	6'-6"	6'–0"	5'–6"	5'–2"	4'–10"	4'–6"	4'–3"	4'–0"

*Linear interpolation within the above table for house widths not given is permitted.
Courtesy BuilderBooks www.builderbooks.com

ONE STORY FLOOR LOADS

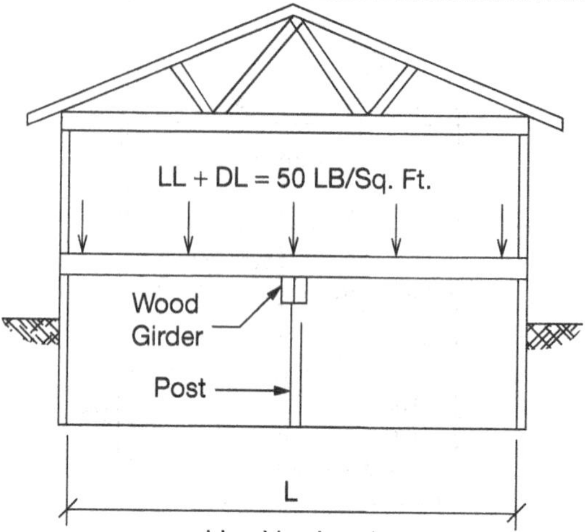

LL + DL = 50 LB/Sq. Ft.

Wood Girder

Post

L

LL – Live Load
DL – Dead Load

Maximum Span Designs for Girders Supporting One-Story Floor Loads Using Lumber Having an Allowable Bending Stress Not Less Than 1500 psi*

Nominal Lumber Sizes	House Widths 5 L (ft)								
	20	22	24	26	28	30	32	34	36
2–2 × 6	5'–3"	4'–10"	4'–5"	4'–1"	—	—	—	—	—
3–2 × 6	6'–9"	6'–5"	6'–2"	5'–11"	5'–8"	5'–3"	4'–11"	4'–8"	4'–5"
2–2 × 8	7'–0"	6'–4"	5'–10"	5'–4"	5'–0"	4'–8"	4'–4"	4'–1"	—
3–2 × 8	8'–11"	8'–6"	8'–1"	7'–9"	7'–5"	7'–0"	6'–6"	6'–2"	5'–10"
2–2 × 10	8'–11"	8'–1"	7'–5"	6'–10"	6'–4"	5'–11"	5'–7"	5'–3"	4'–11"
3–2 × 10	11'–4"	10'–10"	10'–4"	9'–11"	9'–6"	8'–11"	8'–4"	7'–10"	7'–5"
2–2 × 12	10'–10"	9'–10"	9'–0"	8'–4"	7'–9"	7'–2"	6'–9"	6'–4"	6'–0"
3–2 × 12	13'–9"	13'–2"	12'–7"	12'–1"	11'–7"	10'–10"	10'–2"	9'–6"	9'–0"

*Linear interpolation within the above table for house widths not given is permitted.

Courtesy BuilderBooks www.builderbooks.com

SOUND TRANSMISSION AND FIRE RATINGS

The sound transmission class (STC) is a single number that indicates the effectiveness of the material or assembly of materials to resist the transmission of airborne sound through the material into the next area.

Typical Sound Transmission Characteristics for Several STC Ratings

STC Number	
25	Normal speech can be understood easily
35	Loud speech audible but not intelligible
45	You must strain to hear loud speech
50	Loud speech not audible
60 to 70	Practically no sound heard

The noise reduction coefficient (NRC) is a single number rating used to calculate the quantity of sound absorbing material required.

Flame spread ratings are measures of the ability of a material to resist flaming combustion over its surface designated by a single number. The flame spread ratings are:

Class 1: Flame spread, 0-25
Class 2: Flame spread, 26-75
Class 3: Flame spread, 76-200

The fire rating is the time in hours or fractions of an hour that a material or assembly of materials can resist exposure to fire.

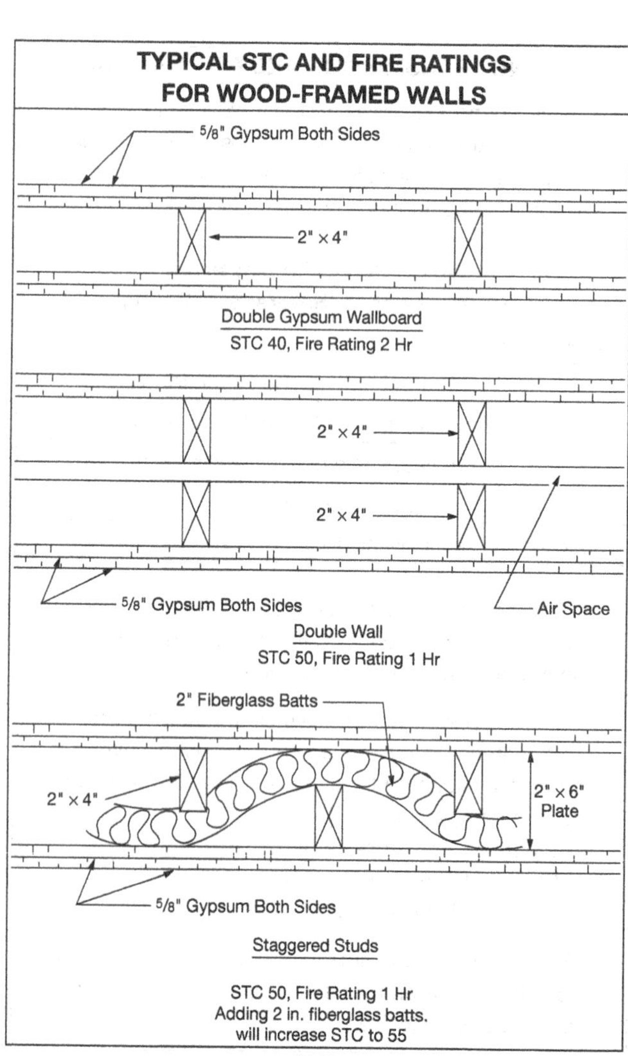
TYPICAL STC AND FIRE RATINGS
FOR WOOD-FRAMED WALLS
5/8" Gypsum Both Sides
2" × 4"
Double Gypsum Wallboard
STC 40, Fire Rating 2 Hr
2" × 4"
2" × 4"
5/8" Gypsum Both Sides
Air Space
Double Wall
STC 50, Fire Rating 1 Hr
2" Fiberglass Batts
2" × 4"
2" × 6"
Plate
5/8" Gypsum Both Sides
Staggered Studs
STC 50, Fire Rating 1 Hr
Adding 2 in. fiberglass batts.
will increase STC to 55

TYPICAL STC AND FIRE RATINGS FOR WOOD-FRAMED WALLS *(cont.)*

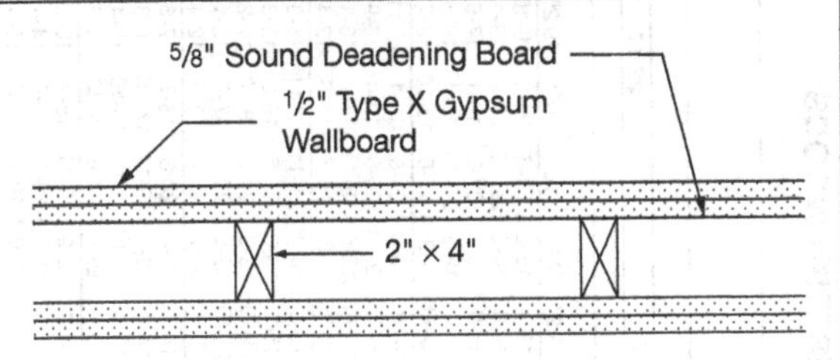

Gypsum Wallboard and Sound Deadening Board

STC 45, Fire Rating 1 hr

Adding 2" Fiberglass Batts will Increase STC to 50

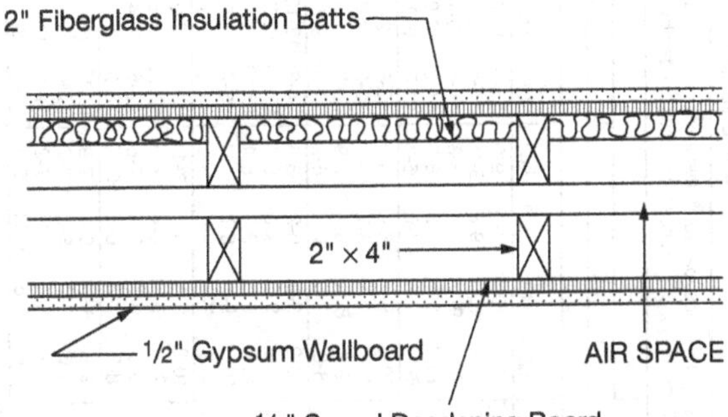

Double Wall with Gypsum Wallboard and Sound Deadening Board

STC 55, Fire Rating 1 hr.

MAXIMUM SPANS FOR FLOOR JOISTS USING WESTERN WOODS

40# Live Load 10# Dead Load L/360

Design Criteria: *Strength* - 40 lbs. per sq. ft. live load, plus 10 lbs. per sq. ft. dead load.
Deflection - Limited in span in inches divided by 360 for live load only.

		Span (feet and inches)															
		2 3 6				2 3 8				2 3 10				2 3 12			
		Spacing on Center															
Species or Group	**Grade**	**12"**	**16"**	**19.2"**	**24"**	**12"**	**16"**	**19.2"**	**24"**	**12"**	**16"**	**19.2"**	**24"**	**12"**	**16"**	**19.2"**	**24"**
Douglas Fir-Larch	Sel. Struc.	11-4	10-4	9-8	9-0	15-0	13-7	12-10	11-11	19-1	17-4	16-4	15-2	23-3	21-1	19-10	18-5
	No. 1 & Btr.	11-2	10-2	9-6	8-10	14-8	13-4	12-7	11-8	18-9	17-0	16-0	14-9	22-10	20-9	19-1	17-1
	No. 1	10-11	9-11	9-4	8-8	14-5	13-1	12-4	11-0	18-5	16-5	15-0	13-5	22-0	19-1	17-5	15-7
	No. 2	10-9	9-9	9-2	8-3	14-2	12-9	11-8	10-5	18-0	15-7	14-3	12-9	20-11	18-1	16-6	14-9
	No. 3	8-11	7-8	7-0	6-3	11-3	9-9	8-11	8-0	13-9	11-11	10-11	9-9	16-0	13-10	12-7	11-3
Douglas Fir-South	Sel. Struc.	10-3	9-4	8-9	8-2	13-6	12-3	11-7	10-9	17-3	15-8	14-9	13-8	21-0	19-1	17-11	16-8
	No. 1	10-0	9-1	8-7	7-11	13-2	12-0	11-3	10-6	16-10	15-3	14-5	12-11	20-6	18-4	16-9	15-0
	No. 2	9-9	8-10	8-4	7-9	12-10	11-8	11-0	10-2	16-5	14-11	13-10	12-5	19-11	17-7	16-1	14-4
	No. 3	8-8	7-6	6-10	6-2	11-0	9-6	8-8	7-9	13-5	11-8	10-7	9-6	15-7	13-6	12-4	11-0
Hem-Fir	Sel. Struc.	10-9	9-9	9-2	8-6	14-2	12-10	12-1	11-3	18-0	16-5	15-5	14-4	21-11	19-11	18-9	17-5
	No. 1 & Btr.	10-6	9-6	9-0	8-4	13-10	12-7	11-10	11-0	17-8	16-0	15-1	14-0	21-6	19-6	18-3	16-4
	No. 1	10-6	9-6	9-0	8-4	13-10	12-7	11-10	10-10	17-8	16-0	14-10	13-3	21-6	18-10	17-2	15-5
	No. 2	10-0	9-1	8-7	7-11	13-2	12-0	11-3	10-2	16-10	15-2	13-10	12-5	20-4	17-7	16-1	14-4
	No. 3	8-8	7-6	6-10	6-2	11-0	9-6	8-8	7-9	13-5	11-8	10-7	9-6	15-7	13-6	12-4	11-0
Spruce-Pine-Fir (South)	Sel. Struc.	10-0	9-1	8-7	7-11	13-2	12-0	11-3	10-6	16-10	15-3	14-5	13-4	20-6	18-7	17-6	16-3
	No. 1	9-9	8-10	8-4	7-9	12-10	11-8	11-10	10-2	16-5	14-11	14-0	12-1	19-11	17-10	16-3	14-7
	No. 2	9-6	8-7	8-1	7-6	12-6	11-4	10-8	9-3	15-11	14-6	13-3	11-10	1-4	16-10	15-4	13-9
	No. 3	8-3	7-2	6-6	5-10	10-5	9-0	8-3	7-5	12-9	11-0	10-1	9-0	14-9	12-10	11-8	10-5
Western Woods	Sel. Struc.	9-9	8-10	8-4	7-9	12-10	11-8	11-0	10-2	16-5	14-11	14-0	12-9	19-11	18-1	16-6	14-9
	No. 1	9-6	8-7	8-0	7-2	12-6	11-1	10-1	9-0	15-7	13-6	12-4	11-0	18-1	15-8	14-4	12-10
	No. 2	9-2	8-4	7-10	7-2	12-1	11-0	10-1	9-0	15-5	13-6	12-4	11-0	18-1	15-8	11-4	12-10
	No. 3	7-6	6-6	5-11	5-4	9-6	8-3	7-6	6-9	11-8	10-1	9-2	8-3	13-6	11-8	10-8	9-6

Courtesy Western Wood Products Association www.wwpa.org Additional information available from WWPA.

MAXIMUM SPANS FOR FLOOR JOISTS USING WESTERN WOODS *(cont.)*

30# Live Load 10# Dead Load L/360

Design Criteria: *Strength* - 30 lbs. per sq. ft. live load plus 10 lbs. per sq. ft. dead load.
Deflection - Limited in span in inches divided by 360 for live load only.

		Span (feet and inches)															
		2 3 6				2 3 8				2 3 10				2 3 12			
		Spacing on Center															
Species or Group	Grade	12"	16"	19.2"	24"	12"	16"	19.2"	24"	12"	16"	19.2"	24"	12"	16"	19.2"	24"
Douglas Fir-Larch	Sel. Struc.	12-6	11-4	10-8	9-11	16-6	15-0	14-1	13-1	21-0	19-1	18-0	16-8	25-7	23-3	21-10	20-3
	No. 1 & Btr.	12-3	11-2	10-6	9-9	16-2	14-8	13-10	12-10	20-8	18-9	17-8	16-5	25-1	22-10	21-4	19-1
	No. 1	12-0	10-11	10-4	9-7	15-10	14-5	13-7	12-4	20-3	18-5	16-9	15-0	24-8	21-4	19-6	17-5
	No. 2	11-10	10-9	10-1	9-3	15-7	14-2	13-0	11-8	19-10	17-5	15-11	14-3	23-4	20-3	18-6	16-6
	No. 3	9-11	8-7	7-10	7-0	12-7	10-11	10-0	8-11	15-5	13-4	12-2	10-11	17-10	15-5	14-1	12-7
Douglas Fir-South	Sel. Struc.	11-3	10-3	9-8	8-11	14-11	13-6	12-9	11-10	19-0	17-3	16-3	15-1	23-1	21-0	19-9	18-4
	No. 1	11-0	10-0	9-5	8-9	14-6	13-2	12-5	11-6	18-6	16-10	15-10	14-5	22-6	20-6	18-9	16-9
	No. 2	10-9	9-9	9-2	8-6	14-2	12-10	12-1	11-3	18-0	16-10	15-10	13-10	21-11	19-8	17-11	16-1
	No. 3	9-8	8-5	7-8	6-10	12-4	10-8	9-9	8-8	15-0	13-0	11-10	10-7	17-5	15-1	13-9	12-4
Hem-Fir	Sel. Struc.	11-10	10-9	10-1	9-4	15-7	14-2	13-4	12-4	19-10	18-0	17-0	15-9	24-2	21-11	20-8	19-2
	No. 1 & Btr.	11-7	10-6	9-10	9-2	15-3	13-10	13-0	12-1	19-5	17-8	16-7	15-5	23-7	21-6	20-2	18-3
	No. 1	11-7	10-6	9-10	9-2	15-3	13-10	13-0	12-1	19-5	17-8	16-7	14-10	23-7	21-1	19-3	17-2
	No. 2	11-0	10-0	9-5	8-9	14-6	13-2	12-5	11-4	18-6	16-10	15-6	13-10	22-6	19-8	17-11	16-1
	No. 3	9-8	8-5	7-8	6-10	12-4	10-8	9-9	8-8	15-0	13-0	11-10	10-7	17-5	15-1	13-9	12-4
Spruce-Pine-Fir (South)	Sel. Struc.	11-0	10-0	9-5	8-9	14-6	13-2	12-5	11-6	18-6	16-10	15-10	14-8	22-6	20-6	19-3	17-11
	No. 1	10-9	9-9	9-2	8-6	14-2	12-10	12-1	11-3	18-0	16-5	15-5	14-1	21-11	19-11	18-3	16-3
	No. 2	10-5	9-6	8-11	8-3	13-9	12-6	11-9	10-10	17-6	15-11	14-9	13-3	21-4	18-9	17-2	15-4
	No. 3	9-3	8-0	7-3	6-6	11-8	10-1	9-3	8-3	14-3	12-4	11-3	10-1	16-6	14-4	13-1	11-8
Western Woods	Sel. Struc.	10-9	9-9	9-2	8-6	14-2	12-10	12-1	11-3	18-0	16-5	15-5	14-3	21-11	19-11	18-6	16-6
	No. 1	10-5	9-6	8-11	8-0	13-9	12-4	11-4	10-1	17-5	15-1	13-10	12-4	20-3	17-6	16-0	14-4
	No. 2	10-1	9-2	8-8	8-0	13-4	12-1	11-4	10-1	17-0	15-1	13-10	12-4	20-3	17-6	16-0	14-4
	No. 3	8-5	7-3	6-8	5-11	10-8	9-3	8-5	7-6	13-0	11-3	10-3	9-2	15-1	13-1	11-11	10-8

Courtesy Western Wood Products Association www.wwpa.org Additional information available from WWPA.

MAXIMUM SPANS FOR CEILING JOISTS USING WESTERN WOODS

20# Live Load 10# Dead Load L/240

Design Criteria: *Strength* - 20 lbs. per sq. ft. live load plus 10 lbs. per sq. ft. dead load.
Deflection - Limited in span in inches divided by 240 for live load only.

		Span (feet and inches)															
		2 3 4				2 3 6				2 3 8				2 3 10			
		Spacing on Center															
Species or Group	**Grade**	**12"**	**16"**	**19.2"**	**24"**	**12"**	**16"**	**19.2"**	**24"**	**12"**	**16"**	**19.2"**	**24"**	**12"**	**16"**	**19.2"**	**24"**
Douglas Fir-Larch	Sel. Struc.	10-5	9-6	8-11	8-3	16-4	14-11	14-0	13-0	21-7	19-7	18-5	17-2	27-6	25-0	23-7	21-3
	No. 1 & Btr.	10-3	9-4	8-9	8-1	16-1	14-7	13-9	12-3	21-2	19-1	17-5	15-7	26-10	23-3	21-3	19-0
	No. 1	10-0	9-1	8-7	7-8	15-9	13-9	12-6	11-2	20-1	17-5	15-10	14-2	24-6	21-3	19-5	17-4
	No. 2	9-10	8-11	8-2	7-3	15-0	13-0	11-11	10-8	19-1	16-6	15-1	13-6	23-3	20-2	18-5	16-5
	No. 3	7-10	6-10	6-2	5-7	11-6	9-11	9-1	8-1	14-7	12-7	11-6	10-3	17-9	15-5	14-1	12-7
Douglas Fir-South	Sel. Struc.	9-5	8-7	8-1	7-6	14-9	13-5	12-8	11-9	19-6	17-9	16-8	15-6	24-10	22-7	21-3	19-9
	No. 1	9-2	8-4	7-10	7-3	14-5	13-1	12-1	10-9	19-0	16-9	15-3	13-8	23-7	20-5	18-8	16-8
	No. 2	8-11	8-1	7-8	7-1	14-1	12-8	11-7	10-4	18-6	16-0	14-8	13-1	22-7	19-7	17-10	16-0
	No. 3	7-8	6-8	6-1	5-5	11-2	9-8	8-10	7-11	14-2	12-4	11-3	10-0	17-4	15-0	13-8	12-3
Hem-Fir	Sel. Struc.	9-10	8-11	8-5	7-10	15-6	14-1	13-3	12-3	20-5	18-6	17-5	16-2	26-0	23-8	22-3	20-6
	No. 1 & Btr.	9-8	8-9	8-3	7-8	15-2	13-9	12-11	11-9	19-11	18-2	16-8	14-11	25-5	22-3	20-4	18-2
	No. 1	9-8	8-9	8-3	7-7	15-2	13-7	12-4	11-1	19-10	17-2	15-8	14-0	24-3	21-0	19-2	17-1
	No. 2	9-2	8-4	7-10	7-1	14-5	12-8	11-7	10-4	18-6	16-0	14-8	13-1	22-7	19-7	17-10	16-0
	No. 3	7-8	6-8	6-1	5-5	11-2	9-8	8-10	7-11	14-2	12-4	11-3	10-0	17-4	15-0	13-8	12-3
Spruce-Pine-Fir (South)	Sel. Struc.	9-2	8-4	7-10	7-3	14-5	13-1	12-4	11-5	19-0	17-3	16-3	15-1	24-3	22-1	20-9	19-3
	No. 1	8-11	8-1	7-8	7-1	14-1	12-9	11-9	10-6	18-6	16-3	14-10	13-3	22-11	19-10	18-2	16-3
	No. 2	8-8	7-11	7-5	6-9	13-8	12-1	11-0	9-10	17-8	15-4	14-0	12-6	21-7	18-8	17-1	15-3
	No. 3	7-3	6-4	5-9	5-2	10-8	9-3	8-5	7-6	13-6	11-8	10-8	9-6	16-5	14-3	13-0	11-8
Western Woods	Sel. Struc.	8-11	8-1	7-8	7-1	14-1	12-9	11-11	10-8	18-6	16-6	15-1	13-6	23-3	20-2	18-5	16-5
	No. 1	8-8	7-9	7-0	6-4	13-0	11-3	10-4	9-3	16-6	14-3	13-0	11-8	20-2	17-5	15-11	14-3
	No. 2	8-5	7-9	7-0	6-4	13-0	11-3	10-4	9-3	16-6	14-3	13-0	11-8	20-2	17-5	15-11	14-3
	No. 3	6-8	5-9	5-3	4-8	9-8	8-5	7-8	6-10	12-4	10-8	9-9	8-8	15-0	13-0	11-10	10-7

Courtesy Western Wood Products Association www.wwpa.org Additional information available from WWPA.

MAXIMUM SPANS FOR ROOF RAFTERS USING WESTERN WOODS

30# Snow Load 15# Dead Load L/180

Design Criteria: *Strength* - 30 lbs. per sq. ft. live load plus 10 lbs. per sq. ft. dead load.
Deflection - Limited in span in inches divided by 180 for live load only.

		Span (feet and inches)															
		2 3 6				2 3 8				2 3 10				2 3 12			
		Spacing on Center															
Species or Group	Grade	12"	16"	19.2"	24"	12"	16"	19.2"	24"	12"	16"	19.2"	24"	12"	16"	19.2"	24"
Douglas Fir-Larch	Sel. Struc.	15-9	14-4	13-5	12-0	20-9	18-8	17-0	15-3	26-4	22-9	20-9	18-7	30-6	26-5	24-1	21-7
	No. 1 & Btr.	15-2	13-2	12-0	10-9	19-3	16-8	15-3	13-7	23-6	20-4	18-7	16-8	27-3	23-7	21-7	19-3
	No. 1	13-11	12-0	11-0	9-10	17-7	15-3	13-11	12-5	21-6	18-7	17-0	15-2	24-11	21-7	19-8	17-7
	No. 2	13-2	11-5	10-5	9-4	16-8	14-5	13-2	11-9	20-4	17-8	16-1	14-5	23-7	20-5	18-8	16-8
	No. 3	10-1	8-9	7-11	7-1	12-9	11-0	10-1	9-0	15-7	13-6	12-4	11-0	18-0	15-7	14-3	12-9
Douglas Fir-South	Sel. Struc.	14-3	12-11	12-2	11-3	18-9	17-0	16-0	14-5	23-11	21-7	19-9	17-8	28-11	25-1	22-10	20-5
	No. 1	13-4	11-7	10-7	9-5	16-11	14-8	13-4	11-11	20-8	17-11	16-4	14-7	23-11	20-9	18-11	16-11
	No. 2	12-10	11-1	10-1	9-1	16-2	14-0	12-10	11-6	19-10	17-2	15-8	14-0	22-11	19-11	18-2	16-3
	No. 3	9-10	8-6	7-9	6-11	12-5	10-9	9-10	8-9	15-2	13-2	12-0	10-9	17-7	15-3	13-11	12-5
Hem-Fir	Sel. Struc.	14-10	13-6	12-9	11-7	19-7	17-10	16-5	14-8	25-0	22-0	20-1	18-0	29-6	25-6	23-3	20-10
	No. 1 & Btr.	14-7	12-7	11-6	10-4	18-5	16-0	14-7	13-0	22-6	19-6	17-10	15-11	26-1	22-7	20-8	18-6
	No. 1	13-8	11-10	10-10	9-8	17-4	15-0	13-9	12-3	21-2	18-4	16-9	15-0	24-7	21-3	19-5	17-5
	No. 2	12-10	11-1	10-1	9-1	16-2	14-0	12-10	11-6	19-10	17-2	15-8	14-0	22-11	19-11	18-2	16-3
	No. 3	9-10	8-6	7-9	6-11	12-5	10-9	9-10	8-9	15-2	13-2	12-0	10-9	17-7	15-3	13-11	12-5
Spruce-Pine-Fir (South)	Sel. Struc.	13-10	12-7	11-10	11-0	18-3	16-7	15-8	14-2	23-4	21-2	19-4	17-4	28-5	24-7	22-5	20-1
	No. 1	13-0	11-3	10-3	9-2	16-5	14-3	13-0	11-8	20-1	17-5	15-11	14-2	23-3	20-2	18-5	16-6
	No. 2	12-3	10-7	9-8	8-8	15-6	13-5	12-3	10-11	18-11	16-4	14-11	13-4	21-11	19-0	17-4	15-6
	No. 3	9-4	8-1	7-4	6-7	11-9	10-3	9-4	8-4	14-5	12-6	11-5	10-2	16-8	14-6	13-2	11-10
Western Woods	Sel. Struc.	13-2	11-5	10-5	9-4	16-8	14-5	13-2	11-9	20-4	17-8	16-1	14-5	23-7	20-5	18-8	16-8
	No. 1	11-5	9-10	9-0	8-1	14-5	12-6	11-5	10-3	17-8	15-3	13-11	12-6	20-5	17-9	16-2	14-6
	No. 2	11-5	9-10	9-0	8-1	14-5	12-6	11-5	10-3	17-8	15-3	13-11	12-6	20-5	17-9	16-2	14-6
	No. 3	8-6	7-4	6-9	6-0	10-9	9-4	8-6	7-7	13-2	11-5	10-5	9-4	15-3	13-2	12-1	10-9

Courtesy Western Wood Products Association www.wwpa.org Additional information available from WWPA.

MAXIMUM SPANS FOR ROOF RAFTERS USING WESTERN WOODS *(cont.)*

20# Live Load 15# Dead Load L/240

Design Criteria: *Strength* - 20 lbs. per sq. ft. live load plus 15 lbs. per sq. ft. dead load.
Deflection - Limited in span in inches divided by 240 for live load only.

		Span (feet and inches)															
		2 3 6				2 3 8				2 3 10				2 3 12			
		Spacing on Center															
Species or Group	**Grade**	**12"**	**16"**	**19.2"**	**24"**	**12"**	**16"**	**19.2"**	**24"**	**12"**	**16"**	**19.2"**	**24"**	**12"**	**16"**	**19.2"**	**24"**
Douglas Fir-Larch	Sel. Struc.	16-4	14-11	14-0	13-0	21-7	19-7	18-5	17-2	27-6	25-0	23-7	21-10	33-6	30-5	28-6	25-6
	No. 1 & Btr.	16-1	14-7	13-9	12-9	21-2	19-3	18-0	16-1	27-1	24-1	22-0	19-8	32-3	27-11	25-6	22-10
	No. 1	15-9	14-3	13-0	11-7	19-9	18-0	16-5	14-8	25-5	22-0	20-1	17-11	29-5	25-6	23-3	20-10
	No. 2	15-6	13-6	12-4	11-0	20-9	17-1	15-7	13-11	24-1	20-10	19-0	17-0	27-11	24-2	22-1	19-9
	No. 3	11-11	10-4	9-5	8-5	15-1	13-0	11-11	10-8	18-5	15-11	14-6	13-0	21-4	18-6	16-10	15-1
Douglas Fir-South	Sel. Struc.	14-9	13-5	12-8	11-9	19-6	17-9	16-8	15-6	24-10	22-7	21-3	19-9	30-3	27-6	25-10	24-0
	No. 1	14-5	13-1	12-4	11-2	19-0	17-3	15-10	14-2	24-3	21-2	19-4	17-3	28-4	24-6	22-5	20-0
	No. 2	14-1	12-9	12-0	10-8	18-6	16-7	15-2	13-7	23-5	20-3	18-6	16-7	27-2	23-6	21-5	19-2
	No. 3	11-7	10-1	9-2	8-2	14-8	12-9	11-7	10-5	17-11	15-7	14-2	12-8	20-10	18-0	16-5	14-9
Hem-Fir	Sel. Struc.	15-6	14-1	13-3	12-3	20-5	18-6	17-5	16-2	26-0	23-8	22-3	20-8	31-8	28-9	27-1	24-8
	No. 1 & Btr.	15-2	13-9	12-11	12-0	19-11	18-2	17-1	15-5	25-5	23-1	21-1	18-10	30-10	26-9	24-5	21-10
	No. 1	15-2	13-9	12-10	11-5	19-11	17-9	16-3	14-6	25-1	21-8	19-10	17-9	29-1	25-2	23-0	20-7
	No. 2	14-5	13-1	12-0	10-8	19-0	16-7	15-2	13-7	23-5	20-3	18-6	16-7	27-2	23-6	21-5	19-2
	No. 3	11-7	10-1	9-2	8-2	14-8	12-9	11-7	10-5	17-11	15-7	14-2	12-8	20-10	18-0	16-5	14-9
Spruce-Pine-Fir (South)	Sel. Struc.	14-5	13-1	12-4	11-5	19-0	17-3	16-3	15-1	24-3	22-1	20-9	19-3	29-6	26-10	25-3	23-5
	No. 1	14-1	12-9	12-0	10-10	18-6	16-10	15-4	13-9	23-8	20-7	18-9	16-9	27-6	23-10	21-9	19-6
	No. 2	13-8	12-5	11-5	10-3	18-0	15-10	14-6	12-11	22-4	19-4	17-8	15-10	25-11	22-5	20-6	18-4
	No. 3	11-0	9-6	8-8	7-9	13-11	12-1	11-0	9-10	17-0	14-9	13-6	12-0	19-9	17-1	15-7	14-0
Western Woods	Sel. Struc.	14-1	12-9	12-0	11-0	18-6	16-10	15-7	13-11	23-8	20-10	19-0	17-0	27-11	24-2	22-1	19-9
	No. 1	13-6	11-8	10-8	9-6	17-1	14-9	13-6	12-1	20-10	18-1	16-6	14-9	24-2	20-11	19-1	17-1
	No. 2	13-3	11-8	10-8	9-6	17-1	14-9	13-6	12-1	20-10	18-1	16-6	14-9	24-2	20-11	19-1	17-1
	No. 3	10-1	8-8	7-11	7-1	12-9	11-0	10-1	9-0	15-7	13-6	12-3	11-0	18-0	15-7	14-3	12-9

[1]A 1.25 Duration of Load adjustment has been applied. See page xxx.

Courtesy Western Wood Products Association www.wwpa.org Additional information available from WWPA.

SOUTHERN PINE SPAN TABLES FLOOR JOISTS

40 psf Live Load, 10 psf Dead Load, 360 Deflection

All Rooms Except Sleeping Rooms and Attic Floors

Size inches	Spacing inches on center	Grade: Visually Graded				Machine Stress Rated (MSR)			Machine Evaluated Lumber (MEL)		
		SS	No. 1	No. 2	No. 3	2400f - 2.0E	2250f - 1.9E	1950f - 1.7E	M23	M14	M29
2 3 6	12.0	11-2	10-11	10-9	9-4	11-7	11-4	10-11	11-2	10-11	10-11
	16.0	10-2	9-11	9-9	8-1	10-6	10-4	9-11	10-2	9-11	9-11
	19.2	9-6	9-4	9-2	7-4	9-10	9-8	9-4	9-6	9-4	9-4
	24.0	8-10	8-8	8-6	6-7	9-2	9-0	8-8	8-10	8-8	8-8
2 3 8	12.0	14-8	14-5	14-2	11-11	15-3	15-0	14-5	14-8	14-5	14-5
	16.0	13-4	13-1	12-10	10-3	13-10	13-7	13-1	13-4	13-1	13-1
	19.2	12-7	12-4	12-1	9-5	13-0	12-10	12-4	12-7	12-4	12-4
	24.0	11-8	11-5	11-0	8-5	12-1	11-11	11-5	11-8	11-5	11-5
2 3 10	12.0	18-9	18-5	18-0	14-0	19-5	19-1	18-5	18-9	18-5	18-5
	16.0	17-0	16-9	16-1	12-2	17-8	17-4	16-9	17-0	16-9	16-9
	19.2	16-0	15-9	14-8	11-1	16-7	16-4	15-9	16-0	15-9	15-9
	24.0	14-11	14-7	13-1	9-11	15-5	15-2	14-7	14-11	14-7	14-7
2 3 12	12.0	22-10	22-5	21-9	16-8	23-7	23-3	22-5	22-10	22-5	22-5
	16.0	20-9	20-4	18-10	14-6	21-6	21-1	20-4	20-9	20-4	20-4
	19.2	19-6	19-2	17-2	13-2	20-2	19-10	19-2	19-6	19-2	19-2
	24.0	18-1	17-5	15-5	11-10	18-9	18-5	17-9	18-1	17-9	17-9

These spans are intended for use in enclosed structures or where the moisture content in use does not exceed 19 percent for an extended period of time unless the table is labled Wet-Service. Applied loads are given in psf (pounds per square foot). Deflection is limited to the span in inches divided by 360, 240, or 180 and is based on live load only. The load duration factor, C_D, is 1.0 unless shown as 1.15 or 1.25. An asterisk (*) indicates the listed span has been limited to 26'0" based on availability; check sources of supply for lumber longer than 20'. Highlighted sizes/grades are NOT commonly produced.

Courtesy Southern Pine Council www.southernpine.com

SOUTHERN PINE SPAN TABLES FLOOR JOISTS *(cont.)*

30 psf Live Load, 10 psf Dead Load, 360 Deflection

Sleeping Rooms and Attic Floors

Size inches	Spacing inches on center	Grade									
		Visually Graded				Machine Stress Rated (MSR)			Machine Evaluated Lumber (MEL)		
		SS	No. 1	No. 2	No. 3	2400f - 2.0E	2250f - 1.9E	1950f - 1.7E	M23	M14	M29
2 3 6	**12.0**	12-3	12-0	11-10	10-5	12-9	12-6	12-0	12-3	12-0	12-0
	16.0	11-2	10-11	10-9	9-0	11-7	11-4	10-11	11-2	10-11	10-11
	19.2	10-6	10-4	10-1	8-3	10-10	10-8	10-4	10-6	10-4	10-4
	24.0	9-9	9-7	9-4	7-4	10-1	9-11	9-7	9-9	9-7	9-7
2 3 8	**12.0**	16-2	15-10	15-7	13-3	16-9	16-6	15-10	16-2	15-10	15-10
	16.0	14-8	14-5	14-2	11-6	15-3	15-0	14-5	14-8	14-5	14-5
	19.2	13-10	13-7	13-4	10-6	14-4	14-1	13-7	13-10	13-7	13-7
	24.0	12-10	12-7	12-4	9-5	13-4	13-1	12-7	12-10	12-7	12-7
2 3 10	**12.0**	20-8	20-3	19-10	15-8	21-5	21-0	20-3	20-8	20-3	20-3
	16.0	18-9	18-5	18-0	13-7	19-5	19-1	18-5	18-9	18-5	18-5
	19.2	17-8	17-4	16-5	12-5	18-3	18-0	17-4	17-8	17-4	17-4
	24.0	16-5	16-1	14-8	11-1	17-0	16-8	16-1	16-5	16-1	16-1
2 3 12	**12.0**	25-1	24-8	24-2	18-8	26-0	25-7	24-8	25-1	24-8	24-8
	16.0	22-10	22-5	21-1	16-2	23-7	23-3	22-5	22-10	22-5	22-5
	19.2	21-6	21-1	19-3	14-9	22-3	21-10	21-1	21-6	21-1	21-1
	24.0	19-11	19-6	17-2	13-2	20-8	20-3	19-7	19-11	19-7	19-7

Maximum spans given in feet and inches inside to inside of bearings.
Courtesy Southern Pine Council www.southernpine.com

SOUTHERN PINE SPAN TABLES RAFTERS

30 psf Live Load, 10 psf Dead Load, 240 Deflection, C_D 5 1.15

Light Roofing; Drywall Ceiling; Snow Load

Size inches	Spacing inches on center	Grade: Visually Graded				Machine Stress Rated (MSR)			Machine Evaluated Lumber (MEL)		
		SS	No.1	No.2	No.3	2400f - 2.0E	2250f - 1.9E	1950f - 1.7E	M23	M14	M29
2 3 6	12.0	14-1	13-9	13-6	11-2	14-7	14-4	13-9	14-1	13-9	13-9
	16.0	12-9	12-6	12-3	9-8	13-3	13-0	12-6	12-9	12-6	12-6
	19.2	12-0	11-9	11-5	8-10	12-5	12-3	11-9	12-0	11-9	11-9
	24.0	11-2	10-11	10-2	7-11	11-7	11-4	10-11	11-2	10-11	10-11
2 3 8	12.0	18-6	18-2	17-10	14-3	19-2	18-10	18-2	18-6	18-2	18-2
	16.0	16-10	16-6	16-2	12-4	17-5	17-2	16-6	16-10	16-6	16-6
	19.2	15-10	15-6	14-9	11-3	16-5	16-1	15-6	15-10	15-6	15-6
	24.0	14-8	14-5	13-2	10-1	15-3	15-0	14-5	14-8	14-5	14-5
2 3 10	12.0	23-8	23-2	22-3	16-10	24-6	24-1	23-2	23-8	23-2	23-2
	16.0	21-6	21-1	19-3	14-7	22-3	21-10	21-1	21-6	21-1	21-1
	19.2	20-2	19-7	17-7	13-4	20-11	20-7	19-10	20-2	19-10	19-10
	24.0	18-9	17-6	15-9	11-11	19-5	19-1	18-5	18-9	18-5	18-5
2 3 12	12.0	26-0*	26-0*	26-0*	20-0	26-0*	26-0*	26-0*	26-0*	26-0*	26-0*
	16.0	26-0*	25-7	22-7	17-4	26-0*	26-0*	25-7	26-0*	25-7	25-7
	19.2	24-7	23-4	20-7	15-10	25-5	25-0	24-1	24-7	24-1	24-1
	24.0	22-10	20-11	18-5	14-2	23-7	23-3	22-5	22-10	22-5	22-5

These spans are intended for use in enclosed structures or where the moisture content in use does not exceed 19 percent for an extended period of time unless the table is labled Wet-Service. Applied loads are given in psf (pounds per square foot). Deflection is limited to the span in inches divided by 360, 240, or 180 and is based on live load only. The load duration factor, C_D, is 1.0 unless shown as 1.15 or 1.25. An asterisk (*) indicates the listed span has been limited to 26'0" based on availability; check sources of supply for lumber longer than 20'. Highlighted sizes/grades are NOT commonly produced.

Courtesy Southern Pine Council www.southernpine.com

SOUTHERN PINE SPAN TABLES RAFTERS *(cont.)*

20 psf Live Load, 10 psf Dead Load, 240 Deflection, C_D 5 1.15

Light Roofing; Drywall Ceiling; Snow Load											
Size inches	Spacing inches on center	Grade									
		Visually Graded				Machine Stress Rated (MSR)			Machine Evaluated Lumber (MEL)		
		SS	No. 1	No. 2	No. 3	2400f - 2.0E	2250f - 1.9E	1950f - 1.7E	M23	M14	M29
2 3 6	12.0	16-1	15-9	15-6	12-11	16-8	16-4	15-9	16-1	15-9	15-9
	16.0	14-7	14-4	14-1	11-2	15-2	14-11	14-4	14-7	14-4	14-4
	19.2	13-9	13-6	13-2	10-2	14-3	14-0	13-6	13-9	13-6	13-6
	24.0	12-9	12-6	11-9	9-2	13-3	13-0	12-6	12-9	12-6	12-6
2 3 8	12.0	21-2	20-10	20-5	16-5	21-11	21-7	20-10	21-2	20-10	20-10
	16.0	19-3	18-11	18-6	14-3	19-11	19-7	18-11	19-3	18-11	18-11
	19.2	18-2	17-9	17-0	13-0	18-9	18-5	17-9	18-2	17-9	17-9
	24.0	16-10	16-6	15-3	11-8	17-5	17-2	16-6	16-10	16-6	16-6
2 3 10	12.0	26-0*	26-0*	25-8	19-5	26-0*	26-0*	26-0*	26-0*	26-0*	26-0*
	16.0	24-7	24-1	22-3	16-10	25-5	25-0	24-1	24-7	24-1	24-1
	19.2	23-2	22-7	20-4	15-4	23-11	23-7	22-8	23-2	22-8	22-8
	24.0	21-6	20-3	18-2	13-9	22-3	21-10	21-1	21-6	21-1	21-1
2 3 12	12.0	26-0*	26-0*	26-0*	23-1	26-0*	26-0*	26-0*	26-0*	26-0*	26-0*
	16.0	26-0*	26-0*	26-0*	20-0	26-0*	26-0*	26-0*	26-0*	26-0*	26-0*
	19.2	26-0*	26-0*	23-10	18-3	26-0*	26-0*	26-0*	26-0*	26-0*	26-0*
	24.0	26-0*	24-1	21-3	16-4	26-0*	26-0*	25-7	26-0*	25-7	25-7

Maximum spans given in feet and inches inside to inside of bearings.

Courtesy Southern Pine Council www.southernpine.com

SOUTHERN PINE SPAN TABLES CEILING JOISTS

20 psf Live Load, 10 psf Dead Load, 240 Deflection

Drywall Ceiling; No Future Room Development, But Limited Attic Storage Available											
Size inches	Spacing inches on center	Grade									
		Visually Graded				Machine Stress Rated (MSR)			Machine Evaluated Lumber (MEL)		
		SS	No. 1	No. 2	No. 3	2400f - 2.0E	2250f - 1.9E	1950f - 1.7E	M23	M14	M29
2 3 6	**12.0**	10-3	10-0	9-10	8-2	10-7	10-5	10-0	10-3	10-0	10-0
	16.0	9-4	9-1	8-11	7-1	9-8	9-6	9-1	9-4	9-1	9-1
	19.2	8-9	8-7	8-5	6-5	9-1	8-11	8-7	8-9	8-7	8-7
	24.0	8-1	8-0	7-8	5-9	8-5	8-3	8-0	8-1	8-0	7-9
2 3 8	**12.0**	16-1	15-9	15-6	12-0	16-8	16-4	15-9	16-1	15-9	15-9
	16.0	14-7	14-4	13-6	10-5	15-2	14-11	14-4	14-7	14-4	14-4
	19.2	13-9	13-6	12-3	9-6	14-3	14-0	13-6	13-9	13-6	13-6
	24.0	12-9	12-6	11-0	8-6	13-3	13-0	12-6	12-9	12-6	12-3
2 3 10	**12.0**	21-2	20-10	20-1	15-4	21-11	21-7	20-10	21-2	20-10	20-10
	16.0	19-3	18-11	17-5	13-3	19-11	19-7	18-11	19-3	18-11	18-11
	19.2	18-2	17-9	15-10	12-1	18-9	18-5	17-9	18-2	17-9	17-9
	24.0	16-10	15-10	14-2	10-10	17-5	17-2	16-6	16-10	16-6	16-2
2 3 12	**12.0**	26-0*	26-0*	23-11	18-1	26-0*	26-0*	26-0*	26-0*	26-0*	26-0*
	16.0	24-7	23-1	20-9	15-8	25-5	25-0	24-1	24-7	24-1	24-1
	19.2	23-2	21-1	18-11	14-4	23-11	23-7	22-8	23-2	22-8	22-8
	24.0	21-6	18-10	16-11	12-10	22-3	21-10	21-1	21-6	21-1	20-

These spans are intended for use in enclosed structures or where the moisture content in use does not exceed 19 percent for an extended period of time unless the table is labled Wet-Service. Applied loads are given in psf (pounds per square foot). Deflection is limited to the span in inches divided by 360, 240, or 180 and is based on live load only. The load duration factor, C_D, is 1.0 unless shown as 1.15 or 1.25. An asterisk (*) indicates the listed span has been limited to 26'0" based on availability; check sources of supply for lumber longer than 20'. Highlighted sizes/grades are NOT commonly produced.

Courtesy Southern Pine Council www.southernpine.com

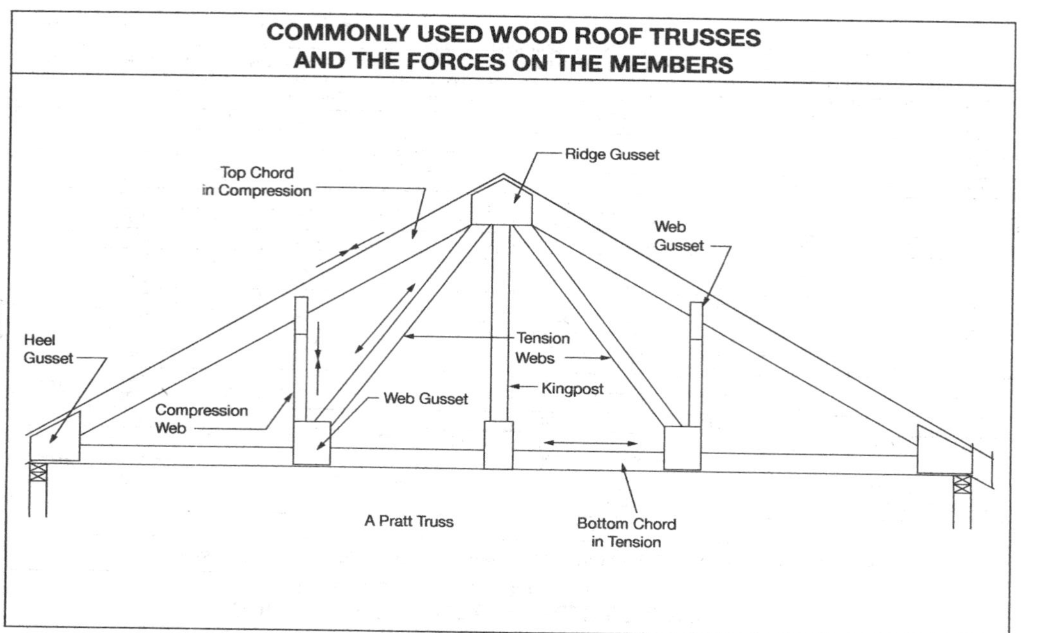
COMMONLY USED WOOD ROOF TRUSSES
AND THE FORCES ON THE MEMBERS
Ridge Gusset
Top Chord
in Compression
Web
Gusset
Heel
Gusset
Tension
Webs
Kingpost
Web Gusset
Compression
Web
A Pratt Truss
Bottom Chord
in Tension

COMMONLY AVAILABLE WOOD-FRAMED ROOF TRUSSES

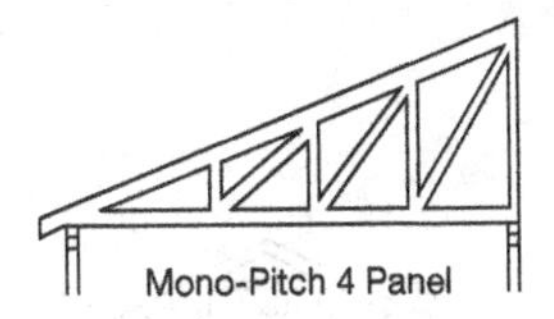
Mono-Pitch 4 Panel

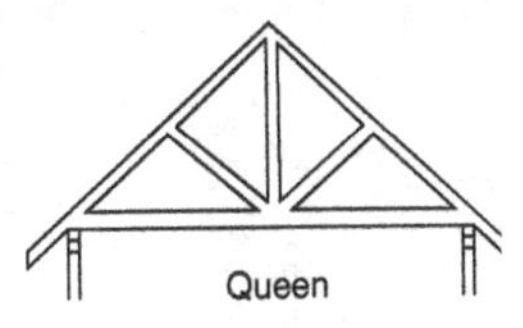
Queen

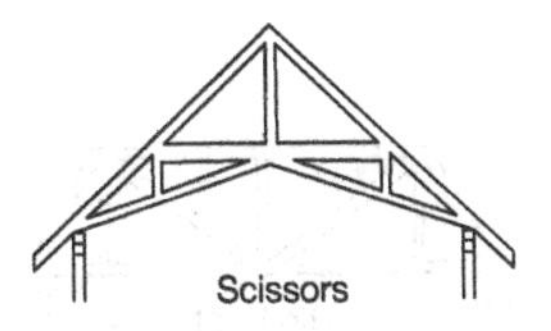
Scissors

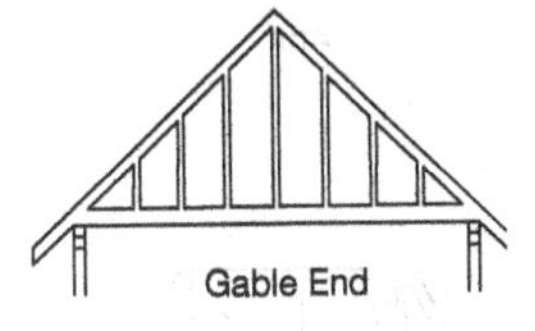
Gable End

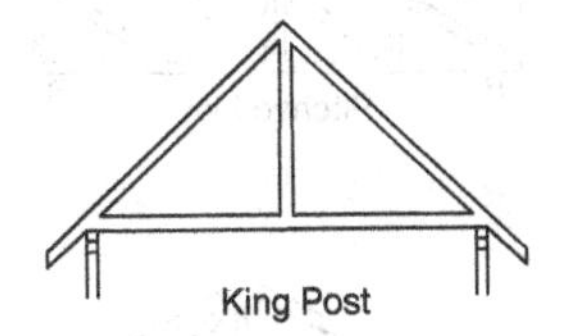
King Post

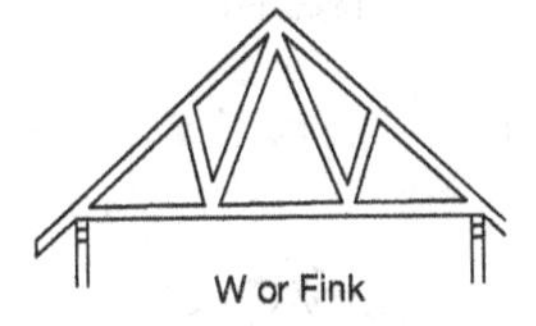
W or Fink

Belgin or Double W

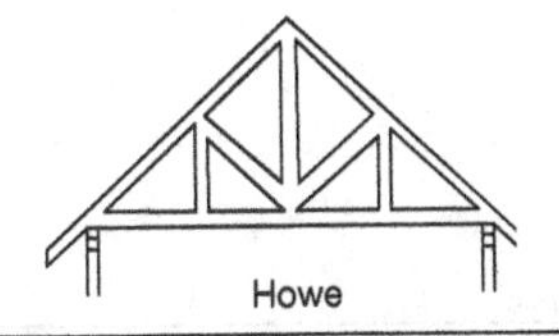
Howe

COMMONLY AVAILABLE WOOD-FRAMED ROOF TRUSSES *(cont.)*

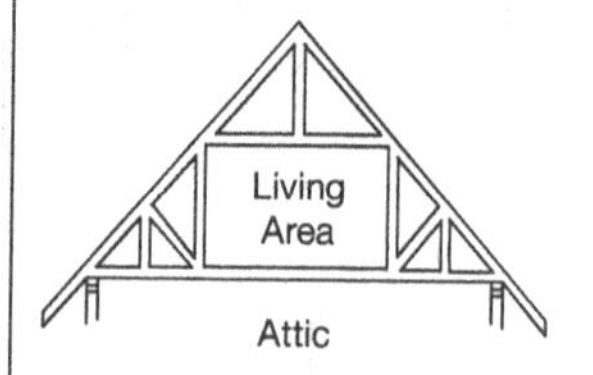

Attic

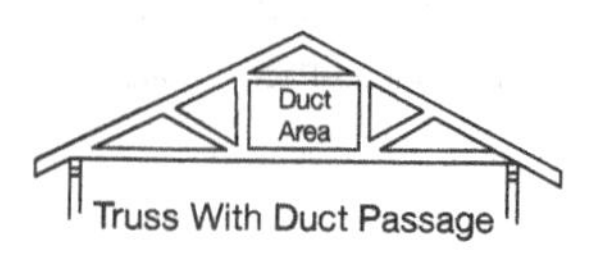

Truss With Duct Passage

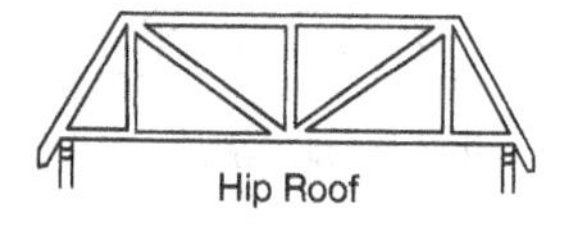
Hip Roof

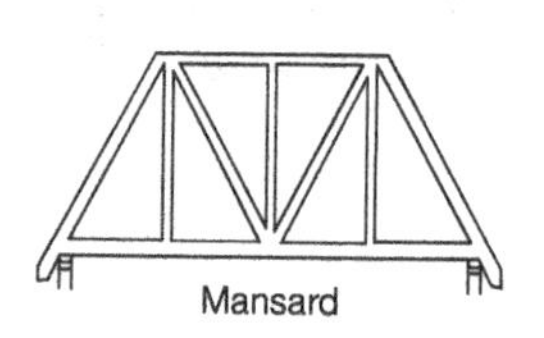
Mansard

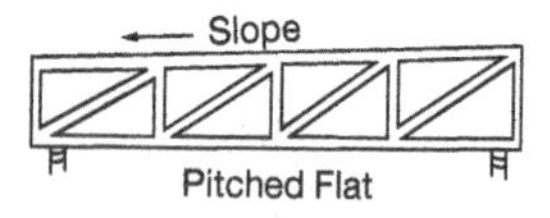

Pitched Flat

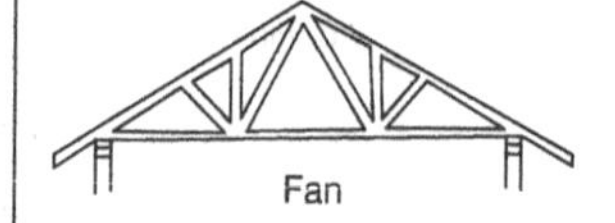
Fan

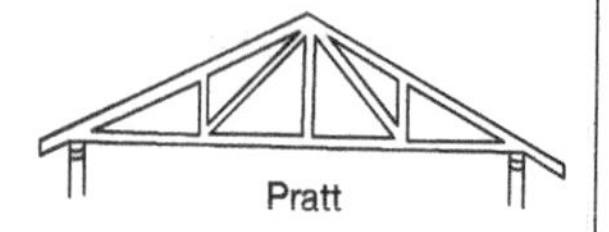
Pratt

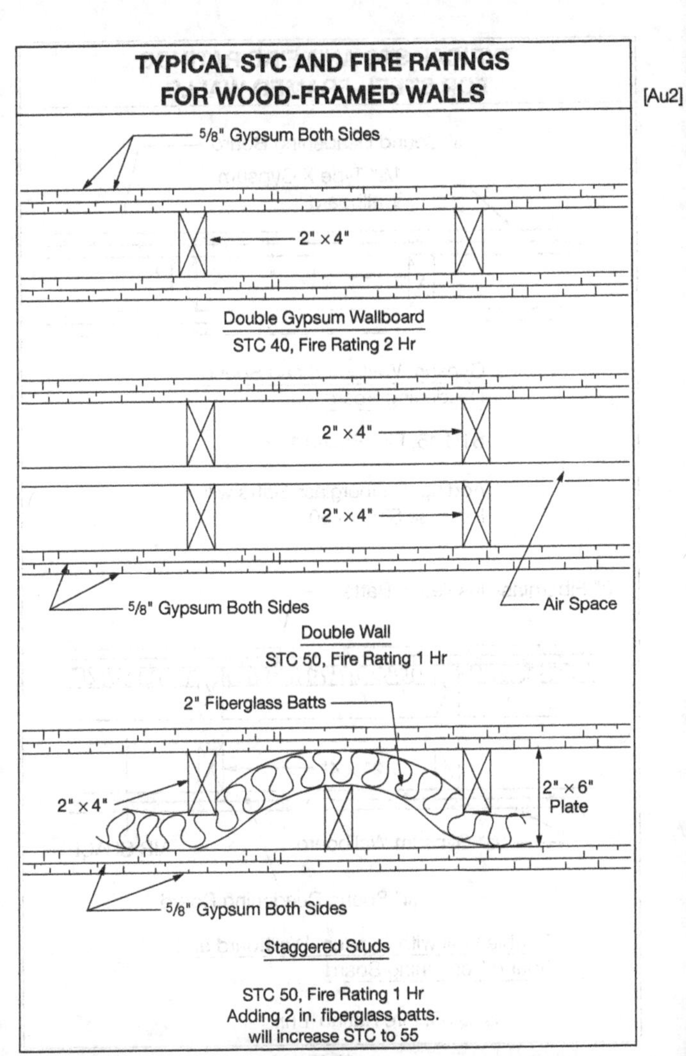
TYPICAL STC AND FIRE RATINGS
FOR WOOD-FRAMED WALLS
[Au2]
5/8" Gypsum Both Sides
2" × 4"
Double Gypsum Wallboard
STC 40, Fire Rating 2 Hr
2" × 4"
2" × 4"
5/8" Gypsum Both Sides
Air Space
Double Wall
STC 50, Fire Rating 1 Hr
2" Fiberglass Batts
2" × 4"
2" × 6"
Plate
5/8" Gypsum Both Sides
Staggered Studs
STC 50, Fire Rating 1 Hr
Adding 2 in. fiberglass batts.
will increase STC to 55

TYPICAL STC AND FIRE RATINGS FOR STEEL-FRAMED WALLS

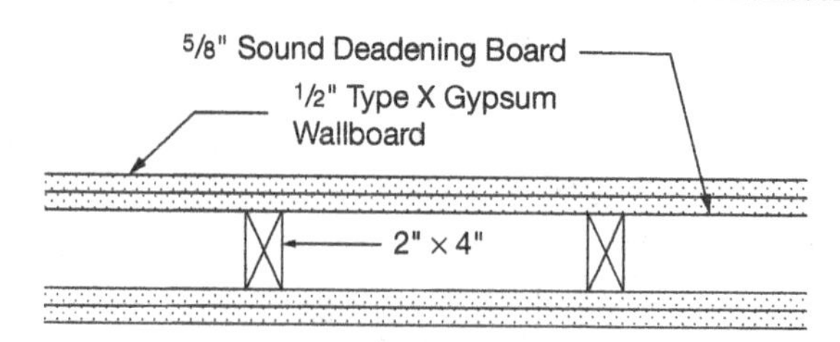

Gypsum Wallboard and Sound Deadening Board

STC 45, Fire Rating 1 hr

Adding 2" Fiberglass Batts will Increase STC to 50

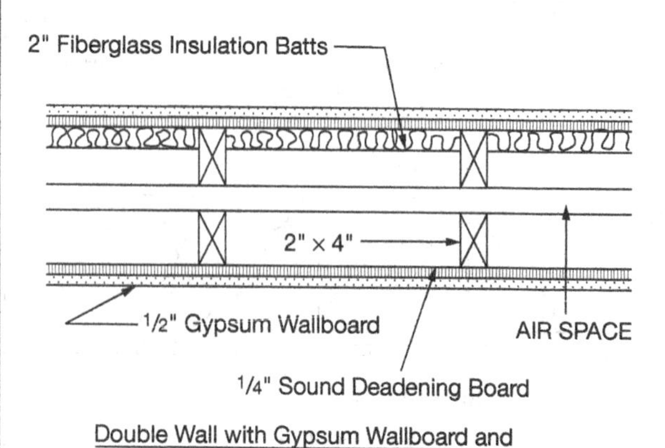

Double Wall with Gypsum Wallboard and Sound Deadening Board

STC 55. Fire Rating 1 hr.

TYPICAL STC AND FIRE RATINGS FOR STEEL-FRAMED WALLS *(cont.)*

Typical Steel Framed Wall Assemblies

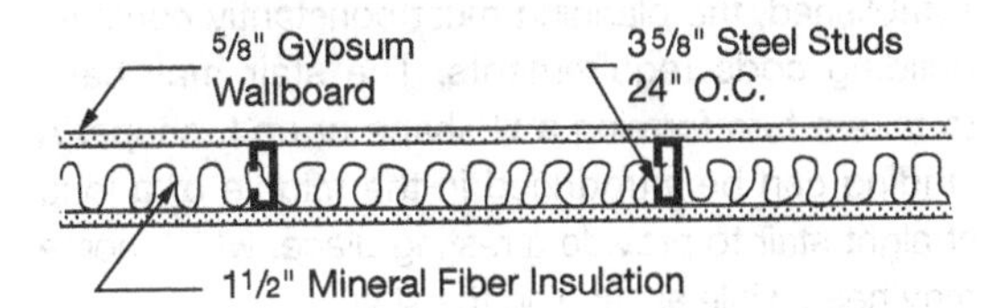

Single Wall with Insulation

STC 45, Fire Rating 1 hr.

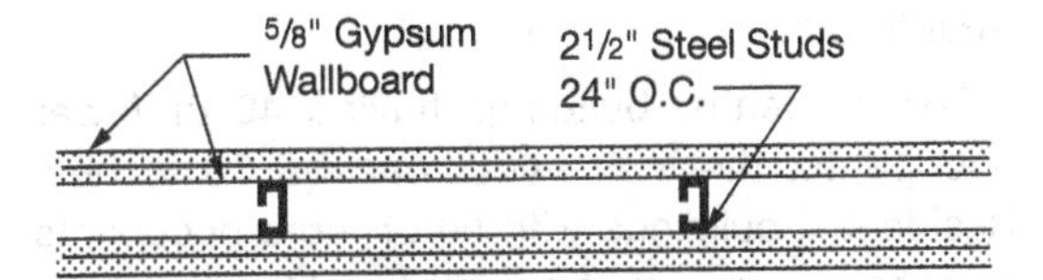

Double Gypsum Wallboard

STC 45, Fire Rating 2 hr.

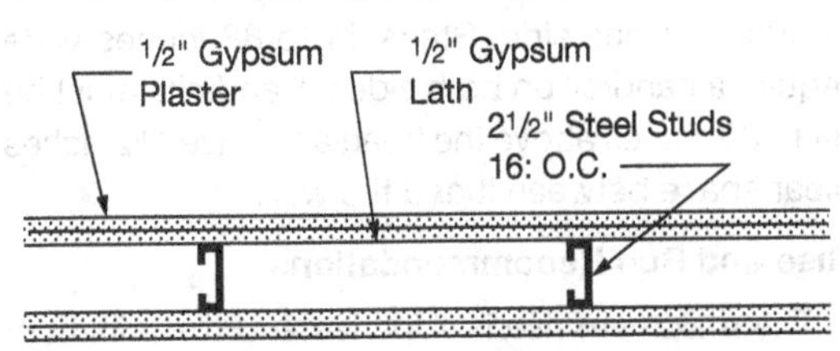

Gypsum Lath and Plaster

STC 40, Fire Rating 2 hr.

STAIRS

Stairs require considerable preliminary planning. After the location on the floor plan has been established, the planning must constantly consider building code requirements. The stair may have to take a turn forming a U-shape or an L-shape. A landing can be introduced in the middle of a long straight stair to provide a resting place, which some may need while ascending the stair.

TYPICAL STAIR CODE REQUIREMENTS

Stair Width

For residential buildings having 49 or fewer occupants, the stair width can be 36 inches. Residential buildings with 50 or more occupants must have minimum stair widths of 44 inches.

Handrails

Stairs less than 44 inches in width can have a handrail on one side. Stairs 44 to 88 inches wide require a handrail on both sides. Handrails must be 34 to 38 inches above the tread and have 1½ inches clear space between it and the wall.

Rise and Run Recommendations

The maximum rise is 7 inches and the minimum is 4 inches in buildings with occupancies below ten people. Some codes permit 7¾- and 8½-inch rises. The maximum run in 11 inches. Some codes permit 10-inch run if the occupancy is ten or fewer.

TYPICAL RISE AND RUN PROPORTIONS

Unit Rise	Unit Run	Approx. Degrees of Slope	Total 2 Risers Plus 1 Unit Run = 24" to 25"
6⅝"	11⅜"	30°-30'	24⅝"
6¾"	11¼"	31°-30'	24¾"
7"	10½"	33°-35'	24½"
7½"	9½"	38°-15'	24½"

The sum of a properly selected rise and run should equal 24 to 25 inches. Check the local code to verify stair requirements.

STAIR HEADROOM

Minimum headroom on a stair is 6'-8".

DIMENSIONS FOR SEVERAL STRAIGHT STAIRS

Height Floor-to Floor	Number of Risers	Height of Risers	Width of Treads	Total Run	Minimum Headroom	Well Opening
8'-0"	14	6⅞"	10⅝"	11'-6⅛"	6'-8"	11'-10³¹⁄₃₂"
8'-0"	16	6"	11½"	14'-4½"	6'-8"	15'-⁵⁄₃₂"
8'-6"	15	6¹³⁄₁₆"	10¹¹⁄₁₆"	12'-5⅝"	6'-8"	11'- 11¹⁵⁄₁₆"
8'-6"	17	6"	11½"	15'-4"	6'-8"	15'-⁵⁄₃₂"
9'-0"	16	6¾"	10¾"	13'-5¼"	6'-8"	12'-11¹¹⁄₁₆"
9'-0"	18	6"	11½"	16'-3½"	6'-8"	15'-⁵⁄₃₂"
9'-6"	17	6¹¹⁄₁₆"	10¹³⁄₁₆"	14'-5"	6'-8"	13'-¾"
9'-6"	19	6"	11½"	17'-3"	6'-8"	15'-⁵⁄₃₂"

TREADS AND RISERS

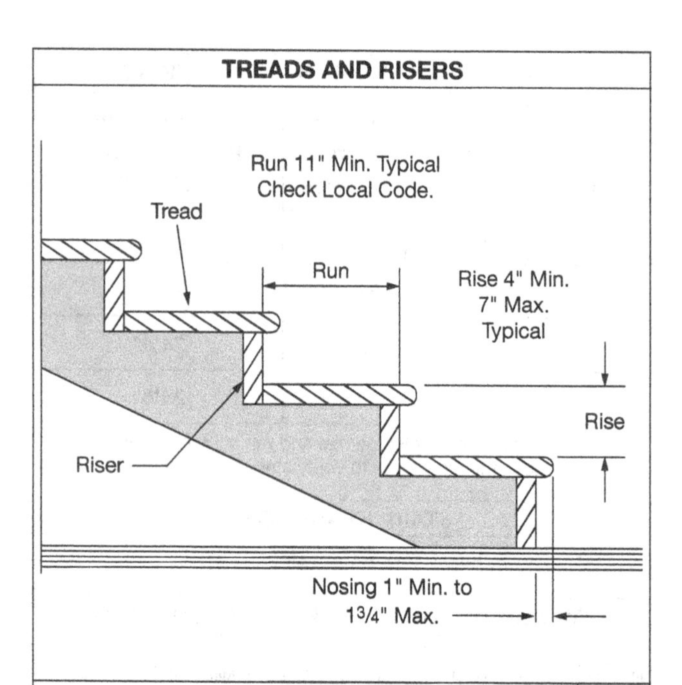

TYPICAL ACCEPTABLE RISE AND RUN PROPORTIONS

Unit Rise	Unit Run	Approx. Degrees of Slope	Total 2 Risers Plus 1 Unit Run 5 24" to 25"
6⅝"	11⅜"	30°-30'	24⅝"
6¾"	11¼"	31°-30'	24¾"
7"	10½"	33°-35'	24½"
7½"	9½"	38°-15'	24½"

THE MANY THINGS THAT MUST BE CONSIDERED WHEN A STAIR IS PLANNED

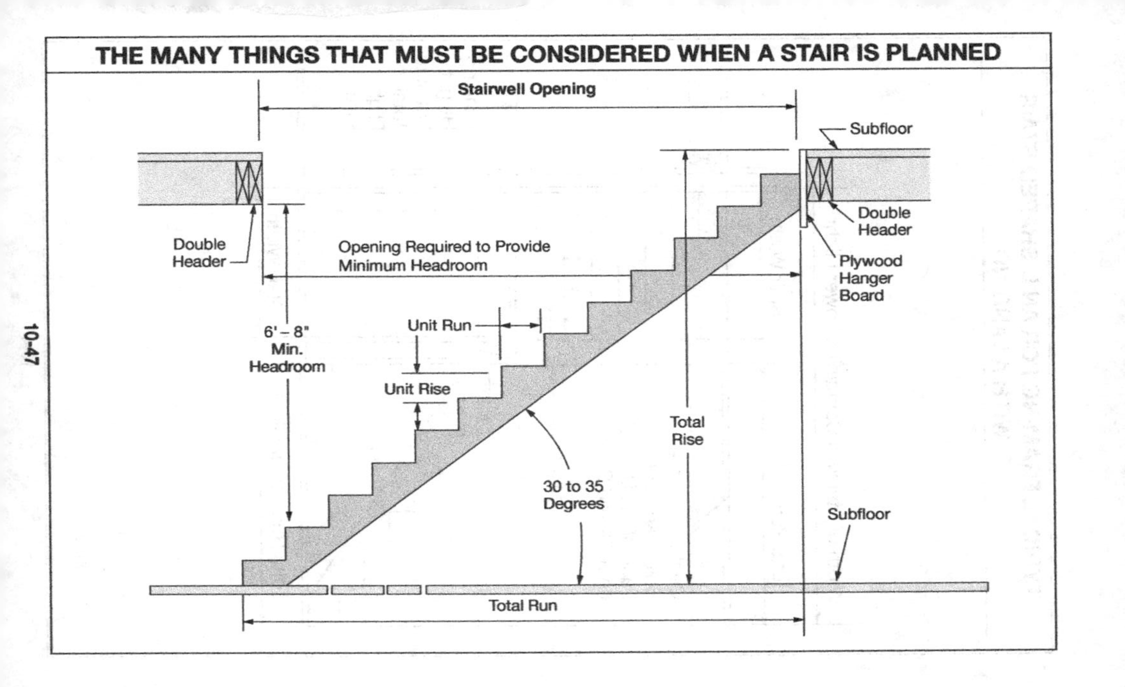

TYPICAL FRAMING FOR AN L-SHAPED STAIR WITH A LANDING

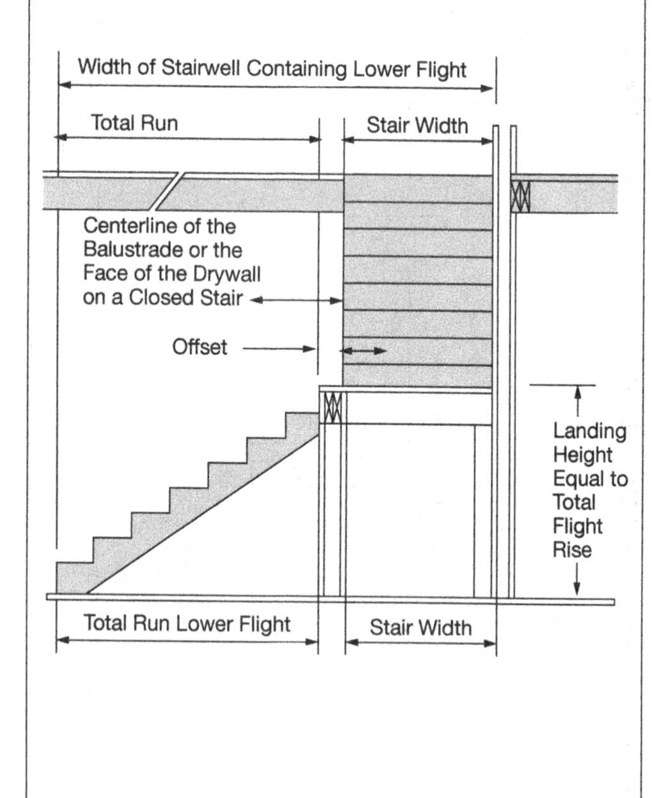

TYPICAL FRAMING FOR AN L-SHAPED STAIR WITH A LANDING *(cont.)*

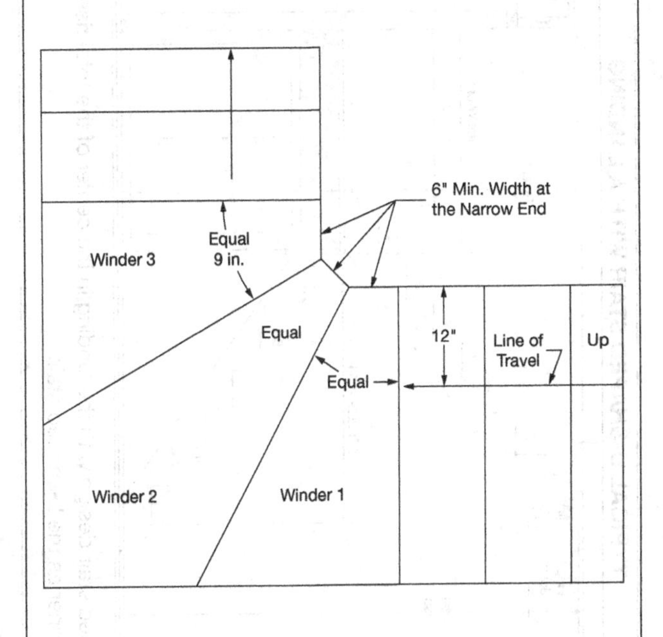

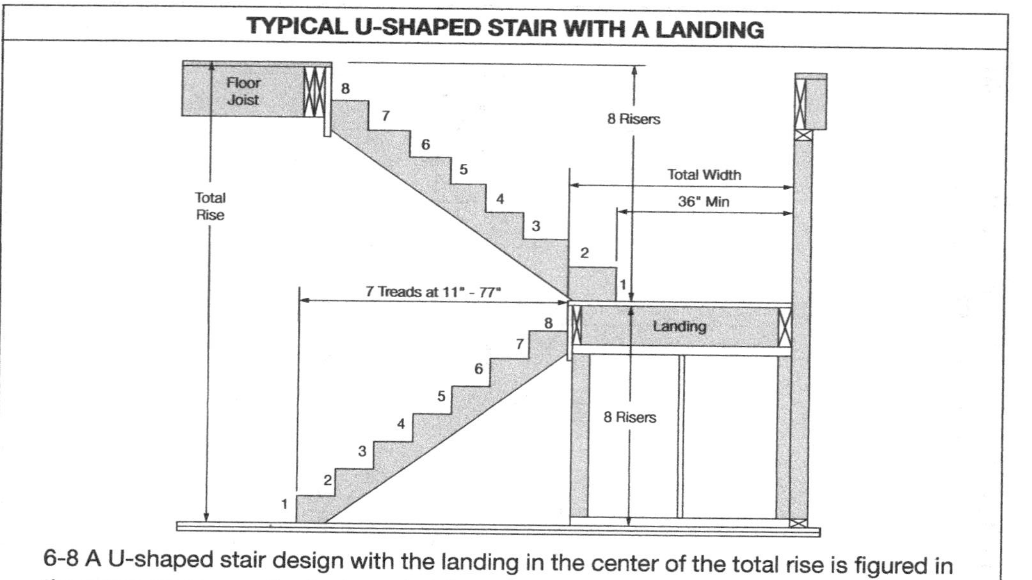

6-8 A U-shaped stair design with the landing in the center of the total rise is figured in the same manner as the L-shaped stair.

TYPICAL FRAMING FOR A STAIRWELL WHEN IT RUNS PARALLEL WITH THE FLOOR JOISTS

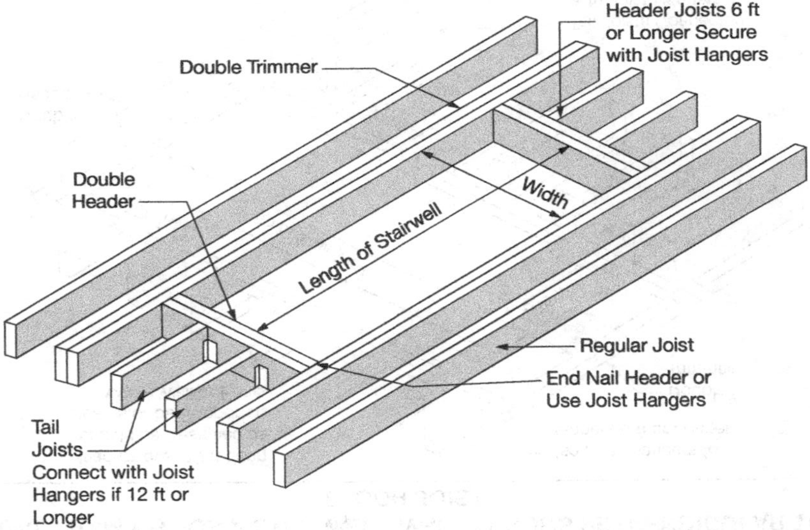

Header Joists 6 ft or Longer Secure with Joist Hangers

Double Trimmer

Double Header

Width

Length of Stairwell

Regular Joist

End Nail Header or Use Joist Hangers

Tail Joists Connect with Joist Hangers if 12 ft or Longer

TYPICAL FRAMING FOR A STAIRWELL WHEN IT RUNS PERPENDICULAR TO THE FLOOR JOISTS

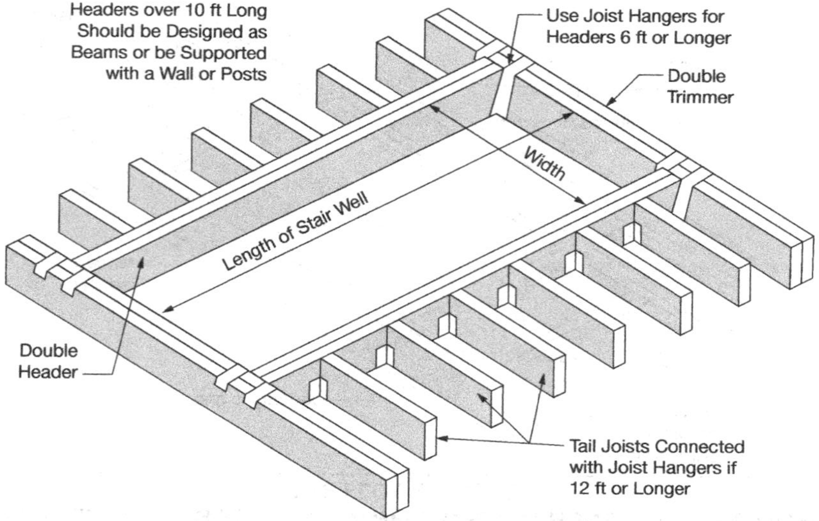

Headers over 10 ft Long Should be Designed as Beams or be Supported with a Wall or Posts

Use Joist Hangers for Headers 6 ft or Longer

Double Trimmer

Width

Length of Stair Well

Double Header

Tail Joists Connected with Joist Hangers if 12 ft or Longer

SLIDING STAIRS

Same sizes as folding stairs. Typical load carrying capacity 400 to 800 pounds.

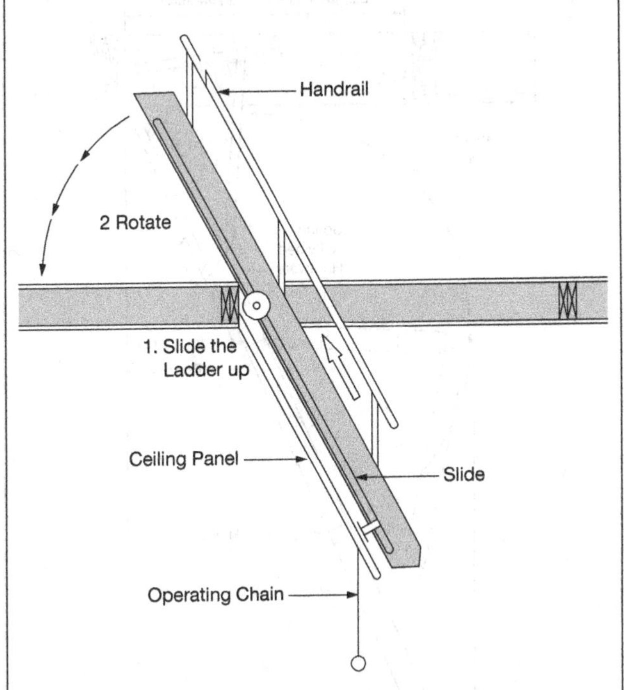

1. To Open Pull Down on the Operating Chain. To Close Slide up and Rotate.

FOLDING STAIRS

Require a ceiling opening 22 in to 30 in wide and 54 in. to 72 in. long. Typical load carrying capacity 250 to 300 pounds.

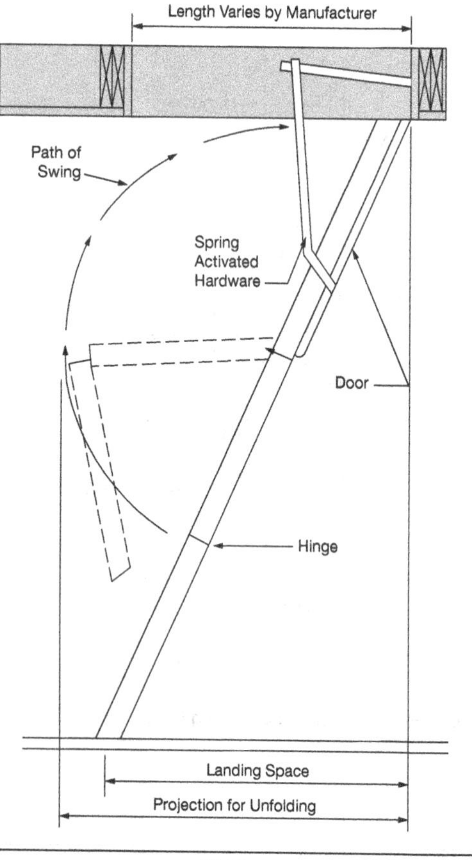

PANEL PRODUCTS

There are a variety of veneered and reconstituted wood panels available. It is important to consult the manufacturer's recommendations when deciding how to use each type. Resistance to moisture and load-carrying capacity are two major considerations.

TYPICAL PANEL THICKNESSES AND SPANS FOR VARIOUS STRUCTURAL PANELS

Panel	Wall Sheathing Studs 16" o.c.	Roof Sheathing rafters 24" o.c.		Subfloor Joists 16" o.c.	Subfloor-Underlayment Joists 24" o.c.
		Edge Support	No Support		
Waterboard	3/8"	7/16"	9/16"	5/8"*	3/4"**
Oriented Strand Board	3/8"	3/8"	7/16"	7/16"	3/4"
Particleboard	3/8"	3/8"	7/16"	1/2"	3/4"
Plywood	5/16"	3/8"	1/2"	1/2"	3/4"
Comply	3/8"	3/8"	1/2"	1/2"	3/4"

These panels may be made to different engineering standards by different manufacturers. Therefore, check with the manufacturer for specific span capabilities of their products. Those carrying the APA stamp have been tested and have code approval.

APA PANEL SUBFLOORING (APA RATED SHEATHING)[a,b]

				Maximum Nail Spacing (in.)	
Panel Span Rating	Panel Thickness (in.)	Maximum Span (in.)	Nail Size & Type[c,g]	Supported Panel Edges[h]	Intermediate Supports
24/16	7/16	16	6d common	6	12
32/16	15/32, 1/2	16	8d common[e]	6	12
40/20	19/32, 5/8	20[d]	8d common	6	12
48/24	23/32, 3/4	24	8d common	6	12
60/32[f]	7/8	32	8d common	6	12

[a] For subfloor recommendations under ceramic tile, refer to Table 14, For subfloor recommendations under gypsum concrete, contact manufacture of floor topping.
[b] APA RATED STURD,1. FLOOR may be substituted when the Span Rating is equal to or greater than tabalated maximum span.
[c] 6d common nail permitted if panel is ½ inch or thinner.
[d] Span may be 24 inches if a minimum 1-½ inches of lightweight concrete is applied over panels.
[e] Other code-approved fasteners may be used.
[f] Check with supplier for availability.
[g] See Table 5 for nail dimensions.
[h] Supported panel joints shall occur approximately along the centerline of framing with a minimum bearing of ½*. Fasteners shall be located ⅜ inch from panel edges.

Courtesy APA-The Engineered Wood Association www.apawood.org

APA PLYWOOD UNDERLAYMENT[c]

Plywood Grades[a]	Application	Minimum Plywood Thickness (in.)	Fastener Size and Type[f]	Maximum Fastener Spacing (in.)[e]	
				Panel Edges[d]	Intermediate
APA Underlayment	Over smooth subfloor	1/4	3d × 1-1/4-in. ring-shank nails[b], min. 12-1/2 gage (0.099 in.) shank dia.	3	6 each way
APA C-C Plugged EXT APA Rated Sturd-I-Floor (19/32" or thicker)	Over lumber sub-floor or uneven surfaces	11/32		6	8 each way

[a]In areas to be finished with resilient floor coverings such as tile or sheet vinyl, or with fully adhered carpet, specify Underlayment, C-C Plugged or veneer-faced STURD-I-FLOOR with "sanded face." Underlayment A-C, Underlayment B-C, Marine EXT or sanded plywood grades marked "Plugged Crossbands Under Face," "Plugged Crossbands (or Core)," "Plugged Inner Plies" or "Meets Underlayment Requirements" may also be used under resilient floor coverings.

[b]Use 4d × 1-1/2-in. ring-shank nails, minimum 12-1/2 gage (0.099 in.) shank diameter, for underlayment panels 19/32 inch to 3/4 inch thick.

[c]For underlayment recommendations under ceramic tile, refer to Table 14.

[d]Fasten panels 3/8 inch from panel edges.

[e]Fasteners for 5-ply plywood underlayment panels and for panels greater than 1/2 inch thick may be spaced 6 inches on center at edges and 12 inches each way intermediate.

[f]See Table 5 for nail dimensions.

Courtesy APA-The Engineered Wood Association www.apawood.org

PLYWOOD PANEL VENEER GRADES	
A	Smooth, paintable. Not more than 18 neatly made repairs, boat, sled, or router type, and parallel to grain, permitted. Wood or synthetic repairs permitted. May be used for natural finish in less demanding applications.
B	Solid surface. Shims, sled or router repairs, and tight knots to 1 inch across grain permitted. Wood or synthetic repairs permitted. Some minor splits permitted.
C Plugged	Improved C veneer with splits limited to ⅛-inch width and knotholes or other open defects limited to ¼ × ½ inch. Wood or synthetic repairs permitted. Admits some broken grain.
C	Tight knots to 1-½ inch. Knotholes to 1 inch across grain and some to 1-½ inch if total width of knots and knotholes is within specified limits. Synthetic or wood repairs. Discoloration and sanding defects that do not impair strength permitted. Limited splits allowed. Stitching permitted.
D	Knots and knotholes to 2-½-inch width across grain and ½ inch larger within specified limits. Limited splits are permitted. Stitching permitted. Limited to Exposure 1 or Interior panels.

Courtesy APA_The Engineered Wood Association www.apawood.org

APA GRADE STAMPS FOR SHEATHING, SIDING AND STURD-I-FLOOR

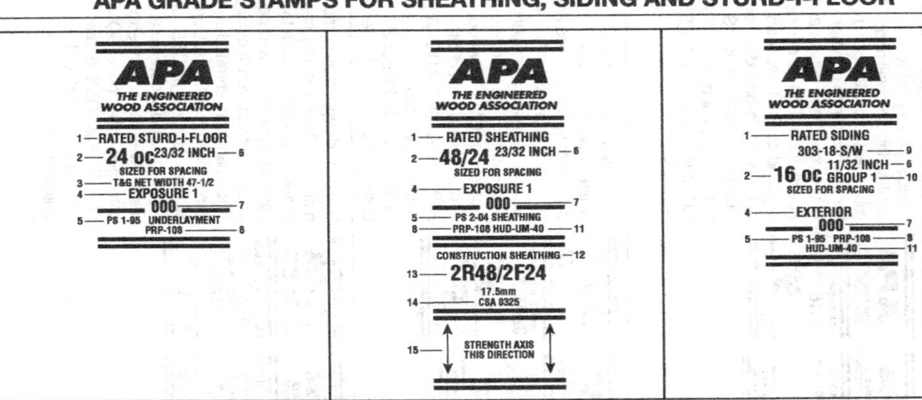

1. Panel grade
2. Span rating
3. Tongue-and-groove
4. Bond classification
5. Product standard
6. Thickness
7. Mill number
8. APA's performance rated panel standard
9. Siding face grade
10. Species group number
11. HUD recognition
12. Panel grade, Canadian standard
13. Panel mark — Rating and end-use designation per the Canadian standard
14. Canadian performance rated panel standard
15. Panel face orientation indicator

Courtesy APA-The Engineered Wood Association www.apawood.org

GUIDE TO APA PERFORMANCE RATED PANELS[a,b]

For Application Recommendations, See Following Pages.

Panel	Typical Trademarks	Description
APA Rated Sheathing Typical Trademark	APA THE ENGINEERED WOOD ASSOCIATION RATED SHEATHING 40/20 19/32 INCH SIZED FOR SPACING EXPOSURE 1 000 PS 2-94 SHEATHING PRP-108 HUD-UM-40 APA THE ENGINEERED WOOD ASSOCIATION RATED SHEATHING 24/16 7/16 INCH SIZED FOR SPACING EXPOSURE 1 000 PRP-108 HUD-UM-40	Specially designed for subflooring and wall and roof sheathing. Also good for a broad range of other construction and industrial applications. Can be manufactured as OSB, plywood, or other wood-based panel. BOND CLASSIFICATIONS: Exterior, Exposure 1, COMMON THICKNESSES: 5/16, 3/8, 7/16, 15/32, 1/2, 19/32, 5/8, 23/32, 3/4.
APA Structural I Rated Sheathing[c] Typical Trademark	APA THE ENGINEERED WOOD ASSOCIATION RATED SHEATHING STRUCTURAL 1 32/16 15/32 INCH SIZED FOR SPACING EXPOSURE 1 000 PS 1-95 C-D PRP-108 APA THE ENGINEERED WOOD ASSOCIATION RATED SHEATHING 32/16 15/32 INCH SIZED FOR SPACING EXPOSURE 1 000 STRUCTURAL I RATED DIAPHRAGMS-SHEAR WALLS PANELIZED ROOFS PRP-108 HUD-UM-40	Unsanded grade for use where shear and cross-panel strength properties are of maximum importance, such as panelized roofs and diaphragms. Can be manufactured as OSB, plywood, or other wood-based panel. BOND CLASSIFICATIONS: Exterior, Exposure 1. COMMON THICKNESSES: 5/16, 3/8, 7/16, 15/32, 1/2, 19/32, 5/8, 23/32, 3/4.
APA Rated Sturd-I-Floor	APA THE ENGINEERED WOOD ASSOCIATION RATED STURD I FLOOR 24 oc 23/32 INCH SIZED FOR SPACING T&G NET WIDTH 47-1/2 EXPOSURE 1 000 PS 2-94 SINGLE FLOOR PRP-108 HUD-UM-40 APA THE ENGINEERED WOOD ASSOCIATION RATED STURD I FLOOR 20 oc 19/32 INCH SIZED FOR SPACING T&G NET WIDTH 47-1/2 EXTERIOR 000 PS 1-95 C-C PLUGGED PRP-108	Specially designed as combination subfloor-underlayment. Provides smooth surface for application of carpet and pad and possesses high Typical Trademark concentrated and impact load resistance. Can be manufactured as OSB, plywood, or other wood-based panel. Available square edge or tongue and-groove. BOND CLASSIFICATIONS: Exterior, Exposure 1, COMMON THICKNESSES: 19/32, 5/8, 23/32, 3/4, 1, 1-1/8.

GUIDE TO APA PERFORMANCE RATED PANELS *(cont.)*

APA Rated Siding Typical Trademark	APA THE ENGINEERED WOOD ASSOCIATION RATED SIDING 24 oc 19/32 INCH SIZED FOR SPACING EXTERIOR 000 PRP-108 HUD-UM-40 APA THE ENGINEERED WOOD ASSOCIATION RATED SIDING 303-18-S/W 11/32 INCH 16 oc GROUP 1 SIZED FOR SPACING EXTERIOR 000 PS 1-95 PRP-108 HUD-UM-40	For exterior siding, fencing, etc. Can be manufactured as plywood, as other wood-based panel or as an overlaid OSB. Both panel and lap siding available. Special surface treatment such as V-groove, channel groove, deep groove (such as APA Texture 1-11), brushed, rough sawn and overlaid (MDO) with smooth- or texture-embossed face. Span Rating (stud spacing for siding qualified for APA Sturd-I-Wall applications) and face grade classification (for veneer-faced siding) indicated in trademark. BOND CLASSIFICATION: Exterior. COMMON THICKNESSES: 11/32, 3/8, 7/16, 15/32, 1/2, 19/32, 5/8.

[a]Specific grades, thicknesses and bond classifications may be in limited supply in some areas. Check with your supplier before specifying.

[b]Specify Performance Rated Panels by thickness and Span Rating. Span-Ratings are based on panel strength and stiffness. Since these properties are a function of panel composition and configuration as well as thickness, the same Span Rating may appear on panels of different thickness. Conversely, panels of the same thickness may be marked with different Span-Ratings.

[c]All plies in Structural I plywood panels are special improved grades and panels marked PS-1 are limited to Group-1 species. Other panels marked Structural-I-Rated qualify through special performance testing.

Courtesy APA-The Engineered Wood Association www.apawood.org

GUIDE TO APA SANDED AND TOUCH-SANDED PLYWOOD PANELS[a,b,c]

APA A-A
Typical Trademark (mark on panel edge)

A-A • G-1 • EXPOSURE 1-APA • 000 • PS1-95

Use where appearance of both sides is important for interior applications such as built-ins, cabinets, furniture, partitions; and exterior applications such as fences, signs, boats, shipping containers, tanks, ducts, etc. Smooth surfaces suitable for painting. BOND CLASSIFICATIONS: Interior, Exposure 1, Exterior. COMMON THICKNESSES: 1/4, 11/32, 3/8, 15/32, 1/2, 19/32, 5/8, 23/32, 3/4.

APA A-B
Typical Trademark (mark on panel edge)

A-B • G-1 • EXPOSURE 1-APA • 000 • PS1-95

For use where appearance of one side is less important but where two solid surfaces are necessary. BOND CLASSIFICATIONS: Interior, Exposure 1, Exterior. COMMON THICKNESSES: 1/4, 11/32, 3/8, 15/32, 1/2, 19/32, 5/8, 23/32, 3/4.

APA A-C
Typical Trademark

For use where appearance of only one side is important in exterior or interior applications, such as soffits, fences, farm buildings, etc.[f] BOND CLASSIFICATION: Exterior. COMMON THICKNESSES: 1/4, 11/32, 3/8, 15/32, 1/2, 19/32, 5/8, 23/32, 3/4.

APA A-D
Typical Trademark

For use where appearance of only one side is important in interior applications, such as paneling, built-ins, shelving, partitions, flow racks, etc[f] BOND CLASSIFICATIONS: Interior, Exposure 1. COMMON THICKNESSES: 1/4, 11/32, 3/8, 15/32, 1/2, 19/32, 5/8, 23/32, 3/4.

GUIDE TO APA SANDED AND TOUCH-SANDED PLYWOOD PANELS *(cont.)*

Panel	Trademark	Description
APA B-B **Typical Trademark (mark on panel edge)**	B-B • G-2 • EXT-APA • 000 • PS1-95	Utility panels with two solid sides. BOND CLASSIFICATIONS: Interior, Exposure 1, Exterior. COMMON THICKNESSES: 1/4, 11/32, 3/8, 15/32, 1/2, 19/32, 5/8, 23/32, 3/4.
APA B-C **Typical Trademark**		Utility panel for farm service and work buildings, boxcar and truck linings, containers, tanks, agricultural equipment, as a base for exterior coatings and other exterior uses or applications subject to high or continuous moisture.[f] BOND CLASSIFICATION: Exterior. COMMON THICKNESSES: 1/4, 11/32, 3/8, 15/32, 1/2, 19/32, 5/8, 23/32, 3/4.
APA B-D **Typical Trademark**		Utility panel for backing, sides of built-ins, industry shelving, slip sheets, separator boards, bins and other interior or protected applications.[f] BOND CLASSIFICATION: Interior, Exposure 1. COMMON THICKNESSES: 1/4, 11/32, 3/8, 15/32, 1/2, 19/32, 5/8, 23/32, 3/4.
APA Underlayment **Typical Trademark**		For application over structural subfloor. Provides smooth surface for application of carpet and pad and possesses high concentrated and impact load resistance. For areas to be covered with resilient flooring, specify panels with "sanded face".[e] BOND CLASSIFICATIONS: Interior, Exposure 1. COMMON THICKNESSES[d]: 1/4, 11/32, 3/8, 15/32, 1/2, 19/32, 5/8, 23/32, 3/4.

GUIDE TO APA SANDED AND TOUCH-SANDED PLYWOOD PANELS *(cont.)*

APA C-C PLUGGED[d]
Typical Trademark

For use as an underlayment over structural subfloor, refrigerated or controlled atmosphere storage rooms, pallet fruit bins, tanks, boxcar and truck floors and linings, open soffits, and other similar applications where continuous or severe moisture may be present. Provides smooth surface for application of carpet and pad and possesses high concentrated and impact load resistance. For areas to be covered with resilient flooring, specify panels with "sanded face."[e] BOND CLASSIFICATION: Exterior. COMMON THICKNESSES[d]: 11/32, 3/8, 15/32, 1/2, 19/32, 5/8, 23/32, 3/4.

APA C-D PLUGGED
Typical Trademark

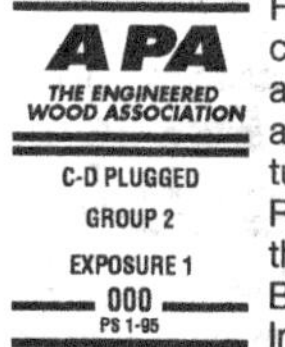

For open soffits, built-ins, cable reels, separator boards and other interior or protected applications. Not a substitute for Underlayment or APA Rated Sturd-I-Floor as it lacks their puncture resistance. BOND CLASSIFICATIONS: Interior, Exposure 1. COMMON THICKNESSES: 3/8, 15/32, 1/2, 19/32, 5/8, 23/32, 3/4.

[a]Specific plywood grades, thicknesses and bond classifications may be in limited supply in some areas. Check with your supplier before specifying.

[b]Sanded Exterior plywood panels, C-C Plugged, C-D Plugged and Underloyment grodes can also be manufactured in Structural I (all plies limited to Group 1 species).

[c]Some manufacturers also produce plywood ponels with premium N-grade veneer on one or both faces. Available only by special order. Check with the manufacturer.

Courtesy APA-The Engineered Wood Association www.apawood.org

[d]Some panels 1/2 inch and thicker are Span Rated and do not contain species group number in trademark.

[e]Also available in Underlayment A-C or Underlayment B-C grades, marked either "touch sanded" or "sanded face."

[f]For nonstructural floor underloyment, or other applications requiring improved inner ply construction, specify ponels marked either "plugged inner plies" (may olso be designated plugged crossbonds under face or plugged crossbands or core); or "meets underlayment requirements."

[g]Also may be designated APA Underloyment C-C Plugged. order.

GUIDE TO APA SPECIALTY PLYWOOD PANELS[a]

APA Decorative **Typical Trademark**  DECORATIVE GROUP 2 EXPOSURE 1 000 PS 1-95	Rough-sawn, brushed, grooved, or striated faces. For paneling, interior accent walls, built-ins, counter facing, exhibit displays. Can also be made by some manufacturers in Exterior for exterior siding, gable ends, fences and other exterior applications. Use recommendations for Exterior panels vary with the particular product. Check with the manufacturer. BOND CLASSIFICATIONS: Interior, Exposure-1, Exterior. COMMON THICKNESSES: 5/16, 3/8, 1/2, 5/8.
APA High Density Overlay (HDO)[b] **Typical Trademark (mark on panel edge)** HDO • A-A • G-1 • EXT-APA • 000 • PS 1-95	Has a hard semi-opaque resin-fiber overlay on both faces. Abrasion resistant. For concrete forms, cabinets, countertops, signs, tanks. Also available with skid-resistant screen-grid surface. BOND CLASSIFICATION: Exterior. COMMON THICKNESSES: 3/8, 1/2, 5/8, 3/4.
APA Medium Density Overlay (MDO)[b] **Typical Trademark**  M.D. OVERLAY GROUP 1 EXTEROR 000 PS 1-95	Smooth, opaque, resin-fiber overlay on one or both faces. Ideal base for paint, both indoors and outdoors. For exterior siding, paneling, shelving, exhibit displays, cabinets, signs. BOND CLASSIFICATION: Exterior. COMMON THICKNESSES: 11/32, 3/8, 15/32, 1/2, 19/32, 5/8, 23/32, 3/4.

GUIDE TO APA SPECIALTY PLYWOOD PANELS *(cont.)*

APA Marine **Typical Trademark (mark on panel edge)** MARINE • A-A • EXT-APA • 000 • PS 1-95	Ideal for boot hulls. Made only with Douglas-fir or western larch. Subject to special limitations on core gaps and face repairs. Also available with HDO or MDO faces. BOND CLASSIFICATION: Exterior. COMMON THICKNESSES: 1/4, 3/8, 1/2, 5/8, 3/4.
APA Plyform Class-I[b] **Typical Trademark**  	Concrete form grades with high reuse factor. Sanded both faces and mill-oiled unless otherwise specified. Special restrictions on species. Also available in HDO for very smooth concrete finish, and with special overlays. BOND CLASSIFICATION: Exterior. COMMON THICKNESSES: 19/32, 5/8, 23/32, 3/4.
APA Plyron **Typical Trademark (mark on panel edge)** PLYRON • EXPOSURE 1 • APA • 000	Hardboard face on both sides. Faces tempered, untempered, smooth or screened. For countertops, shelving, cabinet doors, flooring. BOND CLASSIFICATIONS: Interior, Exposure 1, Exterior. COMMON THICKNESSES: 1/2, 5/8, 3/4.

[a]Specific plywood grades, thicknesses and bond classifications may be in limited supply in some areas. Check with your supplier before specifying.

[b]Can also be manufactured in Structural I (all plies limited to Group-1 species).

Courtesy APA-The Engineered Wood Association www.apawood.org

APA RATED PANEL AND LAP SIDING SIZES

Apa Rated Siding Nominal Thickness	
in.	mm
11⁄32	8.7
3⁄8	9.5
7⁄16	11.1
15⁄32	11.9
1⁄2	12.7
19⁄32	15.1
5⁄8	15.9

Panel Siding Nominal Dimensions (width 3 length)		
ft.	mm	m (approx.)
4 × 8	1219 × 2438	1.22 × 2.44
4 × 9	1219 × 2743	1.22 × 2.74
4 × 10	1219 × 3048	1.22 × 3.05

Lap Siding Nominal Dimensions (width 3 length)		
in. 3 ft.	mm	m (approx.)
6 × 16	152.4 × 4877	0.15 × 4.88
8 × 16	203.2 × 4877	0.20 × 4.88
12 × 16	304.8 × 4877	0.30 × 4.88

Courtesy APA – The Engineered Wood Association www.apawood.org

TYPICAL PLYWOOD PANEL THICKNESSES	
In.	**mm***
¼	6.4
11/32	8.7
⅜	9.5
15/32	11.9
½	12.7
19/32	15.1
⅝	15.9
23/32	18.3
¾	19.1
⅞	22.2
1	25.4
1⅛	28.6

TYPICAL PLYWOOD PANEL SIZES (width 3 length)		
ft.	**mm***	**m***
4 × 8	1,219 × 2,438	1.22 × 2.44
4 × 9	1,219 × 2,743	1.22 × 2.74
4 × 10	1,219 × 3,048	1.22 × 3.05

*Soft metric conversions

CLASSIFICATION OF SOFTWOOD PLYWOOD LUMBER SPECIES FOR STRENGTH AND STIFFNESS

Group 1 has the Strongest Woods.

Group 1	Group 2	Group 3	Group 4	Group 5
Apitong Beech, American Birch Sweet Yellow Douglas-fir 1 Kapur Keruing Larch, Western Maple, Sugar Pine Caribbean Ocole Pine, Southern Loblolly Longleaf Shortleaf Slash Tanoak	Cedar, Port Orford Cypress Douglas-fir 2 Fir Balsam California Red Grand Noble Pacific Silver White Hemlock, Western Lauan Almond Bagtikan Mayapis Red Tangile White Maple, Black Mengkulang Meranti, Red Mersawa Pine Pond Red Virginia Western White Spruce Black Red Sitka Sweetgum Tamarack Yellow-Poplar	Alder, Red Birch, Paper Birch, Paper Cedar, Alaska Fir, Subalpine Hemlock, Eastern Maple, Bigleaf Pine Jack Lodgepole Ponderosa Spruce Redwood Spruce Engelmann White	Aspen Bigtooth Quaking Cativo Cedar Incense Western Red Cottonwood Eastern Black (Western Poplar) Pine Eastern White Sugar	Basswood Poplar, Balsam

Courtesy APA – The Engineered Wood Association.

SUMMARY OF USE CATEGORIES FOR TREATED WOOD THAT APPLY TO PLYWOOD				
Use Category	**Service Conditions**	**Use Environment**	**Common Agents of Deterioration**	**Typical Applications**
UC1	Interior construction, above ground, dry	Continuously protected from weather or other sources of moisture	Insects only	Interior construction and furnishings
UC2	Interior construction, above ground, damp	Protected from weather, but may be subject to sources of moisture	Decay fungi and insects	Interior construction
UC3B	Exterior construction, above ground, uncoated or poor water run-off	Exposed to all weather cycles, including prolonged wetting	Decay fungi and insects	Decking, deck joists, railings, fence pickets, uncoated millwork
UC4A	Ground contact or fresh water, non-critical components	Exposed to all weather cycles, normal exposure conditions	Decay fungi and insects	Fence, deck, and guardrail posts, crossties and utility poles (low decay areas)
UC4B	Ground contact or fresh water, critical components or difficult replacement	Exposed to all weather cycles, high decay potential including salt water splash	Decay fungi and insects with increased potential for biodeterioration	Permanent wood foundations, building poles, horticultural posts, crossties and utility poles (high decay areas)

SUMMARY OF USE CATEGORIES FOR TREATED WOOD THAT APPLY TO PLYWOOD *(cont.)*

Use Category	Service Conditions	Use Environment	Common Agents Of Deterioration	Typical Applications
UC4C	Ground contact or fresh water, critical structural components	Exposed to all weather cycles, severe environments, extreme decay potential	Decay fungi and insects with extreme potential for biodeterioration	Land & freshwater piling, foundation piling, crossties and utility poles severe decay areas)
UC5A	Salt or brackish water and adjacent mud zone northern waters	Continuous marine exposure (salt water)	Salt water organisms	Piling, bulkheads, bracing
UC5B	Salt or brackish water and adjacent mud zone NJ to GA, south of San Francisco	Continuous marine exposure (salt water)	Salt water organisms including creosote tolerant *Limnoria tripunctata*	Piling, bulkheads, bracing
UC5C	Salt or brackish water and adjacent mud zone South of Ga and Gulf Coast, Hawaii, and Puerto Rico	Continuous marine exposure (salt water)	Salt water organisms including creosote tolerant *Martesia, and Sphaeroma*	Piling, bulkheads, bracing

Courtesy APA – The Engineered Wood Association, www.apawood.org

NOMINAL THICKNESSES OF HARDBOARD PANELS		
mm	inch	
2.1	1/12	(0.083)
2.5	1/10	(0.100)
3.2	1/8	(0.125)
4.8	3/16	(0.188)
6.4	1/4	(0.250)
7.9	5/16	(0.312)
9.5	3/8	(0.375)
11.1	7/16	(0.438)
12.7	1/2	(0.500)
15.9	5/8	(0.625)

Courtesy Composite Panel Association, www.pbmdf.com

TYPICAL SIZES OF HARDBOARD PANELS	
Panel width	4 ft.
Panel lengths	4 ft., 6 ft., 8 ft., 10 ft., 12 ft.
Lap Siding	6 in., 8 in., 12 in. wide, 16 ft. long
Width	12 in., length 16 ft.

Other sizes are available by special order.

HARDBOARD IDENTIFICATION STRIPES AND TENSILE STRENGTH

Class	Number and Color of Stripes on Edges of Panels	Tensile Strength psi
Tempered	1 Red	3000 psi
Service Tempered	2 Red	2000 psi
Standard	1 Green	2200 psi
Service	2 Green	1500 psi
Industrialite	1 Blue	1000 psi

Courtesy Composite Panel Association, www.pbmdf.com

CLASSIFICATIONS OF HARDBOARD PANELS

Class 1	Tempered
Class 2	Standard
Class 3	Service-tempered
Class 4	Service
Class 5	Industrialite

Class 1: Impregnated with siccative materials and special additives. Has improved properties of stiffness, hardness and resistance to water.

Class 2: Standard type of hardboard as it comes from the press. Has high strength and water resistance.

Class 3: Impregnated with siccative material and stabilized by heat and additives. Properties better than Service Class 4.

Class 4: Basically the same form as it comes from the press but has less strength than Standard Type Class 2.

Class 5: A medium density hardboard with moderate strength and lower weight than the other classes.

Courtesy National Particleboard Association, www.pdmdf.com

CLASSIFICATION OF HARDBOARD BY THICKNESS AND PHYSICAL PROPERTIES

Class	Nominal Thickness		Water resistance (max. Average Per Panel) Water Absorption Based on Weight	Water resistance (max. Average Per Panel) Thickness Swelling	Modulus of Rupture (min. Average Per Panel)		Tensile Strength (min. Average Per Panel) Parallel to Surface		Tensile Strength (min. Average Per Panel) Perpendicular to Surface	
	mm	inch	percent	percent	MPa	psi	MPa	psi	MPa	psi
1 Tempered	2.1	1/12	30	25	41.4	6000	20.7	3000	0.90	130
	2.5	1/10	25	20						
	3.2	1/8	25	20						
	4.8	3/16	25	20						
	6.4	1/4	20	15						
	7.9	5/16	15	10						
	9.5	3/8	10	9						
2 Standard	2.1	1/12	40	30	31.0	4500	15.2	2200	0.62	90
	2.5	1/10	35	25						
	3.2	1/8	35	25						
	4.8	3/16	35	25						
	6.4	1/4	25	20						
	7.9	5/16	20	15						
	9.5	3/8	15	10						
3 Service-Tempered	3.2	1/8	35	30	31.0	4500	13.8	2000	0.52	75
	4.8	3/16	30	30						
	6.4	1/4	30	25						
	9.5	3/8	20	15						
4 Service	3.2	1/8	45	35	20.7	3000	10.3	1500	0.34	50
	4.8	3/16	40	35						
	6.4	1/4	40	30						
	9.5	3/8	35	25						
	11.1	7/16	35	25						
	12.7	1/2	30	20						
	15.9	5/8	25	20						
5 Industrialite	6.4	1/4	50	30	13.8	2000	6.9	1000	0.17	25
	9.5	3/8	40	25						
	11.1	7/16	40	25						
	12.7	1/2	35	25						
	15.9	5/8	30	20						

Courtesy Composite Panel Association www.pbmnf.com

PARTICLEBOARD GRADES

H	High density (generally above 800 kg/m^3 (50 lb/cu. ft.))
M	Medium density (generally between 640-800 kg/m^3 (40-50 lb/cu. ft.))
LD	Low density (generally less than 640 kg/m^3 (40 lb/cu. ft.))
D	Manufactured home decking
PBU	Underlayment

Courtesy Composite Panel Association, www.pbmdf.com

TYPES, GRADES, AND USES OF PARTICLEBOARD

Type	Grade	Use
High Density	H-1,H-2, H-3 H-1, H-2, H-3 Exterior Glue	High-density industrial High-density exterior industrial
Medium Density	M-1 M-2, M-3 M-1, M-2, M-3 Exterior Glue	Commercial Industrial Exterior construction Exterior industrial
Medium density — specialty grade	M-S	Commercial
Low density	LD-1, LD-2	Door core
Underlayment	PBU	Underlayment
Manufactured home decking	D-2, D-3	Flooring in manufactured homes

Courtesy National Particleboard Association, www.pdmdf.com

PARTICLEBOARD GRADES	
The first letter indicates the following:	
H	High density usually 800 kg/m^3 (50 lb/cu. ft.).
M	Medium density usually 640 to 800 kg/m^3 (40 to 50 lb/cu. ft.)
LD	Low density usually less than 640 kg/m^3 (40 lb/cu. ft.)
D	Manufactured home decking
PDU	Underlayment
The second digit indicates the grade identification within the grade. Example Grade H-2 indicates high density grade 2.	

ORIENTED STRAND BOARD PRODUCT INFORMATION	
APA RATED ORIENTED STRAND BOARD PANELS*	
OSB PANELS	4 × 8 ft. (1.25 × 2.50 m)
OSB LAP SIDING	Widths 6 in. (152 mm), 8 in. (203 mm) Lengths 11 ft. (4.88 m)
*Oversize panels can be manufactured up to 8 × 24 ft.	

ORIENTED STRAND BOARD PRODUCT INFORMATION *(cont.)*

APA RATED ORIENTED STRAND BOARD SHEATHING
Thicknesses 5⁄16" (7.9 mm), 15⁄32" (11.9 mm), ½" (12.7 mm), 19⁄32" (15.1 mm), ⅝" (15.9 mm), 23⁄32" (18.2 mm), ¾" (19.0 mm)

APA RATED ORIENTED STRAND BOARD SUBFLOORING*
Thicknesses 19⁄32" (15.1 mm), ⅝" (15.9 mm), 23⁄32" (18.2 mm), ¾" (19.0 mm), ⅞" (22.2 mm), 1" (25.4 mm), 1⅛" (28.6 mm).
*Can have tongue and groove edges.

APA ORIENTED STRAND BOARD EXPOSURE CLASSIFICATIONS
Exterior. Panels have a fully waterproof bond and are used on applications subject to permanent exposure to weather or moisture.
Exposure 1. Panels have a waterproof bond and are used where long construction delays are expected.
Courtesy APA – The Engineered Wood Association, www.apawood.org

CHAPTER 11
Roofing

STORING ROOFING MATERIALS

Asphalt Shingles

Store in a shady area so heat does not cause adhesive strips to bond. Keep on wood pallets but do not stack over 4 feet high. Cover with plastic sheeting so they are dry when installed. In the winter keep above 40°F (4.5°C).

Wood Shakes and Shingles

Leave in original bundles. Put in a dry place. Put on wood pallets to keep off the ground. Cover with plastic sheeting.

Metal Roofing Panels

Store horizontally on wood pallets to keep flat and off the ground. Be careful to not bend, twist, or scratch. Cover with plastic sheeting.

Clay, Concrete, and Slate Shingles

Store on wood pallets in bundles as received. Cover with plastic sheeting.

WEIGHTS OF SHEATHING MATERIALS

PLYWOOD

Thickness (inches)	lb/sq. ft.
5/16"	1.0
3/8"	1.1
7/16"	1.3
15/32"	1.4
1/2"	1.5
19/32"	1.8
5/8"	1.9
23/32"	2.2
3/4"	2.3
7/8"	2.6
1"	3.0

ORIENTED STRAND BOARD

Thickness (inches)	lb/sq. ft.
5/16"	1.1
3/8"	1.2
7/16"	1.4
15/32"	1.5
1/2"	1.7
19/32"	2.0
5/8"	2.1
23/32"	2.4
3/4"	2.5
7/8"	2.8
1"	3.0

RECOMMENDED MAXIMUM SPANS FOR APA PANEL ROOF DECKS FOR LOW SLOPE ROOFS[a]
(Panel Strength Axis Perpendicular to Supports and Continuous Over Two or More Spans)

Grade	Minimum Nominal Panel Thickness (in.)	Minimum Span Rating	Maximum Span (in.)	Panel Clips Per Span[b] (number)
APA Rated Sheathing	15/32	32/16	24	1
	19/32	40/20	32	1
	23/32	48/24	48	2
	7/8	60/32	60	2

Courtesy APA – The Engineered Wood Association, www.apawood.org.

RECOMMENDED UNIFORM ROOF LIVE LOADS FOR APA RATED SHEATHING[c] AND APA RATED STURD-I-FLOOR WITH STRENGTH AXIS PERPENDICULAR TO SUPPORTS[*]

Panel Span Rating	Minimum Panel Thickness (in.)	Maximum Span (in.)		Allowable Live Loads (psf)[d]							
		With Edge Support[e]	Without Edge Support	Spacing of Supports Center-to-Center (in.)							
				12	16	20	24	32	40	48	60
APA Rated Sheathing[c]											
12/0	5/16	12	12	30							
16/0	5/16	16	16	70	30						
20/0	5/16	19.2	19.2	120	50	30					
24/0	3/8	24	19.2[b]	190	100	60	30				
24/16	7/16	24	24	190	100	65	40				
32/16	15/32, 1/2	32	28	325	180	120	70	30			
40/20	19/32, 5/8	40	32	—	305	205	130	60	30		
48/24	23/30, 3/4	48	36	—	—	280	175	95	45	35	
60/32[a]	7/8	60	40	—	—	—	305	165	100	70	35
60/48[g]	1-1/8	60	48	—	—	—	305	165	100	70	35

APA Rated Sturd-I-Floor[f]											
16 oc	19/32, 5/8	24	24	185	100	65	40				
20 oc	19/32, 5/8	32	32	270	150	100	60	30			
24 oc	23/32, 3/4	48	36	–	240	160	100	50	30	25	
32 oc	7/8	48	40	–	–	295	185	100	60	40	
48 oc	1-3/32, 1-1/8	60	48	–	–	–	290	160	100	65	40

[a]Tongue-and-groove edges, panel edge clips (one midway* between each support, except two equally spaced between supports 48 inches on center or greater), lumber blocking, or other. For low slope roofs, see Table 30.

[b]19.2 inches for 3/8-inch and 7/16-inch panels. 24 inches for 15/32-inch and 1/2-inch panels.

[c]Includes APA rated sheathing/ceiling deck.

Courtesy APA – the Engineered Wood Association, www.apawood.org

[d]10 psf dead load assumed.

[e]Applies to panels 24 inches or wider applied over two or more spans.

[f]Also applies to C-C Plugged grade plywood.

[g]Check with supplier for availability.

*No established tolerance.

ROOF SLOPE AND PITCH

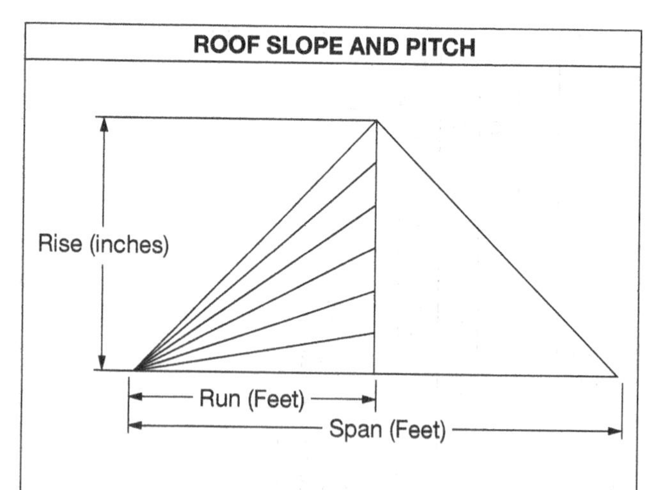

Slope	Pitch
12:12	½
10:12	5/12
8:12	⅓
6:12	¼
4:12	⅙
2:12	1/24

Slope is the rise in inches per foot of run

Pitch is the rise in inches per foot of span

INCLINE ANGLES FOR ROOF SLOPES IN TERMS OF RISE TO RUN AND THEIR EQUIVALENTS IN PERCENTS AND DEGREES			
Slope – Inches Rise to Inches Run	Percent Slope	Ratio-Rise to Run	Angles in Degrees
4:12	33	1:3	18.4
5:12	42	1:2.4	22.6
6:12	50	1:2	26.6
7:12	58	1:1.7	30.3
8:12	67	1:1.5	33.7
9:12	75	1:1.13	36.9
10:12	83	1:1.2	39.8
12:12	100	1:1	45.0
14:12	117	1:2:1	50.2
15:12	125	1:25:1	51.3
16:12	133	1:3:1	52.4
18:12	150	1:5:1	56.3
20:12	167	1:7:1	59.5
24:12	200	2:1	63.4

TYPICAL MINIMUM SLOPES AND UNDERLAYMENT RECOMMENDATIONS FOR STEEP SLOPED ROOFS

Material	Allowable Slope
Asphalt shingles	4:12 if one layer of asphalt-saturated felt underlayment is used. 2:12 if two layers of asphalt-saturated felt underlayment are used.
Clay and concrete tile	4:12 if interlocking tiles are used with one layer of 5 lb cap sheet underlayment. Less than 3:12 when noninterlocking tile are used with two layers of No. 40 asphalt-saturated felt underlayment.
Slate	4:12 with one layer of 30 lb asphalt-saturated felt underlayment. 2:12 with double underlayment.
Wood shingles and shakes	3:12 for shingles and 4:12 for shakes with 30 lb felt underlayment and interlayment with shakes or shingles.
Metal shingles and shakes	3:12 and 4:12 depending on the material and style. Requires 30 lb asphalt-saturated felt and underlayment.

*Consult local building codes and observe the manufacturer's recommendations on acceptable slope and installation details.

ALLOWABLE REROOFS OVER EXISTING ROOFING (Secure Approval of Local Building Official Before Proceeding.)

Existing Roofing	Allowable Reroofing
Built-up	Built-up, wood shingles, asphalt shingles, tile roofing, metal roofing, modified bitumen, polyurethane sprayed foam
Wood shingle	Wood shakes, wood shingles, asphalt shingles, tile roofing, metal roofing
Asphalt shingle	Wood shakes or shingles, asphalt shingles, tile roofs, metal roofs, modified bitumen
Tile roof	No approved reproofing
Metal roof	Metal roofing
Modified bitumen	Built-up, wood shingles, asphalt shingles, tile roofs, metal roofs, modified bitumen

ROOFING MATERIAL FIRE RATINGS*

Class A – Materials have the highest fire-resistance rating.

Class B – Materials will withstand moderate exposure from fire that develops outside the building.

Class C – Materials will withstand light exposure to fire that develops outside the building.

Nonclassified – Materials are untreated wood shingles considered to be combustible.

*Ratings are noted on the bundles of roofing materials.
Check local codes to verify fire ratings acceptable.

FIRE-RATED ROOFING MATERIALS

Brick, masonry, slate, clay, concrete tile are considered Class A.

Asphalt – Have Class A, Class B and Class C ratings, depending upon their construction.

Wood shingles and shakes are classified as unclassified unless treated with fire-retardant chemicals.

Metal roofing rated Class A if the sheathing is covered with a code-specified fire-resistant panel; otherwise is rated Class C.

FREQUENTLY USED BITUMEN MEMBRANE FLASHING

Self-adhered modified bitumen membrane flashing is set in place against the wall or other surfaces where flashing is required. Seal its joint with the roof membrane in the same manner as the laps between the roof membranes.

Hot-mopped modified bitumen membrane flashing is installed by applying hot asphalt to the area to be flashed and to the back of the flashing membrane and then setting it in place. Seal the joint with the roof membrane in the same manner as used to seal laps between the roof membranes.

Torch-applied modified bitumen membrane flashing is used where noncombustible substrates or other joining materials such as concrete walls are flashed. Prime the wall and area to be flashed. Heat the back of the flashing with a torch and set it in place. Seal the joints with the roof membrane in the same manner as laps between the roof membranes.

MATERIALS USED FOR FINISHED ROOFING

Material (kg/m^2)	Type of Roof	Descriptive Factors	Weight per Square Foot (lb/100 sq. ft)	Weight per Square Meter (kg/m^2)
Aluminum (sheet, shingles)	Steep-slope	Fire resistant, long life, range of colors	5–90	2.44–4.39
Asphalt (built-up)	Steep-slope, low-slope	Granular topping applied influences fire class, life 20–30 years	100–600	4.88–29.3
Asphalt shingles (fiberglass, asphalt)	Steep-slope	Fire resistance varies with product, range of colors, life 20–30 years	235–325	11.47–15.86
Cement-fiber tile	Steep-slope	Fire resistance, long life, Heavy, use in warm climates	950	46.4
Clay tile	Steep-slope	Fire resistant, long life, heavy	800–1,600	39–78
Copper (sheet)	Steep-slope low-slope	Fire resistant, long life, can be soldered	0.019" thick 160 0.040" thick 320	7.8 15.6
Lead, copper coated	Steep-slope, low-slope	Fire resistant, long life	$1/32$" thick 200 $1/16$" thick 400	9.76 19.52
Monel (Ni-Cu)	Steep-slope	Fire resistant, long life	22 gauge 1424 26 gauge 827	69.5 40.4
Perlite-portland cement	Steep-slope	High fire rating, lightweight, long life	900–1000	43.9–48.8
Plastic (single-ply Membrane)	Low-slope	Long life, requires careful installation, limited fire classification, several types available	Loose laid ballasted, 1000–1200 Full adhered, 30–55	48.8–58.6 1.5–2.7

MATERIALS USED FOR FINISHED ROOFING *(cont.)*

Material (kg/m^2)	Type of Roof	Descriptive Factors	Weight per Square Foot (lb/100 sq. ft)	Weight per Square Meter (kg/m^2)
Plastic (liquid applied)	Low-slope	Limited fire classification follow manufacturer's directions	20–50	0.98–2.4
Slate	Steep-slope	Fire resistant, heavy long life	¼" thick 900 ⅜" thick 1100 ½" thick 1700 ¾" thick, 2600	43.9 53.7 83.0 126.9
Stainless steel, terne coated	Steep-slope	Fire resistant, long life	90	3.89
Steel (sheet, shingles)	Steep-slope, low-slope	Fire resistant, long life, durable colors	Copper coated 130 Galvanized 130	6.3 6.3
Wood (shingles, shakes)	Steep-slope	No fire resistance unless treated, limited life	200–450	9.8–22.0
Zinc	Steep-slope	Fire resistant, long life, can be painted	9 gauge 670 12 gauge 1050	33.7 51.2
Terneplate copper-bearing sheet	Steep-slope	Fire resistant, long life	30 gauge 540 26 gauge 780	26.4 38.1
Mineral-surfaced cap sheet	Low-slope	Limited fire resistance, limited life	55–60	2.68–2.9
Modified bitumen	Low-slope	Fire resistance, 10 years	100	4.9

ORGANIC AND FIBERGLASS ASPHALT SHINGLES

Asphalt shingles are available made with an organic felt generally composed of wood pulp, waste paper, and waste rags. Organic felt shingles have a Class C rating while fiberglass type have a Class A rating. The fiberglass mat is more stable than the organic felt and has more asphalt used to coat it, so it is more durable and has a longer life than organic felt types.

TYPICAL ASPHALT ROLLS

Product	Approx. Shipping Weight per Roll (lbs.)	Approx. Shipping Weight per Square (lbs.)	Squares per pkg.	Width (in.)	Length (ft.)	Selvage (in.)	Exposure (in.)	ASTM Fire and Wind Ratings
Mineral surface roll	75–90	75–90	1	36–39½	32.7–38	2–4	32–34	Same Class C
Mineral surface roll (double coverage)	55–70	110–140	½	36–39½	32.7–36	19	17	Same Class C
Smooth surface roll	50–86	40–65	1–2	36–39¼	32.7–72	2–4	34–37¼	None
Non-perforated felt underlayment	24–60	6–30	2–8	36	72–288	2–19	17–34	May be a component in a complete fire-rated system. Check with manufacturer for details
Self-adhered eave and flashing membrance	35–82	33-40	1–2¼	36	36–75	2–6	34	May be a component in a complete fire-rated system. Check with manufacturer for details

Courtesy Asphalt Roofing Manufactures Association, www.asphaltroofing.com

TYPICAL ASPHALT SHINGLES

Product	Configuration	Approx. Shipping Weight per Square (pounds)	Shingles per Square	Bundles per Square	Width (inches)	Length (inches)	Exposure (inches)	ASTM Fire & Wine Ratings
Laminated self-sealing random tab shingle	Various edge, surface texture, and application treatments	240-360	64–90	3–5	11½–14¼	36–40	4–6⅛	Class A or C fire rating. Many wind resistant.
Multi-tab self-sealing square tab strip shingle	Various edge, surface texture, and application treatments	240–300	65–80	3–4	12–17	36–40	4–8	Class A or C fire rating. Many wind resistant.
Multi-tab self-sealing square tab strip shingle	Three-tab or four-tab	200–300	48–80	3–4	12–13¼	36–40	5–5⅝	Class A or C fire rating. Many wind resistant.
No-cutout self-sealing square tab strip shingle	Various edge, and surface texture treatments	200–300	65–81	3–4	12–13¼	36–40	5–5⅝	Class A or C fire rating. Many wind resistant.
Individual interlocking shingle (basic design)	Several design variations	180–250	72–120	3–4	18–22¼	20–22½	n/a	Class A or C fire rating. Many wind resistant.

Courtesy Asphalt Roofing Manufactures Association, www.asphaltroofing.com

ASPHALT SHINGLE ROOFING NAILS

1. Minimum shank diameter 12 gauge (0.105 inc.) wire.
2. Minimum head diameter ⅜ in.
3. Smooth shank. Some nails with barbs below the head are acceptable.
4. Corrosion resistance – galvanized steel or aluminum.

TYPICAL NAIL LENGTHS USED WITH ASPHALT SHINGLES*

Application	Length (in.)
Roll roofing on new sheathing	1
Individual or strip asphalt shingles on new sheathing	1¼
Reroof over old asphalt shingles	1½ to 1¾
Reroof over old wood shingles*	1¾

*Nails should penetrate at least ¾ inch into the sheathing. On sheathing less than ¾ inch thick, the nail should penetrate it ⅛ inch.

STAPLE RECOMMENDATIONS FOR INSTALLING ASPHALT SHINGLES*

1. Must have a 1 inch crown.
2. Legs should be long enough to penetrate ¾ inch into the sheathing.
3. Use 6 staples per strip shingle.

*Consult the local codes to see if staples are an approved fastener.

TYPICAL STAPLE SIZES FOR INSTALLING ASPHALT SHINGLES

Wood Sheathing Thickness	Minimum Staple Leg Length	Staple Crown Width
⅜"	⅞"	⅞"
½"	1"	1"
⅝" and thicker	1¼"–1½"	1"

MINIMUM PITCH AND SLOPE REQUIREMENTS FOR VARIOUS ASPHALT ROOFING PRODUCTS

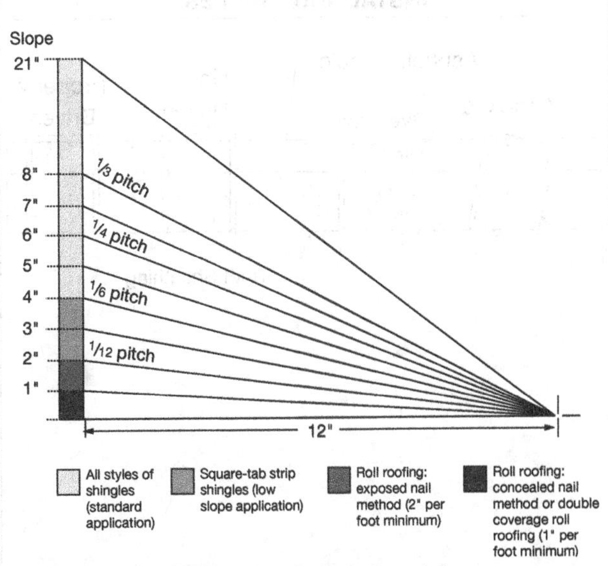

Courtesy Asphalt Roofing Manufactures Association, www.asphaltroofing.com

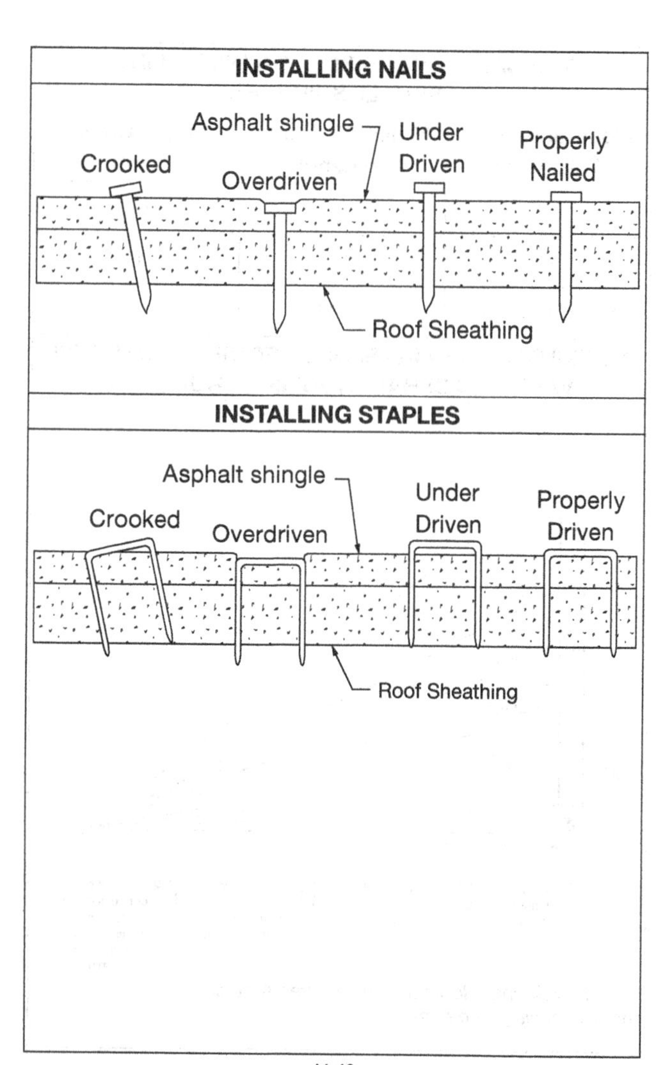
INSTALLING NAILS
Asphalt shingle
Crooked
Overdriven
Under
Driven
Properly
Nailed
Roof Sheathing
INSTALLING STAPLES
Asphalt shingle
Crooked
Overdriven
Under
Driven
Properly
Driven
Roof Sheathing

WOOD SHAKES AND SHINGLES

Grades of Red Cedar Shingles

Number 1 Grade (Blue Label)

Clear heartwood, 100 percent edge grain, no defects. Thickness of a 16-inch shingle is $\frac{5}{2}$, which means the thickness of five shingles at the butt end is 2 inches. Thickness of 18-inch shingle is $\frac{1}{4}$", and 24-inch shingle is $\frac{4}{2}$. Used for roofs 3:12 or greater. Best grade.

Number 2 Grade (Red Label)

Limited sapwood and flat grain are allowed. Free of knots and checks in lower two-thirds from butt.

Number 3 Grade (Black Label)

Utility or economy grade. Unlimited sapwood and flat grain allowed. Clear only in the lower third from the butt.

Undercoursing Grade (Green Label)

Unlimited defects, sapwood and flat grain allowed. Used for underlaying starter course shingles at the eaves where they are not exposed to the weather.

Grades of Red Cedar Shakes

Number 1 Handsplit and Resawn

Has a handsplit face and a smooth sawn back side.

Number 1 Tapersplit

Both faces are handsplit but from opposite ends, producing a tapered shake.

Number 1 Tapersawn

Both faces are sawn and are smooth.

WOOD SHAKES AND SHINGLES *(cont.)*

Number 1 Straightsplit

Both faces are split but from the same end of the shake, producing a shake of uniform thickness from end to end.

Number 1 Machine-Grooved

They are shingles that have the faces machine-grooved, producing striated faces and parallel edges.

WOOD SHAKE WEATHER EXPOSURE AND ROOF SLOPE

Roofing Material	Length (inches)	Grade	Exposure (inches) 4:12 Pitch or Steeper
Shakes of naturally durable wood	18	No. 1	7.5
	24	No. 1	10
Preservative-treated tapersawn shakes of Southern yellow pine	18	No. 1	7.5
	24	No. 1	10
	18	No. 2	5.5
	24	No. 2	7.5
Taper sawn shakes of naturally durable wood	18	No. 1	7.5
	24	No. 1	10
	18	No. 2	5.5
	24	No. 2	7.5

WOOD SHINGLE WEATHER EXPOSURE AND ROOF SLOPE				
			Exposure (inches)	
Roofing Material	Length (inches)	Grade	3:12 Pitch to < 4:12	4:12 Pitch or Steeper
Shingles of naturally durable wood	16	No. 1	3.75	5
		No. 2	3.5	4
		No. 3	3	3.5
	18	No. 1	4.25	5.5
		No. 2	4	4.5
		No. 3	3.5	4
	24	No. 1	5.75	7.5
		No. 2	5.5	6.5
		No. 3	5	5.5

NAIL RECOMMENDATIONS FOR WOOD SHINGLES AND SHAKES FOR NEW ROOFS		
Wood Shingles	**Nail Type and Gauge**	**Min. Length (in.)**
16" and 18" shingles	3d box	1¼
24" shingles	4d box	1½
Wood Shakes		
18" straightsplit	5d box	1¾
18" and 24" hard-split and resawn	6d box	2
24" tapersplit	5d box	1¾
18" and 24" tapersawn	6d box	2

WOOD SHINGLE COVERAGE

Length and Thickness	Approximate Coverage of One Square (4 bundles) of Shingles Based on Following Weather Exposures								
	3½"	4"	4½"	5"	5½"	6"	6½"	7"	7½"
16" × ⅖"	70	80	90	100*					
18" × ⁵⁄₂ ¼"		72½	81½	90½	100*				
24" × ½"					73½	80	86½	93	100*

NOTE: *Maximum exposure recommended for roofs.

WOOD SHAKE EXPOSURE

Slope	Maximum Exposure Recommended for Roofs	
	Length	
	18"	24"
4:12 and steeper	7½"	10"[a]

[a]24" × ⅜" handsplit shakes limited to 7½" maximum weather exposure per UBC and IBC.

WOOD SHINGLE EXPOSURE

Slope	Maximum Exposure Recommended for Roofs Length								
	Number 1 Blue Label			Number 2 Red Label			Number 3 Black Label		
	16"	18"	24"	16"	18"	24"	16"	18"	24"
3:12 to 4:12	3¾"	4¼"	5¾"	3½"	4"	5½"	3"	3½"	5"
4:12 and steeper	5"	5½"	7½"	4"	4½"	6½"	3½"	4"	5½"

WOOD SHAKE COVERAGE

Shake Type, Length and Thickness	Approximate Coverage (in sq. ft.) of One Square, When Shakes are Applied with an Average ½" Spacing, at Following Weather Exposures, in Inches[d]				
	5	5½	7½	8½	10
18" × ½" Handsplit-and-Resawn Mediums[a*]	—	75[b,f]	100[f,c]	—	—
18" × ¾" Handsplit-and-Resawn Heavies[a*]	—	55[b,f]	75[f,c]	—	—
18" × ⅝" Tapersawn*	—	Approx. 75[b]	100[f,c]	—	—
24" × ⅜" Handsplit	—	—	75[e]	—	—
24" × ½" Handsplit-and-Resawn Mediums	—	—	75[b]	85	100[c]
24" × ¾" Handsplit-and-Resawn Heavies	—	—	75[b]	85	100[c]
24" × ⅝" Tapersawn	—	—	75[b]	85	100[c]
24" × ½" Tapersplit	—	—	75[b]	85	100[c]
18" × ⅜" Straight-Split	—	65[b]	90[c]	—	—
24" × ⅜" Straight-Split	—	—	75[b]	85	100[c]
15" Starter-Finish course	Use supplementary with shakes applied not over 10" weather exposure.				

[a]5 bundles will cover 100 sq. ft. roof area when used as starter-finish course at 10" weather exposure; 7 bundles will cover 100 sq. ft. roof area at 7½" weather exposure; see footnote (d).

[b]Maximum recommended weather exposure for 3-ply roof construction.

[c]Maximum recommended weather exposure for 2-ply roof construction.

[d]All coverage based on an average ½" spacing between shakes.

[e]Maximum recommended weather exposure.

*100 sq. ft. coverage is based on 12/12 pack, 5 bundle square, at 7½" exposure.

[f]Note: While most shakes are packed in bundles of 12 courses each side (12/12) they may be packed 9/9. This will after the number of bundles required to cover 1 square.

For example: 18" shake bundles 12/12 should cover 100 square feet at 7½" exposure. 9/9 pack should give 75% coverage of a square. When ordering check with your supplier to confirm bundle size.

RECOMMENDED FASTENERS	
Type of Certi-label Shake or Shingle	**Nail Type and Minimum Length**
Certi-Split & Certi-Sawn Shakes	**Type (in)**
18" Straight-Split	5d Box 1¾
18" and 24" Handsplit-and-Resawn	6d Box 2
24" Tapersplit	5d Box 1¾
18" and 24" Tapersawn	6d Box 2
Certigrade Shingles	**Type (in)**
16" and 18" Shingles	3d Box 1¼
24" shingles	4d Box 1½

BUILT-UP AND MODIFIED BITUMEN ROOFING

Built-up Roofing

It is a multilayer roofing system consisting of layers of reinforcing felt coated with a bitumen. The number of layers can vary. Typically, a top coat of gravel is laid in the last coat of hot asphalt. It has a life expectancy of 15 to 20 years.

Hot-Mopped Modified Bitumen

This is a variation of the built-up roof that consists of fabric-reinforced sheets formulated with a modified bitumen. It can be installed as a single-ply roofing but more likely will be in layers similar to built-up roofing.

SINGLE-PLY MEMBRANES

Roll Roofing

This is single-ply membrane using an asphalt-impregnated organic or fiberglass mat and a mineral aggregate surfacing. It may be installed as a double coverage roofing.

EPDM (ethylene propylene)

This is a single-layer, synthetic, rubber roofing membrane. It is available in large sheets ranging from 50 × 50 feet to 50 × 200 feet and is available in several thicknesses. It can be adhered or mechanically installed to the sheathing. It is installed over foam board that is mechanically secured to the sheathing. It is usually warranted for 15 to 20 years.

TPO (thermoplastic polyolefin)

This membrane has a high rubber content. It is available in rolls 75 inches and 8, 10, and 12 feet wide and 100 feet long and 45- and 60-mil thicknesses. It is available in black and white.

PVC (polyvinyl chloride)

This membrane is a total plastic sheet and not a synthetic rubber. It does not last as long as synthetic rubber membranes. It is in rolls 75 inches wide and 100 feet long. It is available in gray, white, and several colors.

CSPF (chlorosulfonel polyethylene)

This is a thermoplastic membrane that becomes a thermoset material as it ages. It uses a polyester fiber scrim as a reinforcement. It is available in a variety of colors. Sheets 5 feet wide and 100 feet long are typically available.

TYPICALLY RECOMMENDED INCLINES FOR VARIOUS TYPES OF ASPHALT	
	Incline (inches per foot)
Type 1	1.0
Type 2	0.5 to 3.0
Type 3	0.5 to 6.0
Type 4	*

*Requires special considerations.

TYPES AND CLASSES OF MODIFIED BITUMEN MATERIALS	
Types	
Type 1	For exposed roofing applications 1a. Fully adhered 1b. Partially adhered
Type 2	For covered roofing application 2a. Fully adhered 2b. Partially attached 2c. Loose-laid
Classes	
Class A	Granule surfaced
Class B	Metallic surfaced
Class C	Plain surfaced
Grades	
Grade 1	Standard service
Grade 2	Heavy-duty service

Courtesy National Roofing Contractors Association, www.nrca.net

TYPES OF ASPHALT AND THEIR MAXIMUM HEATING TEMPERATURES		
ASTM D-312 Type No.	**Kind of Asphalt**	**Maximum Heating Temperature**
Type I	Dead Level Asphalt	475F
Type II	Flat Asphalt	500F
Type III	Steep Asphalt	525F
Type IV	Special Steep Asphalt	525F

Courtesy ASTM International. Reprinted with permission, www.astm.org

ASPHALT FOR HOT BUILT-UP ROOFING

	Softening Point		Heating Temperature	
	Degrees F	Degrees C	Degrees F	Degrees C
Type 1 (dead level)	135 to 150	52 to 65	475	232
Type 2 (flat)	160 to 175	71 to 79	500	260
Type 3 (steep)	180 to 200	82 to 93	525	260
Type 4 (special steep)	205 to 225	95 to 107	525	260

F. Fahrenheit C. Celsius

COLD APPLIED BITUMENS

1. Asphalt mastic consists of asphalt and carefully graded mineral aggregate. It is troweled to its final finish.
2. Bituminuous grout consists of a bitumen and fine sand. It is troweled to its final finish.
3. Asphalt emulsions consist of fine drops of water distributed in asphalt with an emulsifier. They can be spray applied.
4. Cutback bitumens are those thinned with light oil and organic solvents so they flow easily. They are usually used between plies and as a top flood coat into which aggregate is embedded.

TYPES OF SINGLE-PLY MEMBRANES

Chlorinated Polyethylene (CPE)

Chlorosulfonated Polyethylene (CSPE)

Ethylene Propylene Dienne Monomer (EPDM)

Hypalon Chlorsulfonated Polyethylene (CSPE)

Modified Bitumen

Polyisobutylene (PIB)

Polychloroprene

Nitrile Alloys (NBP)

Polyvinyl Chloride (PVC)

POLYVINYL CHOLORIDE SINGLE-PLY ROOFING CLASSES

Type I	Unreinforced Sheet
Type II	Class 1. Unreinforced Sheet Containing Fibers Class 2. Unreinforced Sheet Containing Fabrics
Type III	Reinforced Sheet Containing Fibers or Fabrics

Courtesy ASTM International. Reprinted with permission, www.astm.org

SLATE ROOFING

SLATE SHINGLE DATA

Face Dimensions[3] in Inches	Minimum Number to Square (3" lap)	Weight per Square	Nails per Square 3d Galvd lb–oz
12 by 6	533	800	5–4
12 by 7	457	800	4–8
12 by 8	400	800	3–15
12 by 10	320	800	3–3
14 by 7	374	780	3–10
14 by 8	327	780	3–3
14 by 10	261	780	2–9
14 by 12	218	780	2–2
16 by 8	277	750	2–12
16 by 9	246	750	2–7
16 by 10	221	750	2–3
16 by 12	185	750	1–13
18 by 9	213	735	2–1
18 by 10	192	735	1–14
18 by 12	160	735	1–9
20 by 10	169	725	
20 by 12	141	725	1–11
20 by 14	121	725	1–7
22 by 11	138	710	1–6
22 by 12	126	710	1–4
24 by 12	115	700	1–2
24 by 14	98	700	1–0

Courtesy Hilltop Slate, Inc. www.hilltopslate.com

EXPOSURE IN INCHES FOR SLOPING SLATE ROOFS	
Length of Slate in Inches	**Slope 8" to 20" per Foot, 3" lap**
24	10½
22	9½
20	8½
18	7½
16	6½
14	5½
12	4½
10	3½

Courtesy Hilltop Slate, Inc., www.hilltopslate.com

SLATE SHINGLE HEADLAP	
Slope	**Headlap (inches)**
4:12 < slope < 8:12	4
8:12 < slope < 20:12	3
Slope > 20:12	2

For SI: 1 inch = 25.4 mm.

2006 International Building Code, copyright 2006, Washington, DC. International Code Council. Reproduced with permission. www.iccsafe.org

SPACING FOR WOOD LATH WHEN SLATE IS LAID WITH A 3-INCH LAP			
Length of Slate (inches)	**Spacing of Lath (inches)**	**Length of Slate (inches)**	**Spacing of Lath (inches)**
24	10½	16	6½
22	9½	14	5½
20	8½	12	4½
18	7½		

Courtesy Hilltop Slate Inc. www.hilltopslate.com

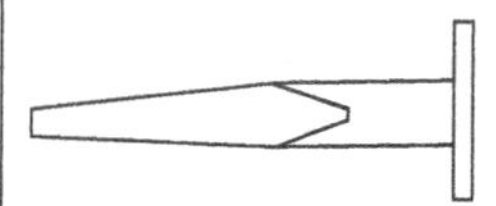

Flathead Cut Copper Slating Nail

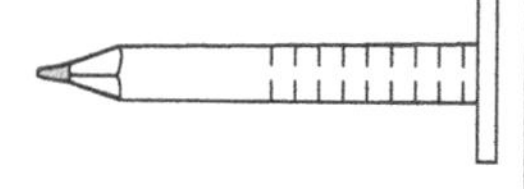

Flathead Copper Wire Slating Nail

COPPER WIRE SLATING NAILS		
Length	**Gage**	**No. per lb***
⅞	12	303
1	12	270
1¼	10	144
1¼	11	196
1¼	12	231
1½	10	134
1½	12	210
1¾	10	112
2	10	104
2½	8	46

*Add 5% to above for Brass Nails.

CUT COPPER SLATING NAILS				
Length	Weight	No. per lb	Length	Weight
1	2d		2½	8d
1¼	3d	190	2¾	9d
1½	4d	135	3	10d
1¾	5d	100	3¼	12d
2	6d			
2¼	7d			

CUT BRASS NAILS			
		No. per lb	
Length	Weight	Slating	Roofing
1⅛		172	
1¼	3d	164	
1⅜		144	
1½	4d	140	216
1¾	5d	108	172
2	6d	88	132
2¼	7d	80	
2½	8d	64	112
2¾	9d	52	
3	10d	48	75
3½	16d		66

Courtesy Hilltop Slate Inc. www.hilltopslate.com

CLAY AND CONCRETE ROOF TILE

UNDERLAYMENT FOR CLAY AND CONCRETE TILE ROOFING*

Roof Slope	Underlayment
2½:12 to 4:12	Minimum of two layers parallel with the eave and secured to the sheathing. Overlap sheets 19 inches.
4½:Greater	Minimum of one layer laid parallel with the eave. Overlap sheets 2 inches.

*Typically roofing felt ASTM 226. Type II or mineral-surfaced roll roofing ASTM D2626 Class M are required. Check local codes for specific requirements.

CLAY AND CONCRETE TILE NAILS

1. Corrosion resistant
2. 10d or larger
3. Head minimum diameter 5⁄16 in (8.0 mm)
4. Length – Must penetrate sheathing at least ¾ in. (19.1 mm)

SOME TYPICAL CONCRETE ROOF TILES*

Type of Tile	Minimum Slope	Tile Size Length × Width (inches)	Exposure Length × Width	Approx. No. of Tiles per Square	Weight lb/sq.
Flat	4:12	17 × 12⅜	14 × 11⅜	89	1050
		11 × 9	8 × 8	220	1025
Low Profile	4:12	16 × 12¾	13 × 10¾	95	950
High Profile	4:12	17 × 12¾	14 × 10¾	92	950

*There are other sizes and designs available. Consult a manufacturer for specific details.

SOME TYPICAL CLAY ROOF TILES*

Type of Tile	Minimum Slope	Tile Size Length × Width (inches)	Exposure Length × Width	Approx. No. of Tiles per Square	Weight lb/sq.
English	3:12	15 × 15	6 × 5	468	2005
		15 × 16	6 × 6	390	2005
		15 × 6.5	6 × 6.5	360	2005
		15 × 7.5	6 × 7½	315	2005
Flat	4:12	12 × 7	5 × 7	412	1840
		15 × 7	6½ × 7	316	1600
		18 × 8	7½ × 8	240	1200
Pan and cover	4:12	16¾ × 9¼	14 × 13¾	75	1480
		19 × 10	16 × 7½	120	1010
		18 × 12	14 × 9½	106	1040
Interlocking	3:12	11 × 8¾	8 × 8	225	900
		14 × 9	11 × 8¼	158	900
		16½ × 13	13½ × 11¾	90	900
Spanish or S-tile	4:12	11 × 8¾	8 × 8	225	900
		14 × 9	11 × 8¼	160	800
		16½ × 13	13½ × 11¾	90	900
		17½ × 13	14 × 12	85	1000

*There are other sizes and designs available. Consult a manufacturer for details.

CLAY AND CONCRETE TILE ATTACHMENT[a,b,c]

GENERAL – CLAY OR CONCRETE ROOF TILE			
Maximum Basic Wind Speed (mph)	Mean Roof Height (feet)	Roof Slope up to < 3:12	Roof Slope 3:12 and Over
85	0-60	One fastener per tile. Flat tile without vertical laps, two fasteners per tile.	Two fasteners per tile. Only one fastener on slopes of 7:12 and less for tiles with installed weight exceeding 7.5 lb/sq. ft. having a width no greater than 16 inches.
100	0:40		
100	>40–60	The head of all tiles shall be nailed. The nose of all eave tiles shall be fastened with approved clips. All rake tiles shall be nailed with two nails. The nose of all ridge, hip, and rake tiles shall be set in a bead of roofer's mastic.	
110	0–60	The fastening system shall resist the wind forces in Section 1609.5.2.	
120	0–60	The fastening system shall resist the wind forces in Section 1609.5.2.	
130	0–60	The fastening system shall resist the wind forces in Section 1609.5.2.	
All	>60	The fastening system shall resist the wind forces in Section 1609.5.2.	

[a]Minimum fastener size. Corrosion resistant nails not less than No. 11 gage with 5/16-inch head. Fasteners should be long enough to penetrate into the sheathing 0.75 inches or through the thickness of the sheathing, whichever is less. Attaching wire for clay and concrete tile shall not be smaller than 0.082 inch.

[b]Snow areas. A minimum of two fasteners per tile are required or battens and one fastener.

[c]Roof slopes greater than 24:12. The nose of all tiles shall be securely fastened.

CLAY AND CONCRETE TILE ATTACHMENT* *(cont.)*

Sheathing	Roof Slope	Number of Fasteners
Solid without battens	All	One per tile
Spaced or solid with battens and slope < 5:12	Fasteners not required	—
Spaced sheathing without battens	5:12 < slope < 12:12	One per tile/ every other row
	12:12 < slope < 24:12	One per tile

*Attach as specified by manufacturer in areas where winds can exceed 100 miles per hour. In areas subject to snow, use two fasteners per tile.

2006 International Residential Code, copyright 2006, Washington, DC. International Code Council. Reproduced with permission. www.iccsafe.org

METAL ROOFING

1. Structural. Has load-carrying properties and does not require a substructure to support it. Often widely spaced wood strips are used to provide some support.
2. Architectural. Must be supported by a substructure, typically plywood; however, closely spaced wood strips are also used.
3. Panels may be copper, aluminum, zinc, stainless steel, weathering steel, galvanized steel, granular-coated steel, or terne metal.

METAL ROOFING *(cont.)*

4. Typical metal roof tile panels are 16 to 18 inches wide and 42 to 46 inches long.
5. Typical standing seam panels are 10, 11, 12, 16 and 18 inches wide and 4 to 50 feet long. Consult the manufacturer for specific data.

INSTALLING METAL ROOF SHINGLES

1. Must be applied over solid sheathing or closely butted deck material.
2. Minimum slope 3:12.
3. Material and fasteners must be corrosion resistant.

INSTALLING METAL ROOF PANELS

1. Must be applied over solid sheathing or closely fitted deck material. Some panels are designed to be installed over spaced decking.
2. Minimum slope 3:12 for lapped nonsoldered panels.
3. Minimum slope for lapped nonsoldered seams that have a lap sealant applied is ½:12.
4. Minimum slope for standing seam roof panels is ¼:12.
5. Panels and fasteners must be corrosion resistant.
6. Steel panel roofing is installed with galvanized fasteners. Copper fasteners are used for stainless steel and all other types of metal roofing.

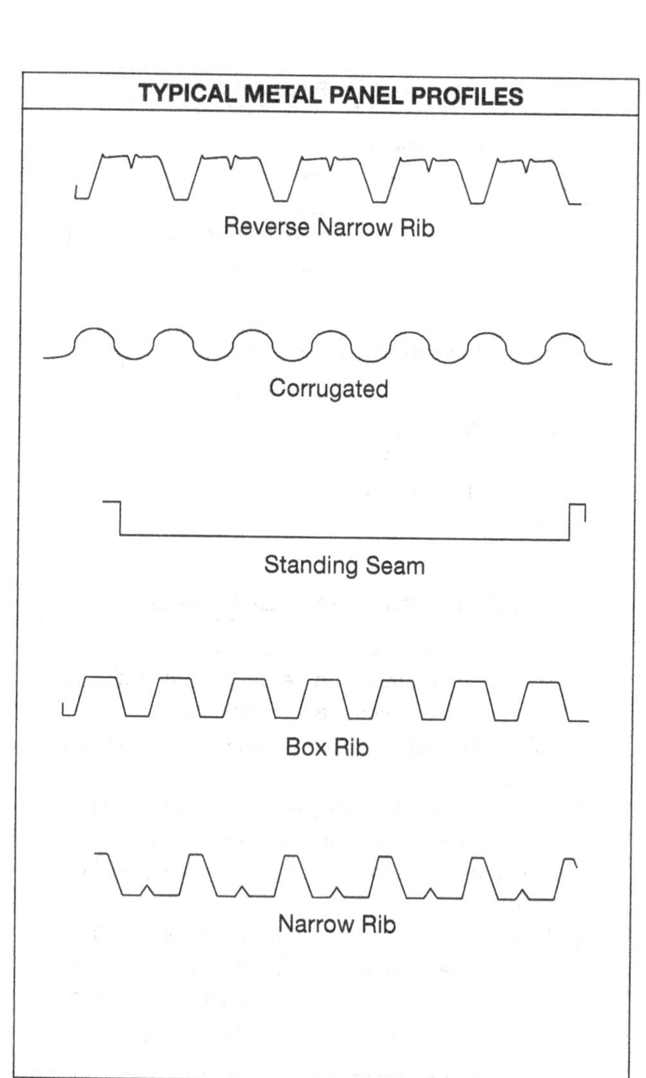
TYPICAL METAL PANEL PROFILES
Reverse Narrow Rib
Corrugated
Standing Seam
Box Rib
Narrow Rib

METAL GAUGES, THICKNESSES, AND WEIGHT PER FOOT

Metal	Gauge	Nominal Thickness (in)	Approximate Pound per Square Foot (lb/sq. ft.*)	Nominal Thickness (mm)	Approximate Kilogram per Square Meter (kg/m^{2}*)
Aluminum		0.024	0.35	0.64	1.72
		0.032	0.45	0.81	2.20
		0.040	0.57	1.02	2.75
		0.050	0.70	1.27	3.50
		0.063	0.89	1.63	4.40
Copper					
16 oz		0.022	1.00	0.56	4.87
20 oz		0.027	1.25	0.69	6.10
24 oz		0.032	1.50	0.81	7.31
32 oz		0.043	2.00	1.09	9.77
Lead sheets					
4 lb		0.062	4.00	1.57	19.53
3 lb		0.047	3.00	1.20	14.65
2½ lb		0.039	2.50	0.99	12.21
2 lb		0.031	2.00	0.79	9.77
Lead-coated copper sheets					
16 oz		0.026	1.07	0.66	5.21
20 oz		0.031	1.31	0.79	6.39
24 oz		0.036	1.54	0.91	7.51
32 oz		0.047	2.03	1.19	9.95

METAL GAUGES, THICKNESSES, AND WEIGHT PER FOOT *(cont.)*

Metal	Gauge	Nominal Thickness (in)	Approximate Pound per Square Foot (lb/sq. ft.*)	Nominal Thickness (mm)	Approximate Kilogram per Square Meter (kg/m²*)
Stainless steel	28	0.015	0.63	0.38	3.19
	26	0.018	0.79	0.46	3.84
	24	0.024	1.05	0.61	5.12
Steel*:	28	0.015	0.63	0.38	3.06
Galvanized Steel	26	0.019	0.91	0.48	4.42
Galvalume*	24	0.025	1.16	0.64	5.64
Aluminized Steel	22	0.031	1.41	0.79	6.85
	20	0.038	1.66	0.97	8.08
	18	0.050	2.16	1.27	10.51
	16	0.063	2.64	1.60	12.83
Terne	26	0.018	0.80**	0.46	3.91***
	28	0.015	0.67**	0.38	3.27***
	30	0.012	0.54**	0.30	2.63***
Terne-coated stainless steel (TCS)	28	0.015	0.66	0.38	3.19
	26	0.018	0.79	0.46	3.84
	24	0.024	1.05	0.61	5.12
Zinc	24	0.020	0.75	0.51	3.66
	21	0.027	1.00	0.69	4.87

*U.S. Standard Gauge

**40-lb coating weight

***88.2-kg coating weight

Courtesy National Roofing Contractors Association, www.nrca.net

COEFFICIENT OF THERMAL EXPANSION AND INCREASE IN LENGTHS OF METAL ROOFING PANELS		
Metal Type	**Coefficient of Thermal Expansion**	**Increase in 10 Foot Lengths per 100°F Temperature Change**
Galvanized Steel	0.0000067 in./in.°F	.080 in.
Steel	0.0000067 in./in.°F	.080 in.
Terne	0.0000067 in./in.°F	.080 in.
Copper	0.0000094 in./in.°F	.113 in.
Stainless Steel	0.0000096 in./in.°F	.115 in.
Aluminum	0.0000129 in./in.°F	.155 in.
Lead	0.0000161 in./in.°F	.193 in.
Zinc	0.0000174 in./in.°F	.209 in.

[Au5]

Courtesy National Roofing Contractors Association. www.nrca.net

SOLDERABLE AND WELDABLE METALS BASED ON THICKNESSES		
Metal Type	**Solderable**	**Weldable**
Aluminum, 0.063" (1.6 mm)	No	Yes
Aluminized steel	No	No
Copper, 16 oz (4.9 kg/m^2)	Yes	No
Galvalume®	No	No
Galvanized steel, 22 ga. (0.031 inch [0.79 mm])	Yes	Yes
Galvanized steel, 26 ga. (0.019 inch [0.48 mm])	Yes	No
Lead, 2 lb (9.8 kg/m^2)	Yes	No
Lead-coated copper, 16 oz (5.2 kg/m^2)	Yes	No
Stainless steel, 22 ga. (0.029 inch [0.74 mm])	Yes	Yes
Stainless steel, 28 ga. (0.015 inch [0.38 mm])	Yes	No
Terne-coated stainless steel, 0.015 inch (0.38 mm)	Yes	No
Zinc, 0.027 inch (0.7 mm)	Yes	No
Courtesy National Roofing Contractors Association. www.nrca.net		

GALVANIC CORROSION

Both ferrous and nonferrous metals are subject to galvanic corrosion. In the presence of an electrolyte (moisture in the air), dissimilar metals in contact will corrode more rapidly than similar metals in contact. For example, aluminum roofing secured with copper nails produces galvanic action at the point of contact. The metal that is higher on the Table of Electrolytic Corrosion Potential will be sacrificed. The types of fasteners used with the metal to be secured should be of the same metal or one close to it on the table.

THE GALVANIC SERIES

Metal	
Aluminum	Highest resistance to galvanic action
Zinc	
Steel	▲
Tin	
Lead	
Brass	▼
Copper	
Bronze	Lowest resistance to galvanic action

ROOF VENTILATION

The most efficient attic ventilation system uses continuous vents in the soffit and ridge vents along the ridge of the roof.

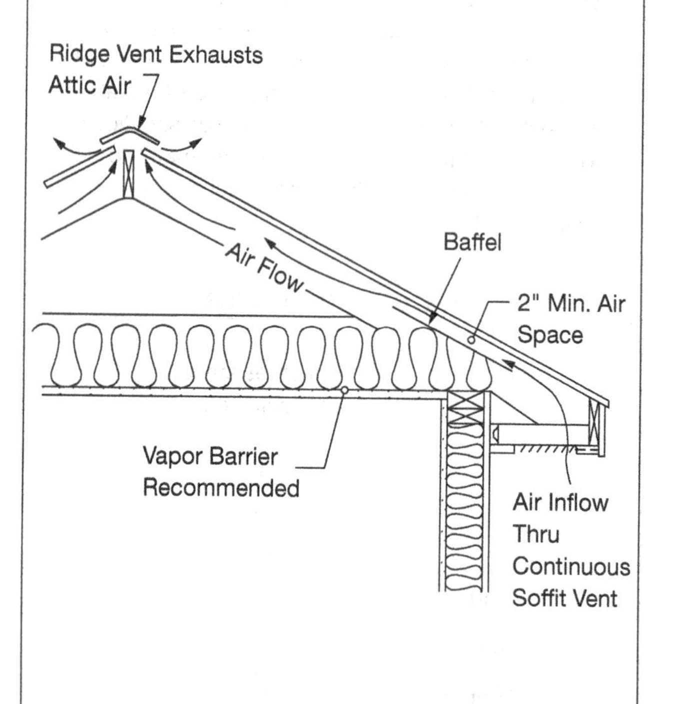

AMOUNT OF ATTIC VENTILATION NEEDED

If the ceiling has a vapor barrier that has a transmission rate not exceeding 1 perm on the warm side of the ceiling, the net free ventilation area can be one square foot of vent for every 300 square feet of ceiling (1 in 300). If there is no vapor barrier, the ratio is 1 in 150.

The net free ventilation area should be divided equally between air inflow at the soffit and the ridge vents. The amount of net free ventilation area is increased depending upon the screen used to keep out insects.

NET FREE VENTILATION ADJUSTMENTS FOR SCREENS

Type of Vent	Increase Needed in Vent Area
¼-inch screen mesh	10%
⅛-inch screen mesh	30%
No. 16 screen mesh	200%

Vent manufacturers cite the net free ventilation area of their products, so check the brand you plan to buy.

CHAPTER 12
Drywall and Plaster

TYPES OF DRYWALL JOINT COMPOUNDS
Taping Compounds. A premixed compound used for the first coat over nails, screws, and to bond the tape to the panel. **Topping Compounds.** A premixed compound used for the second and final finish coat over the taping compound and tape. **All-purpose Compounds.** A premixed compound that can be used for all three coats. It is available in several formulations and used for texturing. **Quick-set Compounds.** They cure rapidly and can be recoated in 30 to 300 minutes depending upon the formulation chosen. They are in powder form and mixed with water on the job.

DRYWALL REINFORCING TAPES
Paper Tape. Typically about 2" wide with a crease down the center to help fold it around corners. It has tiny holes to help the compound move through it as it is laid in the joint compound. **Fiberglass Tape.** Typically 2" wide and is moisture resistant. The plain type (nonadhesive) is stapled to the drywall. The open mesh type has an adhesive on the back and is pressed in place on the drywall panel.

TYPICAL FLAME-SPREAD REQUIREMENTS FOR INTERIOR WALL AND CEIING CONSTRUCTION*			
Use	**Enclosed Vertical Exitways**	**Other Exitways**	**Rooms or Areas**
Assembly buildings	1	2	3
Storage and sales areas for combustible goods	1	2	3
Restaurants less than 50 occupants/ retail gasoline stations	1	2	3
Schools	1	2	3
Factories, warehouses not using highly flammable material	1	2	3
Hospital, nursing home	1	2	2
Hotels, apartments	1	2	3
Residential, single and multi-family	3	3	3
Storage, handling or sale of highly flammable or explosive materials	1	2	3
Auto repair garages	1	2	3
Nurseries	1	2	2
Private garages, carports shed, agricultural buildings	No restrictions		
*See local codes for specific requirements			

TYPES OF GYPSUM PANELS			
Type	Thickness (inches)	Width (feet)	Length (feet)
Regular	1/4, 1/2, 3/8, 5/8	4	6-16
Type X (fire-resistant)	1/2, 5/8	4	6-16
Moisture Resistant	1/2 Reg., 5/8 Type X	4	6-16
Flexible	1/4	4	8 and 10
High-strength Ceiling	1/2	4	6-16
Predecorated	1/2	4	8, 9, and 10
Sound Deadening	1/4	4	8
Sheathing	1/2, 5/8	4	8, 9, 10
Soffit	1/2, 5/8	4	8-12
Paperless	1/2, 5/8	4	8
Sheathing	1/2, 5/8, 1	2, 4	8 to 16
Floor Underlayment	1/4, 5/16, 3/8	4	4
Vinyl-Faced Panels	1/2	4	8, 9, 10
Lay-in Ceiling Tile	1/2	2	2, 4

SIZES OF GYPSUM PANELS

Panels	Thickness (inches)	Length (feet)	Weight per Square Foot
Standard	1/4	8, 10	1.2
	3/8	8, 9, 10, 12, 14	1.4
	1/2	8, 9, 10, 12, 14	1.7
Water-resistant	1/2	8, 10, 12	1.8
	5/8	8, 10, 12	2.2
Interior Ceiling	1/2	8, 12	1.6
1/4" Flexible	1/4	8, 10	1.2
Type X Fire-resistant	1/2	8, 9, 10, 12, 14	1.7
	5/8	8, 9, 10, 12, 14	1.8

GYPSUM BOARD MATERIALS AND ACCESSORIES

Materials	Standard*
Gypsum Sheathing	ASTM C79
Gypsum Wallboard	ASTM C36
Joint Reinforcing Tape and Compound	ASTM C474; C475
Nails for Gypsum Boards	ASTM C514
Steel Screws	ASTM C1002; C954
Water-resistant Gypsum Backing Board	ASTM C630

Courtesy American Society for Testing and Materials, www.astm.org.

MINIMUM BENDING RADII FOR 1/4" HIGH FLEX GYPSUM WALLBOARD				
	Lengthwise		**Widthwise**	
Application	**Bend RADII**	**Max. Stud Spacing**	**Bend RADII**	**Max. Stud Spacing**
Inside (concave) Dry	32"	9" o.c.	20"	9" o.c.
Outside (convex) Dry	30"	9" o.c.	15"	8" o.c.
Inside (concave) Wet	20"	9" o.c.	10"	6" o.c.
Outside (convex) Wet	14"	6" o.c.	7"	5" o.c.

Courtesy National Gypsum Corporation, www.nationalgypsum.com.

MAXIMUM FRAMING SPACING FOR SINGLE-LAYER GYPSUM PANEL PRODUCT		
Gypsum Panel Product Thickness Inches (mm)	**Gypsum Panel Product Orientation to Framing**	**Maximum Framing Spacing Inches (mm) O.C.**
CEILINGS		
3/8 (9.5)[A]	Perpendicular[B]	16 (406)
1/2 (12.7)	Parallel[B]	16 (406)
1/2 (12.7)	Perpendicular[B]	24 (610)
5/8 (15.9)	Parallel	16 (406)
5/8 (15.9)	Perpendicular	24 (610)
WALLS		
3/8 (9.5)	Perpendicular or Parallel	16 (406)
1/2 (12.7)	Perpendicular or Parallel	24 (610)
5/8 (15.9)	Perpendicular or Parallel	24 (610)

[A] Shall not support thermal insulation.

[B] On ceiling to receive hand or spray-applied water-based texture material either i) 1/2" (12.7 mm) gypsum ceiling board (ASTM C 1396/C 1396M) shall be applied perpendicular to framing; or ii) other gypsum panel products shall be applied perpendicular to framing and board thickness shall be increased from 3/8" (9.5 mm) to 1/2" (12.7 mm) for 16" (406 mm) o.c. framing and from 1/2" (12.7 mm) to 5/8" (15.9 mm) for 24" (610 mm) o.c. framing.

Courtesy Gypsum Association, www.gypsum.org.

RECOMMENDED DRYWALL FASTENER LENGTHS		
Fastener Types	Panel Thickness (inches)	Recommended Length (inches)
Annular Ring Nail	3/8	1 1/4
	1/2	1 1/4
	3/8	1 3/8
	3/4	1 1/2
Smooth Shank Nail	3/8	1 1/4
	1/2	1 1/4
	5/8	1 3/8
	3/4	1 1/2
Type W Bugle Head to Wood	3/8, 1/2, 5/8	1 1/4
Type S Bugle Head to Steel Studs	1/2	1
	5/8	1 1/8
	3/4	1 1/4

FASTENER LENGTHS FOR GYPSUM PANEL PRODUCT APPLICATION TO WOOD FRAMING[A]

Gypsum Panel Product Thickness[B] Inches (mm)	Min. Nail Length Inches (mm)	Min. Screw Length Inches (mm)	Min. Staple[C] Length Inches (mm)
1/4 (6.4)	*D*	*D*	*D*
3/8 (9.5)	1 1/4 (32)	1 (25)	1 (25)
1/2 (12.7)	1 3/8 (35)	1 1/8 (28)	1 1/8 (28)
5/8 (15.9)	1 1/2 (38)	1 1/4 (32)	1 1/4 (32)

A Where fire resistance is required for gypsum panel product systems, fasteners of the same or larger length, shank diameter, and head bearing area as those described in the fire-rated design shall be used.

B For other thicknesses, for multi-layer applications, or for application over rigid foam insulation fasteners shall be of sufficient lenth to penetrate framing not less than 3/4" (19 mm) for nails, 5/8" (15 mm) for screws, and 5/8" (16 mm) for staples.

C Staple attachment is restricted to base layers of multi-layer systems only.

D For application over existing solid surfaces or in multi-layer applications, fastener shall be of sufficient length to penetrate framing not less than 3/4" (19 mm) for nails, 5/8" (16 mm) for screws.

Courtesty Gypsum Association, www.gypsum.org

BASE LAYER FASTENER SPACING FOR MULTI-LAYER GYPSUM PANEL PRODUCT APPLICATION*

		Base Layer Nail Spacing Inches (mm)		Base Layer Screw Spacing Inches (mm)		Base Layer Staple Spacing Inches (mm)	
Location	Framing Spacing Inches (mm)	Where Face Layer is Laminated	Where Face Layer is Mechanically Attached	Where Face Layer is Laminated	Where Face Layer is Mechanically Attached	Where Face Layer is Laminated	Where Face Layer is Mechanically Attached
Walls	16 (406)	8 (203)	24 (610)	16 (406)	24 (610)	7 (178)	16 (406)
	24 (610)	8 (203)	24 (610)	12 (305)	24 (610)	7 (178)	16 (406)
Ceilings	16 (406)	7 (178)	16 (406)	12 (305)	24 (610)	7 (178)	16 (406)
	24 (610)	7 (178)	16 (406)	12 (305)	24 (610)	7 (178)	16 (406)

Courtesy Gypsum Association, www.gypsum.org

FASTENER SPACING WITH ADHESIVE OR MASTIC AND SUPPLEMENTAL FASTENING OF GYPSUM PANELS

	Ceilings		Load-Bearing Partitions		Nonload-bearing Partitions	
Framing Spacing Inches (mm)	Nail Spacing Inches (mm)	Screw Spacing Inches (mm)	Nail Spacing Inches (mm)	Screw Spacing Inches (mm)	Nail Spacing Inches (mm)	Screw Spacing Inches (mm)
16 (406)	16 (406)	16 (406)	16 (406)	24 (610)	24 (610)	24 (610)
24 (610)	12 (305)	16 (406)	12 (305	16 (406)	16 (406)	24 (610)

Courtesy Gypsum Association, www.gypsum.org

MULTI-LAYER APPLICATION WITH ADHESIVE BETWEEN LAYERS OF GYPSUM PANELS

Gypsum Panel Product Thickness Inches (mm)		Gypsum Panel Product Orientation to Framing		Max. Framing Spacing Inches (mm) o.c.
Base	Face	Base	Face	
CEILINGS				
3/8 (9.5)	3/8 (9.5)	Perpendicular	Parallel or Perpendicular	16 (406)
1/2 (12.7)	3/8 (9.5)	Parallel or Perpendicular	Parallel or Perpendicular	16 (406)
1/2 (12.7)	1/2 (12.7)	Parallel or Perpendicular	Parallel or Perpendicular	16 (406)
5/8 (15.9)	1/2 (12.7)	Parallel	Parallel or Perpendicular	24 (610)
5/8 (15.9)	5/8 (15.9)	Parallel or Perpendicular	Parallel or Perpendicular	24 (610)
WALLS				
For multi-layer application with adhesive between layers 3/8" (9.5 mm), 1/2" (12.7 mm), or 5/8" (15.9 mm) thick gypsum panel products shall be permitted to be applied either perpendicular or parallel on framing spaced not more than 24" (610 mm) o.c.				

* Adhesive between layers shall be dry or cured prior to the application of any decorative treatment.

Courtesy Gypsum Association, www.gypsum.org

MULTI-LAYER APPLICATION WITHOUT ADHESIVE BETWEEN LAYERS OF GYPSUM PANELS

Gypsum Panel Product Thickness Inches (mm)		Gypsum Panel Product Orientation to Framing		Max. Framing Spacing Inches (mm) o.c.
Base	Face	Base	Face	
CEILINGS				
1/4 (6.4)	3/8 (9.5)	Perpendicular	Perpendicular*	16 (406)
1/4 (6.4)	1/2 (12.7)	Perpendicular	Perpendicular*	16 (406)
3/8 (9.5)	3/8 (9.5)	Perpendicular	Perpendicular*	16 (406)
3/8 (9.5)	1/2 (12.7)	Perpendicular	Perpendicular*	16 (406)
1/2 (12.7)	3/8 (9.5)	Parallel	Perpendicular*	16 (406)
1/2 (12.7)	1/2 (12.7)	Parallel	Perpendicular	16 (406)
1/2 (12.7)	1/2 (12.7)	Perpendicular	Perpendicular*	24 (610)
1/2 (12.7)	5/8 (15.9)	Perpendicular	Perpendicular	24 (610)
5/8 (15.9)	1/2 (12.7)	Perpendicular	Perpendicular*	24 (610)
5/8 (15.9)	5/8 (15.9)	Perpendicular	Perpendicular*	24 (610)
WALLS				
For multi-layer application with no adhesive between layers 1/2" (12.7 mm) or 5/8" (15.9 mm) thick gypsum panel products shall be permitted to be applied either perpendicular or parallel on framing spaced not more than 24" (610 mm) o.c. Framing spacing shall not be more that 16" (406 mm) o.c. when 3/8" (9.5 mm) thick gypsum panel products are used.				

* On ceilings to receive hand or spray-applied water-based texture material either i) 1/2" (12.7 mm) gypsum ceiling board (ASTM C 1396/C1396M) shall be applied perpendicular to framing; or ii) other gypsum panel products shall be applied perpendicular to framing and board thickness shall be increased from 3/8" (9.5 mm) to 1/2" (12.7 mm) for 16" (406 mm) o.c. framing and from 1/2" (12.7 mm) to 5/8" (15.9 mm) for 24" (610 mm) o.c. framing.

Courtesy Gypsum Association, www.gypsum.org

FIGURING THE NUMBER OF POUNDS OF NAILS FOR A JOB*

Panel Size	Number of Nails per Panel	
	Single-Nailed	Double-Nailed
4' × 8'	39	54
4' × 12'	54	78

*Approximately 300 nails per pound

TYPICAL JOINT COMPOUND UNDER TAPE DRYING TIME (HOURS)*

	Degrees F	60	70	80	90	100
	Degrees C	16	21	27	32	38
Relative Humidity	80%	50	38	27	20	15
	70%	36	25	20	15	10
	60%	30	20	13	10	8
	50%	24	16	12	10	7
	40%	20	14	10	8	6
	30%	18	12	10	7	4

* Times will vary depending upon the specific product used.

TYPES OF PLASTERS

Gypsum
Unfibered, fibered, wood fibered, perlite, bond coat, gauging, Keens, molding, acoustical, texture.

Cement
Regular Portland cement plaster, Portland cement-lime plaster, Portland cement glass fiber, enhanced plaster, Portland cement, glass fiber enhanced, acrylic modified brown coat plaster.

TYPES OF PLASTER FINISH COATS

Portland Cement
Lime smooth finish
Lime sand float
Lime dash finish
Manufacturer available Portland cement finishes

Gypsum
Keens cement – lime smooth finish
Gauging – lime smooth finish
Keens cement – lime smooth float
Machine dash
Manufacturer available gypsum finishes

PLASTER BASES

- Gypsum Lath
- Metal Lath
- Brick
- Clay Tile
- Concrete Masonry

GYPSUM LATH SPANS

Thickness	Unsupported Span (inches)
3/8"	16"
1/2"	16"

TYPICAL METAL LATH SPAN RECOMMENDATIONS

Lath Type	Weight lb./sq. ft.	Vertical Support (inches)	Horizontal Support (inches)
Diamond Mesh	0.27	16	12
Diamond Mesh	0.38	16	16
1/8" Flat Rib	0.31	16	12
1/8" Flat Rib	0.38	19	19
3/8" Flat Rib	0.38	24	24
Welded Wire Lath	0.50	24	24

TYPES OF METAL LATH		
Types	**Weight lb./sq. yd.**	**Size**
Diamond Mesh	2.5	27" × 96"
Diamond Mesh with Asphalt Paper Backing	2.5	27" × 96"
1/8" Flat Rib	2.75	27" × 96"
3/8" Flat Rib	3.4	27" × 96"
3/4" Flat Rib	5.4	24" × 96"

TYPES OF GYPSUM LATH

Standard Gypsum Lath
16" wide, 32", 36", and 48" long
Same available up to 12 feet long

Perforated Gypsum Lath
Is a standard sheet with 3/4" diameter holes every 4" both ways.

Insulating Gypsum Lath
Is a standard gypsum lath with aluminum foil on the back.

Veneer Plaster Gypsum Lath
Available in standard gypsum and Type X core.

NAILS AND STAPLES USED TO SECURE GYPSUM LATH TO WOOD SUPPORTS				
Lath Thickness and Width	**Support Spacing**	**No. of Fasteners Per Sheet**	**Nail Length and Head Diameter**	**Staple Crown Width**
3/8" × 16"	16	4	1 1/8" Long 19/64" Head	19/64"
1/2" × 16"	24	4	1 1/4" Long 19/64" Head	19/64"

NAILS AND STAPLES USED TO SECURE METAL LATH TO WOOD FRAMING		
Lath Type	**Nails**	**Staples**
Diamond Mesh, Expanded Metal	1 1/2" Long	1" Long
1/8" Flat Rib	11 Gauge	14 Gauge
Welded Wire	7/16" Head	7/16" Crown
3/8" Flat Rib	Provide 1 3/8" Penetration	Provide 1 3/8" Penetration

PORTLAND CEMENT PLASTER

MAXIMUM VOLUME AGGREGATE PER VOLUME CEMENTITIOUS MATERIAL[a]

Coat	Portland Cement Plaster[a] Maximum Volume Aggregate per Volume Cement	Portland Cement-Lime Plaster[c]			Min. Period Moist Coats	Min. Interval Between
		Max. Volume Lime per Volume Cement	Max. Volume Sand per Volume Cement and Lime	Approx. Min. Thickness[d] Curing (inches)		
First	4	1/4	4	3/8[e]	48 hours[f]	48 hours[g]
Second	5	3/4	5	First and second coats	48 hours	7 days[h]
Finish	3[i]	—	3[i]	1/8	—	Note h

For S1: 1 inch = 25.4 mm, 1 pound = 0.454 kg.

a. When determining the amount of aggregate in set plaster, a tolerance of 10 percent may be allowed.
b. From 10 to 20 pounds of dry hydrated lime (or an equivalent amount of lime putty) may be added as a plasticizing agent to each sack of Type Land Type H standard Portland cement in base coat plaster.
c. No plasticizing agents shall be added.
d. See Table R702.1(1) in the 2006 International Residential Code.
e. Measured from face of support or backing to crest of scored plaster.
f. Twenty-four hour minimum period for moist curing of interior Portland cement plaster.
g. Twenty-four hour minimum interval between coats of interior Portland cement plaster.
h. Finish coat plaster may be applied to interior Portland cement base coats after a 48-hour period.
i. For finish coat, plaster up to an equal port of dry hydrated lime by weight (or an equivalent volume of lime putty) may be added to Type I, Type H and Type III standard Portland cement.

THICKNESS OF PLASTER

Plaster Base	Finished Thickness of Plaster from Face of Lath, Masonry, Concrete (inches)	
	Gypsum Plaster	Portland Cement Mortar
Expanded Metal Lath	5/8 minimum[a]	5/8 minimum[a]
Wire Lath	5/8 minimum[a]	3/4 minimum (interior)[b] 7/8 minimum (exterior)[b]
Gypsum Lath[g]	1/4 minimum	3/4 minimum (interior)[b]
Masonry Walls[c]	1/2 minimum	1/2 minimum
Monolithic Concrete Walls[c,d]	5/8 maximum	7/8 maximum
Monolithic Concrete Ceilings[c,d]	3/8 maximum[e]	1/2 maximum
Gypsum Veneer Base[f,g]	1/16 minimum	3/4 minimum (interior)[b]
Gypsum Sheathing[g]	—	3/4 minimum (interior)[b] 7/8 minimum (exterior)[b]

For S1: 1 inch = 25.4mm.

a. When measured from back plane of expanded metal lath, exclusive of ribs, or self-furring lath, plaster thickness shall be 3/4" minimum.
b. When measured from face of support or backing.
c. Because masonry and concrete surfaces may vary in plane, thickness of plaster need not be uniform.
d. When applied over a liquid bonding agent, finish coat may be applied directly to concrete surface.
e. Approved acoustical plaster may be applied directly to concrete or over base coat plaster, beyond the maximum plaster thickness shown.
f. Attachment shall be in accordance with Table R702.3.5.
g. Where gypsum board is used as base for Portland cement plaster, weather-resistant sheathing paper complying with Section R703.2 shall be provided.

GYPSUM PLASTER PROPORTIONS

Number	Coat	Plaster Base or Lath	Max. Volume Aggregate per 100 lb. Neat Plaster[b] (cubic feet)	
			Damp Loose Sand[a]	Perlite or Vermiculite[e]
Two-coat work	Base coat	Gypsum lath	2.5	2
	Base coat	Masonry	3	3
Three-coat work	First coat	Lath	2[d]	2
	Second coat	Lath	3[d]	2[e]
	First and second coat	Masonry	3	3

For S1: 1 inch = 25.4 mm, 1 cubic foot = 0.0283m3, 1 pound = 0.454 kg.

a. Wood-fibered gypsum plaster may be mixed in the proportions of 100 pounds of gypsum to not more than 1 cubic food of sand where applied on masonry or concrete.
b. When determining the amount of aggregate in set plaster, a tolerance of 10 percent shall be allowed.
c. Combinations of sand and lightweight aggregate may be used, provided the volume and weight relationship of the combined aggregate to gypsum plaster is maintained.
d. If used for both first and second coats, the volume of aggregate may be 2.5 cubic feet.
e. Where plaster is 1/4 inch or more in total thickness, the proportions for the second coat may be increased to 3 cubic feet.

CHAPTER 13
Finish Flooring

UNDERLAYMENTS

Plywood underlayment panels are available in thicknesses of 1/4, 3/8, and 1/2". Thicker plywood panels can be used if the subfloor is unstable. Generally, 1 1/4" ring-shank nails on 1/4" thick panels are spaced 3" apart around the edges of the panel and 6" apart on the interior of the panel. Longer nails are used on thicker panels.

Oriented strandboard underlayment panels are 5/16" thick and 4 × 8 foot panels.

Particleboard underlayment panels are 3/8, 1/2, 5/8, and 3/4" thick and in 4 × 8 foot panels.

Concrete backer board panels are 3/8, 1/2 and 5/8" thick, 3' wide and 4, 5, and 6' long. It is used as an underlayment for ceramic floor tile. It is screwed to the subfloor with 1 5/8" wood screws spaced 8" around the edges of the panel. Leave a 1/8 to 3/8" space between panels and fill it with setting mortar. Then press a fiberglass tape over the joint.

Ditra underlayment is a polyethylene membrane that is much lighter than backerboard or plywood. It provides a stable base for ceramic tile and stone flooring, serves as a waterproof barrier and helps moisture and vapor to escape from the substrate. It is bonded to the subfloor with thin-set mortar. It is 1/8" thick and in sheets 3'-3" wide and up to 95' long.

Builders felt (15 lb.) or red rosin paper is installed over the subfloor before wood flooring is installed.

APA PLYWOOD UNDERLAYMENT

Plywood Grades[a]	Application	Min. Plywood Thickness (inches)	Fastener Size and Type	Max. Fastener Spacing (inches)[d]	
				Panel Edges[c]	Intermediate
APA Underlayment APA C-C Plugged EXT APA Rated STURD-I-FLOOR (19/32" or thicker)	Over Smooth Subfloor	1/4	3d × 1 1/4" Ring-shank Nails (b) min. 12 1/2 gauge (0.099") shank dia.	3	6 each way
	Over Lumber Subfloor or Uneven Surfaces	11/32		6	8 each way

(a) In areas to be finished with resilient floor coverings such as tile or sheet vinyl, or with fully adhered carpet, specify Underlayment, C-C Plugged or veneer-faced STURD-I-FLOOR with "sanded face." Underlayment A-C, Underlayment B-C, Marine EXT or sanded plywood grades marked "Plugged Crossbands Under Face," "Plugged Crossbands (or Core)," "Plugged Inner Plies" or "Meets Underlayment Requirements" may also be used under resilient floor coverings.

(b) Use 4d × 1 1/2" ring-shank nails, minimum 12 1/2 gauge (0.099") shank diameter, for underlayment panels 19/32" to 3/4" thick.

(c) Fasten panels 3/8" from panel edges.

(d) Fasteners for 5-plywood underlayment panels and for panels greater than 1/2" thick may be spaced 6" on center at edges and 12" each way intermediate.

Courtesy APA-The Engineered Wood Association, www.apawood.org.

TYPICAL PANEL FLOOR SPECIFICATIONS BASED ON FINISH FLOOR INSTALLATIONS

All Must Meet Minimum Structural Requirements[a] of IBC or IRC

Finish Floor	Typical Panel Installation[b]	Example Specification[b]
Carpet and Pad	Single Layer of APA Rated STURD-I-FLOOR[c] with T&G Edges	APA STURD-I-FLOOR[c] 24 o.c. Exposure 1 T&G (for joists spaced 24" o.c. or less)
Hardwood Flooring	Single Layer of APA Rated STURD-I-FLOOR[c] or APA Rated Sheathing	APA Rated STURD-I-FLOOR[c] 24 o.c. Exposure 1 **or** 48/24 APA Rated Sheathing Exposure 1 (for joists spaced 24" o.c. or less)[d]
Vinyl (or other thin resilient floor covering) or Glue-down Carpet	APA Rated STURD-I-FLOOR[c] or APA Rated Sheathing Exposure 1 plus min. 1/4" APA Underlayment[e] **Sanded Face** Exposure 1	APA Rated STURD-I-FLOOR[c] 24 o.c. Exposure 1 **or** 48/24 APA Rated Sheathing Exposure 1 (for joists spaced 24" o.c. or less). Cover with 1/4" (or thicker) APA Underlayment[f] **Sanded Face** Exposure 1
Ceramic Tile[g]	Two layers minimum 19/32" APA Rated STURD-I-FLOOR[c] Exposure 1	Two layers of min. 19/32" plywood APA Rated STURD-I-FLOOR[c] 20 o.c. Exposure 1 (for joists spaced 16" o.c. or less)

(a) Floor Span Rating must equal or exceed joist spacing.

(b) www.apawood.org for installation specifics and alternate installation combinations.

(c) **Plywood** APA STURD-I-FLOOR is Underlayment with a Span Rating.

(d) Minimum 19/32" **plywood** APA STURD-I-FLOOR(c) or APA Rated Sheathing. Minimum 23/32" **OSB** APA STURD-I-FLOOR or APA Rated Sheathing. See APA Technical Note: *APA Performance Rated Panel Subfloors Under Hardwood Flooring*, Form R280.

(e) APA Underlayment is always plywood.

(f) For rough floors, specify minimum 11/32" APA Underlayment[c].

(g) For other specialty flooring products, including marble and slate, please refer to the finish floor manufacturer's recommendations.

Courtesy APA-The Engineered Wood Association, www.apawood.org.

SOUTHERN PINE FLOORING GRADE DESCRIPTIONS	
Grade	**Characteristics**
B&B	Highest recognized grade of flooring. Generally clear, although a limited number of pin knots are permitted. Finest quality for natural or staining finish.
C	Excellent for painted or natural finish where requirements are less exacting. Reasonably clear but permits limited number of surface checks and small tight knots.
C & BTR	Combination for B&B and C grades; satisfies requirements for high quality finish.
D	This grade requires a face as good as D Finish grade except scant width of face not permitted and only medium warp allowed. Economical, serviceable grade for natural or painted finish.
No. 1	No. 1 Flooring is not provided under SPIB Grading Rules as a separate grade but, if specified, will be designated and graded as D flooring.
No. 2	This grade requires a face as good as No. 2 Boards which is suitable for high quality sheathing. High utility value where appearance is not a factor.
No. 3	Admits all pieces below No. 2 Flooring if suitable for low-cost sheathing or lathing without wasting over 1/4 length of any piece.
Courtesy Southern Forest Products Association, www.sfpa.org	

SIZES OF SOUTHERN PINE FLOORING

Species	Actual Thickness		Actual Width	
	in.	mm	in.	mm
Longleaf pine, slash pine, shortleaf pine, loblolly pine	5/16	8	1 3/8	35
	7/16	11	2 3/8	60
	9/16	14	3 3/8	86
	3/4	76	4 3/8	101
	1	101	5 3/8	136
	1 1/4	107	—	—

TYPICAL SIZES OF HARDWOOD TONGUE-AND-GROOVE STRIP FLOORING

Species	Actual Thickness		Actual Width	
	in.	mm	in.	mm
Oak, beech, birch, hard maple, hickory	3/4	19	1 1/2, 2, 2 1/4, 3 1/4	38, 51, 57, 82.5
	11/32	8.7	1 1/2, 2	38, 51
	11/32	12	1 1/2, 2	38, 51
	33/32	26	2, 2 1/4, 3 1/4	51, 57, 82.5

Courtesy Southern Forest Products Association, www.sfpa.org

GRADES OF MAPLE FLOORING

- First (highest quality)
- Second and Better (mix of First and Second)
- Second (good quality, some imperfections)
- Third and Better (mix of First, Second, and Third)
- Third (good economy flooring)

Courtesy Maple Flooring Manufacturers Association, www.maplefloor.org

THIN WOOD PARQUET BLOCKS

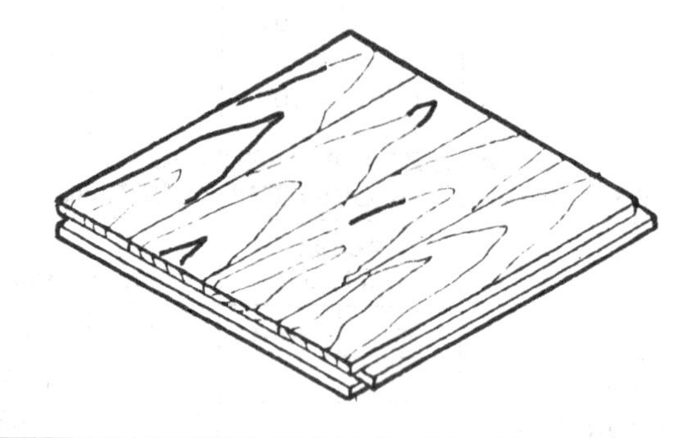

Solid Wood Blocks

Thickness: 3/32", 5/16", 1/2", 25/32", 33/32"

Square Sizes (length/width): 6 3/4" to 9"

Laminated Blocks

Thickness: 5/16", 1/2"

Square Sizes (length/width): 9" to 16"

HEAVY SOLID WOOD PARQUET BLOCKS

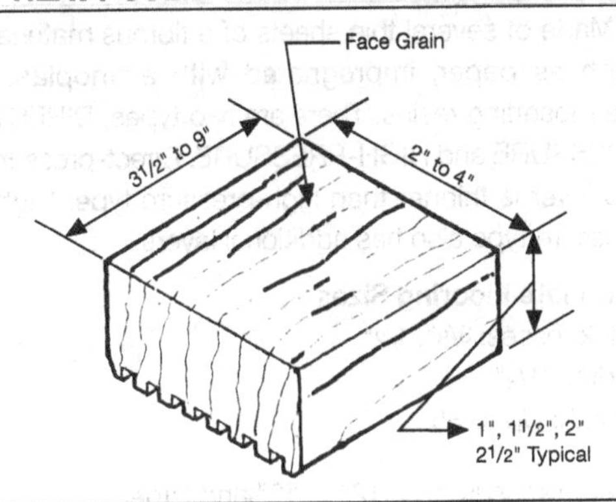

Thickness: 1", 1 1/2", 2", 2 1/2"
Width: 2" to 4"
Length: 3 1/2" to 9"

Consult manufacturer.

ENGINEERED WOOD FLOORING

This is a laminate of three thin layers of solid wood.

Thicknesses: 5/16", 3/8", 1/2"
Widths: 2 1/4" to 7"
Lengths: Random Lengths

BAMBOO FLOORING

Thickness: 3/8"
Widths: 5", 7"
Lengths: Random Lengths

LAMINATE FLOORING

Made of several thin sheets of a fibrous material, such as paper, impregnated with aminoplastic thermosetting resins. There are two types, DIRECT-PRESSURE and HIGH-PRESSURE. Direct-pressure top layer is thinner than high-pressure type. High-pressure type also has additional layers.

Laminate Flooring Sizes

Thicknesses: 3/8", 1/2"

Width: 7 1/2"

Length: Up to 85"

Direct-pressure tiles 12" × 12" and larger.

TYPICAL INSTALLATION OF WOOD STRIP FLOORING

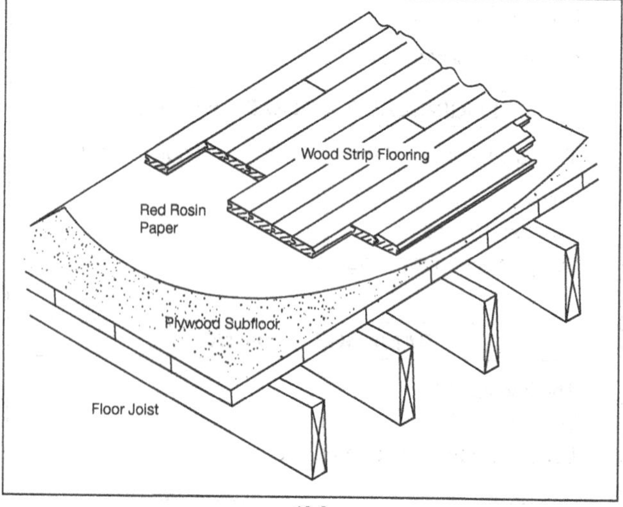

INSTALLING FLOATING LAMINATE FLOORING THAT HAS A SPECIAL GLUELESS JOINT

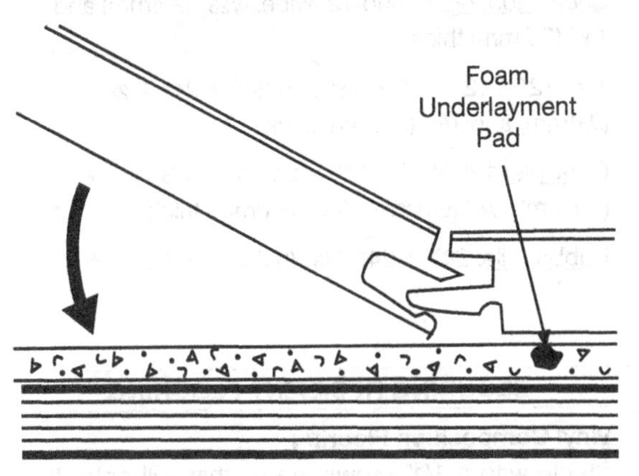

1. Install the butting strip on an angle and press down.

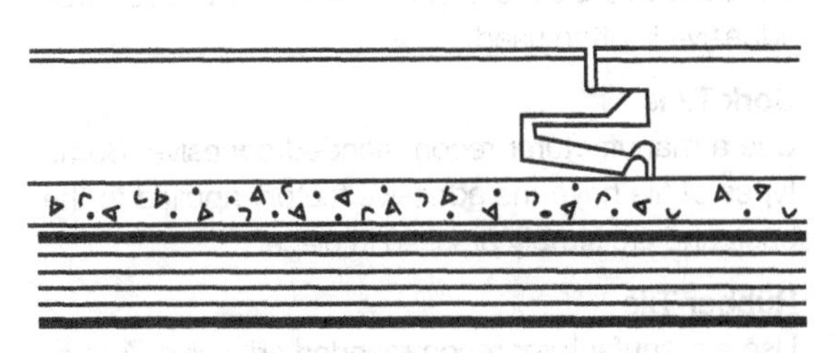

2. This locks the strips together.

RESILIENT FLOORING

Vinyl Composition Flooring

Sheet Flooring: 6' and 12' wide. 3/23" (2.4mm) and 1/8" (3.2mm) thick

Tile: 12" × 12", 18" × 18" and 24" × 24", 3/23" (2.4mm) and 1/8" (3.2mm) thick

Cork Tile: 12" × 12", 12" × 24", 12" × 36", 3/16" (4.7mm), 1/4" (6.0mm), 5/32" (8.0mm) thick

Rubber Tile: 24" × 24", 1/8" (3.2mm) thick

SECURING RESILIENT FLOORING

Vinyl Composition Flooring

Staple with a 1/2" crown staple that will enter the wood subfloor 1/2". Space them every 2" around the perimeter. Some installations require staples and a manufacturer recommended adhesive. Apply the adhesive before stapling. A water-based latex adhesive is often used.

Cork Tiles

Use a manufacturer recommended adhesive. Some types of tile have the adhesive factory applied to the back and are simply pressed in place.

Rubber Tile

Use a manufacturer recommended adhesive. A one-component urethane adhesive prepared especially for rubber tile is widely used.

TYPES OF CARPET FIBERS

Acrylic. Good resistance to abrasion, mildew, sunlight, aging, and some chemicals.

Nylon Fiber. Strong, resists staining, aging, mildew, abrasion, and water absorption.

Wool. A natural fiber that has good resistance to abrasion, aging and damage from mildew and sunlight.

Polyester Fiber. High tensile strength, resists aging, mildew, and abrasion. Not as durable as nylon. Exposure to sunlight over time will cause loss of strength.

Olefin (Polypropylene) Fiber. Has the lowest moisture-absorption rate of all the fibers. Resists mildew, sunlight, abrasion, aging, and some solvents.

TYPES OF CARPET CONSTRUCTION

Tufted construction is most widely used. It has 3 layers, consisting of a woven polypropylene fabric, a layer of woven polypropylene scrim mesh into which the tufting yarn is stitched.

Knitted construction involves looping the elements, stitching, backing yarn, and pile yarn, into a single pad.

Flocked construction bonds the pile yarn to a backing material that is coated with a vinyl adhesive.

Fusion-bonded construction bonds the pile yarn to a backing material with an adhesive.

TYPES OF CARPET CONSTRUCTION *(cont.)*

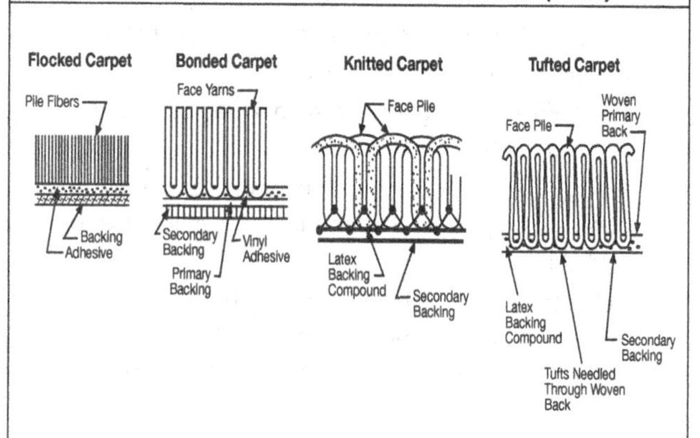

CARPET CUSHION

Urethane foam cushion is formed from recycled foam materials. The denser the foam in the cushion, the better it feels when you walk on it. Four-pound foam cushion is not satisfactory for quality installations. Six or eight pound are much better. It has a vapor barrier on the back. Synthetic felt cushions are made from waste carpet fiber. They provide a firm feeling when walking upon the carpet. They have a vapor barrier on the back.

Rubber cushion is the best to use for quality installations. It also resists moisture. Some types of carpet are made with a cushion bonded on the back.

TYPES AND APPLICATIONS OF TACKLESS STRIPS USED FOR CARPET INSTALLATION	
Type	**Application**
Standard (no anchoring nails)	Use on wood, concrete, and other hard surface floors.
Pre-nailed (wood)	Use on wood subfloors
Pre-nailed (concrete)	Use on concrete
Acoustical	Use on lightweight aggregate concrete flooring that has been designed for absorbing sound
Commercial or Architectural	Use when any area dimensions exceed 30' (9.1 m)
NOTE: Standard and pre-nailed wood tackless strips should be a minimum of 1" (25 mm) wide and 1/4" (6.4 mm) thick.	
Courtesy Carpet and Rug Institute, www.carpet-rug.com	

TROWEL SIZE RECOMMENDATIONS FOR APPLYING CARPET ADHESIVES

For Direct Glue Down Installation					
	Adhesive Type	**Notch Width**	**Notch Depth (in.*)**	**Space Between**	**Notch Shape**
Carpet Backing					
Jute	Latex	3/32 1/8	3/32 1/8	3/32 1/8	V, or U
Rubber (Foam and Sponge	Latex				
Polyurethane Cushion	Latex				
Jute/Vinyl	Vinyl				
Vinyl/Foam	Vinyl				
Vinyl/Slab	Vinyl				
Vinyl/Coated	Vinyl				
Polypropylene Secondary	Latex	1/8 1/8	1/8 1/8	1/16 1/8	V, or U
Unitary	Latex	1/8 1/8	1/8 3/16	1/16 1/8	V, or U
Woven	Latex	1/8 1/8	1/8 3/16	1/16 1/8	V, or U
Hot Melt	Latex	1/8	1/8	1/8	U
For Double Glue Down Installation					
Between Floor and Cushion	Latex	1/16	1/16	1/16	U
Between Cushion and Carpet					
Smooth Back Carpet	Latex	1/8	1/8	1/16	U
Rough Back Carpet	Latex	1/8	3/16	1/8	U

* To convert dimensions to metric: 1/32 = .08mm, 1/16 = 1.6mm, 3/32 = 2.4mm, 1/8 = 3.2mm, 3/16 = 4.8mm

Note: The above guidelines should only be used when specific recommendations are not available from the carpet manufacturer and/or the adhesive supplier. Rough, porous concrete surfaces and heavily textured carpet backs may require a trowel with deeper notches than listed above. A 100% transfer of floor adhesive into the carpet backing while maintaining full coverage of the floor must be attained.

TROWEL SIZE RECOMMENDATIONS FOR APPLYING CARPET ADHESIVES *(cont.)*

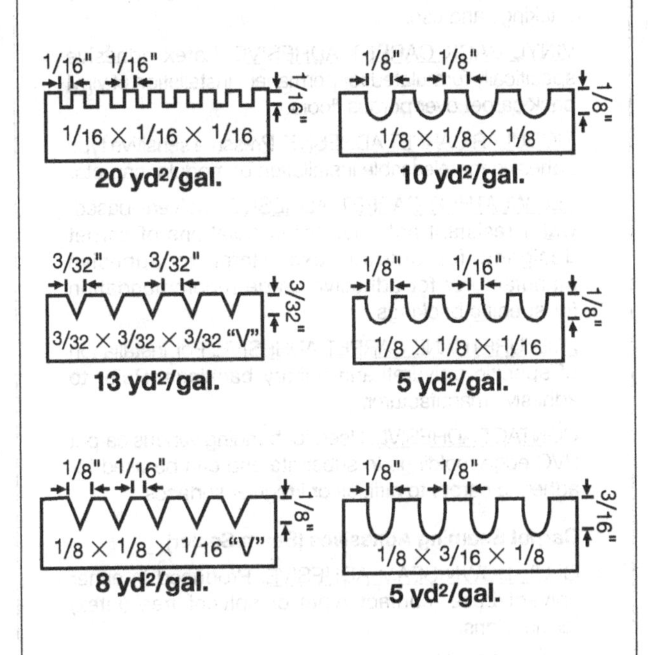

Courtesy Carpet and Rug Institute, www.carpet-rug.com

ADHESIVES USED FOR CARPET INSTALLATION

Carpet Floor Adhesives

LATEX ADHESIVE. For installation of carpet, excluding those with vinyl backing. Refer to adhesive manufacturer for adhesive grade recommendation for specific backings and use.

VINYL BACK CARPET ADHESIVE. Latex adhesive specifically formulated for permanent installation of vinyl back carpet over porous floors.

MODULAR CARPET ADHESIVE. Pressure sensitive type adhesive for releasable installation of modular carpets.

ALL-WEATHER CARPET ADHESIVE. Solvent-based, water resistant adhesive for installations of carpet designed for outdoor use. Refer to adhesive manufacturer for adhesive grade recommendation for specific backings.

POLYURETHANE CARPET ADHESIVE. For installation of specific polyurethane unitary backings. Refer to adhesive manufacturer.

CONTACT ADHESIVE. Used for bonding various carpet PVC edge moldings to substrate and can be used for adhering carpet to difficult or irregular surfaces.

Carpet Seaming Adhesives (Seam Sealer)

GLUE-DOWN SEAM ADHESIVE. Produced in either solvent base (contact type) or solvent free (latex) formulations.

VINYL BACK SEAM ADHESIVE. Solvent base (chemical weld) or solvent free (mechanical bond).

LATEX SEAM ADHESIVE. For applying seaming tapes, reinforcing sewn seams, sealing trimmed edges prior to "hot melt" seaming, securing, binding, etc.

Courtesy Carpet and Rug Institute, www.carpet-rug.com

CERAMIC FLOOR TILE

Quarry Tile

Unglazed, made from shales and fire clay. Impervious to dirt, moisture, and stains. Typical sizes include 4" × 4", 4" × 8", 6" × 6", 6" × 9", 8" × 8".

Paver Tiles

Clay tile, glazed or unglazed, weather and abrasion resistant. Typically 4" × 4" and 4" × 8".

Ceramic Mosaic Tile

Clay tile, glazed or unglazed, available in squares, rectangles and hexagon shapes. They are small, usually 1" × 1" or 1" × 2" and are used on floors, swimming pools, and walls.

Glazed Ceramic Floor Tile

Smooth and textured surfaces available. Typically available in 4" and 8" squares and 4" × 8" rectangles. Used on residential and moderate duty commercial floors.

Porcelain Tile

Available glazed or unglazed. Various types can be used for interior and exterior walls and floors. Typical sizes include 8" × 8", 8" × 18", 12" × 12", 12" × 24", 18" × 18".

TYPICAL SIZES OF CERAMIC FLOOR TILE (INCHES)	
Thickness	**Tile Size**
5/16, 3/8	4 × 4, 6 × 6, 8 × 8, 12 × 16, 24 × 24, 4 × 8, 4 × 12, 4 × 16

TYPICAL SIZES OF QUARRY TILE (INCHES)	
Tile Size	4 × 4, 6 × 6, 8 × 8, 4 × 8, 6 × 9

TYPICALLY USED JOINT WIDTHS BETWEEN CERAMIC FLOOR TILE			
Type of Tile	**Size of Tile**	**Max. Joint Width**	**Min. Joint Width**
Ceramic	2 3/16" to 4 1/4"	1/4"	1/8"
Ceramic	6"×6"	3/4"	1/4"
Ceramic Mosaic	2 3/16" or less	1/8"	1/16"
Quarry	All sizes	3/4"	3/8"

NUMBER OF FLOOR TILES TO COVER VARIOUS FLOOR AREAS			
No. of Sq. Ft.	6" × 6"	9" × 9"	12" × 12"
10	40	18	10
20	80	36	20
30	120	54	30
40	160	72	40
50	200	89	50
60	240	107	60
70	280	125	70
80	320	143	80
90	360	160	90
100	400	178	100
200	800	356	200
300	1200	534	300
400	1600	712	400
500	2000	890	500

ADHESIVES AND MORTARS FOR BONDING CERAMIC FLOOR TILE

Latex Portland cement thin-set mortars are a mixture of Portland cement and sand mixed with a liquid latex. They can be used on interior and exterior installations. They can be used over exterior grade plywood, concrete, old ceramic tile floors, well bonded vinyl flooring, cement terrazzo, and cement backer board. They are laid 3/32 to 1/8" thick.

Epoxy mortars are a mixture of an epoxy resin combined with a hardener, sand, and Portland cement. They are chemical resistant and have a high bond. They are used over concrete floors and cement backer board underlayment. They are laid 3/32 to 1/8" thick.

Nonpolymer thin-set mortars are a mixture of Portland cement, carefully graded aggregates and special chemicals. They are used to bond tile to concrete floors and cement backer board underlayment. They are laid 3/32 to 1/8" thick.

Type I mastics are composed of an organic material. They are used on interior floors only but will work in slightly damp areas. They are suitable for light duty attachment. They are laid 3/32 to 1/16" thick.

Type II mastics are a latex based material that is water-resistant and has a strong bond. They are recommended for interior use. They are laid 3/32 to 1/16" thick.

Always consult the manufacturers recommendations before selecting a clay tile bonding agent.

GROUTING PRODUCTS FOR CERAMIC FLOOR TILE

Epoxy-modified grouts are a two-part epoxy emulsion mixed with a ceramic tile grout. They resist chemicals and stain and can have colors added. They are used for joints 1/8 to 1" wide.

Polymer-modified sanded grouts are a mixture of an acrylic latex resin and a hardener blended with a silica filler. They are available in several grades depending upon the space between the tiles. They are resistant to abrasion, high temperatures, and alkalies.

Unsanded polymer-modified grouts are a mixture of an acrylic latex resin and Portland cement. They are used on narrow joints, typically 1/8" wide, on both exterior and interior floors.

Grout stain is used to stain a light grout darker or to lighten a dark grout.

Grout repellants improve the water repelling and stain resistance of grout.

Grout sealers seal the grout to improve stain resistance and prevent water absorption.

CERAMIC TILE INSTALLATION

Ceramic tile is bonded to cement backerboard which is laid over the plywood subfloor.

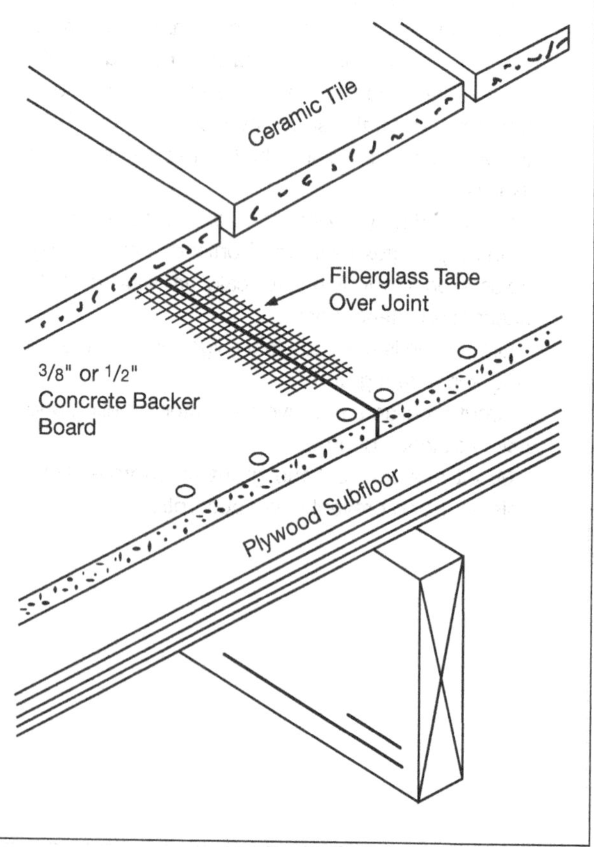

STONE TILES AND PAVERS

Flagstone
Thickness: 1/2" to 4"
Length/Width: 12" × 12" to 24" × 36"

Slate
Thickness: 1/2" to 1"
Length/Width: 12" × 12" to 24" × 54"

Marble Tiles
Thickness: 1/4" to 1/2"
Length/Width: 12" × 12"

Marble Pavers
Thickness: 1 1/4"
Length/Width: 24" × 24"

Limestone Pavers
Thickness: 1 3/4" to 2 1/2"
Length/Width: 24" × 36"

Granite Tiles
Thickness: 3/8" × 12", 3/4", 7/8"
Length/Width: 12" × 24"

Granite Pavers
Thickness: 1 1/4" to 4 1/2"
Length/Width: 15" × 30"

CHAPTER 14
Protective and Decorative Coatings

FEDERAL REGULATIONS

Federal regulations affecting the use of paint and other coatings are in the 1992 Residential Lead-Based Paint Hazard Reduction Act and the 1970 Clean Air Act. For example, lead is prohibited from being used in paints.

Air quality regulations are detailed in the Code of Federal Regulations. The regulation of finishing materials is detailed under the National Volatile Organic Compound (VOC) Emission Standard for Architectural Coatings. It regulates all products used as finishing materials in construction that emit volatile organic compounds. VOC is specified in grams per liter and pounds per gallon.

LIMITS ON VOLATILE ORGANIC COMPOUND EMISSIONS FOR SELECTED FINISHING MATERIALS		
Coating Category	**Grams VOC per Liter**	**Lbs. VOC per Gallon**
Concrete Protective Coating	400	3.3
Clear Fire-retardant Coatings	850	7.1
Flat Exterior Coatings	250	2.1
Flat Interior Coatings	250	2.1
Lacquers	680	5.7
Nonflat Exterior Coatings	380	3.2
Nonflat Interior Coatings	380	3.2
Enamels	450	3.8
Rust Preventative Coatings	400	3.3
Interior Sealers	400	3.3
Clear Shellac	730	6.1
Opaque Stains	350	2.9
Varnishes	450	3.8
Clear Wood Sealer	550	4.6

For more information, see the National Volatile Organic Compound Emission Standards for Architectural Coatings, Code of Federal Regulations, Environmental Protection Agency.

INITIAL APPLICATION AND MAINTENANCE OF EXTERIOR WOOD FINISHES

Finish	Initial Application		Appearance of Wood	Maintenance		
	Process	Cost		Process	Cost	Timing
Water-repellent Preservative	Brushing	Low	Grain Visible; Wood Brown to Black, Fades Slightly with Age	Brush to Remove Surface Dirt	Low	1–3 Years
	Pressure (factory applied)	Medium	Grain Visible; Wood Greenish or Brownish, Fades with Age	Brush to Remove Surface Dirt	Nil, unless Stained, Painted or Varnished	None, unless Stained, Painted or Varnished
Organic Solvent Preservative	Pressure, Steeping, Dipping, and Brushing	Low to Medium	Grain Visible; Color as Desired	Brush and Reapply	Medium	2–3 years or when Preferred
Water Repellent and Oils	One or Two Brush Coats of Clear Material or, Preferably, Dip Application	Low	Grain and Natural Color Visible, Becoming Darker and Rougher Textured with Age	Clean and Reapply	Low to Medium	1–3 Years or when Preferred
Semi-transparent Stain	One or Two Brush Coats	Low to Medium	Grain Visible; Color as Desired	Clean and Reapply	Low to Medium	3–6 Years or when Preferred
Clear Varnish	Three Coats (minimum)	High	Grain and Natural Color Unchanged if Adequately Maintained	Clean, Sand, and Stain Bleached Areas; Apply Two More Coates	High	2 Years or when Break Down Begins
Paint and Solid-color Stain	Brushing; Water Repellent, Prime, and Two Top Coats	Medium to High	Grain and Natural Color Obscured	Clean and Apply Top Coat, or Remove and Repeat Initial Treatment if Damaged	Medium	7–10 Years for Paint; 3–7 Years for Solid Color Stain

U.S. Department of Agriculture, Forest Service, Forest Products Laboratory, Madison, WI

SUITABLE AND EXPECTED SERVICE LIFE OF FINISHES FOR EXTERIOR WOOD SURFACES							
	Water-repellent Preservative and Oil		Semitransparent Stain		Paint and Solid-color Stain	Paint	Solid-color Stain
Type of Exterior Wall Surface	Suitability	Expected Life (years)	Suitability	Expected Life (years)	Suitability	Expected Life (years)	
Siding							
Cedar and Redwood							
Smooth (vertical grain)	High	1–2	Moderate	2–4	High	4–6	3–5
Roughsawn	High	2–3	High	5–8	High	5–7	4–6
Pine, Fir, Spruce							
Smooth (flat-grained)	High	1–2	Low	2–3	Moderate	3–5	3–4
Rough (flat-grained)	High	2–3	High	4–7	Moderate	4–6	4–5
Shingles							
Sawn	High	2–3	High	4–8	Moderate	3–5	3–4
Split	High	1–2	High	4–8	—	3–5	3–4
Plywood (Douglas-fir and Southern Pine)							
Sanded	Low	1–2	Moderate	2–4	Moderate	2–4	2–3
Textured (smooth)	Low	1–2	Moderate	2–4	Moderate	3–4	2–3
Textured (roughsawn)	Low	2–3	High	4–8	Moderate	4–6	3–5
Medium-density Overlay	—	—	—	—	Excellent	6–8	5–7

Plywood (Cedar and Redwood)							
Sanded	Low	1–2	Moderate	2–4	Moderate	2–4	2–3
Textured (smooth)	Low	1–2	Moderate	2–4	Moderate	3–4	2–3
Textured (roughsawn)	Low	2–3	High	5–8	Moderate	4–6	3–5
Hardboard, Medium Density							
Smooth							
Unfinished	—	—	—	—	High	4–6	3–5
Preprimed	—	—	—	—	High	4–6	3–5
Textured							
Unfinished	—	—	—	—	High	4–6	3–5
Preprimed	—	—	—	—	High	4–6	3–5
Millwork (usually Pine)							
Windows, Shutters, Doors, Exterior Trim	High	—	Moderate	2–3	High	3–6	3–4
Decking							
New (smooth)	High	1–2	Moderate	2–3	Low	2–3	1–2
Weathered (rough)	High	2–3	High	3–6	Low	2–3	1–2
Glued-laminated Members							
Smooth	High	1–2	Moderate	3–4	Moderate	3–4	2–3
Rough	High	2–3	High	6–8	Moderate	3–5	3–4
Oriented Strandboard	—	—	Low	1–3	Moderate	2–4	2–3

U.S. Department of Agriculture, Forest Service, Forest Products Laboratory, Madison, WI

INTERIOR WALL AND CEILING FINISH REQUIREMENTS BY OCCUPANCY						
	Sprinklered*			Nonsprinklered*		
Group	Exit Enclosures and Exit Passageways	Corridors	Rooms and Enclosed Spaces	Exit Enclosures and Exit Passageways	Corridors	Rooms and Enclosed Spaces
A-1 and A-2	B	B	C	A	A	B
A-3, A-4, A-5	B	B	C	A	A	C
B, E, M, R-1, R-4	B	C	C	A	B	C
F	C	C	C	B	C	C
H	B	B	C	A	A	B
I-1	B	C	C	A	B	B
I-2	B	B	B	A	A	B
I-3	A	A	C	A	A	B
I-4	B	B	B	A	A	B
R-2	C	C	C	B	B	C
R-3	C	C	C	C	C	C
S	C	C	C	B	B	C
U	No Restrictions			No Restrictions		

*** Classes of Finishing Materials**

Class A Flame Spread 0–25; Smoke Developed 0–450
Class B Flame Spread 26–75; Smoke Developed 0–450
Class C Flame Spread 76–200; Smoke Developed 0–450

Occupancy Groups

A-1: Fixed Seating (i.e. theater)
A-2: Food, Drink Consumption
A-3: Worship, Recreation
A-4: Indoor Sports (i.e. tennis)
A-5: Viewing Outdoor Activities (i.e. bleachers)
B: Business Group (i.e. banks)
E: Education Group
M: Mercantile Group
R-1: Residential (i.e. motel)
R-4: Residential, (i.e. care/assisted living)
F: Factory Industrial Group
H: High Hazard Group
I-1: Institutional Group (i.e. group home)
I-2: Institutional Group (i.e. medical)
I-3: Institutional Group (i.e. persons under restraint)
I-4: Institutional Group (i.e. day care)
R-2: Residential (i.e. apartments)
R-3: Residential (i.e. permanent living)
S: Storage Group
U: Utility and Miscellaneous Group

CLEAR COATINGS

Natural Resin Varnishes

Linseed Oil Varnishes

These include Long Oil, Medium Oil, Short Oil, Tung Oil, and Spirit.

Long Oil Varnishes

Also known as spar varnish. Used on exterior surfaces exposed to occasional moisture but not constantly wetted.

Medium Oil Varnishes

Dry faster than long oil varnishes, have a harder film but are not as water resistant.

Short Oil Varnishes

Dry very rapidly, are brittle, and do not resist abrasion well.

Tung Oil Varnishes

Used in areas where heavy use occurs.

Spirit Varnishes

Known as shellac. Dry rapidly and do not resist moisture. Available in several grades depending upon the amount of resin dissolved in alcohol. Grades are called cuts. For example, a 4 lb. (1.8 kg) cut has 4 lb. of resin dissolved in 1 gallon of denatured alcohol.

Synthetic Resin Varnishes

These include alkyds, polyurethane, silicone, epoxy, acrylics, and phenolics. Properties are shown in the following table.

TYPES AND USES OF SYNTHETIC RESIN VARNISHES

Binder	Base	Uses
Acrylic	Solvent or Water	Waterproofing, Sealing Surface Against Dirt, Used on Concrete, Masonry, Stucco
Alkyd (spar varnish)	Solvent	Used on Interior and Exterior Protected Surfaces
Phenolic (spar varnish)	Solvent	Exterior Wood Exposed to Moisture, Marine Applications
Silicone	Solvent	Waterproofing, Sealing Surface Against Dirt, Used on Concrete, Masonry, Stucco, Wood
Polyurethane (one part)	Solvent	Resists Chemical Attack, Abrasion, Heavy Foot Traffic

Lacquer

Lacquer has a nitrocellulose base, various resins and plasticizers, and a drying oil. The choice of resin in the product you plan to use depends upon where the lacquer will be used. Lacquer is used mainly on interior surfaces. Pigments can be added to give color or opacity. It is usually applied by spraying; however, a brushing lacquer is available.

Opaque Coatings

Opaque coatings obscure the natural color of the surface, give it protection, and provide a range of colors. Metal surfaces must be thoroughly cleaned. Follow the manufacturer's direction. This can include washing with a detergent and water, sanding, using a wire brush, sandblasting, or the use of a dilute acid. Some metals, such as aluminum, should be allowed to weather for 30 days before being primed and painted.

Concrete surfaces should be allowed to completely cure before being primed. Brush off all loose material.

Wood surfaces should have a moisture content of not more than 12 to 15 percent. Interior wood usually should have 6 percent moisture content.

Alkyd Coatings

They have excellent water resistance and weather well; however, regulations on VOC have largely limited their use.

Chlorinated Rubber

They have excellent resistance to alkali and acids, provide some resistance to salt-air and salt water exposure, and excellent water and water vapor resistance, and they resist abrasion. They can be used on masonry, plastic, concrete surfaces, and metal. They are not recommended for use on wood.

Enamel Coatings

Available with a hard, durable gloss or semigloss finish. They are available in a wide range of colors and are hard, washable, and resist alkalis and acids.

Epoxy Coatings

Epoxy coatings are available in several formulations. Epoxy-ester has good resistance to chemical fumes and exposure to water. Epoxy-polyamide has excellent resistance to chemical fumes, oils, atmospheric acids, and alkalis. It bonds well to concrete, metal, and wood. When exposed to weather it may fade but is not damaged.

Latex Coatings

These are formulated with a water base, present no fire hazard, and meet VOC requirements. They dry rapidly and are used on interior and exterior surfaces.

Acrylic Coatings

These are formulated with a water base and are available in clear and pigmented coatings. They protect exterior surfaces from weathering and are used on concrete. One-part emulsions are on interior and exterior surfaces, including wood, masonry, gypsum wallboard, plaster, and metal. Two-part emulsions are also used on interior and exterior surfaces and provide a tough coating that resists stains and can be washed when dirty.

Styrene-butadiene Coatings

These are water based and produce a rubberlike film. They have good resistance to alkali. They can be used on exterior porous concrete surfaces. Some formulations are used on interior masonry, plaster, and gypsum wallboard. It has a film that is washable and resists abrasion. It is not usually used on wood.

Vinyl Coatings

These are available as water-based formulations. One formulation is used for exterior surfaces and another for interior applications. It provides good resistance to soils, alkalies, acids, and salt water and is overall very durable.

Phenolic Coatings

These are solvent based and dry by evaporation, leaving a strong, flexible coating. They are used where resistance to acids, alkalies, and some solvents is required as well as possible immersion in hot water. Phenolic coatings are used on exterior concrete, plaster, wood, metal, and gypsum wallboard. One special type has a catalyst added at the site and hardens due to a chemical reaction. It resists very harsh conditions.

Urethane and Polyurethane Coatings

One-part urethane coatings are moisture cured and clear. They have excellent abrasion resistance. Two-part formulations are available, giving a hard to a rubberlike surface film. They do not adhere well to steel and concrete. They have good resistance to abrasion, water, and solvents. They are used for heavy-duty wall coverings and surfaces such as floors, subject to heavy traffic. They resist abrasion and impact and can be scrubbed clean.

STAINS

Stains are used to color the wood. Some provide protection from moisture. Exterior stains are blends of oil, driers, resins, a coloring pigment, a wood preservative, a mildewcide, and a water repellent. They color the wood and protect it from the weather. They are available in solid and semitransparent types. Some are oil-based and others water-based. Interior stains are available in a range of types. Penetrating stains are made with dyes and do not obscure the grain. Pigmented stains allow some of the pigment to remain on the surface and will obscure some of the grain.

WATER-REPELLENT PRESERVATIVES

Typical water-repellent preservatives contain a preservative (a fungicide), a small amount of wax that is a water repellent, a resin or drying oil, and a solvent such as mineral spirits, turpentine, or paraffinic oil. Waterborne formulations are available. They are not intended for wood to be in contact with the soil. They give short-term protection against decay (1 to 3 years) and mildew growth and prevent water staining. After several years, the surface should be cleaned with a commercial bleach and detergent, rinsed, dried, and retreated.

FIRE-RETARDANT COATINGS

A number of fire-retardant coatings are available from various manufacturers. Pigmented fire-retardant coatings retard the spread of flame along a surface; however, they do not protect the substrate from fire or heat. They are rated for surface burning characteristics on combustible and noncombustible surfaces. Flame spread ratings are CLASS A flame spread 0 to 25; Class B flame spread 26 to 75, and CLASS C flame spread 76 to 200. The building code will specify the requirements for various areas within a building such as rooms, other enclosed spaces, halls, and other exitways. The requirements are different for buildings with and without sprinkler systems.

Intumescent fire-rated coatings develop a thick, rigid foam protective layer that insulates the substrate and prevents the spread of fire.

SEALERS AND PRIMERS

Sealers go over porous surfaces to seal the surface so they eliminate any suction of the finish material into the pores of the surface. The manufacturer of the top coat material will specify the recommended primer for each product.

POLYURETHANE RESIN COATINGS		
ASTM Type Designation	**Curing Agent**	**Chemical Resistance**
ONE COMPONENT		
Type 1	Oxygen	Good
Type 2	Humidity in Air	Very Good
Type 3	Heat	Excellent
TWO COMPONENTS		
Type 4	Amine	Excellent
Type 5	Polyester	Excellent

EXTERIOR PAINT APPLICATIONS

Materials	Finishing Materials									
	Aluminum Paint	Oil-based Exterior Paint	Latex Exterior Paint	Cement Based Paints	Porch and Deck Enamel	Alkyd Enamel	Epoxy Enamel	Varnish	Shellac	Lacquer
METAL										
Aluminum	×	×	×			×	×			
Galvanized Steel	×	×	×			×	×			
Iron	×	×	×			×	×			
Steel	×	×	×			×	×			
MASONRY										
Concrete Block	×	×	×	×		×	×			
Brick	×	×	×	×		×	×			
Concrete Floors			×		×	×	×			
Stucco		×	×	×		×	×			
WOOD										
Unpainted Wood (i.e. trim, siding)		×	×			×	×	×	×	×
Decks					×	×	×	×	×	×
Floors						×	×	×	×	×

COMMONLY USED PRIMERS AND TOPCOATS

Material	Primer	Top Coat	Remarks
Aluminum	Vinyl Red Lead	Vinyl	Exposure to Weather
	Zinc Chromate	Chlorinated Rubber	Exposure to Rain, Salt-water Spray
	Zinc Chromate	Alkyd or Acrylic	Used on Trim, Flashing
	Self-priming	Epoxy Ester	Exposure to Fumes
Ferrous Metal	Self-priming	Phenolic	Exposure to Weather, High Humidity
	Zinc Silicate	Silicate, Alkyd	Exposure to Weather
	Zinc Silicate	Silicone, Aluminum Pigmented	
	Self-priming	Vinyl	Exposure to Rain, Salt-water Spray
	Red Lead	Acrylic	Will not Resist Abrasion
	Self-priming	Urethane	Exposure to Corrosion, Chemicals, Abrasions
	Self-priming	Epoxy	Exposure to Acids, Alkalis, Chemicals
	Self-priming	Coal Tar	Apply Hot to Metal to be Below Ground
	Chlorinated Rubber with Red Lead or Zinc Chromate	Chlorinated Rubber, Oil-based Paints	Exposure to Rain, Salt-water Spray, Chemicals Normal Exterior Conditions
	Red Lead	—	Not Abrasion Resistant, Exposure to Severe Weather
	Red Lead and Alkyd-based Red Lead	Alkyd	
	Zinc-polystyrene	Polystyrene	Chemical Fumes, Exposure to Fresh and Salt-water
Ferrous Metal (galvanized)	Zinc Dust or Zinc Chromate-zinc dust	Alkyd	Does not Require Topcoat
	Zinc Dust or Zinc Oxide	Chlorinated Rubber	Exposure to Rain, Salt-water Spray
Ferrous Metal in Ground	Self-priming	Coal-tar-expoxy	Used on Pipelines, Buried Structural Steel

Gypsum Wallboard	Vinyl	Alkyd	Light Duty
	Self-priming	Acrylic	Heavy Duty
Gypsum Plaster	Self-priming	Acrylic	Plaster Must Be Dry
Concrete and Concrete Masonry (dry), Brick Masonry	Self-priming	Acrylic	Interior Locations, Scrubbable
	Self-priming	Vinyl	Dry Locations
	Self-priming	Epoxy Esters	Exterior Use, Resists Fumes, Scrubbing
	Self-priming	Polychloroprene	Resists Sater, Solvents, Impact, Exterior Uses
	Self-priming	Urethane	Washable, Interior Locations
Concrete Floors, No Moisture Exposure	Self-priming	Urethane	Light to Moderate Traffic
	Self-priming	Epoxy	Moderate to High Traffic
Concrete, Heavy Moisture	Self-priming	Alkali-resistant Chlorinated Rubber	Water Reservoirs, Swimming Pools
Portland Cement Plaster	Self-priming	Vinyl	Dry Locations
	Styrene-butadiene	Alkyd	Dry Locations
Wood, Interior	Self-priming	Vinyl	Walls and Foors
	Self-priming	Alkyd	Doors, Paneling, Trim, Light-duty Floors
	Self-priming	Urethane	Surfaces Subject to Impact, Scrubbing, Heavy-duty Floors
	Self-priming	Acrylic	Surfaces Subject to Impact, Scrubbing
Wood, Exterior	Self-priming	Urethane	Porch Decking, Exterior Stairs
	Self-priming	Alkyd	Siding, Plywood, Cedar Shakes, Trim
	Self-priming	Acrylic	Siding, Plywood
	Self-priming	Phenolic	Siding, Plywood, Trim
	Oil-based Primer	Oil-based Vehicle	Wood Siding, Exterior Trim, Plywood Siding
	Self-priming	Epoxy	Any Exterior Wood

WOOD SEALERS

A sealer is a coating on a porous surface that seals it so the top coats do not soak into the material. Refer to the manufacturer's recommendations for the recommended sealer. Following are some frequently used sealers.

Material	Sealer
Lacquer	Lacquer Sanding Sealer
Water-based Polyurethanes and Acrylics	Water-based Sanding Sealer
Oil-based Polyurethanes and Varnishes	Usually Do Not Need a Sealer
A Clear Exterior Finish as on Decks	Oil-based Clear Sealer Containing Wax

DRIERS

Paints dry through a process of polymerization and oxidation. If these are the only drying properties, the paint will take several days to dry. Driers are added to speed the drying process. Driers frequently used are soluble metallic soaps such as cobalt, manganese, calcium, and zirconium. There are a number of nonmetallic driers used. The specific drier used will most likely not be shown on the paint can, but the actual dry time will be shown.

ADDITIVES

Additives are ingredients added in very small amounts to affect the properties of the paint. Some used are flattening agents, anti-settling agents, and freeze-thaw stabilizers. While these may not be shown on the paint can label, the effects produced may be detailed on the label.

PRIMERS

A primer is a coating applied to the material that helps the top coat to bond to it. The recommended primer for a particular finishing material is usually indicated on the paint can label. The choice of a primer also depends upon the material to be finished. The primer must also be compatible with the top coat to be used. Since there are many possible combinations of finishes and substrate, careful selection of a primer is important.

PRIMER RECOMMENDATIONS		
Material	**Top Coat**	**Primer**
Galvanized Steel	Alkyd and Latex Enamel	Zinc-rich Metal Primer
Aluminum	Alkyd and Latex Enamel	Alkyd Metal Primer
Iron and Steel	Alkyd and Latex Enamel	Alkyd Metal Primer
Concrete Floors	Alkyd, Epoxy Latex Enamel	Thinned First Coat of Enamel or as Directed on Label
Unpainted Gypsum Wallboard	Latex wall	Latex-type Primer-sealer
	Alkyd, Epoxy Enamel	Top Coat Serves as Primer
Plaster	Alkyd, Epoxy Enamel	Alkyd Primer
	Latex Wall Paint	Latex Primer
Concrete Walls, Concrete Block, Clay Brick	Alkyd, Epoxy Latex Enamel	Topcoat Serves as Primer
Interior Woodwork	Alkyd, Latex, Epoxy Enamel	Enamel Undercoat
Exterior Wood Floors, Decks	Alkyd Enamel	Slightly Thinned Alkyd Enamel (thin 10 to 15%)
Exterior Wood Siding	Oil-based Exterior Paint	Manufacturer-recommended Solvent-thinned Primer
	Acrylic Water-thinned Type	Manufacturer-recommended Water-based Primer

DRYING AND NON-DRYING OILS

Drying oils. Linseed oil, Danish oil, tung oil, and walnut oil. When applied directly to the wood they penetrate it and, after several coats, build up a film on the surface that appears much like a varnish surface. These are referred to as penetrating finishes. They are also added to house and trim paints, varnishes, and interior paints to improve the properties.

Non-drying oils. Olive oil, tung oil, and walnut oil. When applied to wood they soak into it but tend to stay wet permanently. They do not form a film.

SOLVENTS

Solvents are liquids that are used to dissolve the ingredients in a finishing material. They are usually toxic, flammable, and poisonous. Following are the most commonly used solvents in finishing materials used in building construction.

Acetone. It will remove or soften epoxy resins, ink, various adhesives, polyester, and contact cement. Used in paint and varnish removers.

Contact Cement Solvent. Used with contact cement adhesives. It dissolves dried contact cement and some plastics.

Citrus Terene. Serves as a cleaner and degreaser, removes oil, tar, grease, ink, and wax.

Ethyl Alcohol. Used in lacquer and will dissolve shellac.

SOLVENTS *(cont.)*

Haxane. Used as a solvent in paint thinners, cleaners, degreasing products, and adhesives. It will soften some plastics and rubber.

Lacquer Thinner. Used to thin lacquers and epoxies. Also a good cleaner. It will soften some plastics.

Methyl Ethyl Keystone. It is used as a solvent in varnishes, lacquers, enamels, and polyurethanes.

Methyl Alcohol. It is used as a solvent and thinner for shellac and shellac-based primers.

Naptha. It is used as an all-purpose thinner.

Mineral Spirits. A solvent for interior and exterior paints.

Turpentine. Used as a solvent and thinner for oil-based paints, stains, varnishes, and enamels.

Water. Used to thin water-based latex paints.

CAULKING COMPOUNDS

Acrylic-Latex. Sets slowly, 24 hours before painting, not as weather resistant as other types, cleanup with mineral spirits, can be painted.

Acrylic-Latex-Silicone. Sets within 1 hour, good weather resistance, water cleanup, good bonding characteristics, widely used for exterior applications, can be painted.

CAULKING COMPOUNDS *(cont.)*

Butyl Rubber. Sets within 2 hours, very elastic, cleanup with mineral spirits, must be painted to resist deterioration by ultraviolet light, bonds to most materials.

Foam. Sprayed polyurethane foam, bonds to most materials, sets within 5 minutes, used to fill wide cracks, and then the crack is topped with one of the more durable caulking compounds.

Latex. Sets within 1 hour, cleanup with water, poor resistance to weathering, best suited for interior applications, can be painted.

Oil-based. Sets slowly, 24 hours before painting, not as weather resistant as other types, cleanup with mineral spirits, can be painted.

Polyurethane. Has good weather resistance, should not be painted, solvent cleanup, good adhesion to most materials, long life.

Rope Caulk. Putty-type material, forced into cracks with a putty knife, does not bond well, not a permanent caulk, loses elasticity and dries out in 6 to 12 months.

Silicone. Excellent moisture resistance, available in colors, bonds well to most materials, some types designed to resist high temperatures as on roof construction, solvent cleanup, cannot be painted.

PAINT FILM THICKNESS

The finishing material must have the proper thickness to assure expected performance. This is given in the wet and dry film thickness. Minimum wet film thicknesses are specified in mils per coat. A mil is 0.001 in. (0.025 mm). Minimum dry film (MDF) thicknesses depend upon the wet film thickness and the solids in the paint.

DRY FILM THICKNESS

The label on the paint can will usually give the volume of solids in the product. If the paint will cover 400 square feet per gallon and has a wet film thickness of 4.0 mil and the solids are 25 percent of the volume, the dry paint film will be 1.0 mil (0.001 inches).

TYPICAL WET FILM THICKNESSES BASED ON THE SPREADING RATE

Spreading Rate (sq. ft./gal.)	Wet Film Thickness (mils)
1,600	1.0
1,000	1.6
700	2.3
400	4.0
200	8.0
160	10.0

INTERIOR FINISH FLAME SPREAD AND SMOKE DEVELOPED INDEXES	
Flame Spread Index	**Smoke Developed Index**
Class A 0-25	Class A 0-450
Class B 26-75	Class B 0-450
Class C 76-200	Class C 0-450

Courtesy National Fire Protection Association, www.nfpa.org

TYPICAL COVERING PROPERTIES OF SELECTED WOOD FINISHING MATERIALS	
Material	**Sq. Ft. per Gal.**
Bleaching Solutions	250-300
Concrete Sealer	75-100
Exterior Stain	250-300
Flat Varnish	300-350
Liquid Wax	600-700
Lacquer	200-300
Lacquer Sealer	250-300
Liquid Filler	250-400
Non-grain-raising Stain	275-325
Oil Stain	300-350
Pigment Oil Stain	350-400
Paint, Acrylic, Alkyd, Latex, Oil	100-900
Rubbing Varnish	450-500
Spirit Stain	250-300
Shellac	300-350
Water Stain	350-400

**Actual coverage depends upon the condition of the surface. For example, rough wood and fiber ceiling tile reduce the coverage per square foot. In paints, the volume of solids in the product establishes the square feet per gallon requirements.*

TYPICAL PAINT COVERAGE ESTIMATES FOR EXTERIOR PRIMER AND PAINT ON ONE STORY WALLS, ONE COAT.*

Total Length of Exterior Walls (ft.)	Wall Height 8'		Wall Height 9'		Wall Height 10'	
	Sq. Ft.	Gallons	Sq. Ft.	Gallons	Sq. Ft.	Gallons
160	1286	3.2	1440	3.6	1600	4.0
190	1520	3.8	1710	4.5	1900	4.8
210	1680	4.2	1890	4.7	2100	5.3
240	1920	4.8	2160	5.5	2400	6.0
270	2160	5.4	2430	6.0	2700	6.8
300	2400	6.0	2700	6.8	3000	7.5

*Based on typical coverage of 400 square feet per gallon. Check paint can label for specific coverage. In addition, add square feet for dormers, gable ends, second floor walls, soffits, exterior trim. Gallons per square foot to be used also can vary depending upon the wet and dry film requirements.

TYPICAL PAINT COVERAGE ESTIMATES FOR INTERIOR PRIMER AND PAINT WALL COVERAGE (ONE COAT)*

Total Length of Room Walls (ft.)	Ceiling Height 8'		Ceiling Height 8 1/2'		Ceiling Height 9'		Ceiling Coverage (gallons)
	Sq. Ft.	Gallons	Sq. Ft.	Gallons	Sq. Ft.	Gallons	
30	240	.60	255	.64	270	.68	.13
40	320	.80	340	.85	360	.90	.25
50	400	1.0	425	1.1	450	1.3	.40
60	480	1.2	510	1.3	540	1.4	.60
70	560	1.4	595	1.5	630	1.6	.50
80	640	1.6	680	1.7	720	1.8	1.00

*Based on typical coverage of 400 square feet per gallon. Check paint can label for specific coverage. Trim coverage not included.

Gallons per square foot will also vary depending upon the wet and dry film requirements.

TYPICAL DRYING TIMES FOR SELECTED FINISHING MATERIALS*	
Material	**Time to Recoat (hours)**
Acrylic Masonry Primer	4 to 5
Acrylic Wood Primer	3 to 4
Acrylic Exterior Paint	4 to 5
Latex Wood Primer	3 to 4
Latex Exterior Paint	4 to 5
Latex Interior Paint	4 to 5
Latex Semi-gloss Enamel	12 to 14
Latex Interior Primer	1 to 2
Latex Flat Enamel	2 to 3
Latex Block Filler	3 to 4
Linseed Based Oil Paint	24
Alkyd Metal Primer	3 to 4
Alkyd Gloss Enamel	16 to 18
Epoxy Enamel	5 to 6
Polyurethane Gloss Enamel	1
Urethanes	24
Lacquer	1 1/2 to 3
Non-grain-raising Stain	3
Oil Stain	24
Paste Wood Filler	24 to 48
Shellac	2
Varnish	18 to 24
Water Stain	—

* Consult the label on the can of finishing material for the manufacturer's recommendations.

SPREAD RATE CHART

Wet film to dry film coverage, square foot/gallon based on percent volume solids. *Courtesy Benjamin Moore Paints, www.benjaminmoore.com*

Dry Film Thickness in Mils	Volume Solids						
	30%	**35%**	**40%**	**45%**	**50%**	**55%**	**60%**
1.0	**3.3**	**2.9**	**2.5**	**2.2**	**2.0**	**1.8**	**1.7**
Sq. Ft./G	*481*	*561*	*642*	*722*	*802*	*882*	*962*
1.5	**5.0**	**4.3**	**3.8**	**3.3**	**3.0**	**2.7**	**2.5**
Sq. Ft./G	*321*	*374*	*428*	*481*	*535*	*588*	*642*
2.0	**6.7**	**5.7**	**5.0**	**4.4**	**4.0**	**3.6**	**3.3**
Sq. Ft./G	*241*	*281*	*321*	*361*	*401*	*441*	*481*
2.5	**8.3**	**7.1**	**6.3**	**5.6**	**5.0**	**4.5**	**4.2**
Sq. Ft./G	*192*	*225*	*257*	*289*	*321*	*353*	*385*
3.0	**10.0**	**8.6**	**7.5**	**6.7**	**6.0**	**5.5**	**5.0**
Sq. Ft./G	*160*	*187*	*214*	*241*	*267*	*294*	*321*
3.5	**11.7**	**10.**	**8.8**	**7.8**	**7.0**	**6.4**	**5.8**
Sq. Ft./G	*137*	*160*	*183*	*206*	*229*	*252*	*275*
4.0	**13.3**	**11.4**	**10.0**	**8.9**	**8.0**	**7.3**	**6.7**
Sq. Ft./G	*120*	*140*	*160*	*180*	*201*	*221*	*241*
4.5	**15.0**	**12.9**	**11.3**	**10.0**	**9.0**	**8.2**	**7.5**
Sq. Ft./G	*107*	*125*	*143*	*160*	*178*	*196*	*214*
5.0	**16.7**	**14.3**	**12.5**	**11.1**	**10.0**	**9.1**	**8.3**
Sq. Ft./G	*96*	*112*	*128*	*144*	*160*	*176*	*192*
5.5	**18.3**	**15.7**	**13.8**	**12.2**	**11.0**	**10.0**	**9.2**
Sq. Ft./G	*87*	*102*	*117*	*131*	*146*	*160*	*175*
6.0	**20.0**	**17.1**	**15.0**	**13.3**	**12.0**	**10.9**	**10.0**
Sq. Ft./G	*80*	*94*	*107*	*120*	*134*	*147*	*160*
6.5	**21.7**	**18.6**	**16.3**	**14.4**	**13.0**	**11.8**	**10.8**
Sq. Ft./G	*74*	*86*	*99*	*111*	*123*	*136*	*148*
7.0	**23.3**	**20.0**	**17.5**	**15.6**	**14.0**	**12.7**	**11.7**
Sq. Ft./G	*69*	*80*	*92*	*103*	*115*	*126*	*137*
7.5	**25.0**	**21.4**	**18.8**	**16.7**	**15.0**	**13.6**	**12.5**
Sq. Ft./G	*64*	*75*	*86*	*96*	*107*	*118*	*128*
8.0	**26.7**	**22.9**	**20.0**	**17.8**	**16.0**	**14.5**	**13.3**
Sq. Ft./G	*60*	*70*	*80*	*90*	*100*	*110*	*120*
8.5	**28.3**	**24.3**	**21.3**	**18.9**	**17.0**	**15.5**	**14.2**
Sq. Ft./G	*57*	*66*	*75*	*85*	*94*	*104*	*113*
9.0	**30.0**	**25.7**	**22.5**	**20.0**	**18.0**	**16.4**	**15.0**
Sq. Ft./G	*53*	*62*	*71*	*80*	*89*	*98*	*107*
9.5	**31.7**	**27.1**	**23.8**	**21.1**	**19.0**	**17.3**	**15.8**
Sq. Ft./G	*51*	*59*	*68*	*76*	*84*	*93*	*101*
10.0	**33.3**	**28.6**	**25.0**	**22.2**	**20.0**	**18.2**	**16.7**
Sq. Ft./G	*48*	*56*	*64*	*72*	*80*	*88*	*96*

GRAIN FILLING REQUIREMENTS FOR SELECTED WOODS		
No Filler Needed	**Liquid Filler**	**Paste Filler**
Basswood	Beech	Ash
Fir	Birch	Butternut
Hemlock	Cherry	Chestnut
Holly	Cottonwood	Elm
Larch	Gum, Red	Hickory/Pecan
Magnolia	Maple, Hard	Locust
Pine, White	Maple, Soft	Mahogany
Pine, Yellow	Sycamore	Red and White Oak
Poplar	—	Walnut
Willow	—	—

WOOD STAINS COMMONLY USED FOR INTERIOR APPLICATION

Type of Stain	Vehicle Solvent	Staining Action	Remarks
Alcohol	Alcohol	Penetrating	Dries quickly but fades easily. Sold in powder form. Will raise grain.
Gelled Wood Stain	Mineral Spirits or Turpentine	Pigmenting	Slow drying and does not fade. Sold in gelled form.
Latex Stain	Water	Pigmenting	Slow drying and does not fade. Sold in liquid form.
Non-grain-raising Stain	Alcohol Glycol	Penetrating	Dries quickly and does not fade or raise the grain. Sold in liquid form.
Oil Stain (penetrating)	Mineral Spirits or Turpentine	Penetrating	Dries quickly and fades easily but does not raise the grain. Sold in liquid form.
Oil Stain (pigmenting)	Mineral Spirits or Turpentine	Pigmenting	Slow drying and does not fade. Sold in liquid form.
Penetrating Resin Stain	Mineral Spirits	Penetrating	Sold in liquid form. Contains stain and protective coating in one coating.
Water	Water	Penetrating	Dries quickly but fades easily. Sold in powder form. Will raise the grain.

WOOD STAINS – CHARACTERISTICS, APPLICATIONS, AND DRYING TIMES			
Stain	**Characteristics**	**Application Methods**	**Drying Time**
Water Stain	Produces one of the clearest colors; nonbleeding; raises grain, which requires light sanding.	Wiping, Spraying	12 Hours
Penetrating Oil Stains	Non-grain-raising; likely to bleed; fades.	Wiping, Brushing, Spraying	24 Hours
Pigmented Oil Stains	Easiest to apply; non-grain-raising; hides grain.	Wiping, Brushing, Spraying	3 to 12 Hours
Non-grain-raising Stains	Dries rapidly; nonbleeding non-grain-raising; clear colors; streaks easily.	Spraying	15 Minutes to 3 Hours

WALLCOVERING MATERIALS	
Lining paper.	Used to smooth walls that are in a rough condition.
Vinyl-coated paper. Has a paper substrate that is sprayed with an acrylic-type vinyl or a polyvinyl chloride.	Any room, especially areas with moisture, such as a kitchen or bath.
Vinyl-coated fabric. The fabric is coated with a liquid vinyl or acrylic. A decorative layer is printed on this coating.	Use in a low moisture area, such as living areas.
Solid decorative sheet laminated to a paper substrate.	Use in most any room but not where it might be subject to abuse.
Decorative vinyl film laminated to a fabric substrate.	Use in any room. Is durable and washable.
OTHER WALLCOVERING MATERIALS	
Foil, plastic film with a metallic finish.	Use in most any room where it will not be subject to abuse and is moisture resistant.
Wood veneers, cork, burlap, hemp, grass cloth.	Use in areas of low moisture and unlikely to be subject to abuse.
Flocks with paper, vinyl, or foil substrate.	Use in most areas not subject to abuse.
Borders.	Use in any area desired.

SIZES OF ROLLS

Widths from 18 to 54" are available in various types of wallcovering materials. Most commonly used widths are 20½ and 27". The 54" roll is used for commercial coverings. Rolls are usually sold two to a package. This is called a bolt.

ADHESIVES

Most paper-backed wallcoverings have prepasted adhesive on the back. Fabric-backed materials should be bonded with the adhesive recommended by the manufacturer.

DIMENSIONS OF WALLCOVERINGS

Width of Roll (in.)	Length of Single Rolls (ft.)	Total Area (sq. ft.)
18	24	36
20.5	21	36
24	18	36
27	15	34
28	15	35
36	12	36

WALLCOVERING ESTIMATING

FIGURING THE NUMBER OF SINGLE METRIC ROLLS NEEDED FOR ROOMS MEASURED IN FEET

Single Metric Rolls				
Distance Around Room in Ft.	Ceiling Height			To Decorate Ceiling
	8 ft.	9 ft.	10 ft.	
36	12	14	16	4
40	14	16	18	6
44	16	16	20	6
48	16	20	22	8
52	18	20	24	8
56	20	22	24	8
60	20	24	26	10
64	22	24	28	10
68	24	26	28	12
72	24	28	30	14

* Metric single rolls cover 2.6 square meters (28 square feet).

WALL COVERING ESTIMATING – NUMBER OF INCH ROLLS FOR VARIOUS ROOM SIZES*

Room Size (m)	Room Size (ft.)	Ceiling Height					
		8' 2.4 m	9' 2.7 m	10' 3.0 m	11' 3.4 m	12' 3.7 m	To Cover Ceiling
		Single 20 1/2" Rolls					
3.4 × 3.0	8 × 10	14	14	16	18	20	4
3.0 × 3.0	10 × 10	14	16	18	20	22	4
3.0 × 3.7	10 × 12	16	18	20	22	24	6
3.0 × 4.3	10 × 14	18	20	22	24	26	6
3.7 × 3.7	12 × 12	18	20	22	24	26	6
3.7 × 4.3	12 × 14	20	22	24	26	28	8
3.7 × 4.1	12 × 16	20	22	26	28	30	8
3.7 × 5.5	12 × 18	22	24	28	30	32	10
3.7 × 6.1	12 × 20	24	26	30	32	34	12
4.3 × 4.3	14 × 14	20	22	26	28	30	10
4.3 × 4.9	14 × 16	22	24	28	30	32	10

4.3 × 5.5	14 × 18	24	26	30	32	34	12
4.3 × 6.1	14 × 20	24	28	32	34	38	12
4.3 × 6.7	14 × 22	26	30	32	36	40	14
4.9 × 4.9	16 × 16	24	26	30	32	34	12
4.9 × 5.5	16 × 18	24	28	32	34	38	14
4.9 × 6.1	16 × 20	26	30	32	36	40	14
4.9 × 6.7	16 × 22	28	32	34	38	42	16
4.9 × 7.3	16 × 24	30	32	36	40	44	18
5.5 × 5.5	18 × 18	26	30	32	36	40	14
5.5 × 6.1	18 × 20	28	32	34	38	42	16
5.5 × 6.7	18 × 22	30	32	36	40	44	18
5.5 × 7.2	18 × 24	30	34	38	42	46	20

* Rolls 20 1/2" wide, 11 yards long (56 square feet).

WALLCOVERING DURABILITY CATEGORIES

Wall Coverings – Noncommercial Installations

- I Decorative Only
- II Decorative, Medium Serviceability
- III Decorative, High Serviceability

Wallcovering – Commercial Installations

- IV Type I Commercial Serviceability
- V Type II Commercial Serviceability
- VI Type III Commercial Serviceability

*ASTM Standard Classification of Wallcovering by Durability Characteristics, ASTM F 793. Courtesy American Society for Testing and Materials. www.astm.org.

CHAPTER 15

Metal Structural Products and Lightweight Steel Construction

FIRE AND SOUND TRANSMISSION RATINGS OF STEEL STUD PARTITIONS

Stud (in.)	Fire Rating	Sound Transmission Class*	Description
SINGLE LAYER GYPSUM WALLBOARD ON BOTH SIDES**			
2 1/2" 24" o.c.	1 hr.	40 45 with insulation	5/6" fire-resistant gypsum wallboard screwed vertically both sides of studs, 2 1/2" fiberglass insulation
3 5/8" 24" o.c.	1 hr.	42 47 with insulation	5/8" fire-resistant gypsum wallboard, screwed to both sides, 2 1/2" fiberglass insulation
DOUBLE LAYER GYPSUM WALLBOARD ON BOTH SIDES**			
2 1/2" 24" o.c.	2 hr.	46 53 with insulation	Two layers of 1/2" fire-resistant gypsum wallboard, screwed to both sides, 3" fiberglass insulation
3 1/2" 24" o.c.	2 hr.	48 53 with insulation	Two layers of 1/2" fire-resistant gypsum wallboard, screwed to both sides, 3" fiberglass insulation

*STC – Sound Transmission Class. The higher the number, the more efficient the material reduces sound transmission.

**Consult local building codes for specific requirements.

RECOMMENDED METAL DRILLING SPEEDS			
		Drill Diameter	
Material	**Speed rpm**	**(inches)**	**(mm)**
Cast Iron	6000 to 6500	1/16	1.59
	3500 to 4500	1/8	3.18
	2500 to 3000	3/16	4.75
	2000 to 2500	1/4	6.35
	1500 to 2000	5/16	7.94
	1500 to 2000	3/8	9.53
Soft Metals (copper)	6000 to 6500	1/16	1.59
	6000 to 6500	1/8	3.18
	5000 to 6000	3/16	4.75
	4500 to 5000	1/4	6.35
	3500 to 4000	5/16	7.94
	3000 to 3500	3/8	9.53
Steel	5000 to 6500	1/16	1.59
	3000 to 4000	1/8	3.18
	2000 to 2500	3/16	4.75
	1500 to 2000	1/4	6.35
	1000 to 1500	5/16	7.94
	1000 to 1500	3/8	9.53

METAL CUTTING FLUIDS

Metals	Power Sawing	Drilling	Reaming	Threading
Carbon Steels Malleable Iron	EMO MO MLO	EMO SUL MLO LARD	MLO SUL EMO	SUL MLO EMO LARD
Stainless Steels	EMO MLO MO	EMO SUL MLO	MOL SUL	SUL MLO EMO
Gray Cast Iron	Dry EMO	Dry EMO	Dry EMO	Dry EMO MLO
Aluminum Alloys	Dry EMO MO	EMO MLO KER MO	MLO MO EMO	MLO MO EMO KER LARD
Copper Brass Bronze	Dry MO EMO MLO	EMO MO MLO KER Dry LARD	MLO MO EMO	MLO MO EMO LARD

KEY

Emulsified (soluble) Oils	EMO
Mineral Oils	MO
Mineral-lard Oils	MLO
Sulfurized Oils	SUL
No Cutting Fluid	DRY
Kerosene	KER
Lard Oil	LARD

FASTENERS USED TO SECURE SHEET METAL TO VARIOUS MATERIALS

Fastener Illustration	Fastener Type	Substrate				
		Sheet Metal (18 gauge or light)	Metal (18 gauge or heavier)	Wood	Concrete	Masonry
	Ring or Annular Shank Nail			X		
	Barbed Shank Nail (smooth)			X		
	Screw Shank Nail			X		
	Pop Rivet	X				
	Shear Rivet	X	X			
	Nail-in Expansion Fastener				X	X
or	Nail-in Expansion Fastener				X	

	Concrete Screw				X	X
	Self-Piercing Screw	X		X		
	Self-drilling Screw	X	X			
	Self-cutting Screw	X		X		
	Self-tapping Screw	X	X			

Hex Washer Head Flat Head Wafer Head Oval Head Pancake Head Pan Head Bugle Head Trumpet Head

Courtesy National Roofing Contractors Association, www.nrca.net

TYPICAL FASTENERS FOR SECURING LIGHTWEIGHT STEEL FRAMING

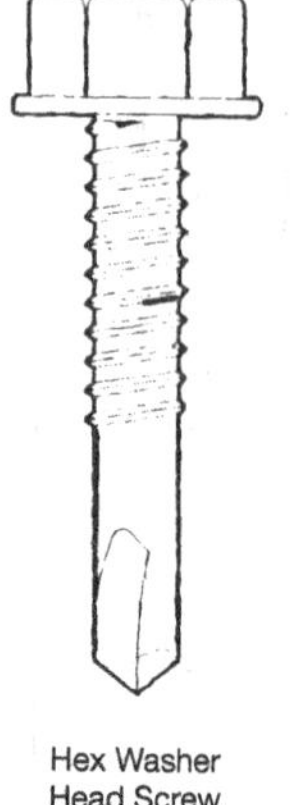

Hex Washer
Head Screw

Pan Head
Screw

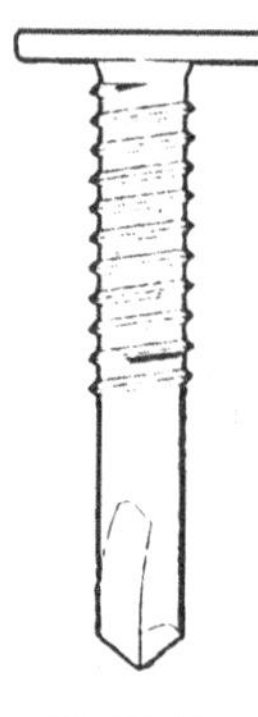

Low Profile
Pan Head Screw

Consult manufacturer for other types and sizes.
See additional information on fasteners in Chapter 7.

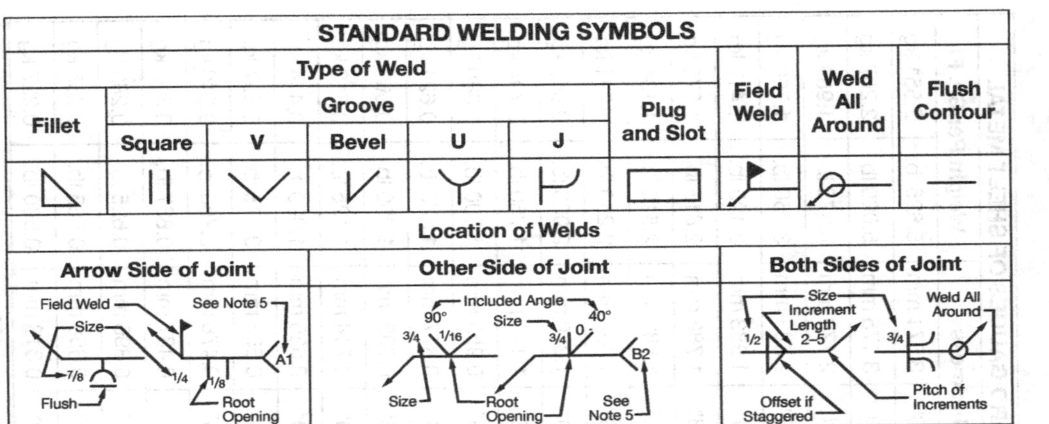

1. The side of the joint to which the arrow points is the arrow side.
2. Both-sides welds of the same type are of same size unless otherwise shown.
3. Symbols apply between abrupt changes in direction of joint or as dimensioned (except where all around symbol is used).
4. All welds are continuous and of user's standard proportions, unless otherwise shown.
5. Tail of arrow used for specification reference. (Tail may be omitted when reference not used.)
6. Dimensions of weld sizes, increment lengths and spacings in inches.

U.S. STANDARD GAUGES OF SHEET METAL				
Gauge	**Thickness**		**Weight Per Sq. Ft.**	
10	0.1406"	3.571 mm	5.625 lb.	2.551 kg
11	0.1250"	3.175 mm	5.000 lb.	2.267 kg
12	0.1094"	2.778 mm	4.375 lb.	1.984 kg
13	0.0938"	2.383 mm	3.750 lb.	1.700 kg
14	0.0781"	1.983 mm	3.125 lb.	1.417 kg
15	0.0703"	1.786 mm	2.813 lb.	1.276 kg
16	0.0625"	1.588 mm	2.510 lb.	1.134 kg
17	0.0563"	1.430 mm	2.250 lb.	1.021 kg
18	0.0500"	1.270 mm	2.000 lb.	0.907 kg
19	0.0438"	1.111 mm	1.750 lb.	0.794 kg
20	0.0375"	0.953 mm	1.500 lb.	0.680 kg
21	0.0344"	0.877 mm	1.375 lb.	0.624 kg
22	0.0313"	0.795 mm	1.250 lb.	0.567 kg
23	0.0280"	0.714 mm	1.125 lb.	0.510 kg
24	0.0250"	0.635 mm	1.000 lb.	0.454 kg
25	0.0219"	0.556 mm	0.875 lb.	0.397 kg
26	0.0188"	0.478 mm	0.750 lb.	0.340 kg
27	0.0172"	0.437 mm	0.687 lb.	0.312 kg
28	0.0156"	0.396 mm	0.625 lb.	0.283 kg
29	0.0141"	0.358 mm	0.563 lb.	0.255 kg
30	0.0120"	0.318 mm	0.500 lb.	0.227 kg

MINIMUM THICKNESS FOR UNCOATED STRUCTURAL-QUALITY SHEET STEEL

Designation (mils)	Minimum Uncoated Thickness Inches (mm)	Reference Gauge Number
18	0.018 (0.455)	25
27	0.027 (0.683)	22
33	0.033 (0.836)	20
43	0.043 (1.087)	18
54	0.054 (1.367)	16
68	0.068 (1.720)	14
97	0.097 (2.454)	12

TYPICAL COLD-FORMED LIGHTWEIGHT STRUCTURAL STEEL SHAPES

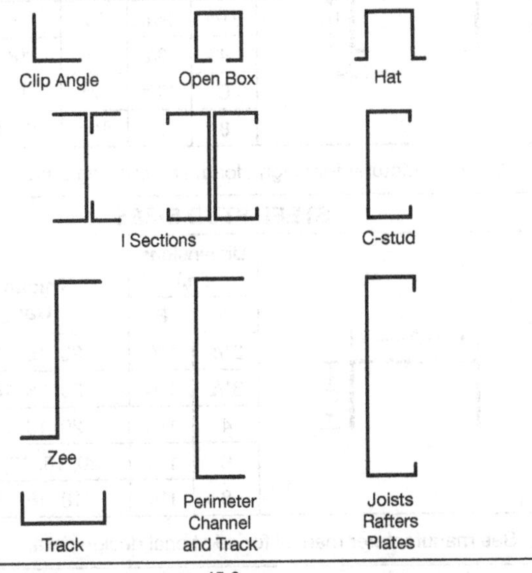

STEEL JOIST SIZES

Dimensions (in.) A	B	C	Thickness Gauge
6	2	9/16	18, 16, 14, 12
8	2	9/16	18, 16, 14, 12
10	2	11/16	16, 14, 12
12	2	11/16	16, 14, 12

See manufacturer manual for additional design data.

STEEL STUD SIZES

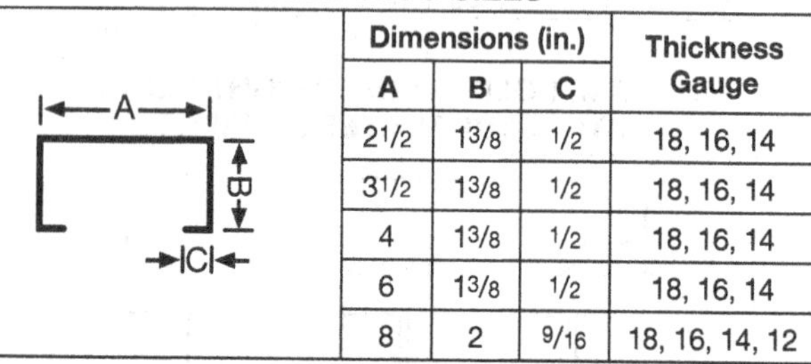

Dimensions (in.) A	B	C	Thickness Gauge
2 1/2	1 3/8	1/2	18, 16, 14
3 1/2	1 3/8	1/2	18, 16, 14
4	1 3/8	1/2	18, 16, 14
6	1 3/8	1/2	18, 16, 14
8	2	9/16	18, 16, 14, 12

See manufacturer for height, load, and spacing data.

STEEL STUD SIZES

Dimensions (in.) A	B	Thickness Gauge
2 1/2	1 1/4	20, 18, 16, 14
3 1/2	1 1/4	20, 18, 16, 14
4	1 1/4	20, 18, 16, 14
6	1 1/4	20, 18, 16, 14, 12
8	1 1/4	18, 16, 14, 12

See manufacturer manual for additional design data.

TYPES OF BRIDGING – STRAP, TRACK, AND SOLID

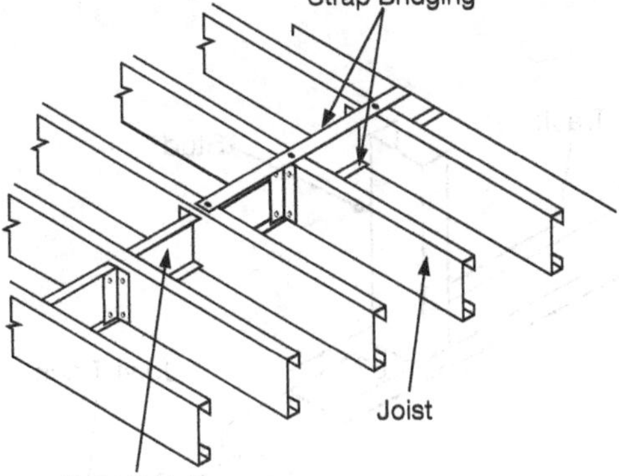

Strap Bridging

Joist

Solid Bridging

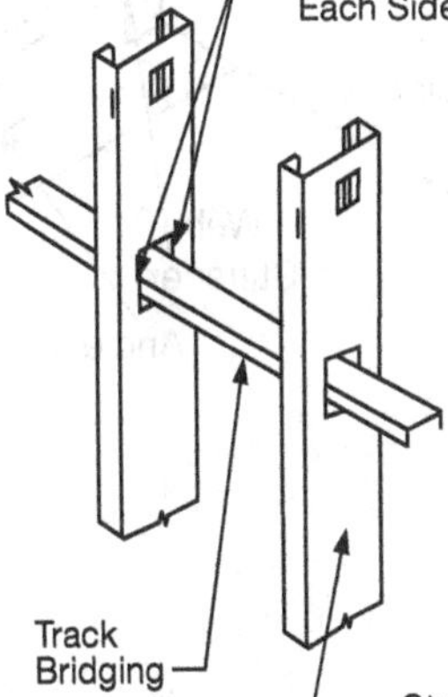

Weld on
Each Side

Track
Bridging

Studs

TYPICAL LIGHT STEEL FLOOR FRAMING

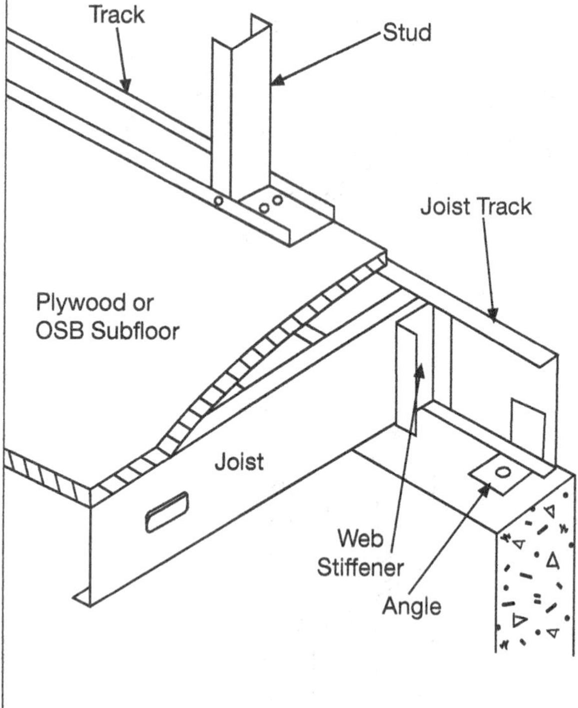

TYPICAL TRUSS ROOF FRAMING DETAIL AT EXTERIOR WALL

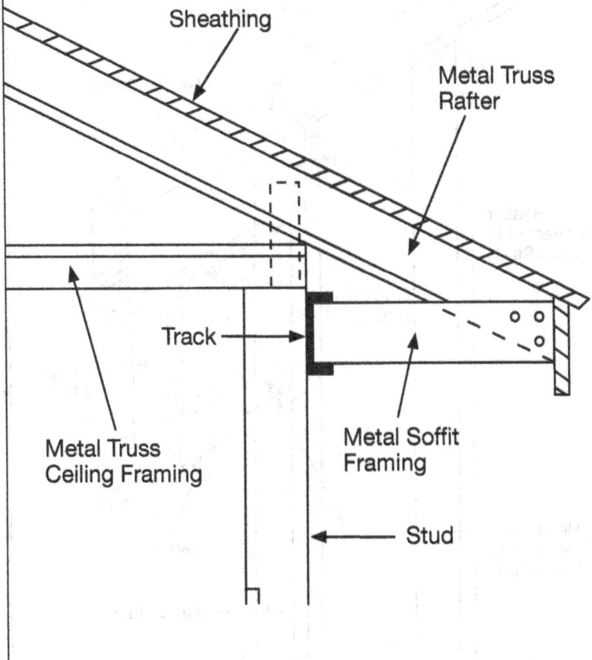

Sheathing

Metal Truss Rafter

Track

Metal Truss Ceiling Framing

Metal Soffit Framing

Stud

TYPICAL WALL OPENING HEADER DETAIL FOR WINDOWS AND DOORS

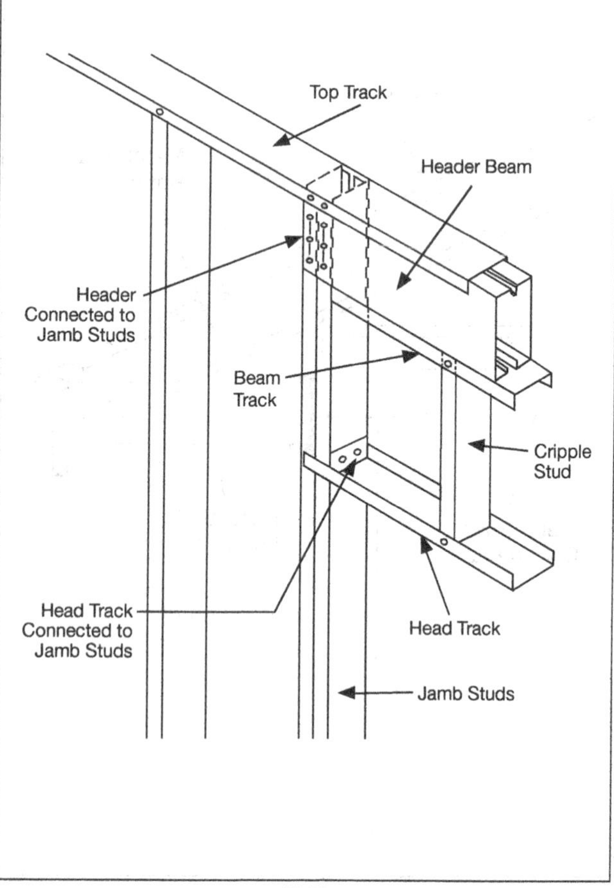

SIZES OF A COLD-FORMED OPEN-WEB STEEL JOIST

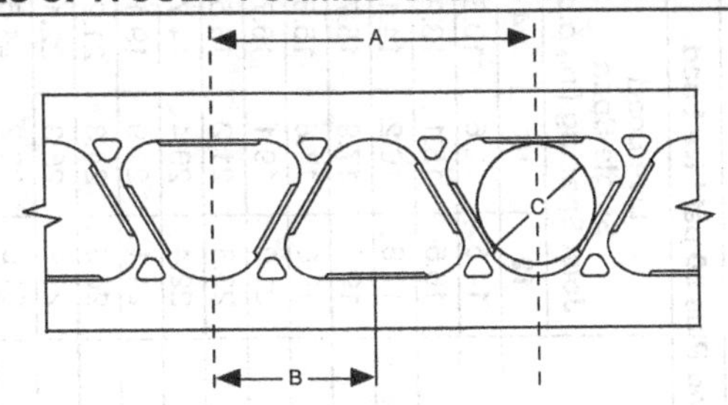

Hole Sizes

Section Depth (in.)	A Pitch (in.)	B 1/2 Pitch (in.)	C Maximum Diameter (in.)
8	14.0	7.00	5.5
10	28.0	14.00	7.2
12	35.0	17.50	9.2
14	35.0	17.50	9.2

Steel Thickness

Product		Design		Minimum		ASTM C955 Color Code
Gauge	Mils	(in.)	(mm)	(in.)	(mm)	
18	43	0.0451	1.15	0.0428	1.09	Yellow
16	54	0.0566	1.44	0.0538	1.37	Green
14	68	0.0713	1.81	0.0677	1.72	Orange
12	97	0.1017	2.58	0.0966	2.45	Red

Courtesy Marino/Ware Industries, www.marinoware.com

SPAN TABLES FOR A COLD-FORMED OPEN-WEB STEEL FLOOR JOIST

JOISTRITE™ FLOOR JOIST SPAN TABLE (FT.)

15 psf Dead Load 40 psf Live Load				15 psf Dead Load Plus 60 psf Live Load			
Section Identification	Deflection L/360 Single Span Joist Spacing (in.) o.c.			Section Identification	Live Load Single Span Joist Spacing (in.) o.c.		
	12	16	24		12	16	24
800JR200-43	17.0	14.8	12.0	800JR200-43	14.6	12.6	10.3
800JR200-54	19.0	17.3	15.1	800JR200-54	16.6	15.1	13.2
800JR200-68	20.4	18.5	16.2	800JR200-68	17.8	16.2	14.1
800JR200-97	22.6	20.5	17.9	800JR200-97	19.7	17.9	15.7
1000JR200-54	22.4	20.3	17.8	1000JR200-54	19.6	17.8	15.5
1000JR200-68	24.0	21.8	19.1	1000JR200-68	21.0	19.1	16.7
1000JR200-97	26.7	24.2	21.2	1000JR200-97	23.3	21.2	18.5
1200JR200-54	25.7	23.3	20.0	1200JR200-54	22.4	20.4	14.7
1200JR200-68	27.6	25.0	21.9	1200JR200-68	24.1	21.9	19.1
1200JR200-97	30.6	27.8	24.3	1200JR200-97	26.8	24.3	21.2
1400JR200-68	31.9	29.0	25.3	1400JR200-68	27.9	25.3	22.1
1400JR200-97	35.5	32.3	28.2	1400JR200-97	31.0	28.2	24.6

Courtesy Marino/Ware Industries, www.marinoware.com

SOME OF THE HOT-ROLLED STRUCTURAL STEEL SHAPES

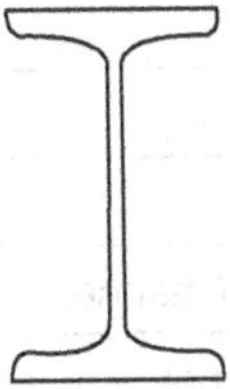

American Standard Beam (S)

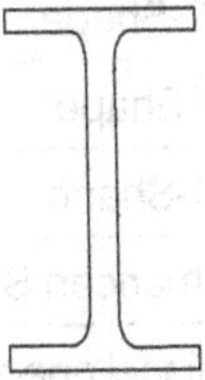

Wide Flange (W) M- and HP-Shapes Similar

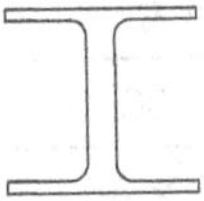

W-Shape

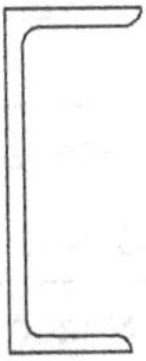

Standard Channel (C)

Angle (L) Equal Legs

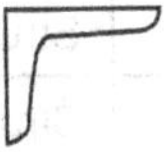

Angle (L) Unequal Legs

Structural Tee (WT, ST, MT)

Structural Tubing (TS) Rectangular

Structural Tubing (ST) Square

Structural Pipe

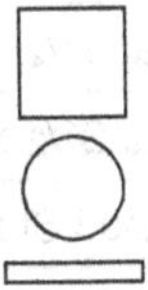

Bars – Square, Round, Flat

Plate (PL)

HOT-ROLLED STRUCTURAL STEEL SHAPE DESIGNATIONS	
Designation	**Type of Shape**
W18 × 70	W-Shape
S18 × 70	S-Shape
M10 × 9	M-Shape
C12 × 30	American Standard Channel
MC12 × 45	Miscellaneous Channel
HP12 × 74	HP Shape
L8 × 8 × 1	Equal-leg Angle
L4 × 3 × 1/2	Unequal-leg Angle
WT8 × 25	Structural Tee Cut from W-Shape
ST12 × 60	Structural Tee Cut from S-Shape
MT4 × 11.25	Structural Tee Cut from M-Shape
PL 1/2 × 24	Plate
Bar 1	Square Bar
Bar 1 1/2	Round Bar
Bar 2 1/2 × 1/2	Flat Bar
Pipe3 Standard	Pipe
Pipe3X–Strong	—
Pipe3XX–Strong	—
TS3 × 3 × 0.375	Structural Tubing: Square
TS4 × 3 × 0.375	Structural Tubing: Rectangular
TS3 OD × 0.250	Structural Tubing: Circular

DATA FOR SELECTED SIZES OF 4" ROUND STRUCTURAL STEEL PIPE

Available Strength in Axial Compression, Kips F_y=35 ksi	Pipe 4

Shape		Pipe 4					
		XXS		XS		Standard	
t Design (in.)		0.628		0.315		0.221	
Weight/Feet		27.6		15.0		10.8	
Design		P_n/Ω_c	$\phi_c P_n$	P_n/Ω_c	$\phi_c P_n$	P_n/Ω_c	$\phi_c P_n$
		ASD	LRFD	ASD	LRFD	ASD	LRFD
Effective Length *KL* (ft.) with Respect to Least Radius of Gyration r_y	0	160	241	86.8	130	62.2	93.6
	6	139	210	76.9	116	55.4	83.3
	7	133	199	73.6	111	53.2	79.9
	8	125	188	70.0	105	50.7	76.2
	9	117	176	66.2	99.4	48.0	72.1
	10	109	164	62.1	93.3	45.1	67.9
	11	101	151	57.9	87.0	42.2	63.4
	12	92.2	139	53.6	80.5	39.2	53.9
	13	83.8	126	49.3	74.1	36.2	54.4
	14	75.6	114	45.0	67.7	33.2	49.9
	15	67.6	102	40.9	61.4	30.2	45.4

ASD	LRFD
Ω_c=1.67	ϕ_c=0.90

DATA FOR SELECTED SIZES OF 8×8" SQUARE STRUCTURAL STEEL TUBING

Available Strength in Axial Compression, Kips F_y=46 ksi							Sq. HSS 8×8
Shape		HSS 8×8"					
Thickness (in.)		1/4		3/16c		1/8c	
t Design (in.)		0.233		0.174		0.116	
Weight/Feet		25.8		19.6		13.3	
Design		P_n/Ω_c	$\phi_c P_n$	P_n/Ω_c	$\phi_c P_n$	P_n/Ω_c	$\phi_c P_n$
		ASD	LRFD	ASD	LRFD	ASD	LRFD
Effective Length KL (ft.) with Respect to Least Radius of Gyration r_y	0	196	294	130	195	63.1	94.9
	6	189	284	127	191	62.3	93.7
	7	186	280	126	190	62.0	93.2
	8	184	276	125	188	61.7	92.7
	9	181	272	124	186	61.3	92.2
	10	177	267	122	184	60.9	91.5
	11	174	261	121	182	60.4	90.8
	12	170	255	119	179	59.9	90.0
	13	166	249	117	177	59.3	89.1
	14	162	243	115	174	58.7	88.2
	15	157	236	113	170	58.0	87.2

ASD	LRFD
Ω_c=1.67	ϕ_c=0.90

ALLOWABLE LOADS AND SPANS FOR STEEL ANGLES*

Angle Size	Weight per Foot	Span (feet)				
		2*	3	4	5	6
		Allowable Uniform Load in Kips				
$5 \times 3^1/2 \times 3/8$	10.4	15.3	10.3	7.7	6.1	5.1
$5 \times 3^1/2 \times 1/2$	13.6	20.0	13.3	10.0	8.0	6.7
$4 \times 3^1/2 \times 5/16$	7.7	8.7	5.8	4.3	3.5	2.9
$3^1/2 \times 3^1/2 \times 1/4$	5.8	5.3	3.5	2.6	2.1	1.8

* A kip is equal to 1000 lbs.

TYPICAL CONCENTRATED LOADS ON ROUND STANDARD WEIGHT STEEL COLUMNS

Diameter (in.)	Column Length (ft.)			
	6	8	10	12
3 1/2	45	40	33	26
4	58	53	47	38
4 1/2	72	67	60	55
5	86	81	75	68

Consult manufacturer for specific data.

CHAPTER 16
Engineering Data

METRIC UNITS AND THEIR SYMBOLS		
METRIC BASE UNITS		
Physical Quality	**Unit**	**Symbol**
Length	Meter	m
Mass	Kilogram	kg
Time	Second	s
Electric Current	Ampere	A
Thermodynamic Temperature	Celsius	C
LuminousIntensity	Candela	cd

METRIC UNITS AND THEIR SYMBOLS *(cont.)*		
METRIC DERIVED UNITS		
Quantity	**Name**	**Symbol**
Frequency	Hertz	Hz
Force	Newton	N
Pressure, Stress	Pascal	Pa
Energy, Work, Quantity of Heat	Joule	J
Power, Radiant Flux	Watt	W
Electric Charge, Quantity	Coulomb	C
Electric Potential	Volt	V
Capacitance	Farad	F
Electric Resistance	Ohm	W
Electric Conductance	Siemens	S
Magnetic Flux	Weber	Wb
Magnetic Flux Density	Tesla	T
Inductance	Henry	H
Luminous Flux	Lumen	lm
Illuminance	Lux	lx

All metric units other than the base units are called derived units. New derived units are developed as new technology requires a means of identification.

SELECTED METRIC CONVERSION FACTORS		
When You Know	**You Can Find**	**If You × By**
LENGTH		
Inches	Millimeters	25.4
Feet	Millimeters	300.48
Yards	Meters	0.91
Miles	Kilometers	1.61
Millimeters	Inches	0.04
Meters	Yards	1.1
Kilometers	Miles	0.6
AREA		
Square Inches	Square Centimeters	6.45
Square Feet	Square Meters	0.09
Square Yards	Square Meters	0.83
Square Miles	Square Kilometers	2.6
Acres	Square Hectometers (Hectares)	0.4
Square Centimeters	Square Inches	0.16
Square Meters	Square Yards	1.2
Square Kilometers	Square Miles	0.4
Hectares	Acres	2.5

SELECTED METRIC CONVERSION FACTORS *(cont.)*		
When You Know	**You Can Find**	**If You × By**
MASS (Weight)		
Ounces	Grams	28.0
Pounds	Kilograms	0.45
Tons (short)	Metric Tons	0.9
Grams	Ounces	0.04
Kilograms	Pounds	2.2
Metric Tons	Tons (Short)	1.1
VOLUME		
Bushels	Cubic Meters	0.04
Cubic Feet	Cubic Meters	0.03
Cubic Inches	Cubic Centimeters	16.4
Cubic Yards	Cubic Meters	0.8
FLUID VOLUME		
Ounces	Milliliters	30.0
Pints	Liters	0.47
Quarts	Liters	0.95
Gallons	Liters	3.8
Milliliters	Ounces	0.03
Liters	Pints	2.1

SELECTED METRIC CONVERSION FACTORS *(cont.)*		
When You Know	**You Can Find**	**If You × By**
FLUID VOLUME *(cont.)*		
Liters	Quarts	1.06
Liters	Gallons	0.26
TEMPERATURE		
Degrees Fahrenheit	Degrees Celsius	0.6 (after subtracting 32)
Degrees Celsius	Degrees Fahrenheit	1.8 (then add 32)
POWER		
Horsepower	Kilowatts	0.75
Kilowatts	Horsepower	1.34
PRESSURE		
Pounds per Square Inch (psi)	Kilopascals	6.9
Kilopascals	Pounds per Square Inch	0.15
VELOCITY (Speed)		
Miles per Hour	Meters per Second	0.45
Miles per Hour	Kilometers per Hour	1.6
Kilometers per Hour	Miles per Hour	0.6

OTHER CONVERSION FACTORS		
Multiply	**By**	**To Obtain**
Acres	43,560	Square Feet
Acres	1.562×10^{-3}	Square Miles
Acre-feet	43,560	Cubic Feet
British Thermal Units	252.0	Calories
British Thermal Units	778.2	Foot-pounds
British Thermal Units	3.960×10^{-4}	Horsepower-hours
British Thermal Units	107.6	Kilogram-meters
British Thermal Units	2.931×10^{-4}	Kilowatt-hours
British Thermal Units	1,055	Watt-seconds
B.t.u. per Hour	2.931×10^{-4}	Kilowatts
B.t.u. per Minute	2.359×10^{-2}	Horsepower
B.t.u. per Minute	1.759×10^{-2}	Kilowatts
Bushels	1.244	Cubic Feet
Cubic Feet	0.02832	Cubic Meters
Cubic Feet	7.481	Gallons
Cubic Feet	28.32	Liters
Cubic Inches	16.39	Cubic Centimeters
Cubic Meters	35.31	Cubic Feet
Cubic Meters	1.308	Cubic Yards
Cubic Yards	0.7646	Cubic Meters
Degrees (angle)	0.01745	Radians
Foot-pounds	1.285×10^{-2}	British Thermal Units
Foot-pounds	5.050×10^{-7}	Horsepower-hours
Foot-pounds	1.356	Joules

OTHER CONVERSION FACTORS *(cont.)*		
Multiply	**By**	**To Obtain**
Foot-Pounds	0.1383	Kilogram-meters
Foot-Pounds	3.766×10^{-7}	Kilowatt-hours
Gallons	0.1337	Cubic Feet
Gallons	231	Cubic Inches
Gallons	3.785×10^{-3}	Cubic Meters
Gallons	3.785	Liters
Gallons per Minute	2.228×10^{-3}	Cubic Feet per Second
Horsepower	42.41	B.t.u. per Minute
Horsepower	2,544	B.t.u per Hour
Horsepower	550	Foot-pounds per Second
Horsepower	33,000	Foot-pounds per Minute
Horsepower	1.014	Horsepower (metric)
Horsepower	10.70	Kg. Calories per Minute
Horsepower	0.7457	Kilowatts
Horsepower (boiler)	33,520	B.t.u. per Hour
Horsepower-hours	2,544	British Thermal Units
Horsepower-hours	1.98×10^{6}	Foot-pounds
Horsepower-hours	2.737×10^{5}	Kilogram-meters
Horsepower-hours	0.7457	Kilowatt-hours
Inches of Mercury	1.133	Feet of Water
Inches of Mercury	70.73	Pounds per Square Foot
Inches of Mercury	0.4912	Pounds per Square Inch
Inches of Water	25.40	Kilograms per Square Meter
Inches of Water	0.5781	Ounces per Square Inch

OTHER CONVERSION FACTORS *(cont.)*		
Multiply	**By**	**To Obtain**
Inches of Water	5.204	Pounds per Square Foot
Joules	9.478×10^{-4}	British Thermal Units
Joules	0.2388	Calories
Joules	10^{7}	Ergs
Joules	0.7376	Foot-Pounds
Joules	2.778×10^{-7}	Kilowatt-hours
Joules	0.1020	Kilogram-meters
Joules	1	Watt-seconds
Kilograms	2.205	Pounds
Kilogram-calories	3.968	British Thermal Units
Kilogram Meters	7.233	Foot-pounds
Kilograms per Square Meter	3.281×10^{-3}	Feet of Water
Kilograms per Square Meter	0.2048	Pounds per Square Foot
Kilograms per Square Meter	1.422×10^{-3}	Pounds per Square Inch
Kilometers	3.281	Feet
Kilometers	0.6214	Miles
Kilowatts	56.87	B.t.u. per Minute
Kilowatts	737.6	Foot-pounds per Second
Kilowatts	1.341	Horsepower
Kilowatts-hours	3409.5	British Thermal Units
Kilowatts-Hours	2.655×10^{6}	Foot-pounds
Knots	1.152	Miles

OTHER CONVERSION FACTORS *(cont.)*		
Multiply	**By**	**To Obtain**
Lumens per Square Ft.	1	Footcandles
Miles	5,280	Feet
Miles	1.609	Kilometers
Ohms	10^{-6}	Megohms
Ohms	10^{6}	Microhms
Ohms per Mil. Foot	0.1662	Microhms per Cm. Cube
Ohms per Mil. Foot	0.06524	Microhms per In. Cube
Pounds	32.17	Poundals
Pound-feet	0.1383	Meter-kilograms
Pounds of Water	0.01602	Cubic Feet
Pounds of Water	0.1198	Gallons
Pounds per Cubic Ft.	16.02	Kg. per Cubic Meter
Pounds per Cubic Ft.	5.787×10^{-4}	Pounds per Cubic Inch
Pounds per Cubic In.	27.68	Grams per Cubic Cm.
Pounds per Cubic In.	2.768×10^{-4}	Kg. per Cubic Meter
Pounds per Cubic In.	1.728	Pounds per Cubic Foot
Pounds per Square Ft.	0.01602	Feet of Water
Pounds per Square Ft.	4.882	Kg. per Square Meter
Pounds per Square Ft.	6.944×10^{-3}	Pounds per Square Inch
Pounds per Square In.	2.307	Feet of Water
Pounds per Square In.	2.036	Inches of Mercury
Pounds per Square In.	703.1	Kg. per Square Meter
Radians	57.30	Degrees
Square Feet	2.296×10^{-5}	Acres

OTHER CONVERSION FACTORS *(cont.)*		
Multiply	**By**	**To Obtain**
Square Feet	0.09290	Square Meters
Square Inches	1.273×10^6	Circular Mils
Square Inches	6.452	Square Centimeters
Square Kilometers	0.3861	Square Miles
Square Meters	10.76	Square Feet
Square Miles	640	Acres
Square Miles	2.590	Square Kilometers
Tons (Long)	2,240	Pounds
Tons (Metric)	2,205	Pounds
Tons (Short)	2,000	Pounds
Watts	0.05686	B.t.u. per Minute
Watts	10^7	Ergs per Second
Watts	44.26	Foot-pounds per Minute
Watts	1.341×10^{-3}	Horsepower
Watts	14.34	Calories per Minute
Watts-hours	3.412	British Thermal Units
Watts-hours	2,655	Footpounds
Watts-hours	1.341×10^{-3}	Horsepower-hours
Watts-hours	0.8605	Kilogram-calories
Watts-hours	376.1	Kilogram-meters

INCH TO MILLIMETERS LENGTH CONVERSIONS

Inches	Millimeters	Inches	Millimeters
1/16	1.59	8	203.20
1/8	3.18	10	254.00
3/16	4.76	12	304.80
1/4	6.35	16	406.40
3/8	9.53	18	457.20
1/2	12.70	21	533.40
5/8	15.88	24	609.60
3/4	19.05	27	685.80
7/8	22.23	30	762.00
1	25.40	33	838.20
2	50.80	36	914.40
4	101.60	39	990.60
6	152.40	48	1219.20

MILLIMETER TO INCHES LENGTH CONVERSIONS

Millimeters	Inches	Millimeters	Inches
1	1/16	22	7/8
3	1/8	25.4	1
6	1/4	32	1 1/4
8	5/16	100	4
10	3/8	305	12
12	1/2	500	19 3/4
16	5/8	1000	39 3/8
19	3/4	—	—

*Inches are expressed to the nearest 16th.

METERS TO FEET LENGTH CONVERSIONS	
Meters	**Feet**
1 meter	3'-3 3/8"
2 meters	6'-6 3/4"
2.5 meters	8'-2 1/2"
4 meters	13'-1 1/2"
6 meters	19'-8 1/4"

VOLUME CONVERSIONS LITERS TO GALLONS	
Liters	**Gallons**
1 liter	0.26 gallons
4 liters	1 gallon
15 liters	4 gallons

VOLUME CONVERSIONS GALLONS TO LITERS	
Gallons	**Liters**
1 gallon	3.8 liters
2 gallons	7.6 liters
3 gallons	11.4 liters
4 gallons	15.2 liters

WEIGHT CONVERSIONS KILOGRAMS TO POUNDS	
Kilograms	**Pounds**
1 kilogram	2.2 pounds
10 kilograms	22 pounds
50 kilograms	110 pounds

WEIGHT CONVERSIONS POUNDS TO KILOGRAMS	
Pounds	**Kilograms**
1 pound	0.45 kilograms
5 pounds	2.25 kilograms
10 pounds	4.50 kilograms
15 pounds	6.75 kilograms

CONVERSIONS OF FRACTIONS OF AN INCH TO DECIMALS OF AN INCH AND A FOOT

Fraction	Decimal/Inch	Decimal/Foot
1/64	0.015625	0.0052
1/32	0.3125	
3/64	0.046875	
1/16	0.0625	
5/64	0.078125	0.0104
3/32	0.09325	
7/64	0.109375	
1/8	0.125	
9/64	0.140625	0.0156
5/32	0.15625	
11/64	0.171875	
3/16	0.1875	
13/64	0.203125	0.0208
7/32	0.21875	
15/64	0.234375	
1/4	0.250	
17/64	0.265625	0.0260
9/32	0.28125	
19/64	0.296875	
5/16	0.3125	
21/64	0.328125	0.0313
11/32	0.34375	
23/64	0.359375	
3/8	0.375	
25/64	0.390625	0.0365
13/32	0.40625	
27/64	0.421875	
7/16	0.4375	
29/64	0.453125	0.0417
15/32	0.46875	
31/64	0.484375	
1/2	0.500	

CONVERSIONS OF FRACTIONS OF AN INCH TO DECIMALS OF AN INCH AND A FOOT *(cont.)*

Fraction	Decimal/Inch	Decimal/Foot
33/64	0.515625	0.0469
17/32	0.53125	
35/64	0.546875	
9/16	0.5625	
37/64	0.578125	0.0521
19/32	0.59375	
39/64	0.609375	
5/8	0.625	
41/64	0.640625	0.0573
21/32	0.65625	
43/64	0.671875	
11/16	0.6875	
45/64	0.703125	0.0625
23/32	0.71875	
47/64	0.734375	
3/4	0.750	
49/64	0.765625	0.0677
25/32	0.78125	
51/64	0.796875	
13/16	0.8125	
53/64	0.828125	0.0729
27/32	0.84375	
55/64	0.859375	
7/8	0.875	
57/64	0.890625	0.0781
29/32	0.90625	
59/64	0.921875	
15/16	0.9375	
61/64	0.953125	0.0833
31/32	0.96875	
63/64	0.984375	
1	1.00	

TYPICAL SOUND TRANSMISSION CLASS RATINGS FOR SELECTED OCCUPANCIES*	
Type	**Rating**
Offices	45-52
Conference room	45
Hotel and motel rooms	48-50
Hotel and motel bathrooms	52-55
Apartments	48-55
Single family residence bedroom	40-50
Single family bathroom	45-50
Single family exterior wall	45-40
*See local codes for specific requirements.	

FAHRENHEIT TO CENTIGRADE TEMPERATURE CONVERSIONS

°F	°C	°F	°C	°F	°C	°F	°C	°F	°C
-459.4	-273	-22	-30	35.6	2	93.2	34	150.8	66
-418	-250	-18.4	-28	39.2	4	96	36	154.4	68
-328	-200	-14.8	-26	42.8	6	100.4	38	158	70
-238	-150	-11.2	-24	46.4	8	104	40	161.6	72
-193	-125	-7.6	-22	50	10	107.6	42	165.2	74
-148	-100	-4	-20	53.6	12	111.2	44	168.8	76
-130	-90	-0.4	-18	57.2	14	114.8	46	172.4	78
-112	-80	3.2	-16	60.8	16	118.4	48	176	80
-94	-70	6.8	-14	64.4	18	122	50	179.6	82
-76	-60	10.4	-12	68	20	125.6	52	183.2	84
-58	-50	14	-10	71.6	22	129.2	54	186.8	86
-40	-40	17.6	-8	75.2	24	132.8	56	190.4	88
-36.4	-38	21.2	-6	78.8	26	136.4	58	194	90
-32.8	-36	24.8	-4	82.4	28	140	60	197.6	92
-29.2	-34	28.4	-2	86	30	143.6	62	201.2	94
-25.6	-32	32	0	89.6	32	147.2	64	204.8	96

°F	°C	°F	°C	°F	°C	°F	°C	°F	°C
208.4	98	347	175	590	310	1004	540	6332	3500
212	100	356	180	608	320	1040	560	7232	4000
221	105	365	185	626	330	1076	580	4500	8132
230	110	374	190	644	340	1112	600	9032	5000
239	115	383	195	662	350	1202	650	9932	5500
248	120	392	200	680	360	1292	700	10832	6000
257	125	410	210	698	370	1382	750	11732	6500
266	130	428	220	716	380	1472	800	12632	7000
275	135	446	230	734	390	1562	850	13532	7500
284	140	464	240	752	400	1652	900	14432	8000
293	145	482	250	788	420	1742	950	15332	8500
302	150	500	260	824	440	1832	1000	16232	9000
311	155	518	270	860	460	2732	1500	17132	9500
320	160	536	280	896	480	3632	2000	18032	10000
329	165	554	290	932	500	4532	2500		
338	170	572	300	968	520	5432	3000		

1 degree F is 1/180 of the difference between the temperature of melting ice and boiling water.
1 degree C is 1/100 of the difference between the temperature of melting ice and boiling water.
Absolute Zero = 273.16°C = -459.69°F; Freezing = 32°F, 0°C; Boiling = 212°F, 100°C

THERMAL COEFFICIENTS OF LINEAR EXPANSION OF SELECTED BUILDING MATERIALS	
Material	**Inches/Inch °F**
Aluminum	0.000012
Bronze, Commercial	0.00001
Brass	0.0000104
Brick	0.0000033
Concrete	0.0000065
Concrete Block	0.0000062
Copper	0.0000094
Float Glass	0.0000051
Glass, Soda Lime	0.0000047
Granite	0.0000047
Gypsum Wallboard	0.0000090
Gypsum Plaster	0.0000070
Limestone	0.0000044
Marble	0.0000073
Stainless Steel	0.0000087
Structural Steel	0.0000064
Terne	0.0000065
Wood, Parallel with the Grain	0.0000017 to 0.0000025

FINDING AREAS, VOLUME, AND CIRCUMFERENCE

Finding Areas

Square or Rectangle

Length × Height

Triangle

One-half of Height × the Length of the Base

Circle

Diameter × Diameter × 0.7854

Trapezoid

One-half the Sum of the Parallel Sides × the Height

Finding Volume

Cube or Rectangular Prisms

Width × Height × Length

Sphere

Diameter × Diameter × Diameter × 0.5236

Pyramid

Area of Base × 1/3 of the Altitude

Cone

Diameter × Diameter × 0.785 × One-third of the Altitude

Cylinder

Radius × Radius × Height × 3.1416

Finding Circumference of a Circle

Circumference

Diameter × 3.1416

Radius × 6.2832

SIMPLE GEOMETRIC CALCULATIONS

Area of a Right Triangle Area = $\frac{\text{Base}}{2} \times$ Height	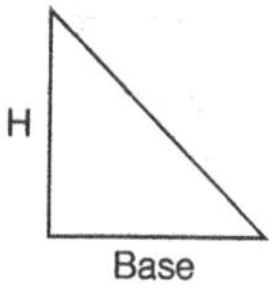 H Base
Area of Any Triangle Area = Base $\times \frac{\text{Altitude}}{2}$	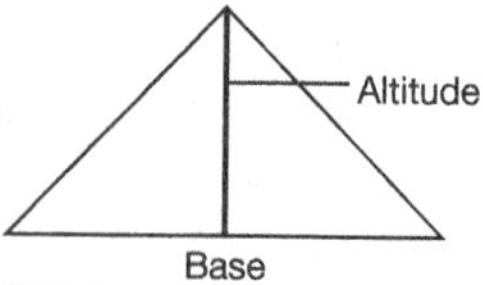 Altitude Base
Area of a Circle Area = Radius Squared × 3.1416	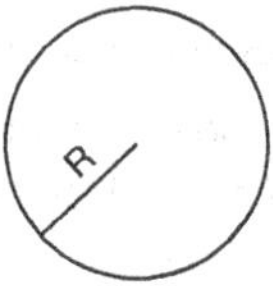 R
Circumference of a Circle Circumference = Diameter × 3.1416	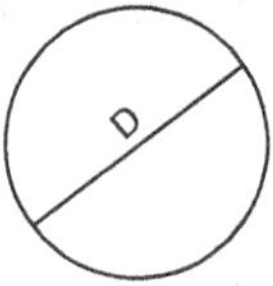 D
Surface of a Sphere Surface + Diameter Squared × 3.1416	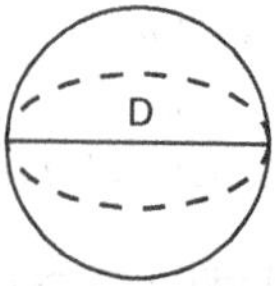 D
Volume of a Cone or Pyramid (Round, Square, or Rectangular) Volume = Area of Base $\times \frac{\text{Altitude}}{3}$	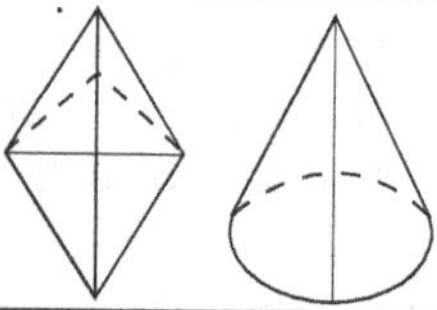

SIMPLE GEOMETRIC CALCULATIONS *(cont.)*	
Area of a Cube Area = Area of One Side × 6	
Area of a Rectangle Area = Long Side × Short Side Perimeter = Sum of the Four Sides	
Area of a Trapezoid Area = 1/2 Sum of Parallel Sides × Altitude	Altitude
Area of a Trapezum Area + Divide into Two Triangles and Figure Area of Each Triangle	
Area of an Ellipse Area + .7854 × D1 × D2	D1 D2

ROOF SLOPE	
Rise and Run	**Degrees**
1 in 12	4.76
2 in 12	9.46
3 in 12	14.04
4 in 12	18.43
5 in 12	22.62
6 in 12	26.57
7 in 12	30.26
8 in 12	33.69
9 in 12	36.87
10 in 12	39.81
11 in 12	42.51
12 in 12	45.00
13 in 12	47.29
14 in 12	49.40
15 in 12	51.34
16 in 12	53.13
17 in 12	54.78
18 in 12	56.31

WEIGHTS OF BUILDING MATERIALS		
Brick and Block Masonry	**lb./ft.^{2}**	**kg/m^2**
4" Brickwall	40	196
4" Concrete Brick, Stone or Gravel	46	225
4" Concrete Brick, Lightweight	33	161
4" Concrete Block, Stone or Gravel	34	167
4" Concrete Block, Lightweight	22	108
6" Concrete, Stone or Gravel	50	245
6" Concrete Block, Lightweight	31	152
8" Concrete Block, Stone or Gravel	55	270
8" Concrete Block, Lightweight	35	172
12" Concrete Block, Stone or Gravel	85	417
12" Concrete Block, Lightweight	55	270
Concrete	**lb./ft.^{3}**	**kg/m^3**
Plain, Slag	132	2,155
Plain, Stone	144	2,307
Reinforced, Slag	138	2,211
Reinforced, Stone	150	2,403
Lightweight Concrete	**lb./ft.^{3}**	**kg/m^3**
Concrete, Perlite	35-50	561-801
Concrete, Pumice	60-90	961-1,442
Concrete, Vermiculite	25-60	400-961
Wall, Ceiling, and Floor	**lb./ft.^{2}**	**kg/m^2**
Acoustical Tile, 1/2"	0.8	3.9
Gypsum Wallboard, 1/2"	2	9.8
Plaster, 2" Partition	20	98
Plaster, 4" Partition	32	157
Plaster, 1/2"	4.5	22
Plaster on Lath	10	49
Tile, Glazed, 3/8"	3	14.7
Tile, Quarry, 1/2"	5.8	28.4
Terrazzo, 1"	25	122.5

WEIGHTS OF BUILDING MATERIALS *(cont.)*		
Wall, Ceiling, and Floor *(cont.)*	**lb./ft.^{2}**	**kg/m^2**
Vinyl Composition Floor Tile	1.4	69
Hardwood Flooring, 25/32"	4	19.6
Flexicore 6", Lightweight Concrete	30	14.7
Structural Facing Tile	**lb./ft.^{2}**	**kg/m^3**
2" Facing Tile	14	68.6
4" Facing Tile	24	118
6" Facing Tile	34	167
Wood	**lb./ft.^{2}**	**kg/m^2**
Ash, White	40.5	198
Birch	44	202
Cedar	22	108
Cypress	33	162
Douglas Fir	32	157
White Pine	27	132
Pine, Southern Yellow	26	127
Redwood	26	127
Plywood, 1/2"	1.5	7.4
Residential Assemblies	**lb./ft.^{2}**	**kg/m^2**
Wood, Framed Floor	10	49
Ceiling	10	49
Frame Exterior Wall, 4" Studs	10	49
Frame Exterior Wall, 6" Studs	13	64
Brick Veneer of 4" Frame	50	245
Brick Veneer over 4" Concrete Block	74	363
Interior Partitions with Gypsum Both Sides *(Allowance per sq. ft. of floor area—not weight of material)*	20	320
Wall, Ceiling, and Floor	**lb./ft.^{2}**	**kg/m^2**
Flexicore 6", Stone Concrete	40	196
Plank, Cinder Concrete, 2"	15	73.5
Plank, Gypsum, 2"	12	58.8

WEIGHTS OF BUILDING MATERIALS *(cont.)*		
Wall, Ceiling, and Floor *(cont.)*	**lb./ft.^{2}**	**kg/m^2**
Concrete Reinforced, Stone, 1"	12.5	61.3
Concrete Reinforced, Lightweight, 1"	6-10	29.4-49
Concrete Plain, Stone, 1"	12	58.8
Concrete Plain, Lightweight, 1"	3-9	14.7-44.1
Partitions	**lb./ft.^{2}**	**kg/m^2**
2 × 4 Wood Studs, Gypsum Wallboard 2 Sides	8	39.2
4" Metal Stud, Gypsum Wallboard 2 Sides	6	29.4
6" Concrete Block, Gypsum Wallboard 2 Sides	35	171.5
Roofing	**lb./ft.^{2}**	**kg/m^2**
Built-up	6.5	31.9
Concrete Roof Tile	9.5	46.6
Copper	1.5-2.5	7.4-12.3
Steel Deck Alone	2.5	12.3
Shingles, Asphalt	1.7-2.8	8.3-13.7
Shingles, Wood	2-3	9.8-14.7
Slate, 1/2"	14-18	68.6-88.2
Tile, Clay	8-16	39.2-78.4
Stone Veneer	**lb./ft.^{3}**	**kg/m^3**
2" Granite, 1/2" Parging	30	481
4" Limestone, 1/2" Parging	36	577
4" Sandstone, 1/2" Parging	49	785
1" Marble	13	208
Structural Clay Tile	**lb./ft.^{2}**	**kg/m^2**
4" Hollow	23	368
6" Hollow	38	609
8" Hollow	45	721
Suspended Ceilings	**lb./ft.^{2}**	**kg/m^2**
Acoustic Plaster on Gypsum Lath	10-11	49-54

WEIGHTS OF BUILDING MATERIALS *(cont.)*		
Suspended Ceilings *(cont.)*	**lb./ft.^{2}**	**kg/m^2**
Mineral Fiberboard	1.4	6.9
Metals	**lb./ft.^{3}**	**kg/m^3**
Aluminum	165	2,643
Copper	556	8,907
Iron, Cast	450	7,209
Steel	490	7,850
Steel, Stainless	490-510	7,850-8,170
Glass	**lb./ft.^{2}**	**kg/m^2**
1/4" (6.3 mm) Plate or Float	3.3	16.2
1/2" (12.7 mm) Plate or Float	6.6	32.3
1/32" (0.79 mm) Sheet	2.8	13.7
1/4" (6.3 mm) Sheet	3.5	17.2
1/8" (3.2 mm) Double Strength	1.6	7.8
7/32" (5.6 mm) Sheet	2.85	14
1/4" (6.3 mm) Laminated	3.3	16.2
1/2" (12.7 mm) Laminated	6.35	31.1
2" (50.2 mm) Bullet Resistant	26.2	128.4
1" (25.4 mm) Insulating, 1/2", (6.3 mm) Air Space	6.54	32
1/4" (6.3 mm) Wired	3.5	17.1
3/8" (9.5 mm) Wired	5	24.4
3 7/8" × 5 3/4" Square, (98.0 × 146 mm) Glass Block	16	78.4

CHAPTER 17
Electrical and Plumbing Installations

ELECTRICAL INSTALLATIONS

The National Electric Code provides details on the technology and practices required in the electrical field. For information, contact the National Fire Protection Association, www.nfpacatalog.org.

Electrical symbols used on architectural drawings are found in Chapter 3.

ELECTRIC TERMS AND FORMULAS

OHMS LAW

The current in an electric circuit is proportional to the voltage and inversely proportional to the resistance. This is a relationship between voltage, current, and resistance.

VOLTAGE, CURRENT, RESISTANCE, AND POWER SYMBOLS

E represents the current in volts.
R represents the resistance in ohms.
I represents amperes, the unit of electric flow.
P represents watts, the electric flow rate.

ELECTRIC TERMS AND FORMULAS *(cont.)*

FORMULAS TO CALCULATE ELECTRIC PROPERTIES

$P = E \times I$ Watts = Volts × Amps

$I = \frac{E}{R}$ $\text{Amperes} = \frac{\text{Volts}}{\text{Ohms}}$

$E = I \times R$ Volts = Amperes × Ohms

To Find Ohms

$P = \frac{E^2}{R}$ Solve for R

SYMBOLS USED TO IDENTIFY WIRE INSULATION

Symbol	Meaning
AC	Armored Cable
C	Corrosion-resistant
F	Feeder
H	Heat-resistant
NM	Non-metallic
R	Rubber
SE	Service Entrance Cable
T	Thermoplastic
U	Underground
W	Water-resistant

WIRE COLOR CODES		
Wire Color	**Purpose**	**Device Connection**
Black	Hot, power carrying wire	Connected to dark screw
Green or Bare Copper	Ground, conducting connection between electric circuit and the earth	Connected to green screw
Red	Hot, 240 volt power carrying wire	The second hot wire connection in a 240 volt circuit
White	Neutral, carries zero voltage	Connected to silver screw
White with Black Tape or Ink Marks	Hot power carrying wire	Connected to dark screw

ALLOWABLE WIRING METHODS	
Allowable Wiring Method	**Designated Abbreviation**
Armored Cable	AC
Electrical Metallic Tubing	EMT
Electrical Non-metallic Tubing	ENT
Flexible Metal Conduit	FMC
Intermediate Metal Conduit	IMC
Liquidtight Flexible Conduit	LFC
Metal-clad Cable	MC
Non-metallic Sheathed Cable	NM
Rigid Non-metallic Conduit	RNC
Rigid Metallic Conduit	RMC
Service Entrance Cable	SE
Surface Raceways	SR
Underground Feeder Cable	UF
Underground Service Cable	USE

COMMONLY USED AMERICAN WIRE GAUGE SIZES

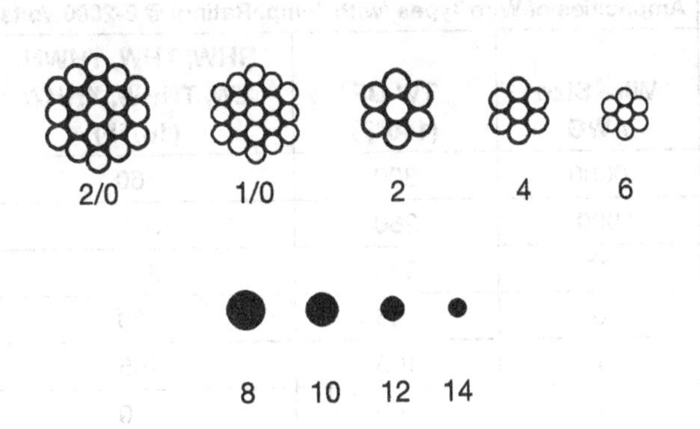

HOUSEHOLD ELECTRIC WIRE COLOR CODING*

Green – Grounding wire
Black – The "hot" wire carrying the current
Red – A second "hot" wire
White/Gray – The neutral wire

*Wires used for electronic applications have a color coding system specified by the Electronic Industries Association.

COPPER WIRE CURRENT CAPACITY – SINGLE WIRE IN OPEN AIR, AMBIENT TEMP. 86°F

Ampacities of Wire Types (with Temp. Rating) @ 0-2000 Volts

Wire Size AWG	TW UF (140°F)	RHW, THW, THWN, ZW, THHW, XHHW (167°F)
0000	300	360
000	260	310
00	225	265
0	195	230
1	165	195
2	140	170
3	120	145
4	105	125
6	80	95
8	60	70
10	40	50
12	30	3
14	25	30
16	–	–
18	–	–

ALUMINUM WIRE AMP CAPACITY — SINGLE WIRE IN OPEN AIR, AMBIENT TEMP. 86°F		
Ampacities of Wire Types (with Temp. Rating) @ 0-2000 Volts		
Wire Size AWG	**UF TW (140°F)**	**RHW, THW, THWN, XHHW, THHW (167°F)**
500kcmil	405	485
400kcmil	355	425
300kcmil	290	350
0000	235	280
000	200	240
00	175	210
0	150	180
1	130	155
2	110	135
3	95	115
4	80	100
6	60	75
8	45	55
10	35	40
12	25	30

CABLES

Cable has two or more wires enclosed in a metal, rubber, or plastic sheath. Cable with two insulated wires is identified by wire-gauge and the number of wires. A wire identified as 14-2 has two 14 gauge wires. The white wire is the neutral wire and carries current at zero voltage, and the black wire is the hot wire and carries full voltage. If it has a ground wire included, that is either bare or has green insulation and is identified on the wire as 14-2G. If the cable has a third insulated wire, it will be red and is a hot wire.

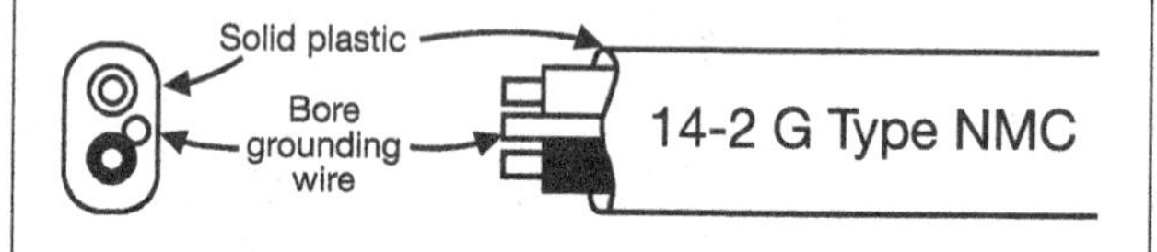

IDENTIFICATION ON CABLE INSULATION

Cables

Electric cable design and use are strictly regulated by codes such as the National Electrical Code published by the National Fire Protection Association. The types of conductors rated by this code are:

- **Type AC.** Insulated conductors are wrapped in paper and enclosed in a flexible, spiral-wrapped metal covering. An internal copper bonding strip in contact with the metal covering provides a means of grounding. It is used only in dry locations. This wire is referred to as BX cable.

IDENTIFICATION ON CABLE INSULATION *(cont.)*

- **Type ACL.** This type has the same insulation and covering as Type AC, but it has lead-covered conductors that make it useful in wet applications.
- **Type ACT.** The individual copper conductors have a moisture-resistant fibrous covering and run inside a spiral metal sheath.

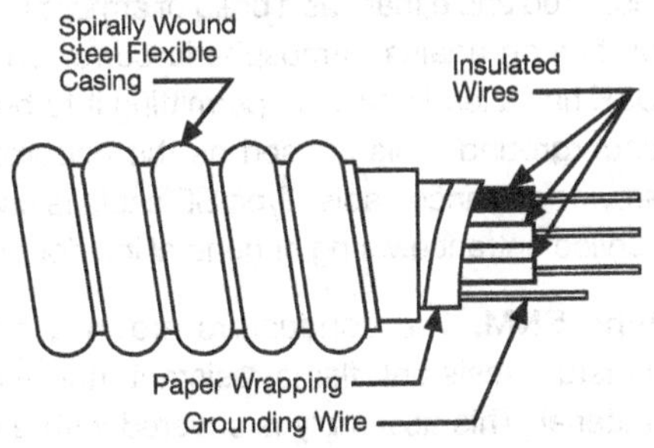

- **Type MC.** Insulated copper conductors are sheathed in a flexible metal casing. If it has a lead sheath, it can be used in wet locations. It is a heavy-duty industrial feeder cable.
- **Type MI.** The conductors are mineral insulated and sheathed in a gas-tight and watertight metal tube. This type can be used in hazardous locations and underground. It can be fire-rated.

IDENTIFICATION ON CABLE INSULATION *(cont.)*

- **Type NM or NMC.** A non-metallic-sheathed cable used in protected areas. It is also called Romex. NM has a flame-retardant and moisture-resistant outer casing and is restricted to interior use. NMC is also fungus resistant and corrosion resistant and can be used on exterior applications.
- **Type SE or USE.** A moisture-resistant, fire-resistant insulated cable that has a braid of armor providing protection against atmospheric corrosion. Type USE has a lead covering, permitting it to be used underground. This is used as the underground service entrance cable. Type SE cable is used for service entrance wiring or general interior use.
- **Type SNM.** The conductors are in a core of moisture-resistant, flame-resistant, non-metallic material. This assembly is covered with a metal tape and a wire shield and is sheathed in an extruded non-metallic material impervious to oil, moisture, fire, sunlight, corrosion, and fungus. It can be used for hazardous applications.
- **Type UF.** The conductors are enclosed in a sheath resistant to corrosion, fire, fungus, and moisture and can be directly buried in the earth.

TYPICAL NONMETALLIC CABLES

Non-metallic Type NMC

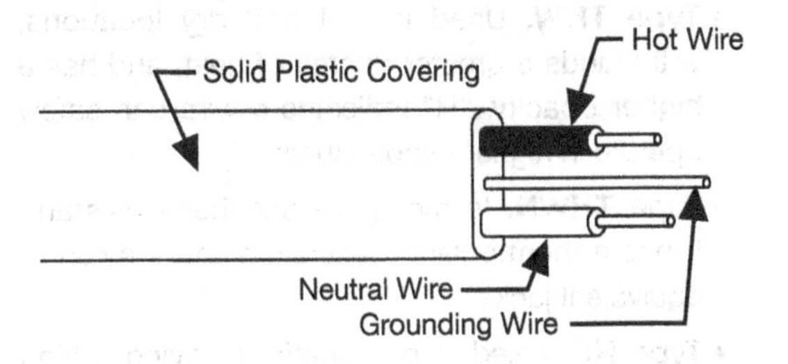

Non-metallic Type NM

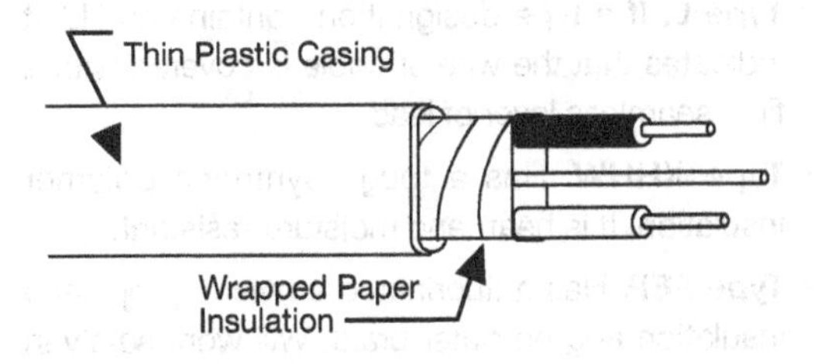

PLASTIC INSULATED WIRES

- **Type T.** Used in dry locations. No longer widely used. Replaced with Type TW.
- **Type TW.** Used in wet and dry locations. "W" indicates a wire that can be used in wet locations.
- **Type THW.** Used in wet and dry locations, withstands a greater degree of heat, and has a higher opacity. "H" indicates a wire can safely operate in higher temperatures.
- **Type THWN.** Is moisture- and heat-resistant. It has a thermoplastic insulation and a nylon or equivalent jacket.
- **Type HH.** Used in dry locations having a high temperature.
- **Type THHN.** Used in dry locations having high temperatures.
- **Type L.** If a type designation contains an "L" it indicates that the wire or cable is covered with a final seamless layer of lead.
- **Type XHHW.** Has a tough synthetic polymer insulation. It is heat- and moisture-resistant.
- **Type FEP.** Has a fluorinated ethylene propylene insulation and no outer braid. Will work safely in elevated temperatures.
- **Type FEPB.** Is the same as Type FEP but has an outer braid of fiberglass or asbestos. Will work in higher temperatures than Type FEP.

WIRE IDENTIFICATION

Wires are identified by having the type designated on their surface. Other designations may include the UL (Underwriters Laboratory) mark, AL if it is aluminum, and the date of manufacture may be given.

IDENTIFICATION ON WIRE INSULATION

(UL) (AWG 12) 4.307 mm^2 Type THWN or THHN

WIRE INSULATION IDENTIFICATION

A "T" prefix (as TW) indicates the wire has a thermoplastic type insulation. An "R" prefix (as RHH) indicates a rubber insulation. Frequently used wires in the thermoplastic group include THHN, THW, THWN, and TW. Frequently used wires with rubber insulation include RH, RHH, and RHW.

INSULATED ELECTRICAL CONDUCTORS USED FOR GENERAL WIRING APPLICATIONS

Wire Types	Insulation	Maximum Operating Temperature	Locations
TW	Moisture-resistant Plastic	60°C 140°F	Wet or dry
THHW	Moisture- and Heat-resistant Plastic	75°C 167° F	Wet or Dry
THHN	Heat-resistant Plastic	90°C 194°F	Damp or Dry
THW	Moisture and Heat	75°C	Wet or Dry
XHHW	Moisture- and Heat-resistant Plastic	90°C 194°F	Damp or Dry
RH	Rubber Thermoset Flame-retardant	75°C 167°F	Damp or Dry
RHH	Rubber Thermoset Flame-retardant	90°C 194°F	Damp or Dry
RHW	Rubber Thermoset Flame-retardant	75°C 167°F	Wet or Dry
UF	Moisture-resistant	60°C 140°F	Underground Feeder

From the National Electric Code, Courtesy The National Fire Protection Association, www.nfpa.org.

ELECTRIC CABLE, TUBING, AND RACEWAY SYMBOLS	
Wiring Product	**Abbreviations**
Armored Cable	AC
Electric Metallic Tubing	EMT
Electric Non-metallic Tubing	ENT
Flexible Metal Conduit	FMC
Intermediate Metal Conduit	IMC
Liquidtight Flexible Conduit	LFC
Metal Clad Cable	MC
Non-metallic Sheathed Cable	NM, RNC
Rigid Metallic Conduit	RMC
Service Entrance Cable	SE
Surface Raceways	SR
Underground Feeder Cable	UF
Underground Service Entrance Cable	USE

CONDUIT

Rigid metal conduit provides a metal enclosure to protect the wiring from damage and the surroundings from hazards such as a fire or overheating of the wire in the conduit. Galvanized steel conduit is most widely used, however, plastic and enamel coatings are used. The types of steel conduit are heavy-metal (also called rigid steel conduit), intermediate metal, which has thinner walls, and a lightweight metal tubing identified as EMT and referred to as thin-wall conduit. They are available in diameters from 1/2" to 4".

WIRE CONNECTOR CAPACITIES

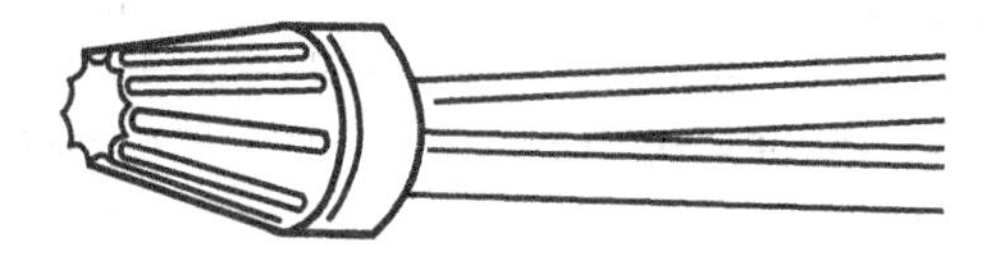

Three sizes for two-wire connections
14, 16, 18 gauge wires

Two sizes for four-wire connections
14 and 18 gauge wires
12 gauge connector can be used to connect three 10 gauge wires

DETERMINING THE SAFE CAPACITY FOR AN ELECTRIC CIRCUIT		
Calculate Wattage Amps × Volts	**Wattage Capacity**	**Maximum Safe Wattage**
120V CIRCUITS		
15A × 120V=	1800 Watts	1440 Watts
20A × 120V=	2400 Watts	1920 Watts
30A × 120V=	3600 Watts	2880 Watts
240V CIRCUITS		
20A × 240V=	4800 Watts	3840 Watts
30A × 240V=	7200 Watts	5760 Watts

CAPACITY OF METAL BOXES

Protective metal boxes designed to house conductor junctions, switches, or outlets should be of sufficient size to provide free space for any electrical components and all conductors to be enclosed in the box. (NEC 370-16.)

Box Dimension, Inches Trade Size or Type	Min. Cu. In. Cap.	Maximum Number of Conductors						
		18	16	14	12	10	8	6
4 × 1 1/4 Round or Octagonal	12.5	8	7	6	5	5	4	2
4 × 1 1/2 Round or Octagonal	15.5	10	8	7	6	6	5	3
4 × 2 1/8 Round or Octagonal	21.5	14	12	10	9	8	7	4
4 × 1 1/4 Square	18.0	12	10	9	8	7	6	3
4 × 1 1/2 Square	21.0	14	12	10	9	8	7	4
4 × 2 1/8 Square	30.3	20	17	15	13	12	10	6
4 11/16 × 1 1/4 Square	25.5	17	14	12	11	10	8	5
4 11/16 × 1 1/2 Square	29.5	19	16	14	13	11	9	5
4 11/16 × 2 1/8 Square	42.0	28	24	21	18	16	14	8
3 × 2 × 1 1/2 Device	7.5	5	4	3	3	3	2	1
3 × 2 × 2 Device	10.0	6	5	5	4	4	3	2
3 × 2 × 2 1/4 Device	10.5	7	6	5	4	4	3	2

3 × 2 × 2 1/2 Device	12.5	8	7	6	5	5	4	2
3 × 2 × 2 3/4 Device	14.0	9	8	7	6	5	4	2
3 × 2 × 3 1/2 Device	18.0	12	10	9	8	7	6	3
4 × 2 1/8 × 1 1/2 Device	10.3	6	5	5	4	4	3	2
4 × 2 1/8 × 1 7/8 Device	13.0	8	7	6	5	5	4	2
4 × 2 1/8 × 2 1/8 Device	14.5	9	8	7	6	5	4	2
3 3/4 × 2 × 2 1/2 Masonry, Box/Gang	14.0	9	8	7	6	5	4	2
3 3/4 × 2 × 3 1/2 Masonry, Box/Gang	21.0	14	12	10	9	8	7	4
FS—Minimum Internal Depth 1 3/4 Single Cover/Gang	13.5	9	7	6	6	5	4	2
FD—Minimum Internal Depth 2 3/8 Single Cover/Gang	18.0	12	10	9	8	7	6	3
FS—Minimum Internal Depth 1 3/4 Multiple Cover/Gang	18.0	12	10	9	8	7	8	3
FD—Minimum Internal Depth 2 3/8 Multiple Cover/Gang	24.0	16	13	12	10	9	8	4

Courtesy National Safety Council. Source National Electric Code, 370-16(a).

MAXIMUM SUPPORT INTERVALS FOR VARIOUS WIRING METHODS

Wiring Method	Support Intervals	NEC Ref.
Cable Tray	Fasten cables security in all nonhorizontal runs	318-8(b)
Open Wiring	4 1/2 ft. and 6 in. from splice or tap. 12 in. from dead-end connection	320-6
Mineral-Insulated, Metal-Sheathed Cables	6 ft.	330-12
Electrical Non-metallic Tubing	3 ft.	331-11
Armored Cable	4 1/2 ft. and 12 in. from box or fitting	333-7
Metal Clad Cable	6 ft.	334-10(a)
Non-metallic Sheathed Cable	4 1/2 ft. and 12 in. from box or fitting	336-15
Non-metallic Extensions	8 in.	342-7(a)(2)
Intermediate Metal Conduit	10 ft. and 3 ft. from box or fitting	345-12
Rigid Metal Conduit	10 ft. and 3 ft. from box or fitting	346-12
Electrical Metallic Tubing	10 ft. and 3 ft. from box or fitting	348-12

Rigid Non-metallic Conduit			347-8
1/2 to 1 in.	3 ft.	and 3 ft. from box or fitting	
1 1/4 to 2 in.	5 ft.		
2 1/2 to 3 in.	6 ft.		
3 1/2 to 5 in.	7 ft.		
6 in.	8 ft.		
Flexible Metal Conduit	4 1/2 ft. and 12 in. from box or fitting		350-4
Liquidtight Flexible Conduit	4 1/2 ft. and 12 in. from box or fitting		351-8
Wireway	5 ft.		362-8
Busway	5 ft.		364-5
Cablebus	12 ft.		365-6(a)

Courtesy National Safety Council, Source National Electric Code, 370-16(a)

HOLES IN JOISTS*	
Joist	Maximum Diameter
2 × 8	2 7/16"
2 × 10	3 1/8"
2 × 12	3 3/4"

*Drilled holes must be less than 33 percent of the depth. They must be at least 2" from the top and bottom of the joist. Diameters are figured using the actual size of the joist.

NOTCHES IN JOISTS*		
Joist	Maximum Joist Depth	Maximum Joist Width
2 × 8	1 1/4"	2 1/2"
2 × 10	1 1/2"	3 1/8"
2 × 12	1 7/8"	3 13/16"

*Maximum notch depth 1/6 of the depth.
Maximum notch width 1/3 of the depth.
No notches permitted in the center third of the joist
Sizes calculated using the actual sizes of the joists.

See the drawing related to holes and notches in joists on page 17-41.

DRILLING HOLES IN STUDS*	
Stud	**Maximum Hole Diameter**
2 × 4 Non-load-bearing	2 1/8"
2 × 4 Load-bearing	1 3/8"
2 × 6 Non-load-bearing	3 5/16"
2 × 6 Load-bearing	2 3/16"

*Load-bearing Studs – Holes must be less than 40 percent of depth.

Non-load-bearing Studs – Holes must be less than 60 percent of the depth.

Sizes calculated using the actual sizes of the studs.

DRILLING NOTCHES IN STUDS*	
Stud	**Maximum Notch Depth**
2 × 4 Non-load-bearing	1 3/16"
2 × 4 Load-bearing Stud	7/8"
2 × 6 Non-load-bearing	2 13/16"
2 × 6 Load-bearing	1 3/8"

*Load-bearing Studs – Notches less than 25 percent of depth.

Non-load-bearing Studs – Notches less than 40 percent of depth.

Sizes calculated using the actual sizes of the studs.

ALLOWABLE APPLICATIONS FOR WIRING METHODS

Allowable Applications (application allowed where marked with an "A")	AC	EMT	ENT	FMC	IMC RMC RNC	LFC*	MC	NM	SR	SE	UF	USE
Services	—	A	A[h]	A[i]	A	A[i]	A	—	—	A	—	A
Feeders	A	A	A	A	A	A	A	A	—	A[b]	A	A[b]
Branch Circuits	A	A	A	A	A	A	A	A	A	A[c]	A	—
Inside a Building	A	A	A	A	A	A	A	A	A	A	A	—
Wet Locations Exposed to Sunlight	—	A	A[h]	A[d]	A	A	A	—	—	A	A[c]	A[c]
Damp Locations	—	A	A	A[d]	A	A	A	—	—	A	A	A
Embedded in Noncinder Concrete in Dry Location	—	A	A	—	A	—	—	—	—	—	—	—
In Noncinder Concrete in Contact with Grade	—	A[f]	A	—	A[f]	—	—	—	—	—	—	—
Embedded in Plaster not Exposed to Dampness	A	A	A	A	A	A	A	—	—	A	A	—
Embedded in Masonry	—	A	A	—	A[f]	A	A	—	—	—	—	—

In Masonry Voids and Cells Exposed to Dampness or Below Grade Line	—	A[f]	A	A[d]	A[f]	A	A	—	—	A	A	—
Fished in Masonry Voids	A	—	—	A	—	A	A	A	—	A	A	—
In Masonry Voids and Cells not Exposed to Dampness	A	A	A	A	A	A	A	A	—	A	A	—
Run Exposed	A	A	A	A	A	A	A	A	A	A	A	A
Run Exposed and Subject to Physical Damage	—	—	—	—	A[g]	—	—	—	—	—	—	—
For Direct Burial	—	A[f]	—	—	A[f]	A	A[f]	—	—	—	A	A

For S1: 1 foot = 304.8 mm

a. Liquid-tight flexible non-metallic conduit without integral reinforcement within the conduit wall shall not exceed 6 feet in length.

b. The grounded conductor shall be insulated except where used to supply other buildings on the same premises. Type USE cable shall not be used inside buildings.

c. The grounded conductor shall be insulated.

d. Conductors shall be a type approved for wet locations and the installation shall prevent water from entering the raceways.

e. Shall be listed as "Sunlight Resistant."

f. Metal raceways shall be protected from corrosion and approved for the application.

g. RNC shall be Schedule 80.

h. Shall be listed as "Sunlight Resistant" where exposed to the direct rays of the sun.

i. Conduit shall not exceed 6 feet in length.

GENERAL INSTALLATION AND SUPPORT REQUIREMENTS FOR WIRING METHODS

Installation Requirements (requirement applicable only to wiring methods marked "A")	AC MC	EMT IMC RMC	ENT	FMC LFC	NM UF	RNC	SE	SR*	USE
When run parallel with the framing member, the wiring shall be 1.25 inches from the edge of a framing member such as a joist, rafter or stud, or shall be physically protected.	A	—	A	A	A	—	A	—	—
Bored holes in studs and vertical framing members for wiring shall be located 1.25 inches from the edge or shall be protected with a minimum 0.0625 inch steel plate or sleeve or other physical protection.	A	—	A	A	A	—	A	—	—
Where installed in grooves, to be covered by wallboard, siding, paneling, carpeting, or similar finish, wiring methods shall be protected by 0.0625 inch thick steel plate, sleeve, or equivalent or by not less than 1.25 inch free space for the full length of the groove in which the cable or raceway is installed.	A	—	A	A	A	—	A	A	A
Bored holes in joists, rafters, beams, and other horiztonal framing members shall be 2 inches from the edge of the structural framing member.	A	A	A	A	A	A	A	—	—

Securely fastened bushings or grommets shall be provided to protect wiring run through openings in metal framing members.	—	—	A	—	A	—	A	—	—
The maximum number of 90 degree bends shall not exceed four between junction boxes.	—	A	A	A	—	A	—	—	—
Bushings shall be provided where entering a box, fitting, or enclosure unless the box or fitting is designed to afford equivalent protection.	A	A	A	A	—	A	—	A	—
Ends of raceways shall be reamed to remove rough edges.	—	A	A	A	—	A	—	A	—
Maximum allowable on center support spacing for the wiring method in feet.	4.5[b,c]	10	3[b]	4.5[b]	4.5[l]	3[d]	2.5[e]	—	2.5[e]
Maximum support distance in inches from box or other terminations.	12[b,f]	36	36	12[b,g]	12[h,i]	36	12	—	12

For S1:
1 inch = 25.4 mm, 1 foot = 304.8 mm, 1 degree = 0.009 rad.

a. Installed in accordance with listing requirements.
b. Supports not required in accessible ceiling spaces between light fixtures where lengths do not exceed 6 feet.
c. Six feet for MC cable.
d. Five feet for trade sizes greater than 1 inch.
e. Two and one-half feet where used for service or outdoor feeder and 4.5 feet where used for branch circuit or indoor feeder.
f. 24 inches where flexibilitiy is necessary.
g. 36 inches where flexibility is necessary
h. Within eight nches of boxes without cable clamps.
i. Flat cables shall not be stapled on edge.

MINIMUM COVER REQUIREMENTS, BURIAL IN INCHES

Location of Wiring Method or Circuit	1 Direct Burial Cables or Conductors	2 Rigid Metal Conduit or Intermediate Metal Conduit	3 Non-metallic Raceways Listed for Direct Burial Without Concrete Encasement or Other Approved Raceways	4 Residential Branch Circuits Rated 120V or Less with GFCI Protection and Maximum Overcurrent Protection of 20 Amperes	5 Circuts for Control of Irrigation and Landscape Lighting Limited to Not More Than 30V and Installed with Type UF or in Other Identified Cable or Raceway
All locations not specified below	24	6	18	12	6
In trench below 2 inch thick concrete or equivalent	18	6	12	6	6
Under a building	0 (in race-way only)	0	0	0 (in raceway only)	0 (in raceway only)
Under minimum of 4 inch thick concrete exterior slab with no vehicular traffic and the slab extending not less than 6 inches beyond the underground installation	18	4	4	6 (direct burial) 4 (in raceway)	6 (direct burial) 4 (in raceway)

Under streets, highways, roads, alleys, driveways, and partking lots	24	24	24	24	24
One- and two-family dwelling drive-ways and outdoor parking areas, and used only for dwelling-related purposes	18	18	18	12	18
In solid rock where covered by mini-mum of 2 inches concrete extending down to rock	2 (in race-way only)	2	2	2 (in raceway only)	2 (in raceway only)

For S1: 1 inch = 25.4 mm

a. Raceways approved for burial only where encased concrete shall require concrete envelope not less than 2 inches thick.

b. Lesser depths shall be permitted where cables and conductors rise for terminations or splices, or where access is otherwise required.

c. Where one of the wiring method types listed in columns 1 to 3 is combined with one of the circuit types in columns 4 and 5, the shallower depth of burial shall be permitted.

d. Where solid rock prevents compliance with the cover depths specified in this table, the wiring shall be installed in metal or non-metallic raceway permitted for direct burial. The raceways shall be covered by a minimum of 2 inches of concrete extending down to the rock.

PLUMBING INFORMATION

Plumbing design and installation are regulated by local codes, which often adopt the International Plumbing Code published by the International Code Council, www.iccsafe.org. Following are tables giving information about materials, processes, and procedures that frequently arise.

Plumbing installation requires notching and boring holes in studs and joists. This information is shown in the electric section of this chapter.

THREADED JOINTS

Threaded joints on all types of pipe should conform to ASME B1.20.1.

SIZES OF COPPER PIPE

Nominal Pipe Size (inches)	Outside Diameter (inches)
3/8	1/2
1/2	5/8
5/8	3/4
3/4	7/8
1	1 1/8
1 1/4	1 3/8
1 1/2	1 5/8
2	2 1/8
2 1/2	2 5/8
3	3 1/8
3 1/2	3 5/8
4	4 1/8

COPPER TUBING, TYPES, USES, AND SIZES

Type	Color	Uses	Nominal Diameter
K	Green	Potable water, fire protection, solar, fuel/fuel oil, HVAC, snow melting	Straight lengths 1/4 – 12 in. (6–305 mm) Coils 1/4 – 2 in. (6-50.8 mm)
L	Blue	Potable water, fire protection, solar, fuel/fuel oil, HVAC, snow melting	Straight lengths 1/4 – 12 in. (6–305 mm)
M	Red	Potable water, fire protection, solar, fuel/fuel oil, HVAC, snow melting	Straight lengths 1/4 – 12 in. (6–305 mm)
DWV	Yellow	Drain, waste, vent, HVAC, solar	Straight lengths 1 1/4 – 8 in. (32–203 mm)
ACR	Blue Soft ACR not marked	Air-conditioning, refrigeration, natural gas, liquified petroleum gas	Straight lengths 3/8 – 4 1/8 in. (9.5–105 mm)
OXY, MED, OXY/MED, OXY/ACR, OXY/MED	K-Green L-Blue	Medical gas	Straight lengths 1/4 – 8 in. (6–203mm)
G	Yellow	Natural gas, liquified petroleum gas	Straight lengths 3/8 – 1 1/8 in. (9.5–28.5 mm) Coils 3/8–7/8 in. (9.5–22 mm)

COPPER TUBING FOR WATER SERVICE AND DISTRIBUTION		
Type	Identification Code Color	Sizes (inches)
K	Green	1/4, 1/2, 3/4, 1, 1 1/2, 2, 4,
L	Blue	1/8, 1/4, 3/8, 1/2, 3/4, 1, 1 1/4
M	Red	1 1/2, 2, 2 1/2, 3, 3 1/2

SIZES OF PLASTIC PIPE	
Nominal Pipe Size (in.)	Outside Diameter (in.)
1/8	0.40
1/4	0.54
3/8	0.67
1/2	0.84
3/4	1.05
1	1.31
1 1/4	1.66
1 1/2	1.90
2	2.37
2 1/2	2.87
3	3.50
3 1/2	4.00
4	4.50
5	5.56

MAJOR TYPES AND USES OF PLASTIC PIPES

Type	Condition	Connections	Maximum Operating Temperature		Typical Uses
			°F	°C	
Acrylonitrile Butadiene Styrene (ABS)	Rigid	Threaded, Serrated, Fittings, Solvent	100 Pressure 180 Nonpressure	38 82.9	Cold Water, Waste, Vent, Sewer, Drain, Conduit, Gas
Chlorinated Polyvinyl Chloride (CPVC)	Rigid	Threaded, Couplings, Serrated Fittings, Solvent	180 at 100 psig (Type 11)	82.9	Cold and Hot Water, Chemical Piping
Polyethylene (PE)	Flexible	Serrated Fittings, Fusion in Socket, Butt Fusion	100 Pressure 180 Nonpressure	38 82.9	Cold Water, Gas, Waste, Chemicals
Polypropylene (PP)	Rigid	Mechanical Couplings, Butt Fusion, Socket Fusion	100 Pressure 180 Nonpressure	38 82.9	Chemical Piping, Chemical Drainage
Polyvinyl Chloride (PVC)	Rigid	Solvent, Threading, Mechanical Couplings, Serrated Fittings	100 Pressure 180 Nonpressure	38 82.9	Cold Water, Gas, Waste, Vents, Drains, Sewers, Conduit
Styrene Rubber Plastic (SRP)	Rigid	Solvent, Serrated Fittings, Elastomer Seal	150 Nonpressure (not used under pressure)	66	Sewage Disposal Field, Storm Drainage, Soil Drainage

IDENTIFICATION SYMBOLS USED ON PLASTIC PIPE

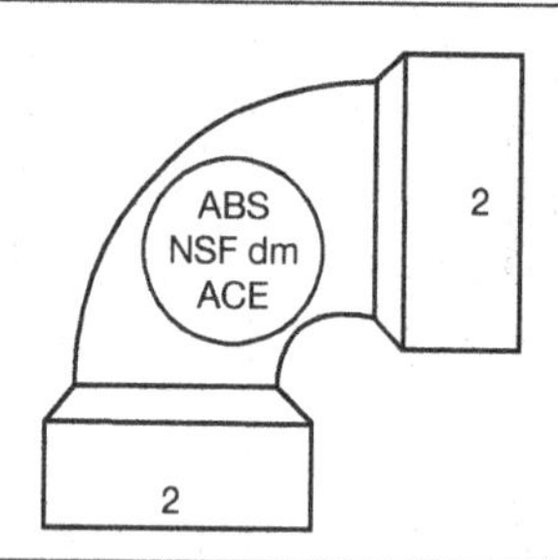

ACE – The name of the manufacturer

ABS	Acrylonitrile Butadiene Styrene, the material.
NSF	Tested by the National Sanitation.
ACE	The name of the manufacturer.
DWV	Foundation Testing Laboratory. The pipe meets or exceeds the current standards for sanitary service.
DWV	Suitable for drainage waste and vent.
SCH 40	Schedule 40. This identifies the wall of thickness of the pipe.
ASTM	"Standards Number" assigned by the American Society for Testing and Materials.

SIZES OF GALVANIZED STEEL PIPE		
Nominal Pipe Size (in.)	**Outside Diameter (in.)**	**Threads per Inch**
1/8	0.40	27
1/4	0.54	18
3/8	0.67	18
1/2	0.84	14
3/4	1.05	14
1	1.31	11 1/2
1 1/4	1.66	11 1/2
1 1/2	1.90	11 1/2
2	2.37	11 1/2
2 1/2	2.87	8
3	3.50	8
3 1/2	4.00	8
4	4.50	8
5	5.56	8

CAST IRON SOIL PIPE USES AND SIZES		
Type	**Uses**	**Nominal Diameter**
Hubless Soil Pipe, Standard and Extra Heavy	Sanitary and Storm Drains, Waste, Vents	2 – 15 in. (50.8–381 mm)
Hub Type Soil Pipe, Service and Extra Heavy	Sanitary and Storm Drains, Waste, Vents	2 – 15 in. (50.8–381 mm)

PIPE AND TUBING USED FOR WATER SUPPLY SYSTEMS			
Type	**Connections**	**Diameters**	**Special Qualities**
Galvanized Steel	Threaded	1/8 – 4 in. (3–102 mm)	Strong, Long Life
Welded Steel	Threaded	1/8 – 4 in. (3–102 mm)	Strong, Long Life
Copper Tube	Soldered, Brazed	1/4 – 6 in. (6–152 mm)	Corrosion Resistant
Red Brass	Threaded, Brazed	1/8 – 6 in. (3–152 mm)	Corrosion Resistant
Plastic	Solvent Joined, Heat Fusion, Serrated Inserts	1/4 – 6 in. (6–152 mm)	Lightweight, Corrosion Resistant

PIPING JOINTS FOR SANITARY DRAINAGE CONNECTIONS	
Types of Pipe	**Types of Joint Connections**
ABS plastic PVC plastic	Mechanical in underground systems, solvent cemented and threaded
Asbestos-cement	Sleeve coupling of same material and an elastomeric ring
Brass	Brazed, mechanical, threaded, welded
Cast iron	Caulk with oakum and fill with lead, use compression gaskets, mechanical coupling
Concrete	Elastometric seal
Copper	Brazed, mechanical, soldered, threaded, welded
Galvanized Steel	Threaded, mechanical
Lead	Electomeric seal
Joints Between Pipes of Different Materials	Compression-type mechanical connectors, and mechanical sealing connectors

DRAINAGE FIXTURE UNIT (DFU) VALUES FOR INDIVIDUAL DWELLING UNITS	
Type	**DFU Values**
Half-bath	3
One Bathroom	5 to 6
Two Bathrooms	7 to 10
Three Bathrooms	9 to 12
Bathtub or Combination Bath/Shower, 1 1/2" Trap	2.0
Bidet, 1 1/4" Trap	1.0
Clothes Washer, Domestic, 2" Standpipe	3.0
Dishwasher, Domestic, with Independent Drain	2.0
Kitchen Sink, Domestic, with One 1 1/2" Trap	2.0
Kitchen Sink, Domestic, with Food-waste-grinder	2.0
Kitchen Sink, Domestic, with Dishwasher	3.0
Kitchen Sink, Domestic, with Grinder and Dishwasher	3.0
Laundry Sink, One or Two Compartments, 1 1/2" Waste	2.0
Laundry Sink, with Discharge from Clothes Washer	2.0
Lavatory, 1 1/4" Waste	1.0
Shower Stall, 2" Trap	2.0
Sink, 1 1/2" Trap	2.0
Sink, 2" Trap	3.0
Trap Size, 1 1/4" (other)	1.0
Trap Size, 1 1/2" (other)	2.0
Trap Size, 2" (other)	3.0
Water Closet, 1.6 GPF Gravity or Pressure Tank	3.0
Water Closet, 1.6 GPF Plushometer Valve	3.0
Water Closet, 3.5 GPF Gravity Tank	4.0
Water Closet, 3.5 GPF Flushometer Valve	4.0
Whirlpool, Bath or Combination Bath/Shower 1 1/2" Trap	2.0

From *2006 National Standard Plumbing Code* published by the Plumbing-Heating-Cooling Contractors National Association. www.naphcc.org

PIPING JOINTS FOR WATER SUPPLY SYSTEMS	
Type of Pipe	**Types of Joint Connections**
ABS Plastic CPVC Plastic	Mechanical, Solvent, Cement, Threaded
Asbestos Cement	Sleeve Couplings of Same Material as Pipe and an Elastomeric Ring
Brass	Brazed, Mechanical, Threaded, Welded
Gravy and Ductile Iron	Follow Manufacturers Instructions
Copper and Copper Tubing	Brazed, Mechanical, Soldered, Threaded, Welded
Galvanized Steel	Threaded, Mechanical
Polybutylene and Polyethylene Plastic	Flared Joint, Heat-fusion, Mechanical

CONNECTIONS FOR PIPE OF DIFFERENT MATERIALS	
Type of Pipe	**Types of Joint Connections**
Copper and Copper Alloy to Galvanized Steel	Brass Converter Connections
Plastic Pipe	Use a Code-approved Adapter Fitting

REQUIREMENTS FOR NOTCHNG AND BORING SOLID WOOD JOISTS				
Joist Size*	Max. Hole D/3	Max. Notch Depth D/6	Max. Notch Width D/3	Max. End Notch D/4
2 × 4	None	None	None	None
2 × 6	1 13/16	7/8	1 13/16	1 3/8
2 × 8	2 1/2	1 1/4	2 1/2	1 7/8
2 × 10	3 1/8	1 1/2	3 1/8	2 3/8
2 × 12	3 13/16	1 7/8	3 13/16	2 7/8
*D is the actual size of the depth of the joist.				

REQUIREMENTS FOR NOTCHNG AND BORING SOLID WOOD JOISTS *(cont.)*

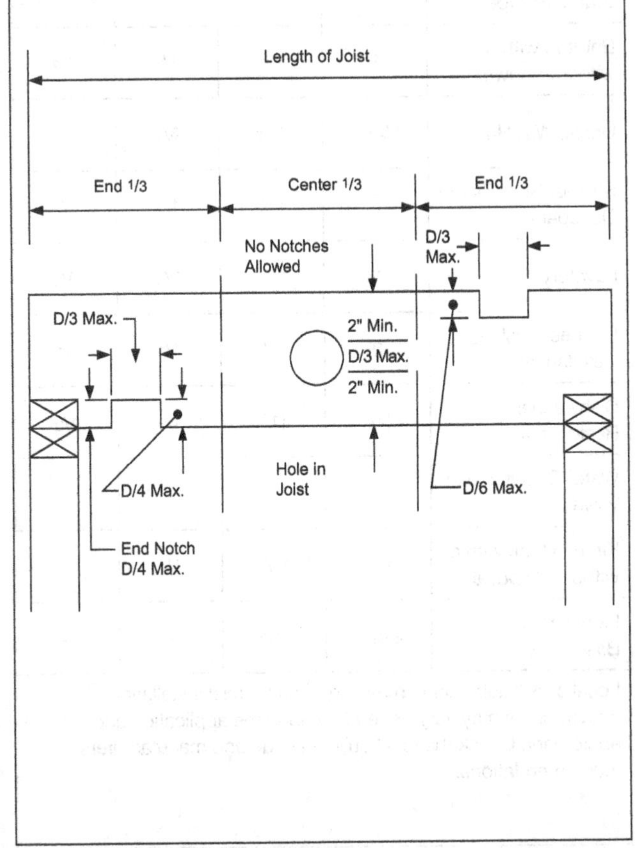

TYPICAL, TRAP, VENT, AND WATER SUPPLY PIPE SIZES				
Fixture	Min. Trap Size (in.)	Vent Size (in.)	Water Supply (in.)	
			Cold	Hot
Water Closet, Standard Type	3	1 1/2	1/2	–
Bathtub with or without Shower	1 1/2	1 1/2	1/2	1/2
Urinal, Wall Hung	1 1/2 or 2	1 1/2	3/4	–
Shower Stall, Single Occupant	2	1 1/2	1/2	1/2
Lavatory	1 1/4	1 1/4	3/8	3/8
Clothes Washer, Residential	2	1 1/2	1/2	1/2
Dishwasher, Residential	1 1/2	1 1/2	1/2	1/2
Water Closet, Flush Valve	1 1/2	1 1/2	1	–
Kitchen Sink with or without Disposal	1 1/2	1 1/2	1/2	1/2
Floor Drain, Basement	2 or 3	1/12	–	–

Local size requirements may vary. Commercial installation requirements may vary depending upon the application and equipment. Check the local building code and manufacturers recommendations.

MINIMUM FLOW AND PRESSURE REQUIRED BY TYPICAL FIXTURES

Fixture	Flow Pressure		Flow Rate	
	(psi)	(kPa)	(gpm)	(L/s)
Ordinary Basin Faucet	8	55	2.0	0.13
Self-closing Basin Faucet	8	55	2.5	0.16
Sink Faucet, 3/8 in. (9.5 mm)	8	55	4.5	0.28
Sink Faucet, 1/2 in. (12.7 mm)	8	55	4.5	0.28
Bathtub Faucet	8	55	6.0	0.38
Laundry Tub Faucet, 1/2 in. (12.7 mm)	8	55	5.0	0.32
Shower	8	55	5.0	0.32
Ball-cock for Closet	8	55	3.0	0.19
Flush Valve for Closet	15	103	15–40	0.95–2.52
Flushometer Valve for Urinal	15	103	15.0	0.95
Garden Hose (50 ft., 3/4 in. Sill Cock) (15 m, 19 mm)	30	207	5.0	0.32
Garden Hose (50 ft., 3/8 in. Outlet) (15 m, 16 mm)	15	103	3.33	0.21
Drinking Fountains	15	103	0.75	0.05
Fire Hose 1 1/2 in. (38 mm), 1/2 in. Nozzle (12.7 mm)	30	207	40.0	2.52

Reproduced from Manual of Individual Water Supply Systems, U.S. Environmental Protection Agency.

MINIMUM INSULATION FOR PIPES (INCHES)

Fluid Design Operating Temp. Range (°F)	Insulation Conductivity: Conductivity Range [BTU in. (h–ft.–°F)]	Insulation Conductivity: Mean Rating Temp. (°F)	Nominal Pipe Diameter (in.): Runouts[a] Up to 2	1 and Less	1–1 1/4 to 2	2 1/2 to 4	5 and 6	8 and Up
Heating Systems (Steam, Steam Condensate, and Hot Water)								
Above 350	0.32–0.34	250	1.5	2.5	2.5	3.0	3.5	3.5
250-350	0.29–0.31	200	1.5	2.0	2.5	2.5	3.5	3.5
201–250	0.27–0.30	150	1.0	1.5	1.5	2.0	2.0	3.5
141–200	0.25–0.29	125	0.5	1.5	1.5	1.5	1.5	1.5
105–140	0.24–0.28	100	0.5	1.0	1.0	1.0	1.0	1.5
Domestic and Service Hot Water Systems[b]								
105 and Greater	0.24–0.28	100	0.5	1.0	1.0	1.5	1.5	1.5
Cooling System (Chilled Water, Brine, and Refrigerant)[c]								
40–55	0.23–0.27	75	0.5	0.5	0.75	1.0	1.0	1.0
Below 40	0.23–0.27	75	1.0	1.0	1.5	1.5	1.5	1.5

a. Runouts to individual terminal units not exceeding 12 ft. in length.

b. Applies to recirculating sections of service or domestic hot water systems and first 8 ft. from storage tank for nonrecirculating systems.

c. The required minimum thicknesses do not consider water vapor transmission and condensation. Additional insulation, vapor retarders, or both, may be required to limit water vapor transmission and condensation.

ASTM STANDARDS FOR WATER SERVICE PIPE	
Material	**Standard**
Acrylonitrile Butadiene Styrene (ABS) Plastic Pipe	ASTM D 1527; ASTM D 2282
Asbestos-cement Pipe	ASTM C 296
Brass Pipe	ASTM B 43
Copper or Copper-alloy Pipe	ASTM B 42; ASTM B 302
Copper or Copper-alloy Tubing (Type K, WK, L, WL, M, or WM)	ASTM B 75; ASTM B 88; ASTM B 251; ASTM B 447
Chlorinated Polyvinyl Chloride (CPVC) Plastic Pipe	ASTM D 2846; ASTM F 441; ASTM F 442; CSA B137.6
Ductile Iron Water Pipe	AWWA C151; AWWA C115
Galvanized Steel Pipe	ASTM A 53
Polybutylene (PB) Plastic Pipe and Tubing	ASTM D 2662; ASTM D 2666; ASTM D 3309; CSA B137.8
Polyethylene (PE) Plastic Pipe	ASTM D 2239; CSA CAN/ CSA-B137.1
Polyethylene (PE) Plastic Tubing	ASTM D 2737; CSA B137.1
Cross-linked Polyethylene (PEX) Plastic Tubing	ASTM F 876; ASTM 877; CSA CAN/CSA-B137.5
Cross-linked Polyethylene/ Aluminum/Cross-linked Polyethylene (PEX-AL-PEX) Pipe	ASTM F 1281; CSA CAN/ CSA B137.10
Polyethylene/Aluminum/ Polyethylene (PE-AL-PE) Pipe	ASTM F 1282; CSA CAN/ CSA-B137.9
Polyvinyl Chloride (PVC) Plastic Pipe	ASTM D 1785; ASTM D 2241; ASTM D 2672; CSA CAN/ CSA-B137.3

The water service pipe is installed underground and outside the structure and runs to the water distribution pipe.

ASTM STANDARDS FOR WATER DISTRIBUTION PIPE	
Material	**Standard**
Brass pipe	ASTM B 43
Chlorinated Polyvinyl Chloride (CPVC) Plastic Pipe and Tubing	ASTM D 2846; ASTM F 441; ASTM F 442; CSA-B137.6
Copper or Copper-alloy Pipe	ASTM B 42; ASTM B 302
Copper or Copper-alloy Tubing (Type K, WK, L, WL, M, or WM)	ASTM B 75; ASTM B 88; ASTM B 251; ASTM B 447
Cross-linked Polyethylene (PEX) Plastic Tubing	ASTM F 877; CSA CAN/ CSA-B137.5
Cross-linked Polyethylene/ Aluminum/Cross-linked Polyethylene (PEX-AL-PEX) Pipe	ASTM F 1281; CSA CAN/ CSA-B137.10
Galvanized Steel Pipe	ASTM A 53
Polybutylene (PB) Plastic Pipe and Tubing	ASTM D 3309; CSA CAN-B137.8

The water distribution pipe is installed within the structure.

MINIMUM SIZES OF FIXTURE WATER SUPPLY PIPES	
Fixture	**Minimum Pipe Size (inch)**
Bathtubs (60" × 32" and smaller)[a]	1/2
Bathtubs (larger than 60" × 32")	1/2
Bidet	3/8
Combination Sink and Tray	1/2
Dishwasher, Domestic[a]	1/2
Drinking Fountain	3/8
Hose Bibbs	1/2
Kitchen Sink[a]	1/2
Laundry, 1, 2 or 3 Compartments[a]	1/2
Lavatory	3/8
Shower, Single Head[a]	1/2
Sinks, Flushing Rim	3/4
Sinks, Service	1/2
Urinal, Flush Tank	1/2
Urinal, Flush Valve	3/4
Wall Hydrant	1/2
Water Closet, Flush Tank	3/8
Water Closet, Flush Valve	1
Water Closet, Flushometer Tank	3/8
Water Closet, One Piece[a]	1/2

For SI: 1 inch = 25.4 mm. 1 foot = 304.8 mm,
1 pound per square inch = 6.895 kPa

a. Where the developed length of the distribution line is 60 feet or less, and the available pressure at the meter is a minimum of 35 psi, the minimum size of an individual distribution line supplied from a manifold and installed as part of a parallel water distribution system shall be one nominal tube size smaller than the sizes indicated.

WATER DISTRIBUTION SYSTEM DESIGN CRITERIA REQUIRED CAPACITIES AT FIXTURE SUPPLY PIPE OUTLETS

Fixture Supply Outlet Serving	Flow Rate (gpm)	Flow Pressure (psi)
Bathtub	4	8
Bidet	2	4
Combination Fixture	4	8
Dishwasher, Residential	2.75	8
Drinking Fountain	0.75	8
Laundry Tray	4	8
Lavatory	2	8
Shower	3	8
Shower, Temperature Controlled	3	20
Sillcock, Hose Bibb	5	8
Sink, Residential	2.5	8
Sink, Service	3	8
Urinal, Valve	15	15
Water Closet, Blow Out, Flushometer Valve	35	25
Water Closet, Flushometer Tank	1.6	15
Water Closet, Siphonic, Flushometer Valve	25	15
Water Closet, Tank, Close Coupled	3	8
Water Closet, Tank, One Piece	6	20

For SI: 1 pound per square inch = 6.895 kPa., 1 gallon per minute (gpm) = 3.785 L/m.

For additional requirements for flow rates and quantities see the International Plumbing Code.

HANGER SPACING DESIGNED TO PROVIDE FOR EXPANSION AND CONTRACTION

Piping Material	Maximum Horizontal Spacing (feet)	Maximum Vertical Spacing (feet)
ABS Pipe	4	10[b]
Aluminum Tubing	10	15
Brass Pipe	10	10
Cast Iron Pipe	5[a]	15
Copper or Copper-alloy Pipe	12	10
Copper or Copper-alloy Tubing, 1 1/4 inch Diameter and Smaller	6	10
Copper or Copper-alloy Tubing, 1 1/2 inch Diameter and Larger	10	10
Cross-linked Polyethylene (PEX) Pipe	2.67 (32 inches)	10[b]
Cross-linked Polyethylene/ Aluminum/Cross-linked Polyethylene (PEX-AL-PEX) Pipe	2 2/3 (32 inches)	4
CPVC Pipe or Tubing, 1 inch or Smaller	3	10[b]
CPVC Pipe or Tubing, 1 1/4 Inches or Larger	4	10[b]
Steel Pipe	12	15
Lead Pipe	Continuous	4
PB Pipe or Tubing	2.67 (32 inches)	4
Polyethylene/Aluminum/ Polyethylene (PE-AL-PE) Pipe	2.67 (32 inches)	4
PVC Pipe	4	10[b]
Stainless Steel Drainage Systems	10	10[b]

For SI: 1 inch = 25.4 mm. 1 foot = 304.8 mm.

a. The maximum horizontal spacing of cast-iron pipe hangers shall be increased to 10 feet where 10-foot lengths of pipe are installed.

b. Midstory guide for sizes 2 inches and smaller.

DISTANCE FROM SOURCES OF CONTAMINATION TO PRIVATE WATER SUPPLIES AND PUMP SUCTION LINES	
Source of Contamination	**Distance (feet)**
Barnyard	100
Farm Silo	25
Pasture	100
Pumphouse Floor Drain of Cast Iron Draining to Ground Surface	2
Seepage Pits	50
Septic Tank	25
Sewer	10
Subsurface Disposal Fields	50
Subsurface Pits	50

For SI: 1 foot = 304.8 mm.

MAXIMUM DISTANCE OF FIXTURE TRAP FROM VENT

Size of Trap (inches)	Size of Fixture Drain (inches)	Slope (inch per foot)	Distance From Trap (feet)
1 1/4	1 1/4	1/4	3 1/2
1 1/4	1 1/2	1/4	5
1 1/2	1 1/2	1/4	5
1 1/2	2	1/4	6
2	2	1/4	6
3	3	1/8	10
4	4	1/8	12

For SI: 1 inch = 25.4 mm, 1 foot = 304.8 mm, 1 inch per foot = 83.3 mm/m

SLOPE OF HORIZONTAL DRAINAGE PIPE

Size (inches)	Minimum Slope (inch per foot)
2 1/2 or less	1/4
3 to 6	1/8
8 or larger	1/16

For SI: 1 inch = 25.4 mm, 1 inch per foot = 83.3 mm/m.

ASTM STANDARDS FOR BUILDING SEWER PIPE	
Material	**Standard**
Acrylonitrile Butadiene Styrene (ABS) Plastic Pipe	ASTM D 2661; ASTM D 2751; ASTM F 628
Asbestos-cement Pipe	ASTM C 428
Cast-iron Pipe	ASTM A 74; ASTM A 888; CISPI 301
Coextruded Composite ABS DWV Schedule 40 IPS Pipe (solid)	ASTM F 1488
Coextruded Composite ABS DWV Schedule 40 IPS Pipe (cellular core)	ASTM F 1488
Coextruded Composite PVC DWV Schedule 40 IPS Pipe (solid)	ASTM F 1488
Coextruded Composite PVC DWV Schedule 40 IPS Pipe (cellular core)	ASTM F 1488, ASTM F 891
Coextruded Composite PVC IPS DR-PS DWV, PS140, PS200	ASTM F 1488
Coextruded Composite ABS Sewer and Drain DR-PS in PS35, PS50, PS100, PS140, PS200	ASTM F 1488

ASTM STANDARDS FOR BUILDING SEWER PIPE *(cont.)*	
Material	**Standard**
Coextruded Composite PVC Sewer and Drain DR-PS in PS35, PS50, PS100, PS140, PS200	ASTM F 1488
Coextruded PVC Sewer and Drain – PS25, PS50, PS100 (cellular core)	ASTM F 891
Concrete Pipe	ASTM C 14; ASTM C 76; CSA A257.1M; CSA CAN/CSA-A257.2M
Copper or Copper-alloy Tubing (Type K or L)	ASTM B 75; ASTM B 88; ASTM B 251
Polyvinyl Chloride (PVC) Plastic Pipe (Type DWV, SDR26, SDR35, SDR41, PS50 or PS100)	ASTM D 2665; ASTM D 2949; ASTM D 3034; CSA-B182.2; CSA CAN/CSA-B182.4
Stainless Steel Drainage Systems, Type 316L	ASME A112.3.1
Vitrified Clay Pipe	ASTM C 4; ASTM C 700

ASTM STANDARDS FOR UNDERGROUND BUILDING DRAINAGE AND VENT PIPE

Material	Standard
Acrylonitrite Butadiene Styrene (ABS) Plastic Pipe	ASTM D 2661; ASTM F 628; CSA-B181.1
Asbestos-cement Pipe	ASTM C 428
Cast-iron Pipe	ASTM A 74; CISPI 301; ASTM A 888
Coextruded Composite ABS DWV Schedule 40 IPS Pipe (solid)	ASTM F 1488
Coextruded Composite ABS DWV Schedule 40 IPS Pipe (cellular core)	ASTM F 1488
Coextruded Composite PVC DWV Schedule 40 IPS Pipe (solid)	ASTM F 1488
Coextruded Composite PVC DWV Schedule 40 IPS Pipe (cellular core)	ASTM F 891; ASTM F 1488
Coextruded Composite PVC IPS-DR, PS140, PS200 DWV	ASTM F 1488
Copper or Copper-alloy Tubing (Type K, L, M, or DWV)	ASTM B 75; ASTM B 88; ASTM B 251; ASTM B 306
Polyolelin Pipe	CSA CAN/CSA-B181.3
Polyvinyl Chloride (PVC) Plastic Pipe (Type DWV)	ASTM D 2665; ASTM D 2949; CSA CAN/CSA-B181.2
Stainless Steel Drainage Systems, Type 316L	ASME/ANSI A112.3.1